Land Use Transitions and Rural Restructuring in China

Hualou Long

Land Use Transitions and Rural Restructuring in China

Springer

Hualou Long
Institute of Geographic Sciences and Natural
Resources Research
Chinese Academy of Sciences
Beijing, Chaoyang, China

National Natural Science Foundation of China, No. 41731286

ISBN 978-981-15-4923-6 ISBN 978-981-15-4924-3 (eBook)
https://doi.org/10.1007/978-981-15-4924-3

This Springer imprint is published by the registered company Springer Nature Singapore Pte Ltd.
The registered company address is: 152 Beach Road, #21-01/04 Gateway East, Singapore 189721, Singapore

Foreword

The restructuring of rural China is one of the most important elements of global social and economic change in the early twenty-first century. From a traditionally rural country, with deep cultural roots in the 'soil', China has been transformed into a highly advanced and largely urbanized society within the course of a few decades. The scale and significance of the transformation is equivalent to the industrialization and urbanization of Europe in the nineteenth and early twentieth centuries, and like the European experience a century earlier, its impacts reverberate around the globe. Yet, contemporary China is not Victorian-era Europe and whilst there are parallels, we cannot assume that the trajectories, consequences, and responses are the same.

The shifting rural-urban dynamics of contemporary China are not just a question of population movement or land use change, but are entangled with a much wider series of social, economic, cultural, environmental, and political issues. Urbanization and rural restructuring raise particular challenges for the governance of an increasingly mobile population and asymmetrical workforce, the allocation of public resources, the reduction of territorial inequalities, the maintenance of food security, and the management of the environmental impacts of development. These challenges have been recognized by the Chinese government, with the policies of 'Building the New Socialist Countryside' since 2006 and subsequent 'Rural Vitalization' since 2017 adopted as a top priority for the Communist Party of China and forming arguably the most ambitious program of rural development and regeneration ever attempted anywhere in the world.

China's rural-urban transformation therefore inevitably demands scientific investigation. Initially, geographers and other social scientists were drawn to China's rapidly expanding cities as the focus of their studies; however, over the last decade researchers have increasingly explored the problems of the rural districts of China and the processes of restructuring that are experienced there. Accordingly, there has been an explosion in the number of published papers on rural China, by both Chinese and international scholars, appearing in leading international journals. Indeed, so extensive now is this literature that there is a desperate need for a coherent and accessible overview text to help students and researchers to navigate the field.

In *Land Use Transitions and Rural Restructuring in China*, Hualou Long provides just such a guide. The book discusses the considerable evidence base accumulated by Long's extensive body of research on rural China over more than a decade and draws on his training in both human and physical geography, his experience of working in both policy and academic contexts, and his detailed knowledge of international debates and concepts in rural studies. It presents a rich blend of national and regional scale statistical analysis with detailed local case studies, combining quantitative and qualitative data and adding sophisticated critical and theoretically-informed analysis that both acknowledge historical trends and look ahead to future projections. Together these elements make for an authoritative and important volume.

The chapters of this book emphasize the key components of China's rural transition, starting with the central significance of land. Land is important as a contested resource, with the conversion of rural land for urban expansion and housing development needing to be balanced against the preservation of farmland to ensure food security for China's expanding population. Achieving this balance requires new approaches to land management and may in places involve changes in land tenure and the consolidation of traditional units into larger holdings. It also leaves little room for the neglect or abandonment of land, especially in densely-packed Eastern Coastal China. In 'hollowed villages' such as those discussed in Chap. 3, land consolidation and reorganization is frequently a vital response to out-migration and the disuse of properties, necessary to ensure the continuing coherence and viability of the communities and to keep good agricultural land in productive use.

At the same time, the place of people in China's rural transition is not over-looked, with Chap. 6 for example drawing attention to the changing profile of the rural population and rural labour force, with out-migration of working-age individuals contributing both to a social problem of 'left-behind' aged communities and an economic problem of a lack of skilled workers. The third key dimension is therefore that of the economy. Revitalizing rural China requires the modernization and diversification of its economy and as the examples in Chap. 8 show this is taking many different forms, from industrial agriculture to eco-tourism to the growing opportunities offered by e-commerce and the proliferation of Taobao villages. Characteristically, Long's analysis of localities following these different trajectories not only unpicks the different drivers but also focuses in detail on the impacts, including changes in land use in the villages concerned. As such, the various case studies presented in this volume aggregate into an integrated and holistic picture of change in rural China.

Underlying the three dimensions of land, people, and economy is the significance of policy. As noted above, the restructuring of rural China described in this book has been shaped not only by the centripetal force of urbanization, but also the Chinese state's response to urbanization in the policies of 'Building the New Socialist Countryside' and 'Rural Vitalization'. As a recurrent thread through the chapters, Long describes and evaluates policy interventions, but also discusses the ways in which policy and the roles and practices of central, provincial, and local governance have evolved, noting for instance the changing contours of state intervention in Chap. 13. Neither is this the perspective of a disinterested bystander, as Long reflects on the role of science—including his own research—in informing the development and application of evidence-based policy for rural China.

As a geographical account, *Land Use Transitions and Rural Restructuring in China* also emphasizes the spatial variations in the experiences of restructuring in different parts of the Chinese countryside. Rural restructuring is different in densely-populated Eastern Coastal China than in more sparsely populated regions of Northern or Western China; and it is different in the peri-urban zones on the edge of mega-cities than in remote rural settlements. These differences are demonstrated in the book through GIS analysis and mapping of trends and outcomes across the country as well as in comparisons between individual case studies. For explanation, we need to recall the vast size of China and the difficult nature of much of its physical terrain, as well as to examine and consider the continuing importance of local traditions, cultural variations between provinces, and the role of local agency and the degree of discretion of local governance actors—which sometimes surprises international observers who perceive China as a centralized state.

At the same time, the book reinforces the embeddedness of all parts of rural China in a complex web of interconnections that link the rural and the urban, and indeed, the rural and the rural—a map of e-commerce sales from Xiaying village in Hubei Province in Chap. 8 being one striking example. Moreover, in the final chapter Long positions rural restructuring in China in the broader context of globalization. This is a dynamic that I have myself examined in research in China for the European Research Council GLOBAL-RURAL, in which I was privileged to collaborate with Hualou and his colleagues at Institute of Geographic Sciences and Natural Resources Research. In case studies of craft revitalization in Shandong Province, labour out-migration and return investment in Henan Province, and manufacturing and tourism in Zhejiang Province, I observed dramatic transformations of rural communities that were taking place against the backdrop of a globalizing economy and society, but which represented, to adapt Jeffrey Henderson's term, 'globalization with Chinese characteristics'. Rural communities in China are, as yet, protected from the more aggressive aspects of globalization, such as competition from imports in unregulated markets and speculative foreign investment, whilst China's unique form of community capitalism and labour force management have enabled some rural localities to exploit opportunities in the global economy for community gain. Equally, if globalization is understood as the multiplication, intensification, and acceleration of relations between places, then increased interconnectivity within China might be the form that it is most commonly experienced by rural communities rather than through international connections—inter-provincial connections in China being arguably the equivalent of cross-border links in Europe, which are often uncritically regarded as expressions of globalization.

These last observations highlight the paradoxical nature of contemporary rural restructuring in China—that it must be understood in global context and has parallels with both present-day and historic processes elsewhere in the world, but is also distinctive and different—which has implications for how we study it as social scientists. Concepts derived from western literature, such as 'rural restructuring' and 'governance', can help to illuminate aspects of the experience of change in rural China and have been valuably employed to develop a more critical perspective and to engage with international debates; yet not all concepts are transferable, and we urgently need rural geography that is *written from* China articulating explanations and formulating concepts that are rooted in empirical research on the Chinese experience. This book is an important contribution to building this new literature.

Michael Woods
Aberystwyth University, UK

Preface

Rural China has experienced a rapid and far-reaching transition since Deng Xiaoping launched economic reforms in 1978. The traditional centrally planned economy has been transformed into a market-based economy, with accompanying processes of rapid industrialization and urbanization based on China's special urban-rural dual institutional barriers in the aspects of household registration, land use policy and social security. Consequently, tremendous land use transitions and obvious rural restructuring have been triggered in China and brought about a series of challenges to sustainable rural development. As China emerges as a global economic super-power, and as it edges towards becoming a majority-urban population, so challenges arise for the Chinese government in ensuring that rural regions can adequately supply resources including food, land and labour; and in avoiding political instability by tackling the growing prosperity gap between urban and rural regions. With rural China is increasingly integrated into global social and economic networks, rural transformation in China is a hybrid and contested process, which involves actors and forces operating at multiple scales, and which echoes elements of rural restructuring in both the 'developed world' of Europe and North America and the 'developing world' of the global south, yet has distinctively different characteristics. How have the changes in the institutional and socio-economic context brought significant impacts to the land, people and economy in rural China?

Since obtaining my Ph.D. in physical geography in 1999, I have devoted my career to rural studies, especially research on land use and rural development, which belong more to the sphere of human geography and social science than physical geography. I do appreciate my lucky experiences without which I would not be able to develop the concepts and connotations of land use transition and rural restructuring that fit in studies over China's specific rural-urban dichotomy. The two-year post-doctoral research experience at the Chinese Academy of Sciences and one-year Research Scholar experience at the International Institute for Applied Systems Analysis (IIASA), made me access to the concepts of land use morphology and land use transition and fueled my research work to a great extent. The five-year work at the Ministry of Land and Resources of China (MLRC) gave me opportunities to link the theories of land use and rural development with practices. The visit to the

Aberystwyth University of UK introduced me the idea of rural restructuring, which has been linked by me with land use transition to study rural China from then on. While there are numerous books depicting urban China, those concerning rural China, especially from a geographical perspective, have found much less. The aim of this book is to systematically review my two-decade research findings on land use transitions and rural restructuring under the context of urbanization, industrialization and rural vitalization strategies in China, by emphasizing the key role of geography in the understanding of rural issues and their solutions.

Land Use Transitions and Rural Restructuring in China is divided into four parts, each containing three to four Chapters. Part I, *Land Use Transitions*, covers the theoretical approach for understanding land use transitions, farmland use transition and rural housing land transition in China and their coupling analysis in four Chapters. It is worth mentioning that, in this part, the concept and connotations of land use morphology are developed and expanded as dominant morphology (e.g., quantity, structure and spatial pattern) and recessive morphology (i.e., quality, property rights, management mode, fixed input, productive ability and function), and the theoretical model of regional land use transitions is established. In addition, three innovative approaches to study land use transitions in China are put forward: multidisciplinary research framework involving geography, management, economics, and sociology; horizontal comparison research method with space to exchange for time; and transect research method taking socio-economic development level and land use as key gradient factors.

Part II, *Rural Restructuring*, treats rural restructuring in four Chapters with particular emphasis upon the evolution of 'elements-structure-function' of rural territory affecting rural restructuring including three aspects of spatial restructuring, economic restructuring and social restructuring, which aims at optimizing the structure and promoting the function within rural territorial system as well as realizing the coordination of structure and complementation of function between urban and rural territorial systems. Part III, *Rural Restructuring Promoted by Land Use Transitions*, with three Chapters, places special emphasis on land consolidation, as an indispensable way of rural restructuring and vitalization promoted by land use transitions, including three types, i.e., agricultural land consolidation, hollowed villages consolidation and industrial and mining land consolidation, as it can provide space and supporting platform for rural restructuring and vitalization and realizing urban-rural integrated development in China.

Part IV, *Reflections and Futures*, devotes three Chapters to the adjusting and controlling land use transitions, the policies and countermeasures ensuring rural restructuring, and probing the future of rural China. Recessive land use morphology and its changes are the key points of land use transitions and land management. Adjusting the changes of recessive land use morphology can be taken as an efficient way to innovate land management policies and institutions in China, also a way to promote the changes of land management mode. It is necessary to restructure the contours of government intervention framework in rural economy and society to realize the optimal allocation and effective management of rural critical resource elements, but more attentions should be paid to the scientific evaluation on the role of industrial and commercial capital invested into the countryside as well as the rational view on the socio-economic and ecological values of land resource allocation. It also needs to fully understand the impacts of globalization on rural restructuring in China.

Domestic peers and I feel very grateful when we noticed that the research of land use transition has sparked great concern in both academic communities and the governmental authorities since it was introduced into China in 2001. To date, there have been 62 Ph.D. and 166 M. S. dissertations on the topic of 'land use transition' in China. Since 2011, 'land use transition' has been selected as one of the keywords of human geography by the National Natural Science Foundation of China (NSFC). During 2002–2019, the NSFC funded 48 research programs on the theme of 'land use transition', which included one key program. In 2015, a Non-profit Industry Financial Program concerning land use transition was granted by the MLRC. We have organized special issues on 'Land Use Policy in China' and 'Land Use and Rural Sustainability in China' with the journal of *Land Use Policy,* and 'Rural Restructuring in China' and 'Rural Revitalisation and Sustainability' with *Journal of Rural Studies*. Although gratifying achievements have been made in the researches of land use transition and rural restructuring in China in recent two decades, ongoing changes in land use and rural development in China will trigger more theoretical and practical research needs. I hope this book will be helpful for attracting more young Chinese scholars to carry out further studies on land use transition and rural restructuring, as well as inspiring discussions among international audiences.

Beijing, China Hualou Long

Acknowledgements

A deep debt of gratitude is owed to the Department of Agricultural Geography and Rural Development at the Institute of Geographic Sciences and Natural Resources Research (IGSNRR), Chinese Academy of Sciences (CAS). We have successfully finished a series of essential national research projects, such as four key programs funded by the NSFC, two national key research and development programs funded by the Ministry of Science and Technology of China, annual third-party assessment of the results of national targeted poverty alleviation for five consecutive years entrusted by the State Council Leading Group Office of Poverty Alleviation and Development, and so on. Working with such a big group of supportive colleagues and also excellent experts in rural studies is always a delightful thing. I have drawn inspiration and gained insights from them, especially the team leader Prof. Yansui Liu, the Fellow of The World Academy of Sciences (TWAS) and the Chairman of the Commission on Agricultural Geography and Land Engineering of the International Geographical Union (IGU-AGLE).

My greatest debt of gratitude is owed to my supervisors. They are Prof. Yunlong Cai (Ph.D. supervisor) at Peking University, Prof. Xiubin Li (post-doctoral research supervisor) at IGSNRR of CAS, and Prof. Huiqing Liu (master supervisor) at Northeast Normal University, who led me on the road of land use and rural development research, in particular, Prof. Li made me access to the concepts of land use morphology and land use transition.

I am grateful to the following international experts and friends: Prof. Gerhard K. Heilig who enlightened me on the topic of sustainable rural development; Prof. Michael Woods, the former Editor of *Journal of Rural Studies*, the Fellow of the UK Academy of Social Sciences, who gave me new insights, suggested fresh ways of approaching the topic of rural restructuring, and invited me to the editorial board of *Journal of Rural Studies*; Prof. Guy M. Robinson, the former Editor of *Land Use Policy* invited me to join the editorial board; and Prof. Charles Choguill, the former Editor of *Habitat International* invited me to join the editorial board, then as an Associate Editor. These editorial jobs have broadened my horizon in the field of land use and urban-rural development.

I would like to acknowledge the supports from my students, including Yurui Li, Shuangshuang Tu, Tingting Li, Dazhuan Ge, Yi Qu, Yongqiang Liu, Yingnan Zhang, Yanfeng Jiang, Li Ma, Kunqiu Chen, Liuwen Liao, Yuhang Zheng, Guipeng Zhou, Jian Zou, Zhichao Hu, Xingna Zhang, Pengcan Fan, Fang Fang, Lingzhi Wang, Xingying Zhou, Xiaoli Liang, Shimeng Wan, Xuegang Hou, Cong Chen, Shangao Xiong, Xiaosong Tu, Yingbin Feng, Liguo Yang, and Ligang Lyu. Some of them assisted me to prepare the draft chapters of the book. Shuangshuang Tu edited Chaps. 5, 7 and 8; Yingnan Zhang edited Chaps. 1, 3 and 4; Kunqiu Chen edited Chaps. 6, 10 and 14; Yanfeng Jiang edited Chaps. 9, 11 and 13; Dazhuan Ge edited Chap. 2; and Yi Qu edited Chap. 12.

Many thanks for the great support by the Key Program of National Natural Science Foundation of China entitled 'Land Use Transitions Driven by Urbanization and Their Effects on Local Environment in the Farming Areas of China: The Case of Huang-Huai-Hai Plain' (Grant No. 41731286). Special thanks are due to Prof. Fang Wang at Peking University, Prof. Weiguo Jiang at Beijing Normal University, and Dr. Jing Zhou at Beijing Institute of Technology who contributed to the publication of this book. Last but not least, thanks are also due to the Journal Publishers, Elsevier, and Springer Nature, the reprinting/adapting permissions from whom have been granted for reusing my published materials in this book.

Contents

Part IV Reflections and Futures

About the Author

Hualou Long is Professor and Ph.D. supervisor of Human Geography at the University of Chinese Academy of Sciences, Director of the Department of Agricultural Geography and Rural Development at the Institute of Geographic Sciences and Natural Resources Research (IGSNRR), Chinese Academy of Sciences (CAS), and Deputy Director of the Center for Assessment and Research on Targeted Poverty Alleviation, CAS.

He is also the Director of the Specialty Committee of Agricultural Geography and Rural Development in the Geographical Society of China, the Director of the Working Committee on Youth in China Land Science Society, and the Deputy Director and concurrent Secretary-General of the Specialty Committee of Land Resources Research in China Society of Natural Resources. He obtained his Ph.D. in physical geography from Peking University and carried out a post-doctoral research at IGSNRR, CAS. Thereafter, he had a five-year working experience at Land Consolidation and Rehabilitation Center, the Ministry of Natural Resources of China. He worked as Research Scholar at the section of Sustainable Rural Development, International Institute for Applied Systems Analysis (IIASA) in 2005, and visited the Institute of Geography and Earth Sciences, Aberystwyth University in 2009.

His research interests include China's rural restructuring, urban–rural development, land use transition, and sustainable land use. He has published eight books including one translation on land use and urban–rural development and over 270 original research papers. He was awarded Highly Cited Researcher in the field of Social

Sciences in 2019. He has guided 10 Ph.D. students as supervisor and 6 as a co-supervisor in the area of human geography.

Currently, he is serving as Associate Editor of *Habitat International*, and Member of Editorial Board of *Journal of Rural Studies* and *Land Use Policy*.

Part I
Land Use Transitions

Chapter 1
Understanding Land Use Transitions: A Theoretical Approach

Abstract Land system science has developed over the past decades. Land provides essential resources to the society, and its changes have large consequences for the local environment and human well-being. Land use change demonstrates non-linear process and is related to other societal and biophysical disturbances through a series of transitions. This chapter aims to theorize land use transitions and construct a theoretical foundation of this book. With the in-depth research of land use transitions, the concept and connotations of land use morphology were further developed and expanded as two kinds, i.e., dominant morphology and recessive morphology. The dominant morphology refers to the land use structure of a certain region over a certain period of time, with features such as the quantity and spatial pattern of land use types. While the recessive morphology includes the land use features in the aspects of quality, property rights, management mode, input, output and function. Accordingly, the concept of land use transition is further developed, and the theoretical model of regional land use transitions is established. Thereafter, three innovative integrated approaches to study land use transitions are put forward, i.e., multidisciplinary research framework, horizontal comparison research method, and transect research method. With the introduction of land use transition research into China by the author, related researches concerning farmland transition, rural housing land transition and their relationships with rural restructuring under the context of urbanization, industrialization and rural vitalization in China have been carried out extensively, which enrich the knowledge of land system science.

1.1 Introduction

Land system science has developed over the past decades (TurnerII and Munroe 2020). Land provides essential resources to the society, and its changes have large consequences for the local environment and human well-being. The past, current and projected state and dynamics of land use is the major content of land use science, which is influenced by long-term anthropogenic changes. As a result, land use change demonstrates non-linear process and is related to other societal and biophysical disturbances through a series of transitions. Land use transition was firstly put

H. Long, *Land Use Transitions and Rural Restructuring in China*,
https://doi.org/10.1007/978-981-15-4924-3_1

forward based on the researches of forest transition. From then on, the research of forest transition as the core of land use transition research has been pushed forward, mainly focusing on the theoretical development and empirical studies in the Europe countries as well as countries in Asia and America.

Accelerated urbanization and subsequent increase of human activities are triggering tremendous land use transitions in China. Land use transitions can be seen as primary forces of driving the transformation of rural system, and bring about direct socio-economic and environmental effects on rural sustainability, e.g., resulting in farmland loss and soil degradation, affecting biodiversity and the ability of ecosystems to serve human needs, polluting the rural environment, influencing agricultural production and food security, and causing the socio-economic and spatial restructuring of rural area. With the introduction of land use transition research into China, related researches combining land use transitions with the sustainability of rural China have been carried out extensively, which enrich the knowledge of land system science.

The objective of this chapter is to advance our theoretical understanding of land use transitions based on literature review. What is land use morphology is firstly clarified to underpin the enrichment of the connotation of land use transitions. Then, the theoretical model and research methods of regional land use transitions are discussed. Finally, this chapter reviews the research of land use transitions in China.

1.2 Land Use Morphology and Land Use Transitions

Land use transition is one of the manifestations of land-use change (Long et al. 2007), and is also one of the major research contents of Global Land Project (GLP), a joint research agenda of the International Geosphere-Biosphere Programme (IGBP) and the International Human Dimensions Programme on Global Environmental Change (IHDP) to improve the understanding of land system dynamics (GLP 2005). Land use transition was put forward by Grainger (1986, 1995) based on his study on the land use in forested countries. From then on, the research concerning land use transition has been carried out vigorously, but which mainly focused on forest transition (Barbier et al. 2010; Mather et al. 1999; Meyfroidt and Lambin 2008).

Although pieces of evidence supported the original forest transition hypothesis and trajectories, there was still some researches that showed different opinions. For example, it showed that forest cover change involves complex trajectories, some of which are cyclical and reversible, while others are linear and permanent. Barbier et al. (2010) developed a more comprehensive theory of the forest transition and argued that long-run changes in forest cover in a country or region cannot be separated from the national or regional pattern of land use changes, taking into account of the competition among different land use. The research of land use transitions including forest transition has been the foci of studying land change science since the turn of the new millennium (Lambin and Meyfroidt 2010; Long et al. 2007; Rudel et al. 2010), ranging from those who favor the development of analysis methods and detection technologies (Ferreira et al. 2015), the dynamic driving mechanism (Jadin et al. 2016; Liu et al. 2008), the environmental effects (Long 2014a; Nuissl et al. 2009), to those who try to understand the relationship between land use transitions and socio-economic development (Chen et al. 2014; Lambin and Meyfroidt 2010). The restoration process of woodland ecosystems can be considered from other three dimensions in addition to demand-induced perspective (Meyfroidt et al. 2018): (1) The intensification of farmland and forest land can result in the abandonment of farmland and the restoration of forest land; (2) The reorganization of the land use spatial layout will fully consider the suitability and sustainability of land use, and the increasing level of market integration, which will also increase the degree of intensive land use and lead to the conversion of farmland to forest land; and (3) International trade activities will cause a decrease in the demand for forest land in one area, and put greater pressure on another.

In addition, Foley (2005) pointed out that land use transition is not only a regional issue but also of global significance. Different regions are in different stages of transformation development due to historical, socio-economic and ecological conditions, and political institutions. Not all regions follow a linear transformation development, and some regions also exhibit a non-linear trajectory. A certain area may stay longer in a certain stage, and some areas may be shorter in the corresponding stage. Meyfroidt et al. (2013) and Lambin (2012) argued that the driving forces of land use transition are not only confined to socio-economic disturbances and natural background factors in one region, but also affected by the displacement effect produced by international trade. Globalization and urbanization are two important global trends since the middle of the 20th century (Long and Woods 2011; Long et al. 2015; Robinson and Carson 2015; Liu 2019; Wei 2019; Fang and Yu 2020). As a result of their interaction, a complex trade network has gradually formed between countries, thus forming an important tele-coupling context of land use transitions.

Land-use and land-cover change, as the core of coupled human-environment systems, has become a potential field of land change science (LCS) in the study of global environmental change (Li 2002; Liu et al. 2010). The goal of GLP is to measure, model and understand the coupled socio-environmental terrestrial system, which is also referred to as "land system" (GLP 2005). Land use transition is the changes in land use morphology, the overall pattern of actual land cover in a region at a given time, over time (Grainger 1995). Land use morphology is the key content of land use transition research. Driven by socio-economic change and innovation, only focusing on the quantity and spatial structure characteristics of land use morphology cannot meet the demands of the research on land use transition (Lambin and Meyfroidt 2010). As such, the concept and connotations of land use morphology need to be developed continuously with in-depth research and socio-economic change and innovation.

Based on the analysis of the evolvement of the concept and connotations of land use morphology, we argue that there are two formats for depicting land use morphology: one is dominant morphology, another is recessive morphology. The dominant land use morphology refers to the quantity, structure and spatial pattern of land use, and the recessive land use morphology includes the land-use features in the aspects of quality, property rights, management mode, fixed input, productive ability and function, etc. So, the conception of land use transition may be further developed as follows: land use transition refers to the changes in land use morphology, including dominant morphology and recessive morphology, of a certain region over a certain period of time driven by socio-economic change and innovation, and it usually corresponds to the transition of socio-economic development stage (Long and Li 2012). Figure 1.1 shows the conceptualization of the models of land use transitions. The land use morphology at a time point (e.g., time point B) is not only the result of land use transition in last period of time (from A to B) but also the beginning of land use transition in the next period of time (from B to C) (Fig. 1.1). Land use transition is a continuous evolution process of land use morphology instead of aiming at forming a fixed or deterministic land-use pattern.

Farmland and rural housing land[1] are the two major land-use types closely related to human livelihood and production activities. It is well known that the urban is originally developed from the rural, and its formation and further development is closely

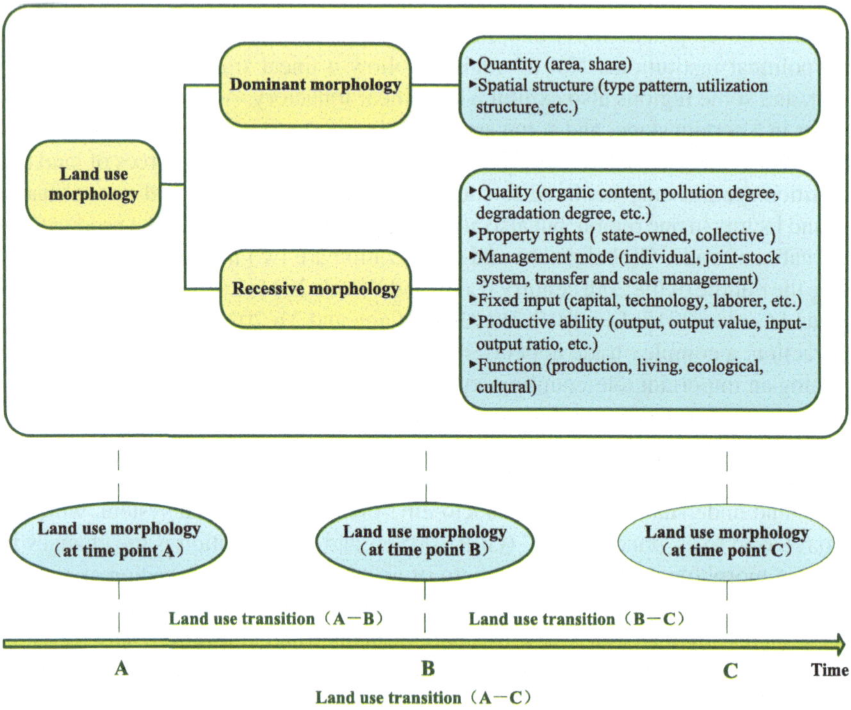

Fig. 1.1 A conceptualization of the models of land use transitions (Adapted from Long 2012a)

[1]Rural housing land refers to the land utilized by rural residents for dwelling and living, i.e. land for building house and other structures or affiliated facilities. According to China's farmers' living customs, which have been formulated for a long time, rural housing land usually includes: (1) land for dwelling and living, such as living house, kitchen, room for livestock (e.g., pigsty, sheepfold, stable and cowshed), warehouse, room for storing farm machinery, toilet; (2) surrounding afforested land, such as bamboo forest, forest tree, flower nursery; and (3) other land for service facilities of living, such as cellar, well, methane-generating pit (usually for lighting and cooking).

linked with the transition of farmland and rural housing land. In general, socio-economic development affects land use morphology, which also makes a counter-action to socio-economic development. The interaction between land use morphology and socio-economic development facilitates the land use transition. Therefore, the changes in the regional land use morphology may be basically reflected by the changes in farmland and rural housing land, two major land-use types affecting human production activities. Farmland and rural housing land, as the two major sources of land use transitions, play an important role in the process of facilitating land use transition. That is also the main reason why this book focuses on the analysis of farmland transition and rural housing land transition, which undoubtedly will be helpful in promoting the understanding of the coupled "land system".

1.3 Theoretical Model

Land use transition is a new theme of the comprehensive research of land-use/land-cover change (LUCC) (Long 2003), the concept of which was first introduced into China with an initial meaning of temporal changes in land use morphology corresponding to socio-economic development transition, and the land use morphology means mainly the quantity and spatial structure characteristics of land use in a region at a given time (Grainger 1995; Long and Li 2002). The national/regional land use morphology refers to the structure of a country/region composed of the main land use types in a specific period, which corresponds to its economic and social development stage, that is, the specific economic and social development stage is associated with a certain land use morphology-the structure of national/regional land use types. It is considered as a basic geographic concept and one of the key elements of the theory of man-land interrelations, as well as the simulation of global environmental changes (Grainger 1995). National/regional land use patterns have spatial and non-spatial representation: Its spatial representation refers to a map of a country/region consisting of cultivated land, grassland, forest land, urban land, abandoned land, etc.; non-spatial expressions refer to the share of each land use type in the country/region's territory (Grainger 1995).

National/regional land use patterns are not static, they vary with socio-economic development. For example, land use patterns in the U.S. in the late 20th century indicate that the proportion of forest land and grassland is almost the same as the world average, but it has changed greatly compared with the previous stages (Grainger 1995): From the founding of the nation to the development stage of the early 20th century, the predatory use of resources in the United States has put a lot of pressure on the fragile ecological environment in the west of the nation. Subsequently, the federal government began to pay attention to the conservation and protection of natural resources. From the 1970s to the 1990s, the government put much concern agriculture subsidies, environmental pollution, and farmland transfers to protect strategic agricultural resources and the natural environment. It can be seen that the above-mentioned land policies corresponding to different stages of economic and social development have greatly influenced the land use morphology in the United States (Grainger 1995).

Grainger's (1986) assumption about national forest land transition is that a nation generally enlarges its cultivated land at the expense of deforestation until the balance between agricultural and forest sector. Because national land use patterns cannot remain static for a long time, the end of the transition does not mean that land use changes between the forestry and agricultural sectors will not continue. Forest land can even increase again due to its own regeneration and plantation, as it did in the United States and Britain in the 20th century. Mather (1992) once referred to this turning point of forest land area change curve as the transformation of forest land, at which the national forestry cover stopped decreasing and began to increase. However, this increase at the end of the land-use transition is not an inevitable process. Even if it increases, the rate of increase must be constrained by socio-economic levels and relevant national policies (Grainger 1995).

To some extent, the regional land use transitions are essentially a process during which different land use types representing the benefits of different departments conflict in space and try to alleviate these conflicts by changing the morphologies in time. Accordingly, the theoretical model of regional land use transitions is established as follows: with the socio-economic development, transformations between different land use types during a certain period of time cause the change of the conflicts resulted from regional land use morphology pattern from strong to weak, i.e., a trend towards coordination; these transformations will lead to a new balance of regional land use morphology pattern consists of different land use types reflecting the development trend of corresponding economic departments, respectively, and finally realize the qualitative transformation of urban-rural land use system (Fig. 1.2).

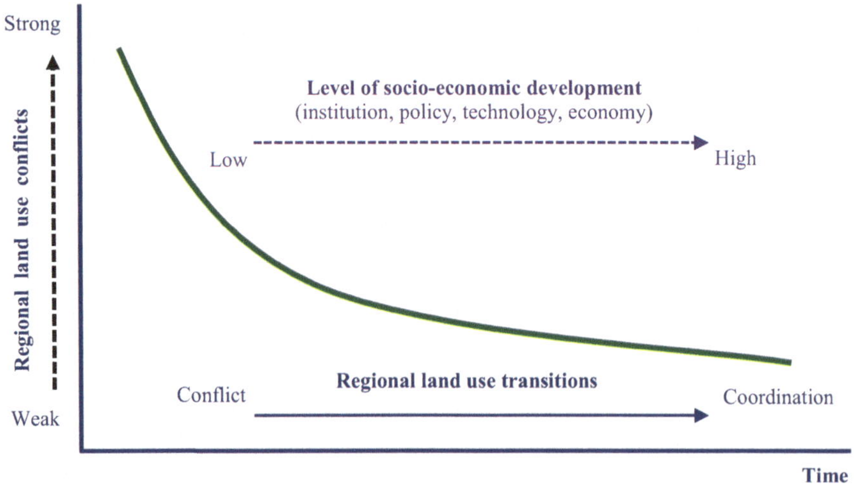

Fig. 1.2 The theoretical model of regional land use transitions (Long 2015; Long and Qu 2018)

Land use transition refers to the changes in regional land use morphology, and the regional feature is an important aspect of land use transition research. Strictly speaking, the transformation of single land use type, e.g., farmland was changed to forest land, can not be treated as land use transition, which can only be treated when it is put into the context of regional land use structure and function to analyze the changes of land use morphology.

Usually, the process of land use transitions comprises long-term and trend changes in regional land use morphology. The fundamental change of land use morphology or the turn of its changing direction indicates an accomplishment of land use transitions in a certain period. New land use issues arise with the socio-economic development will bring about new conflicts of regional land use morphology patterns, thereafter a new regional land use transition process will be triggered. Land use transition can be examined from both quantitative and qualitative aspects. The quantitative aspect is mainly reflected by the change of dominant land use morphology, i.e., changes in quantity and spatial pattern of land use. However, the qualitative aspect is reflected by the change of recessive land use morphology, especially the changes of functions of land use system driven by the comprehensive changes in the aspects of quality, property rights, management mode, input and output of land use. Currently, there are abundant researches on the dominant morphology of regional land use but less on the recessive morphology. However, land use transitions in the aspect of recessive morphology are most closely relative to land use management. The process and outcome of land use transitions may improve land management policies and measures. On the contrary, land use transitions may be deeply affected by the adjustment of existing policy or the implementation of a new land use policy.

1.4 Research Methods

In the aspect of qualitative research of land use transitions, hypothesis-deduction method is commonly used since the trend of land use transitions is uncertain, to some extent. Harvey (1969), a famous British geographer, mentioned that in his book “Explanation in Geography”, there are two approaches to scientific interpretation: inductive and deductive methods. Among them, deductive methods are more commonly used, that is, they are understood through perceptual experience. The structure of the real world, based on the prior model (the formal representation of the image), is abstracted, and then a theory can be assumed. Afterward, the functional-perceived data can be compared with a posteriori model (Another way of expressing the concepts contained in the theory, such as using mathematical symbols) and used to test the hypotheses (Harvey 1969).

Another common method is corresponding systematic comprehensive analysis, which was put forward by Chen (1998) who emphasized the corresponding transformation analysis of regional natural structure and productivity layout, as well as the production, sales, and consumption structure, which pioneered a new way of formulating regional development strategy at county level (Chen 1998). This research method can provide a scientific basis through the structural correspondence analysis and comparative study of physical-ecological structure, technical-economic structure, social-political structure in a region, which can be used for reference in the research of land use transitions and rural transformation development. Temporal changes and historical backgrounds of social and environmental variations can be integrated through comprehensive system analysis. The rural human-land system can be measured and simulated, which is helpful for identifying the stage of land use transitions and rural development. Furthermore, the mechanism and regional patterns can be summarized, thus laying a solid foundation for proposing scientific managerial countermeasures (Long 2012a).

It is generally known that mathematical models and 3S (GIS/RS/GPS) technologies have the advantages for carrying out quantative research of land use transitions, especially in the aspect of dominant morphology of land use. However, it is pivotal to have a tendency to turn to multidisciplinary integration method for the research of recessive morphology of land use. Based on above analysis, the author put forward three innovative integrated approaches to study land use transitions, i.e., multidisciplinary research framework for recessive land use transition (Qu and Long 2018a), horizontal comparison research method (Long 2003; Long et al. 2007), and transect research method (Long et al. 2007).

1.4.1 Multidisciplinary Integration

1.4.1.1 The Deficiency of Single Discipline Research

At present, studies on recessive land use transition are scattered across different disciplines, but no sufficient studies can be found on the regionality, comprehensiveness, and multi-scale aspects specialized by geographical studies.

(1) Regionality. Land has multiple attributes. The production management of land use involves many natural factors, such as soil, temperature, precipitation, daylight, and many economic and social factors like capital, labor, and technology investment. The above natural factors have regional differences, while the social and economic factors vary with the different stages of regional economic development. Neglecting the regional characteristics can lead to biased research conclusions. For example, in the research on the "inverse relationship" between farmland management scale and productivity (Sen 1962; Bardhan1973), some studies argued that this relationship was due to the omission of farmland quality variables and a lack of consideration of market factors. In other words, the existing research failed to fully consider the differences in land quality of different regions and ignore the differences in labor costs of different regions, resulting in the conclusion of inverse relationship. This problem stems from a neglect of regionality in the geographical research.

(2) Comprehensiveness. Land use and agricultural activities involve many systems, such as nature, economy, and society. Land use transition is a complex process affected by many factors. However, current research on recessive land use transition is limited to the economic system and the social system, and even the social system has not been fully considered. In the analysis of management performance and moderate-scale management, the indicators selected were relatively limited, focusing on economic indicators, such as the input-output. However, it failed to consider social factors, such as the status of the labor force and the burden of land for social security functions. At present, unreasonable land use has caused serious agricultural pollution. However, environmental factors were not considered in the selection of management mode and the determination of moderate scale. To some extent, the above shortcomings have weakened the sustainability of the existing research results being used in practice.
(3) Multi-scale. At present, the basic data for quantitative analysis are mostly farmer household survey data at the micro scale; therefore, the evaluation results and analysis conclusions obtained mostly serve management needs at the farmer household level. However, the introduction and implementation of land resource management decisions are hierarchical, and the objectives of land resource management vary at different levels of the administrative units. Management at the national level focuses on macro objectives and overall control. With the decrease of management scale, the management objectives are gradually refined, and the depth of management increases. Farmers belong to the most micro-scale management type. The characteristics of resources management at the level of farmer households are mainly to implement management objectives at the higher level according to local conditions. Therefore, with an improvement in management scale, the conclusions based on the micro farmer household survey need to combine with other information from different scales (economic development level, agricultural labor status, basic agricultural conditions, etc.) for comprehensive extraction to meet the actual needs of land resource management.

In view of the above shortcomings, we can try to introduce geographical analysis methods based on existing researches to make up for the shortcomings on space, regionality, and scale. Geography is a discipline with strong comprehensiveness and regionality (Cai et al 2004; Lu 2011). Geographical research pays attention to systematization. It analyzes the spatial and temporal pattern and process mechanism of the research object, and also focuses on the revelation of the law at different scales. The geographical analysis method can solve the research problems with consideration of regionality, comprehensiveness, and multi-scale (Cai et al 2004; Fang et al 2019). It is worth noting that in geography, regional features are represented by "patterns" through geographical differentiation, and the "geographical process" is depicted by the temporal and spatial evolution of geographical phenomena. The comprehensive study in geography has its advantages in the pattern study, while the process study is insufficient (Fu 2014). Therefore, in-depth discussion of specific research topics should be based on comprehensive research with multi-disciplinary complementarity.

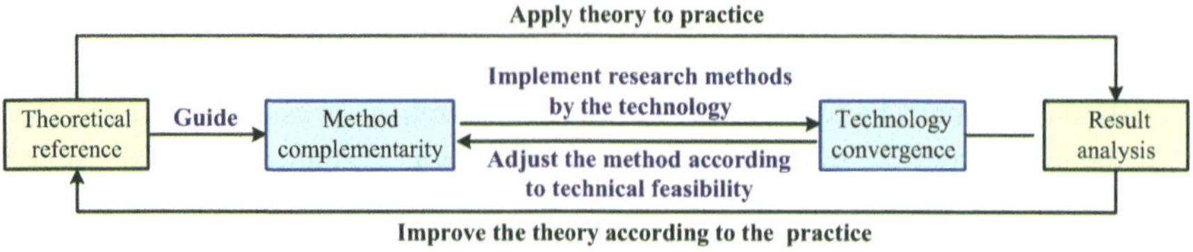

Fig. 1.3 The conceptual framework of multidisciplinary research (Qu and Long 2018a; Long et al. 2020)

1.4.1.2 Conceptual Framework of Multidisciplinary Research

The connection between different subjects should be based on theoretical reference, by means of method connection and bridged by software and data format connection. The purpose is to analyze and mine the results from different perspectives. Theory is the basis of discipline development. The mutual reference of different discipline theories can provide direction and guidance for problem analysis (Cai 2001; Fan 2019). In terms of methods, different disciplines have their own preferred research methods, which come with both strengths and weaknesses. For example, spatial analysis and system analysis are frequently used in geography, but there is a lack of a validation mechanism for the analysis results. Econometrics is good at statistical analysis, but it lacks the comprehensive analysis mechanism of multiple factors at different spatial scales. Therefore, they can complement one another. Technically, the established research methods need to be realized by corresponding software technology, data acquisition technology, and analysis technology, and the selected research methods need to be adjusted appropriately, according to the technical feasibility of the research process. In terms of results, multi-dimensional analysis results can be obtained through the connection of analysis technologies of different disciplines (Cai 2001). Meanwhile, from the perspectives of different disciplines, these results can be further mined and analyzed. In the overall analysis process, the mutual reference and development of theories are the basis of the result analysis. Without theoretical guidance, the data analysis results could be easily biased or even wrong. Meanwhile, rigorous analysis results and research conclusions play an important role in promoting the development and improvement of theory, because the development of theory comes from practice, and theory is a general summary of many individual phenomena in practice (Fig. 1.3).

1.4.1.3 Overall Design of Multidisciplinary Research

Under the guidance of the above conceptual framework, the comprehensive research framework of the disciplines involved in the research into the recessive land use transition was specifically designed (Fig. 1.4). At the theoretical level, research into the recessive transition of land use mainly involves disciplines including geography, management, economics, and sociology. With the research problem-oriented approach, preliminary reference and integration of different theories can

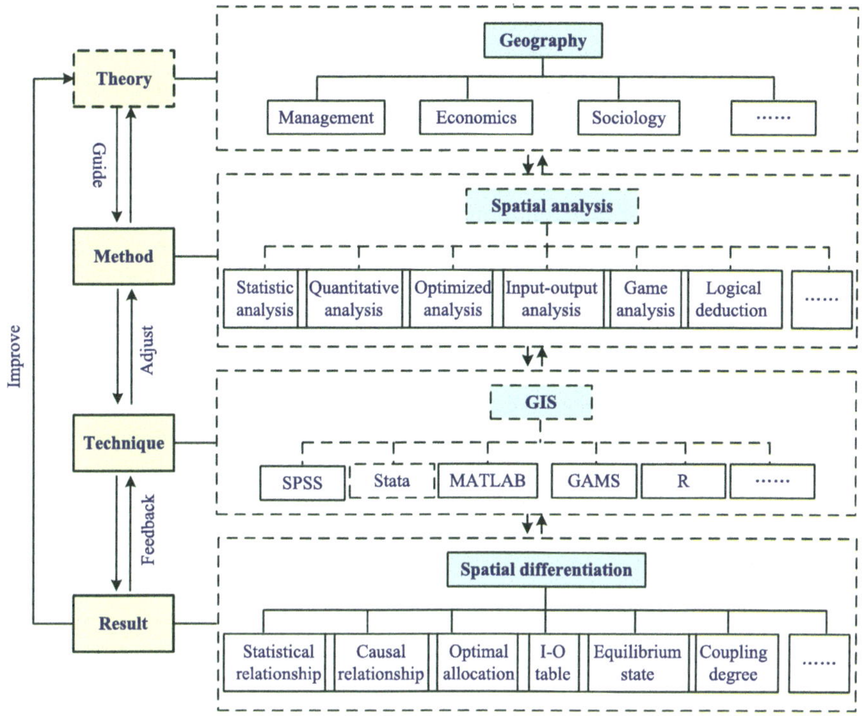

Fig. 1.4 The specific design of multidisciplinary research (Qu and Long 2018a; Long et al. 2020)

be attempted. For example, to study moderate-scale land use management in China, based on "the man-land interrelations territorial system theory" in geography, the regional differentiation rules of natural factors and the differentiation characteristics of socio-economic factors in different regions are outlined. Then, based on the scale economy theory and production theory in economics, quantitative analysis of specific regions can be carried out to explore moderate-scale management in accordance with local conditions.

At the methodological level, spatial analysis is frequently used in geography, and measurement and input-output analysis are common methods in economics. Qualitative logic deduction and quantitative statistical analysis methods are commonly used in management and sociology. The above methods can complement each other to meet the needs of multi-dimensional and multi-angle analysis in terms of time (stage), space, and quantity. For example, to analyze the choice of management mode in a specific region, we can perform qualitative analysis and make logical deductions about local policies, resource endowments, and mainstream characteristics to obtain plan options from the perspectives of management and sociology. Then, based on the field survey data, a quantitative measurement can be carried out to determine the final plan from the perspective of economics.

At the technical level, there needs a dual connection between different analysis software and different analysis data. Through spatial analysis using GIS related software, a preliminary analysis results can be obtained. The results could be exported to the SAV format (read by SPSS) to realize the statistical analysis of spatial information, to the DTA format (read by Stata) to realize the quantitative analysis of spatial information, to the DAT format (read by MATLAB) to realize the mathematical optimization analysis of spatial information.

At the result level, we can obtain the statistical relationship (such as correlation and duality) between different elements through statistical analysis and the causal relationship between different elements through quantitative analysis. Also, through mathematical optimization analysis, we can analyze the optimization relationship among the input ratios of various elements under different restrictions. Through the input-output analysis, the correlation relationship between resource input and economic output can be obtained. Through game theory analysis, an equilibrium state of the decision-making among different activity types can be obtained. Based on the above multi-means, multi-perspective, and multi-angle analysis, we can obtain the comprehensive cognition of the research object from development status, evolution laws, and future trends.

1.4.1.4 A Multidisciplinary Research Framework

The following questions need to be answered in the study of the recessive transition of land use: what is the recessive morphology of land use at different times or stages? What are the recessive transition patterns of land use in different regions? What is the transition mechanism between different recessive morphologies of land use? What kinds of economic effect, social effect, and environmental effect will be caused by different recessive transitions of land use? The above problems involve different disciplines, such as geography, management, economics, and so on. So only with comprehensive analysis from different dimensions with the integration of various disciplines, could these research be successfully completed. In order to determine the type of land use transition to be taken in a given region, and the measures to be taken to guide the implementation, a comprehensive analysis should be conducted from multiple dimensions of time (stage), space, and quantity (Fig. 1.5). The analysis process is similar to finding a point in a cube that satisfies three-dimensional conditions. When only a single dimension is considered, even if the information in this dimension is very accurate, the conclusions can only be specified to one plane. When considering the information in two dimensions, the conclusions could be narrowed down to a line, but there is still great uncertainty. Only by considering comprehensive information in three dimensions can we obtain pertinent and reliable research conclusions.

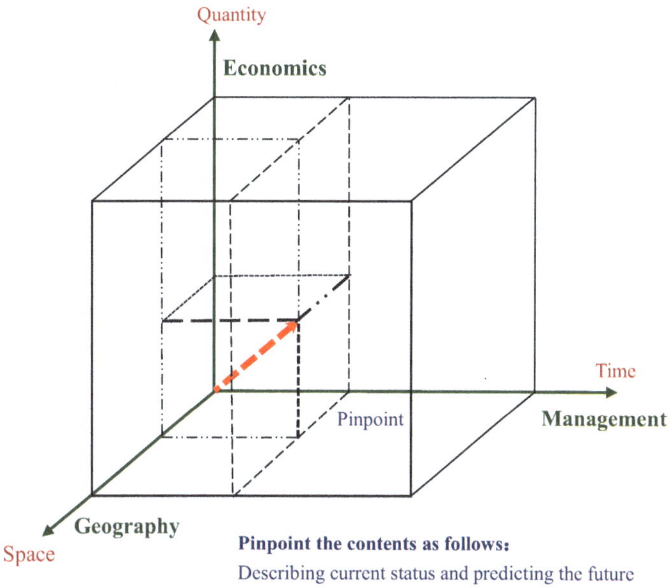

Fig. 1.5 A multidisciplinary research framework for recessive land use transition (Qu and Long 2018a; Long et al. 2020)

For example, quantitative analysis is frequently used in econometrics. However, if the production efficiency of different management scales is only analyzed from the quantitative dimension, and factors such as the differences in the natural conditions of different regions and the market levels of factors at different economic stages are ignored, it will lead to the coexistence of "positive," "reverse," and "non-linear" conclusions. At present, even the "inverse relationship" with a strong empirical basis is controversial. Some scholars have argued that the "inverse relationship" did not follow the objective law, as the imperfection of the factor market was ignored in the analysis (Heltberg 1998). Some scholars believed that it was caused by the omission of important natural difference variables (Chen et al. 2011). And others denied the above reasons to support the existence of such a law (Calogero et al. 2013). The reason why different views exist is that the analysis process cannot be comprehensively investigated within a unified framework, which leads to biased conclusions. For another example, spatial pattern analysis is commonly used in geography. However, if we only analyze the spatial pattern of the climate, soil, labor, and crop planting, the specific process of land use transition cannot be effectively detected. Meanwhile, it is also impossible to extract a transition mode that accords with the quantitative law of production and could be incorporated it into practice effectively. Therefore, research into the recessive transition of land use needs comprehensive research from multiple disciplines and perspectives. Details of this are described as follows.

Management can provide a macro time (stage) background for research from the time (stage) axis. For example, with technological development and institutional changes, what kinds of changes have taken place in the policies for land use and agricultural production? What changes have taken place in the property rights of land? How will the land use management mode and type evolve? From the time (stage) dimension, the research problem can be specified to a plane, which clarifies the problem definition.

Geography can provide regional background for research from the spatial axis. For example, with the change of natural zone, what is the spatial evolution law of natural factors, such as daylight, temperature, soil, and water? What are the spatial differences in the abundance of land resources, farmland quality, and cropping system? What is the spatial distribution of the land production function and ecological function? What is the spatial transfer pattern of the agricultural labor force? What is the spatial differentiation of the agricultural development level? As the regional characteristics of research objects were figured out, the research problem can be further specified from a plane to a straight line.

Economics can give more accurate positioning from the quantitative axis. For example, in a given region with a relatively uniform policy background, development stage, and natural economic characteristics, analysis can be conducted on the variation characteristics of efficiency indicators, such as land (or household, labor) production efficiency (physical or value), technical efficiency, total factor productivity with different conditions of management scale, management mode, and management type. After defining the specific position in the quantitative axis, the research problem is further specified to a certain point on the straight line, so the conclusion is more specific and reliable.

1.4.1.5 Summary

Based on the above shortcomings, through the continuous deepening and refinement at three levels, a multidisciplinary research framework for the recessive transition of land use was proposed in this section. At the level of conceptual framework, the multidisciplinary integration path with "theoretical reference, complementary methods, technical connection, and result analysis" as the main line was clarified. At the level of overall design, this section attempted to take the problem-oriented approach and discussed the potential integration of geography, management, and economics in research theory, analysis method, software connection, data fusion, and result mining. At the level of practical application, this section constructed a multidisciplinary research framework for the recessive transition of land use from the three dimensions of time, space, and quantity, and specifically expounded the theoretical logic of multi-dimensional comprehensive analysis, as well as the questions that different disciplines may answer in the corresponding dimensions.

The multiple attributes of the recessive morphology of land use form a multi-dimensional information space. The specific description of the recessive morphology of land use is similar to positioning a point satisfying different dimensional conditions in the multi-dimensional information space. When the recessive morphology of land use in the same region at different times (stages) is located in turn, the revealing process of the recessive transition law of land use is like seeking a specific curve connecting the above points in this space. The more information dimensions that can be obtained, the more accurate the positioning becomes. The description of information in different dimensions relies on different research from different disciplines, which justifies the inevitability of building a multidisciplinary research framework for recessive land use transition (Song 2017).

1.4.2 Horizontal Comparison

Land use transitions are often studied with long time series statistics. Although global environmental data have grown in abundance in recent decades, time-series data for China are unreliable, because the statistical system was disrupted several times since it was founded in 1949 by several campaigns that hampered economic development, such as Great Leap Forward (1958–1960) and Cultural Revolution (1966–1975). However, China has a vast territory with obvious regional differentiation in the level of socio-economic development (Fan et al. 2005; Lu and Fan 2010; Dunford and Liu 2015), which makes it possible to apply a spatial comparative research method to study land use transitions. This method is called as horizontal comparison research method, i.e., using the spatial differentiation in regional development to compensate data deficiencies in the long time series, reflecting the change process of land use with space to exchange for time (Fig. 1.6). The application of horizontal comparison research method combining with the following transect research method to the research of rural housing land transition can be seen in Chap. 3 (Long et al. 2007).

1.4.3 Transect

The terrestrial transects of the International Geosphere-Biosphere Programme (IGBP) emphasize on explaining issues concerning global change through gradient methods, and temperature, precipitation and land use are three key gradient factors (Koch et al. 1995). We can use this Transect as a coherent set of study sites to determine how land-use associated land-cover is changing as a result of social and environmental factors, and how these changes interacts with biophysical factors (Koch et al. 1995). Usually, land-use change is triggered by socio-economic factors (Li 1996; Heilig 1997; Cai 2001; Krausmann et al. 2003). Therefore, the socio-economic developmental level was considered the most important factor to define a study area for studying land use transitions.

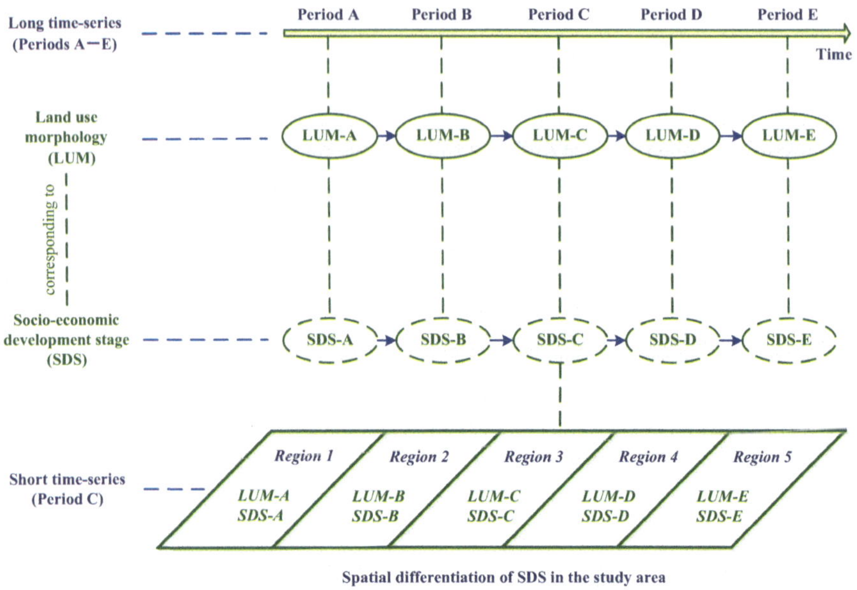

Fig. 1.6 A sketch map of horizontal comparison research method with space to exchange for time (Long et al. 2020)

Here, Transect of the Yangtse River (TYR) is taken as an example. In the Yangtse River Valley, we have identified a Transect in which the underlying gradient is the changing pattern of land use. The level of socio-economic development is sharply declining from the East of the Yangtse Transect to the West. TYR is depicted in Fig. 1.7. It comprises an area of about 1.12 M km^2, containing 312 counties (or cities at county level), which belong to the following eleven provinces—Shanghai, Jiangsu, Anhui, Jiangxi, Hubei, Hunan, Chongqing, Sichuan, Yunnan, Qinghai and Tibet. Along the Transect, there are several of the biggest cities in China—Shanghai (to the East), Nanjing (the capital of Jiangsu province), Wuhan (the capital of Hubei province) and Chongqing. TYR is about 3600 km in length, with an average width of 200 km or so (Fig. 1.7).

Recent years have witnessed the widespread use of novel methods incorporating quantitative analysis, integrated "3S" technology, and terrestrial transect. Introducing the transect to longitudinal comparison can overcome the deficiencies of time series data and combining with the above-mentioned horizontal comparison research method, which is of great significance for the researches of land use transitions and rural development in China. For example, the research team of agriculture geography and rural development of Institute of Geographic Sciences and Natural Resources Research, Chinese Academy of Sciences, has selected Transect of the Yangtse River, "Southern Jiangsu-Northern Shaanxi" Transect, and Transect along No.106 State Road to carry out related studies (Long et al. 2007, 2012b; Lu and Liu 2013).

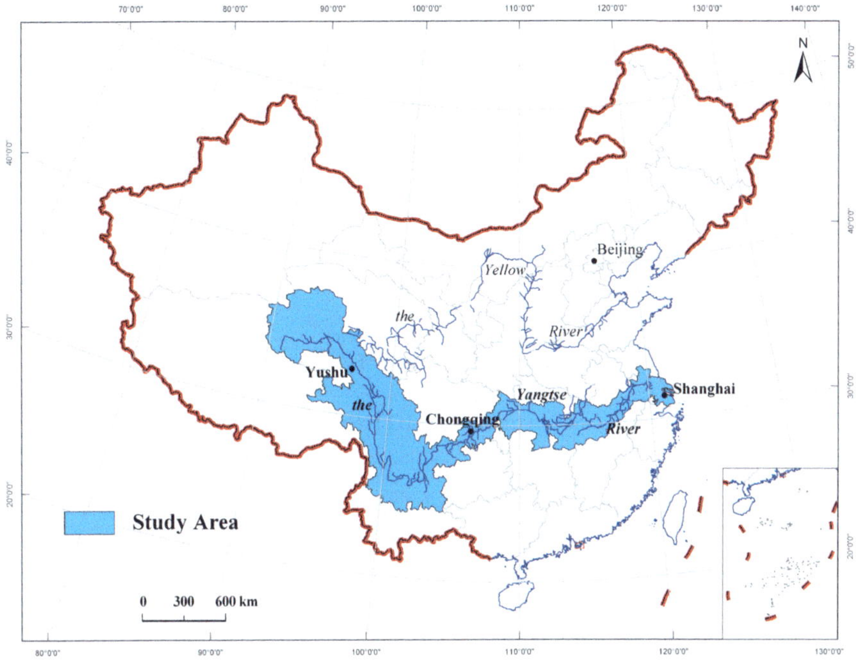

Fig. 1.7 Location of Transect of the Yangtse River (Long et al. 2007)

1.5 Research of Land Use Transitions in China

1.5.1 Overview

Recently, related researches on land use transitions combined with the characteristics of China's socio-economic development have been carried out rapidly since the research field of land use transition was introduced into China (Long 2001, 2003; Long and Li 2002). Over the last decade, the research on land use transitions has flourished in China and involved various aspects, e.g., related concept (Lu et al. 2006), the laws (Cai et al. 2009; Song et al. 2015), and the research framework (Song 2017; Hu et al. 2019a; Long 2012a), theory and hypotheses (Long et al. 2007; Song et al. 2019) of land use transition, the transition of the land use mode and structure (Liu and Li 2006; Long et al. 2018a), the transition of a particular land-use type (Long et al. 2007), the economic and environmental effects of land use transitions (Long et al. 2018b; Qu and Long 2018b), and the relationship between land use transitions and other land-use activities (Long 2003; Chen et al. 2014, 2020; Zhang et al. 2019a). The editor set up a special column of 'Land use transition' in the special issue of "Land Use Policy in China" published by the journal of *Land Use Policy* in 2014 (Long 2014a).

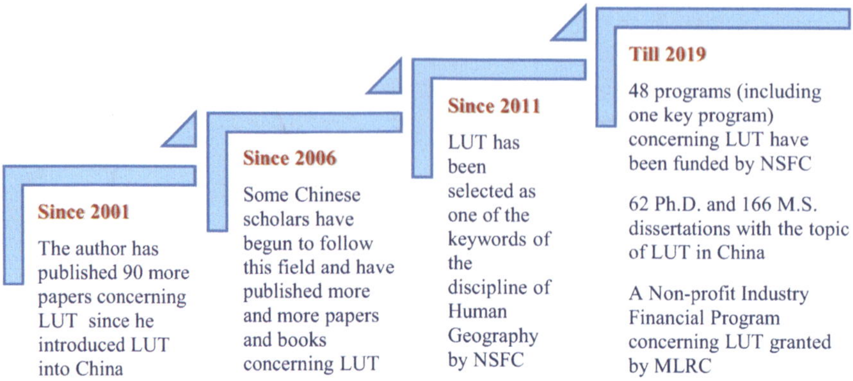

Fig. 1.8 The development of land use transition (LUT) research in China (Long et al. 2020)

To date, there have been 62 Ph.D and 166 M.S. dissertations on the topic of "land use transition" in China. Since 2011, "land use transition" has been selected as one of the keywords in the research field of "rural land use and allocation" under the discipline of Human Geography, which is a significant research brunch of the National Natural Science Foundation of China (NSFC). During the period of 2002–2019, the NSFC funded 48 research programs on the theme of "land use transition", which included one key program. In 2015, A Non-profit Industry Financial Program entitled "Management and Control Technologies and Policy Innovation of Land Use Transitions in the Middle Reaches of the Yangtze River Economic Belt" was granted by the Ministry of Land and Resources of China (MLRC), which demonstrated that land use transition has sparked great concern in both academic communities and the governmental authorities since the author introduced it into China in 2001 (Fig. 1.8). Chinese scholars have paid much attention to farmland transition, rural housing land transition and their relationships with rural restructuring under the policies context of urbanization, industrialization, 'Building the New Socialist Countryside' and subsequent 'Rural Vitalization'.[2]

[2]In 2006, the State Council of the People's Republic of China launched an epochmaking strategy on "Building the New Socialist Countryside" with the targets of 'advanced production, improved livelihood, clean and tidy villages, a civilized social atmosphere and efficient management' (Long et al. 2010). In 2017, the 19th National Congress of the Communist Party of China (CPC) put forward the "Rural Vitalization" strategy to alleviate the issues of uncoordinated urban-rural development with the basic principles concerning 'thriving industry, pleasant living environment, refined rural civilization, effective governance, and prosperous life' (Long et al. 2019).

1.5.2 Farmland Use Transition

When considering the quantity change of farmland, if, at a certain time, the change in the area of farmland presents a turning trend, indicating that the spatial morphology of farmland has changed. After socio-economic development reached a certain stage, the total farmland of a country or region may not necessarily show a recoverable net increase, but the net consumption rate of farmland will slow down or even stagnate significantly (Ge et al. 2018; Long et al. 2012a). Therefore, the transition of farmland can be understood as the process of a country or region's farmland changing from continuous and rapid consumption in the relatively low-level stage of socio-economic development to low-speed consumption and even stagnation in the relatively advanced stage of socio-economic development (Ge et al. 2018; Long et al. 2012a).

The total farmland and the annual net consumption rate are two major indicators for describing the dominant transition of farmland (Ge et al. 2018; Tan 2014). Diagnosing the transformation of the spatial morphology of farmland can start from exploring the turning point of the total area of farmland (Bertoni et al. 2018; Ge et al. 2018; You et al. 2018), which always fluctuates with policy reform, economic growth, and demographic change (Tan et al. 2009). The mutational point of farmland change includes the above-mentioned turning point. If after a certain time, the annual net consumption rate of farmland significantly slows down or even stagnates, indicating that the spatial transformation of farmland morphology occurred at that point. From the perspective of the spatial pattern, farmland is often fragmented and scattered at a lower-level of socio-economic development. With the socio-economic advancement, it tends to be concentrated. The management pattern and landscape pattern are proved to be two major indicators for depicting farmland transition (Long et al. 2007). The farmland management pattern refers to the concentration of plots managed by farmers or other users (Long et al. 2016; Ma et al. 2020). It can be measured by the management scale, such as the area of farmland per agricultural employee, the number of plots per household, and farm size. The landscape pattern can be indicated by multiple landscape attributes, such as the number of patches, the density of patches, average areas of the patches, etc. Land use function also provides a meaningful perspective of examining farmland transition. The definition of farmland function links farmland with multiple demands. When considering the formation process, it is deduced from various human needs through the combined effects of land characteristics, human senses, and socio-economic systems (Lu et al. 2013). Generally, it can be classified into three categories-production function, ecological function and social function (Zhang et al. 2018).

1.5.3 Rural Housing Land Transition

The connotations of rural housing land transition includes dominant transformations in quantity, space, and structure, and it should also fully reflect the changes in employment structure, lifestyle, and family structure corresponding with urban and rural development (Long et al. 2012a; Su et al. 2011). The contradiction between humans and land is gradually highlighted, which facilitates relevant research to take a human into the framework (Qu et al. 2019). The research focusing on per capita rural housing land is the direct reflection, which is simultaneously affected by changes in the total area of housing sites and population migration (Qu et al. 2019). Contrary to the fact that rural housing land area has generally increased, induced by urbanization and industrialization, farmers have migrated to urban areas temporarily or completely, thus causing the increase of rural housing land and population decline (Long et al. 2019). Existing researches on rural housing land transition is relatively rare in China, which mainly concentrates on the spatial distribution, utilization status, mode for land consolidation, and its functional transition process (Jiang et al. 2016). Several types of the spatial distribution of rural settlements (Qu et al. 2017, 2018), such as clustering, random, and uniform discrete distribution, were found in China with significant regional differences (Yang et al. 2016). The spatial distribution studies involve various scales, national/regional level, city level, county level, and village level, and massive empirical researches have been finished in China (Ma et al. 2018). The utilization intensity of rural settlements is also highlighted under the context of the huge demand for land. Much attention has been given to the abandoned rural residential land (Zhang et al. 2019b), which attempts to explore the spatial heterogeneity, influential factors, formation mechanism, and consolidation potential and mode. There are several pieces of research tried to theorize rural housing land transition by linking it with population mobility and urbanization process (Qu et al. 2019). Besides, it has been revealed that the function of rural housing land tends to be diversified (Ma et al. 2019; Zhu et al. 2014).

1.5.4 Land Use Transitions and Rural Restructuring

Rural restructuring, which is internationally prevalent, has the most overlapped meanings with rural transformation development (Liu 2011; Long 2012b; Long and Liu 2016). In the process of rural restructuring, the functional and structural transformation of the rural territorial system constitutes the interaction of land use transition and rural restructuring (Long and Liu 2016; Tu and Long 2017). The research on land use transitions aims to reveal the non-linear change process of regional land use patterns (including dominant and recessive morphologies) (Long and Qu 2018), while the rural transformation development focuses on exploring the trend of the restructuring process (Hoggart and Paniagua 2001). At different stages of socio-economic development, the characteristics of the man-land interrelations in the rural territorial system are significantly different (Hu et al. 2019b). Regional land use patterns and rural development status also show significant spatial and temporal heterogeneity. Researches have revealed that the change process of regional land use pattern conflict is the internal mechanism that drives the evolution of land use pattern (Long and Qu 2018). The regional pattern of rural man-land interrelations corresponding to land use conflicts will also change, which has also become a key factor in determining the process of rural restructuring (Chen et al. 2012). Therefore, the conflict of regional land use patterns is closely related to the driving force of rural restructuring (Long et al. 2016). The response and feedback process of the regional pattern of rural man-land interrelations to the two forces constitutes a bridge between land use transitions and rural restructuring (Long and Tu 2017, 2018). Land use transition and its coupling and coordination process with rural restructuring is the key to rural vitalization (Liu and Li 2017; Liu 2018). Land use transition studies provide a new perspective for an in-depth understanding of land use changes by characterizing the trend variation of land use patterns in both dominant and recessive states (Long 2014b; Long and Qu 2018). At the same time, significant changes have taken place in the rural population, land, industry, culture and society. In the process of urban-rural migration, land allocation between urban and rural, industry transformation and upgrading, and the evolution of a socio-cultural system are all deeply affected by the rural restructuring process.

Land use transition, a dynamic process, is driven by a set of forces, such as the input of capital and labor, industry development, employment and population mobility, which have tight associations with rural restructuring (Fig. 1.9). In turn,

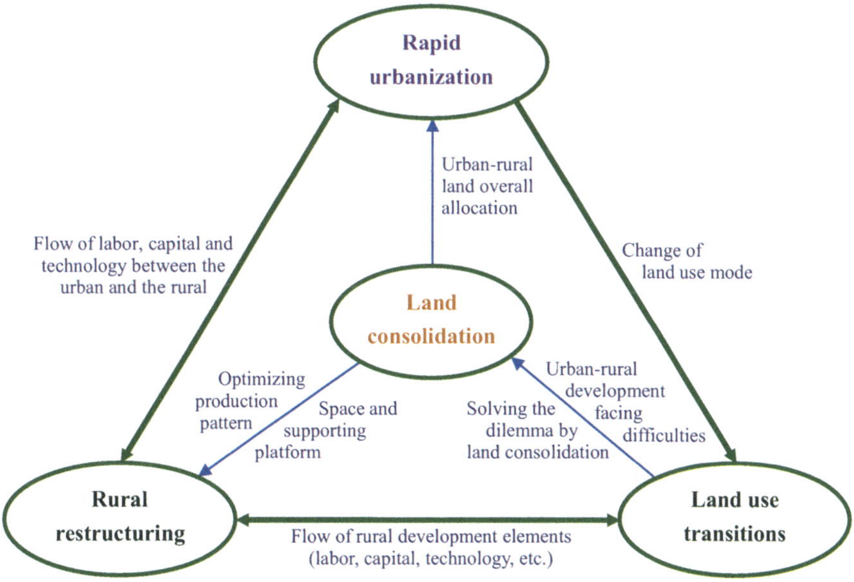

Fig. 1.9 Land use transitions and rural restructuring under the pressure of rapid urbanization (Long 2014a)

the status quo of land use can mirror various issues in the process of rural development. Usually, the spatio-temporal inseparability of socio-economic advancement is transferred to land, which is the main carrier of socio-economic activities, creating a very complex land use pattern. While, changes in this land use pattern will affect the process of regional natural, ecological and social development. In conclusion, land use transitions and rural restructuring are mutual feedback processes. Coordinating the coupling relationship between land use transitions and rural restructuring is of great significance for optimizing the rural transformation development process and the integrated development of the urban and the rural (Tu et al. 2018; Zhang et al. 2016; Fang and Yu 2016).

1.6 Conclusions

This chapter aims to theorize land use transitions and construct a theoretical foundation of this book. With the in-depth research of land use transitions, the concept and connotations of land use morphology were further developed and expanded as two kinds, i.e., dominant morphology and recessive morphology. The dominant morphology refers to the land use structure of a certain region over a certain period of time, with features such as the quantity (area, proportion) and spatial pattern of land use types. While the recessive morphology is a special morphology which relies on the dominant morphology but can only be observed by the means of analyzing, testing, monitoring and surveying, includes the land use features in the aspects of quality (nutrient, pollution, degradation), property rights (state-owed, collective-owed), management mode (individual, joint-stock system, transfer and large-scale management), input (capital, technology, labor), output (yield, output value, input-output ratio) and function (production, living, ecology, culture). Accordingly, the concept of land use transition may be further developed as the changes in land use morphologies, including dominant morphology and recessive morphology, of a certain region over a certain period of time driven by socio-economic change and innovation, and it usually corresponds to the transformation of the socio-economic development stage.

Accordingly, the theoretical model of regional land use transitions is established as follows: with the socio-economic development, transformations between different land use types during a certain period of time cause the change of the conflicts resulted from regional land use morphology pattern from strong to weak, i.e., a trend towards coordination; these transformations will lead to a new balance of regional land use morphology pattern consists of different land use types reflecting the development trend of corresponding economic departments, respectively, and finally, realize the qualitative transformation of urban-rural land use system.

The extension of land use morphology brings about opportunities and challenges as the qualitative aspect of land use transitions is reflected by the changes of recessive land use morphology, especially the changes of functions of land use system driven by the comprehensive variations in the aspects of quality, property rights, management mode, input and output of land use, which is difficult to be measured or represented.

As such, this chapter puts forward three innovative integrated approaches to study land use transitions, i.e., multidisciplinary research framework for recessive land use transition, horizontal comparison research method, and transect research method.

With the introduction of land use transition research into China by the author, related researches combining land use transitions with the sustainability of rural China have been carried out extensively, which enrich the knowledge of land system science. The research fields majorly focus on farmland transition, rural housing land transition and their relationships with rural restructuring under the context of urbanization, industrialization and rural vitalization.

References

Barbier EB, Burgess JC, Grainger A (2010) The forest transition: towards a more comprehensive theoretical framework. Land Use Pol 27(2):98–107

Bardhan PK (1973) Size, productivity, and returns to scale: an analysis of farm-level data in Indian agriculture. J Polit Econ 81(6):1370–1386

Bertoni D, Aletti G, Ferrandi G et al (2018) Farmland use transitions after the cap greening: a preliminary analysis using markov chains approach. Land Use Pol 79:789–800

Cai Y (2001) 土地利用/土地覆被变化研究: 寻求新的综合途径 (A study on land use/cover change: The need for a new integrated approach). Geographic Res 20(6):645–652

Cai Y, Lu D, Zhou Y et al (2004) 中国地理科学的国家需求与发展战略 (National demands for and development strategies of Chinese geography). Acta Geographica Sinica 59(6):811–819

Cai Y, Wang Y, Li Y (2009) 中国耕地供需变化规律研究 (Study on changing relationship of demand and supply of cultivated land in China). China Land Sci 23(3):11–18, 31

Calogero C, Savastano S, Zezza A et al (2013) Fact or artifact: The impact of measurement errors on the farm size-productivity relationship. J Dev Econ 103:254–261

Chen C (1998) 区域综合开发的理论与案例 (Theories and empirical studies of regional comprehensive development). Science Press, Beijing

Chen K, Long H, Liao L et al (2020) Land use transitions and urban-rural integrated development: Theoretical framework and China's evidence. Land Use Pol 92:104465

Chen R, Ye C, Cai Y et al (2014) The impact of rural outmigration on land use transition in China: Past, present and trend. Land Use Pol 40:101–110

Chen Z, Huffman WE, Rozelle S (2011) Invrese relationship between productivity and farm size: the case of China. Contemp Econ Policy 29(4):580–592

Chen Y, Zhang Y, Liu Y (2012) 江苏省典型样带区农业发展的地域差异 (Regional differences of agricultural development in a typical transect in Jiangsu Province). Geographic Res 29(7):1317–1323

Dunford M, Liu W (eds) (2015) The Geographical Transformation of China. Routledge, Abingdon

Fan J (2019) 中国人文地理学70年创新发展与学术特色 (The progress and characteristics of Chinese human geography over the past 70 years). Scientia Sinica Terrae 49(11):1697–1719

Fan J, Heberer T, Taubmann W (2005) Rural China: Economic and Social Change in the Late Twentieth Century. M. E. Sharpe, Armonk

Fang C, Yu D (2016) China's New Urbanization. Science Press & Spriner Press, Beijing

Fang C, Yu D (2020) China's Urban Agglomerations. Science Press and Springer Nature Singapore

Fang C, Liang L, Wang Z (2019) Quantitative simulation and verification of upgrade law of sustainable development in Beijing-Tianjin-Hebei urban agglomeration. Sci China Earth Sci 62(12):2031–2049

Ferreira MP, Alves DS, Shimabukuro YE (2015) Forest dynamics and land-use transitions in the Brazilian Atlantic Forest: the case of sugarcane expansion. Reg Envir Chang 15(2):365–377

Foley JA, DeFries R, Asner GP et al (2005) Global consequences of land use. Science 309(5734):570–574
Fu B (2014) 地理学综合研究的途径与方法: 格局与过程耦合 (The integrated studies of geography: Coupling of patterns and processes). Acta Geographica Sinica 69(8):1052–1059
Ge D, Long H, Zhang Y et al (2018) Farmland transition and its influences on grain production in China. Land Use Pol 70:94–105
GLP (*Global Land Project*) (2005) Science plan and implementation strategy (0264–8377). http://www.sciencedirect.com/science/article/pii/S0264837705000591
Grainger A (1986) The future role of the tropical rain forests in the world forest economy. Dissertation, University of Oxford
Grainger A (1995) National land use morphology: patterns and possibilities. Geography 80(3):235–245
Harvey D (1969) Explanation in geography. Edward Arnold, London
Heilig GK (1997) Anthropogenic factors in land-use change in China. Popul Dev Rev 23(1):139–168
Heltberg R (1998) Rural market imperfections and the farm size—productivity relationship: evidence from Pakistan. World Dev 26(10):1807–1826
Hoggart K, Paniagua A (2001) What rural restructuring? J Rural Stud 17(1):41–62
Hu S, Tong L, Long H (2019a) 论土地利用转型潜力及其评价的理论框架 (Land use transition potential and its assessment framework. Geographical Research). Geographic Res 38(6):1367–1377
Hu X, Li H, Zhang X et al (2019b) Multi-dimensionality and the totality of rural spatial restructuring from the perspective of the rural space system: a case study of traditional villages in the ancient Huizhou region, China. Habitat Int 94:102062
Jadin I, Meyfroidt P, Lambin E (2016) International trade, and land use intensification and spatial reorganization explain Costa Rica's forest transition. Environ Res Lett 11(3):035005
Jiang G, He X, Qu Y, Zhang R et al (2016) Functional evolution of rural housing land: a comparative analysis across four typical areas representing different stages of industrialization in China. Land Use Pol 57:645–654
Koch GW, Scholes RJ, Steffen WL et al (1995) The IGBP terrestrial transects. Science plan. http://www.scopus.com/inward/record.url?scp=85041152131partnerID=8YFLogxK
Krausmann F, Haberl H, Schulz NB et al (2003) Land-use change and socio-economic metabolism in Austria—part I: driving forces of land-use change: 1950–1995. Land Use Pol 20(1):1–20
Lambin EF (2012) Global land availability: malthus versus Ricardo. Glob Food Secur 1(2):83–87
Lambin EF, Meyfroidt P (2010) Land use transitions: socio-ecological feedback versus socio-economic change. Land Use Pol 27(2):108–118
Li X (1996) 全球环境变化研究的核心领域—土地利用/土地覆被变化的国际研究动向 (A review of the international researches on land use/land cover change). Acta Geographica Sinica 63(6):553–558
Li X (2002) 土地利用变化的解释 (Explanation of land use changes). Progress in Geography 21(3):195–203
Liu W (2019) The Belt and Road Initiative: A Pathway Towards Inclusive Globalization. Routledge, London
Liu Y (2011) 中国新农村建设地理论 (Geography of New Countryside Construction in China). Science Press, Beijing
Liu Y (2018) 新时代城乡融合与乡村振兴 (Research on the urban-rural integration and rural revitalization in the new era in China). Acta Geographica Sinica 73(4):637–650
Liu C, Li X (2006) 1980年以来中国农地利用变化的区域差异 (Regional differences in the changes of the agricultural land use in China during 1980–2002). Acta Geographica Sinica 61(2):139–145
Liu Y, Li Y (2017) Revitalize the world's countryside. Nature 548(7667):275–277
Liu Y, Wang L, Long H (2008) Spatio-temporal analysis of land-use conversion in the eastern coastal China during 1996–2005. J Geogr Sci 18(3):274–282

Liu Y, Liu Y, Chen Y et al (2010) The process and driving forces of rural hollowing in China under rapid urbanization. J Geogr Sci 20(6):876–888
Long H (2001) 长江沿线样带土地利用变化与土地可持续利用 (Land Use Changes and Sustainable Land Management in Transect of the Yangtse River). Post-doctoral Research Report, IGSNRR, CAS
Long H (2003) 土地利用转型——土地利用/覆被变化综合研究的新途径 (Land use transition: a new integrated approach of land use/cover change study). Geograph Geo-infor Sci 19(1):87–90
Long H (2012a) 论土地利用转型与乡村转型发展 (Land use transition and rural transformation development). Progress in Geography 31(2):131–138
Long H (2012b) 中国乡村转型发展与土地利用 (Land use and rural transformation development in China). Science Press, Beijing
Long H (2014a) Land consolidation: an indispensable way of spatial restructuring in rural China. J Geogr Sci 24(2):211–225
Long H (ed) (2014b) Land use policy in China. Land Use Pol 40:1–146
Long H (2015) 论土地利用转型与土地资源管理 (Land use transition and land management). Geographic Res 34(9):1607–1618
Long H, Li X (2002) 区域土地利用转型分析——以长江沿线样带为例 (Analysis on regional land use transition: a case study in transect of the Yangtze River). J Nat Res 17(2):144–149
Long H, Li T (2012) The coupling characteristics and mechanism of farmland and rural housing land transition in China. J Geogr Sci 22(3):548–562
Long H, Liu Y (eds) (2016) Rural restructuring in China. J Rural Stud 47(Part B):387–676
Long H, Tu S (2017) 论乡村重构 (On rural restructuring). Acta Geogr Sin 72(4):563–576
Long H, Tu S (2018) 土地利用转型与乡村振兴 (Land use transition and rural vitalization). China Land Science 32(7):1–6
Long H, Qu Y (2018) Land use transitions and land management: A mutual feedback perspective. Land Use Pol 74:111–120
Long H, Woods M (2011) Rural restructuring under globalization in eastern coastal China: what can be learned from Wales? J Rural Community Dev 6(1):70–94
Long H, Heilig GK, Li X et al (2007) Socio-economic development and land-use change: analysis of rural housing land transition in the Transect of the Yangtse River, China. Land Use Pol 24:141–153
Long H, Liu Y, Li X et al (2010) Building new countryside in China: a geographical perspective. Land Use Pol 27:457–470
Long H, Liu Y, Li T (2015) Agricultural production in China under globalization. In: Robinson GM, Carson DA (eds) Handbook on the Globalisation of Agriculture. Edward Elgar, pp 214–234
Long H, Li Y, Liu Y, Woods M et al (2012a) Accelerated restructuring in rural China fueled by 'increasing vs. decreasing balance' land-use policy for dealing with hollowed villages. Land Use Pol 29(1):11–22
Long H, Zou J, Li T et al (2012b) 乡村转型发展特征评价及地域类型划分——以"苏南-陕北"样带为例 (Study on the characteristics and territorial types of rural transformation development: the case of "Southern Jiangsu-Northern Shaanxi" transect). Geographic Res 31(3):144–149
Long H, Tu S, Ge D et al (2016) The allocation and management of critical resources in rural China under restructuring: Problems and prospects. J Rural Stud 47:392–412
Long H, Ge D, Zhang Y et al (2018a) Changing man-land interrelations in China's farming area under urbanization and its implications for food security. J Environ Manage 209:440–451
Long H, Qu Y, Tu S et al (2018b) 城镇化背景下中国农区土地利用转型及其环境效应研究: 进展与展望 (Land use transitions under urbanization and their environmental effects in the farming areas of China: research progress and prospect). Adv Earth Sci 33(5):455–463
Long H, Qu Y, Tu S et al (2020) Development of land use transitions research in China. J Geogr Sci 30(7):1195–1214
Long H, Zhang Y, Tu S (2019) Rural vitalization in China: a perspective of land consolidation. J Geogr Sci 29(4):517–530
Lu D (2011) 中国地理学的发展与全球变化研究 (Development of geographical sciences and research on global change in China). Acta Geogr Sin 66(2):147–156

Lu S, Liu Y (2013) 106国道沿线样带区农村空心化土地整治潜力研究 (Rural land consolidation potential of typical transect along No. 106 state road). J Nat Res 28(4):537–549
Lu D, Fan J (eds) (2010) Regional Development Research in China: A Roadmap to 2050. Science Press and Springer Press, Beijing
Lu D, Shi P, Dong G (2006) 体现土地利用转型的重要理念(Important ideas embodying land use transition). China Land and Resources News 2006/07/17/005
Lu S, Guan X, Liu Y et al (2013) 农业地域类型的识别及其演进特征——以106国道沿线典型样带区为例 (Agricultural region type identification and its spatio-temporal evolution characteristics: a case study of sampling belt along G106 in China). Progress in Geography 32(4):637–648
Ma W, Jiang G, Zhang R et al (2018) Achieving rural spatial restructuring in China: A suitable framework to understand how structural transitions in rural residential land differ across peri-urban interface? Land Use Pol 75:583–593
Ma W, Jiang G, Li W et al (2019) Multifunctionality assessment of the land use system in rural residential areas: confronting land use supply with rural sustainability demand. J Environ Manage 231:73–85
Ma L, Long H, Tu S et al (2020) Farmland transition in China and its policy implications. Land Use Pol 92:104470
Mather AS (1992) The forest transition. Area 24(4):367–379
Mather AS, Fairbairn J, Needle CL (1999) The course and drivers of the forest transition: the case of France. J Rural Stud 15(1):65–90
Meyfroidt P, Lambin EF (2008) The causes of the reforestation in Vietnam. Land Use Pol 25(2):182–197
Meyfroidt P, Lambin EF, Erb KH et al (2013) Globalization of land use: distant drivers of land change and geographic displacement of land use. Curr Opin Environ Sustain 5(5):438–444
Meyfroidt P, Roy Chowdhury R, de Bremond A et al (2018) Middle-range theories of land system change. Glob Environ Change 53:52–67
Nuissl H, Haase D, Lanzendorf M et al (2009) Environmental impact assessment of urban land use transitions—a context-sensitive approach. Land Use Pol 26(2):414–424
Qu Y, Long H (2018a) 中国耕地利用隐性形态转型的多学科综合研究框架 (A framework of multi-disciplinary comprehensive research on recessive farmland transition in China). Acta Geogr Sin 73(7):1226–1241
Qu Y, Long H (2018b) The economic and environmental effects of land use transitions under rapid urbanization and the implications for land use management. Habitat Int 82:113–121
Qu Y, Jiang G, Zhao Q et al (2017) Geographic identification, spatial differentiation, and formation mechanism of multifunction of rural settlements: a case study of 804 typical villages in Shandong province, China. J Clean Prod 166:1202–1215
Qu Y, Jiang G, Yang Y et al (2018) Multi-scale analysis on spatial morphology differentiation and formation mechanism of rural residential land: a case study in Shandong province, China. Habitat Int 71:135–146
Qu Y, Jiang G, Li Z et al (2019) Understanding rural land use transition and regional consolidation implications in China. Land Use Pol 82:742–753
Robinson GM, Carson DA (2015) Handbook on the Globalisation of Agriculture. Edward Elgar, Cheltenham • Northampton
Rudel TK, Schneider L, Uriarte M (2010) Forest transitions: an introduction. Land Use Pol 27(2):95–97
Sen AK (1962) An aspect of Indian agriculture. Economic Weekly 14(4–6):243–246
Song X (2017) 论土地利用转型的研究框架 (Discussion on land use transition research framework). Acta Geogr Sin 72(3):471–487
Song X, Li X (2019) 区域耕地利用功能转型的理论解释与实证 (Theoretical explanation and case study of regional cultivated land use function transition). Acta Geogr Sin 75(4):992–1010
Song X, Huang Y, Wu Z, Ouyang Z (2015) Does cultivated land function transition occur in China? J Geogr Sci 25(7):817–835

Su S, Zhang Q, Zhang Z et al (2011) Rural settlement expansion and paddy soil loss across an ex-urbanizing watershed in eastern coastal China during market transition. Reg Envir Chang 11(3):651–662
Tan M (2014) The transition of farmland production functions in metropolitan areas in China. Sustainability 6(7):4028–4041
Tan R, Beckmann V, van den Berg L et al (2009) Governing farmland conversion: comparing China with the Netherlands and Germany. Land Use Pol 26(4):961–974
Tu S, Long H (2017) Rural restructuring in China: theory, approaches and research prospect. J Geogr Sci 27(10):1169–1184
Tu S, Long H, Zhang Y et al (2018) Rural restructuring at village level under rapid urbanization in metropolitan suburbs of China and its implications for innovations in land use policy. Habitat Int 77:143–152
TurnerII BL, Munroe DK (2020) Land Change Science/Land System Science. In: Kobayashi A (ed) International Encyclopedia of Human Geography (Second Edition). Elsevier, 87–92
Wei H (2019) Urbanization in China: The Path to Harmony and Prosperity. Social Sciences Academic Press and Springer Nature Singapore
Yang R, Xu Q, Long H (2016) Spatial distribution characteristics and optimized reconstruction analysis of China's rural settlements during the process of rapid urbanization. J Rural Stud 47:413–424
You H, Hu X, Wu Y (2018) Farmland use intensity changes in response to rural transition in Zhejiang province, China. Land Use Pol 79:350–361
Zhang Q, Sun Z, Wu F et al (2016) Understanding rural restructuring in China: the impact of changes in labor and capital productivity on domestic agricultural production and trade. J Rural Stud 47:552–562
Zhang Y, Long H, Ma L et al (2018) Farmland function evolution in the Huang-Huai-Hai Plain: processes, patterns and mechanisms. J Geogr Sci 28(6):759–777
Zhang B, Sun P, Jiang G et al (2019a) Rural land use transition of mountainous areas and policy implications for land consolidation in China. J Geogr Sci 29(10):1713–1730
Zhang R, Jiang G, Zhang Q et al (2019b) Does urbanization always lead to rural hollowing? Assessing the spatio-temporal variations in this relationship at the county level in China 2000–2015. J Clean Prod 220:9–22
Zhu F, Zhang F, Li C et al (2014) Functional transition of the rural settlement: analysis of land-use differentiation in a transect of Beijing, China. Habitat Int 41:262–271

Chapter 2
Farmland Use Transition in China

Abstract Farmland is the key resource for sustaining rural development and agricultural production. Farmland use morphology is closely related to rural development status. This chapter establishes a theoretical analysis framework for farmland use transition in China from the perspectives of "structure-function" and "supply-demand", and analyzes the process and mechanism of farmland use transition from the perspectives of intensive use level, grain production ability, changes of agricultural labor and function evolution of farmland. Presently, China's land use structure is closely related to economic development, and the farmland use in China has experienced several phases from slow transition, rapid transition to steady transition and innovation transition. In view of the current challenges of farmland, this chapter puts forward a series of policy recommendations, which include deepening the reform of the land system, combining the utilization of farmland with the optimization of land use space, promoting land consolidation, strengthening the zoning planning and governance of farmland, and promoting the construction of ecological civilization. In the new era, the use of farmland tends to be a high-quality transformation and innovation, and the functions of farmland tend to be more diversified and coordinated. Therefore, in the future, we should focus on the transformation of farmland utilization caused by the change in human demand and new policy instruments, as well as the impact of farmland use transition on farmers' livelihoods, ecological environment and agricultural structure change around the world.

2.1 Introduction

Farmland is an important part of rural land use structure, and the farmland use transition (FUT) is a vital content of land use transitions, and it is also an important basis for presenting the process of rural transformation development. FUT not only changes the way of farmland use, but also the important indication of the process of rural transformation. Whether the evolution process of farmland use quantity and structure characteristics, or the evolution of farmland use efficiency, ownership and function characteristics, the role of farmland use in rural development can be

H. Long, *Land Use Transitions and Rural Restructuring in China*,
https://doi.org/10.1007/978-981-15-4924-3_2

demonstrated from different aspects. The process of FUT is closely related to rural transformation development, and it is also the core component driving the evolution of the rural man-land interrelations. The change in the quantity and the intensive level of farmland use has a direct influence on the pattern of food security. FUT is important to the protection of farmland use, and further research on the FUT could provide policy support for ensuring food security.

From the perspectives of "structure-function" and "supply-demand", this chapter tries to establish a theoretical analysis framework for FUT research under the background of rural transformation development, with emphasis on the analytical path of FUT from the perspective of multi-disciplinary. It also tries to construct the analysis scheme of FUT from the perspectives of intensive use (intensive use of farmland and grain security, intensive use of greenhouse-led farmland and rural development), grain productive ability (the relationship between per capita farmland area and grain production, the matching relationship between farmland resources and grain production), agricultural labor changes (the change of agricultural labor force and the relationship between the aging of rural population and grain production), and the evolution of farmland function. Finally, a integrated perspective concerning dominant transition (focusing on spatial structure and amount of farmland) and recessive transition (focusing on the social, economic and ecological functions of farmland) of farmland use is conducted.

2.2 Theoretical Framework

2.2.1 Farmland Use Transition and Rural Transformation Development

Farmland has not only the production function (economic benefit), but also the ecological service function (ecological benefit) and social security function (social benefit). As a Chinese saying goes, "Food is essential to the people, and farmland is the foundation of food." farmland is the carrier of grain production and the fundamental guarantee of food security. In the absence of breakthroughs in agricultural science and technology, there must be a steady supply of farmland with a certain quantity and quality to stabilize and improve grain production ability and ensure food security. China has a population of more than 1.4 billion, half of whom live in rural areas. To some extent, farmland meets the needs of subsistence security such as employment, old-age care of farmers. Therefore, farmland is an important foundation of agricultural stability and rural development.

With the rapid process of urban-rural transformation development (URTD) in China, the farmland use is also in constant transformation, not only reflected in the quantity, but also reflected in the three functions mentioned above: (1) Under the market economy system, the less efficient farmland has the tendency to be converted into more efficient land, and the farmland will be further lost; (2) The living standard of farmers is changing from "adequate food and clothing" to "well-off". The level of per capita consumption is rising rapidly, and demand for food and other agricultural products will expand further and farmland will be even more deficient; (3) The shift from extensive to intensive management of farmland is conducive to increasing the yield of per unit area and making up for the negative impact on food security, but the potential is not unlimited due to the law of diminishing returns; (4) Aagricultural and rural areas have changed from "be deprived" by industrialization to "be fed back" by industry and cities. Industry and cities have gained a certain strength and can replace scarce land resources with capital and labor input; (5) The urbanization of the rural population is conducive to the concentration of farmland into skilled farmers, thus gradually realizing the scale management and improving agricultural efficiency; (6) At present, the prices of major grain varieties in China are already higher than those of the international market. However, agricultural products such as live livestock, meat and its products, tea and textile fibers still have comparative advantages, so we can use their comparative advantages to carry out international trade in selected agricultural products, but we cannot rely on imported grain to ensure the food security of China.

The history of land use activities is, to a large extent, the expansion and contraction of farmland in space. The change of farmland and its related rural labor force employment transfer, urbanization, local economic development and other influences on rural development are also mainly reflected in the change of farmland management mode and rural industrial structure. Since the reform and opening up, the rapid development of China's economy has driven profound changes in the rural industrial structure, especially in the developed areas. Taking the eastern coastal areas of China as an example, after farmland was expropriated for urban and industrial use, the low-benefit grain production is being replaced by high-quality and efficient vegetable and seedling horticultural economy driven by market demand, which promotes the development of high-efficiency agriculture.

However, after more than four decades of large-scale development, the man-land interrelations in the eastern coastal areas have undergone great changes. The once fertile commodity grain producing base is now an urbanized area, and the problem of farmland loss and agricultural declining is increasingly prominent. With the FUT, the mode of production of farmers and the ways to obtain economic income have changed, farmland operation is no longer the major or the only means to make a living, the function of guaranteeing farmers' survival and development has been weakened, and their production and life styles have certain characteristics of non-agriculturalization and urbanization. Therefore, how to deal with the changing characteristics of different stakeholders and the relationship between them and FUT will directly affect rural transformation development.

With the rapid expansion of urban-rural construction land, resulting in a large amount of farmland occupied, the contradiction between man and land is increasingly sharp. According to the statistics of the Ministry of Land and Resources, from 1996 to 2008, China's farmland area decreased from 1.95 billion mu to 1.83 billion mu (1 ha = 15 mu). At present, the farmland per capita is less than 1.37 mu, less than 40% of the world average. Farmland continues to decrease, and the pressure to stick to the "red line" of 1.8 billion mu farmland is increasing. At the same time, to ensure national ecological security, we also need to strengthen the protection of agricultural land with ecological functions, especially farmland. However, with the intensification of farmland protection and ecological construction, the land resources available for new construction land in China are very limited, and the supply of construction land will face unprecedented pressure. The rapid advance of industrialization and urbanization will inevitably occupy part of farmland. The reserve resources of farmland in China are small, and the constraint of the ecological environment is large, which seriously restricts the capacity of land development and the quantity of farmland supplements. This will bring a profound impact on farmland protection and FUT.

2.2.2 *Recessive Transition of Farmland Use: A Managerial Perspective*

2.2.2.1 Research on the Recessive Transition of Farmland Use

Although the quantity and space of farmland in a specific area remains unchanged for a long time, land use transition still occurs with the development of the economy and society. Such transitions manifest more as part of the evolutionary trend for the recessive morphology of land use. The recessive transition of farmland use involves changes in various recessive morphology attributes. The first example, the change in farmland property rights: the ownership of farmland changes from collective to the state, the use right is separated from the ownership, the contracting right of the farmland is separated from the use right, and the management right of the farmland is separated from the contracting right. Second, the changes in the farmland management modes: single-household farming in the early days of the People's Republic of China, cooperative management in the period of the People's Commune, household contract management in the early days of reform and opening-up, and farmland circulation and scale management with continuous development in recent years. Third, the change in the input-output of farmland, which mainly manifests in the change of land use efficiency accompanied by the change in management modes and management scale, such as the change of per capita grain output, per capita income, per capita production cost or profit margin, and so on. Fourth, the change in the farmland functions. Farmland has the functions of grain production, production activity bearing, and the social security of farmers. The primary and secondary functions change with the evolution of regional economic development.

Land use is a mirror of society (Tuan 1971). The status and problems of farmland use can reflect those of China's agricultural and rural development. Since the founding of the People's Republic of China, the quantity of farmland resources in China has gone through three stages: fluctuating growth before the reform and opening up, a significant increase after the reform and opening up, and the rapid decrease that followed (Ge et al. 2000). In order to ensure food security and alleviate farmland loss, the Fourth Session of the Tenth National People's Congress passed the Outline of the Eleventh Five-year Plan for National Economic and Social Development in 2006, which takes 1.8 billion mu of farmland as a binding index with legal effect. The establishment of a farmland protection red line and the planning of basic farmland[1] have protected farmland use in terms of quantity and space. However, in order to achieve the strategic goal of ensuring food security and increasing farmers' income, more research should be conducted into the recessive transition of farmland use. In 1987, the No. 5 Central Document put forward clearly for the first time that we should actively and steadily promote proper scale management of farmland where conditions permit. Since then, the recessive morphology transition of farmland use has gained more attentions. On the basis of the original farmland, these transitions affect the current agricultural development and rural appearance in terms of property right structures, management modes, resource utilization efficiency, and land function evolution, among other things.

Farmland has multiple attributes. It has natural attributes as a natural resource, economic attributes as an agricultural production factor, asset attributes as a land use right carrier, and social attributes as a type of rural social security. Therefore, the research into the recessive transition of farmland use includes many subjects that involve quality, property rights, management modes, inputs and outputs, efficiency and benefits, and so on. At present, studies on the dominant morphology of farmland use are more common (Long and Li 2012; Lu et al. 2017; Ge et al. 2018a), while studies on the recessive morphology are less common (Qu and Long 2016, 2017). However, the recessive morphology is more closely related to land resource management (Long 2015). The discovery and characterization of its evolution rules play important roles in the innovation and practice of land resource management and are also likely to be of great significance to the future development and improvement of land use transition theory.

The existing research on the recessive transition of farmland use is scattered in relevant studies of management, economics, and sociology. Due to the implementation of a household contract responsibility system for rural land in China, farmland has particularities of ownership, function, and management scale. Meanwhile, farmland resources are scarce and regional differences are significant, which combinedly contribute to the Chinese characteristics and problems of farmland use. In terms of research topics, the existing research into the recessive transition of farmland use involves four main aspects: change of management mode, change of management type, evaluation of management performance, and exploration of moderate-scale management. Among these, evolution of management mode in China is often based on the promulgation of relevant policies and is accompanied by the development of management types. The evaluation of management performance comes from the

---

[1]Basic farmland consists of (Ding 2003): (1) agricultural production bases (such as crops, cotton, edible oils, and other high quality agricultural products) approved by government; (2) farmland with high productivity and a good irrigation system and that has been exploited; (3) vegetation production bases for large and middle cities; and (4) experimental fields for science and educational purposes.

needs of agricultural production management, while the exploration of moderate-scale management mostly originates from the decision-making needs of relevant farmland use policies. Management performance is an important reference point for determining the moderate-scale management, which still needs to consider many other factors, such as quantity of farmland, rural surplus labor force, economic level, social development stage, and so on.

2.2.2.2 Changes in Farmland Management Modes in China

The changes in farmland management modes involve changes in property rights (use rights, contracting rights, management rights, etc.) and management modes (individual, collective, transfer, scale management, etc.). These changes come from the joint action of many factors in the process of technological progress, economic development, and social evolution. Currently, the emergence conditions of farmland management modes in China can be classified as endogenous dynamic conditions and external environmental conditions. With the development of the urban and rural economies, a large number of township enterprises have emerged in developed areas. To a certain extent, the rural labor force has been transferred from agriculture to other industries. With the comprehensive promotion of the socialist market economy, the opportunity cost of agricultural labor continues to increase, and then the non-agricultural transfer is realized spontaneously at a larger regional scale, which provides the endogenous dynamic conditions for the evolution of management modes. The extensive application of agricultural machinery and biochemical technology in agricultural production, the adjustment of the rural land system, and the improvement of rural land rights have further enhanced agricultural production efficiency, providing external environmental conditions for the emergence and promotion of changes in management modes. Since the founding of the People's Republic of China, the transition of farmland management modes has been mostly stepped with the introduction of relevant policies, as shown in Table 2.1.

The implementation of moderate-scale management requires clear regulations on land transfer, and since 2000, many documents have been issued in this regard (Zhang et al. 2010). In 2001, the Central Government issued the regulations on the transfer of rights to land contractual management, which clearly pointed out that land transfer must adhere to the principle of "voluntariness, legality, and compensation." The Law on Rural Land Contract of the People's Republic of China, issued in 2002, proposed that the transfer of the right to contracted management of land can be carried out by subcontracting, leasing, exchanging, transferring, or other means. In 2005, the Ministry of Agriculture issued the Administrative Measures for the Transfer of the Right to Contracted Management of Rural Land, which legally regulated the transfer of the right to the contracted management of rural land. In 2008, the Third Plenary Session of the Seventeenth Central Committee of the Communist Party of China (CPC) passed "the Decision on Several Major Issues Concerning the Promotion of Rural Reform and Development", which proposed to "strengthen the management and service of the transfer of the right to contracted management of land, establish

Table 2.1 The evolution of management modes of Chinese farmland and policy nodes since 1949 (Adapted from Qu and Long 2018)

Time	Event	Farmland management mode
The early days since the founding of PRC	Rural land reform	Decentralized management mode based on "land to the tiller"
The period of agricultural cooperation	Establishment of people's communes	The mode of collective ownership and centralized management of land (Ji and Qian 2010)
1982	Farmers in Xiaogang Village, Fengyang county, Anhui province tried to contract production to each household; 1st No. 1 Document on Rural Work was issued by the Central Committee of the Communist Party of China (CPC)	The legitimacy of the management mode of contracted production to each household was determined, and the household contract responsibility system was determined
1986	5th No. 1 Document on Rural Work	The pilot areas of rural reform were established, and the moderate-scale management in rural areas of developed areas and suburbs of large cities was attempted
1993	Some Policy Measures of the Central Committee of CPC and the State Council on the Current Development of Agriculture and Rural Economy was issued; The 3rd Plenary Session of the 14th CPC Central Committee passed Decisions on Several Issues Concerning the Establishment of a Socialist Market Economic System	The farmland contract would be extended for another 30 years after expiry. A few areas were allowed to carry out moderate-scale management based on the farmers' willing; Developing moderate-scale management in the form of subcontracting, equity participation, etc. was permitted
Middle to late 1990s	The Central Government emphasized agricultural industrialization	The agricultural management mode driven by industrial and commercial enterprises was proposed and developed

(continued)

and improve the transfer of the right to contracted management of land, and develop scale management mainstream, such as large farming household, family farms, and agricultural cooperatives where conditions permit" on the basis of retaining the above provisions.

Table 2.1 (continued)

Time	Event	Farmland management mode
2013	No. 1 Central Document	The policy of guiding and restricting agriculture-related industrial and commercial enterprises was put forward, and the development of various forms of moderate-scale management was encouraged
2014	No. 1 Central Document	The large-scale farmland previously managed by industrial and commercial enterprises would be cut and converted into small-scale family farms, and moderate-scale management would be developed
2017	Speech by General Secretary Xi Jinping at the 19th National Congress of CPC	The development of various forms of moderate-scale management, cultivation of new agricultural management types, and improvement of the socialized agricultural service system were proposed

The introduction of relevant policies often leads to the fine adjustment of the corresponding farmland property rights. A country's overall system of farmland property rights is relatively stable. However, since the reform and opening up, economic and social development have been keeping a high speed, and the demand for adjusting the structure of farmland property rights to release the efficiency of agricultural production has become increasingly strong and urgent in actual farmland production and operation activities. In order to meet this demand, relevant policies have been issued successively, promoting the continuous improvement of farmland property rights. Since China's reform and opening up, there have been two major institutional innovations in rural reform. The first is "the household contract responsibility system", which has been widely implemented in rural areas since 1978. It divides land rights into ownership and contracted management rights. Ownership belongs to the collective, and contracted management rights belong to the farmers. This division greatly mobilized farmers' enthusiasm for production and solved the subsistence problems at that time. The second is the implementation of the "Separation of Three Rights" of rural land. Specifically, the General Office of the CPC Central Committee and the General Office of the State Council issued the Opinions on Improving the Approach to Separation of Rural Land Ownership, Contracting Rights, and Management Rights in 2016. "The right to contracted management of land is divided into contracting rights and management rights. Ownership, contracting rights, and management rights are separated and implemented in parallel to push forward the agriculture modernization." The aim is to "optimize the allocation of land resources, cultivate new management types, and promote the development of moderate-scale management." This adjustment plays a great role in promoting agricultural development and increasing farmers' income. Between these two land property right system reforms, there were many fine adjustments, the goals of which involved stabilizing the farmers' contracting rights, ensuring the farmers' income, removing obstacles for the circulation of farmland, and improving the efficiency of farmland use and agricultural production.

2.2.2.3 Changes in Farmland Management Types in China

With the evolution of farmland management modes, management type has gradually evolved from small-scale farmers, communes, and farmer households to the current stage, where a variety of new agricultural management types coexist. Those new mainstream mainly include family farms, backbone farmers, cooperatives, and agriculture-related enterprises.

The emergence and development of agriculture-related enterprises originated from the policy of "agricultural industrialization" in 1990. However, in practice, due to the difficulties of solving labor supervision problems, coping with natural risks, and making profits, this kind of management type gradually evolved into a form of sub-management by entrusted households in the later stage of its development. As time has gone on, it has further evolved into the form of "company plus family farm," which indirectly promotes the development of the "family farm," a new agricultural management type.

Family farms are supported and promoted by a variety of supportive policies, so their number, as well as their relative research is increasing (Wu 2013). The formation and growth of family farms must be based on the maturity of the labor market system, rural land system, and socialized agricultural service system (Gao et al. 2013). Some scholars have put forward the advantages of family farms from the perspectives of production and management. They argued that family farms can not only take full advantage of the unique production enthusiasm of family members, but also give full play to the market operation of enterprises. It is a new type of production and management form that integrates both the advantages of family management and those of enterprise management (Gao et al. 2013). Some scholars have also analyzed the inevitability of family farms from a sociological perspective. They believed that farmers, village collectives, local governments, and central governments have different logics for the formation of a reasonable scale of farmland. The above logic constitutes a tension and points to family farms that can balance the interests of different mainstream. Meanwhile, the family farm represents the inheritance and improvement of the family management system in China, which is in line with the common interests of the government and farmers and is more likely to increase the income of farmers. Therefore, it is sustainable.

Compared with agriculture-related enterprises and family farms, the mainstream of backbone farmers is much smaller in terms of management scale, but it has already broken through the category of "small-scale farmers" in the early days since the founding of the People's Republic of China. From the perspectives of production and management, scholars argued that compared with small-scale farmers, backbone farmers are more active in purchasing agricultural machinery. Its management scale can guarantee the full load working efficiency of agricultural machinery. In addition, backbone farmers pay more attention to the application of new technology during their agricultural production process. In the process of urbanization, a variety of resource elements continue to flow out of rural areas and into urban areas. The growth of the backbone farmer class helps to maintain the basic order of the rural areas. Although scholars have affirmed the importance of backbone farmers from an academic perspective, this management type has not received enough support in the actual development process. In order to foster the mainstream of rural development and stabilize the rural social development environment, more attention should be paid to support this management type.

Compared with enterprises, family farms, and backbone farmers, cooperatives are relatively vague in terms of management scale and management mode. In terms of scale, cooperatives can be as small as a natural village and as large as a multi-community group. Some scholars have summarized cooperative-based management modes into three types: farmer professional cooperatives plus farmers, farmer professional cooperatives plus leasing management, and shares plus cooperation. Cooperative management often takes the farmland and labor resources within the scope of the village collective as the organization mainstream. Its formation and development require higher leadership and organization ability, as well as higher service awareness. Its operation process should be based on a strong grassroots leadership team.

2.2.2.4 Evaluation on Management Performance and Discussion on Moderate-Scale Management

With the evolution of management modes and management types, the demand for evaluating management performance emerged in farmland use. In order to meet this practical demand, scholars have carried out a wide range of evaluation research into the performance of farmland management. These studies mainly focused on the differences in production efficiency (land productivity, labor productivity, cost-profit ratio, total factor productivity, and technical productivity) or the scale benefit caused by different farmland management scales.

Chinese scholars have studied the performance evaluation of farmland management from various perspectives based on the change of farmland management scale. More studies were conducted on the evaluation of scale efficiency, land productivity, and labor productivity, but less on cost-profit ratio, total factor productivity, and technical productivity. The evaluation of farmland management performance abroad started many years earlier than that in China. The studies focused on the empirical test and the cause exploration of the "inverse relationship" between farmland management scale and productivity. For example, Sen (1962) found that small-scale farmers had higher yield per unit area than large-scale farmers from the production practice in India, so he raised the question of whether small-scale farmers were more efficient (Sen 1962). Bardhan (1973) empirically tested this phenomenon from the perspectives of land productivity and returns to scale and found that there was an inverse relationship between farmland management scale and production efficiency. Bardhan speculated that the reason for this relationship may be the imperfection of the land and labor factor market (Bardhan 1973). From then on, research into an "inverse relationship" was initiated, and scholars have expanded this research topic from various perspectives. Some scholars have questioned the persuasiveness of market imperfection and land quality difference as the causes of the "inverse relationship" (Barrett et al. 2010). However, most studies have shown that compared with other reasons, such as farmers' differences (Carter 1984; Heltberg 1998; Assunção et al. 2007) and measurement errors (Calogero et al. 2013), their influences on the "inverse relationship" have a stronger theoretical basis and practical interpretation ability (Bhalla and Roy 1988; Lamb 2003; Henderson 2015).

Based on long-term research and development, research into the evaluation of farmland management performance is relatively mature. Therefore, scholars often learnt from the methodology of performance evaluation to analyze new research topics, such as exploring moderate-scale management. The moderate-scale management of farmland is a unique research topic under China's conditions. In terms of the endowment of farmland resources, China has more people and less land. Farmland resources are scarce and per capita farmland resources are small. In terms of farmland management and utilization, there is not only the phenomena that local governments blindly adopt administrative means to promote scale management, develop modern agriculture, and cause land-deprived farmers, but also the phenomena of large-scale non-agriculturalization of the agricultural labor force, abandonment of farmland, and threats to food security. Scholars have carried out extensive qualitative analysis and quantitative measurement of moderate-scale farmland management. However, it should be noted that moderate-scale management in China is not only a quantitative study based on economic production efficiency, but also a comprehensive study based on the current economic development background that considers aspects of agricultural development, technological level, and social security.

The moderate management scales obtained from the existing research are not the same, which is not only due to different analysis angles, selected indicators, and data sources, but is also caused by the characteristics of farmland management. Moderate-scale farmland management is related to various factors, such as nature, the economy, society, and technology, and its size is influenced by resources endowment, economic development level, productivity level, quality of labor force, perfection of socialized agricultural service system, development level of relevant policies, etc. Therefore, the discussion of moderate-scale management needs to be based on time and local conditions.

2.2.2.5 Summary

At present, the 1.8 billion mu farmland protection red line in China has played a positive role in protecting the quantity of farmland resources. However, in order to achieve the strategic goal of ensuring food security and increasing farmers' incomes, more research should be conducted on the recessive transition of farmland use under the given resources conditions. The recessive morphology of land use has multiple attributes, such as property rights, management modes, input-output, efficiency, so the research into the recessive transition of farmland use involves many disciplines.

Combined with the practice of farmland management in China, this section combed the recessive transition of farmland use since the founding of China and summarized the related research progress scattered across different disciplines from four aspects of management mode, management type, management performance, and moderate-scale management. Currently, research into the recessive transition of farmland use is characterized by the diversity of perspectives and the decentralization of contents, and it is insufficient in terms of regionality, comprehensiveness, and multi-scale. Lacking regionality, the existing research ignores the differences in natural background and socio-economic development of different samples. Therefore, the conclusions vary with different samples, indicators, and perspectives. Due to the lack of comprehensiveness, existing studies focus on the economic benefits of farmland, but ignore its social security functions and the environmental effects in the process of farmland utilization, so the research results cannot be sustained in practice. Due to the lack of multi-scale, most existing studies provide microanalysis conclusions at the level of farmer households. It is difficult to meet the needs of different levels of farmland resources management.

2.2.3 Theoretical Analysis Framework

The functions of farmland can be divided into two levels: basic functions and derivative functions. The basic functions include production, ecology, and landscape culture, which are inherent attributes of farmland because they exist due to the agricultural production process and natural materiality of farmland. Based on the basic functions of farmland, the derivative functions change with human beings' needs, degree of understanding and ability to use farmland (Jiang et al. 2017). Farmland function transitions can be understood as a process of the mutual transitions of basic farmland functions and the gradual enhancement in farmland derivative functions. From the structure morphology of farmland, FUT can be divided into dominant and recessive transitions, where the former mainly refers to the changes in the quantity, patch area and crop planting structure of farmland, and the latter is mainly reflected in the transition of the functions, property rights, and management modes of farmland. Different stages of socio-economic development often correspond to different

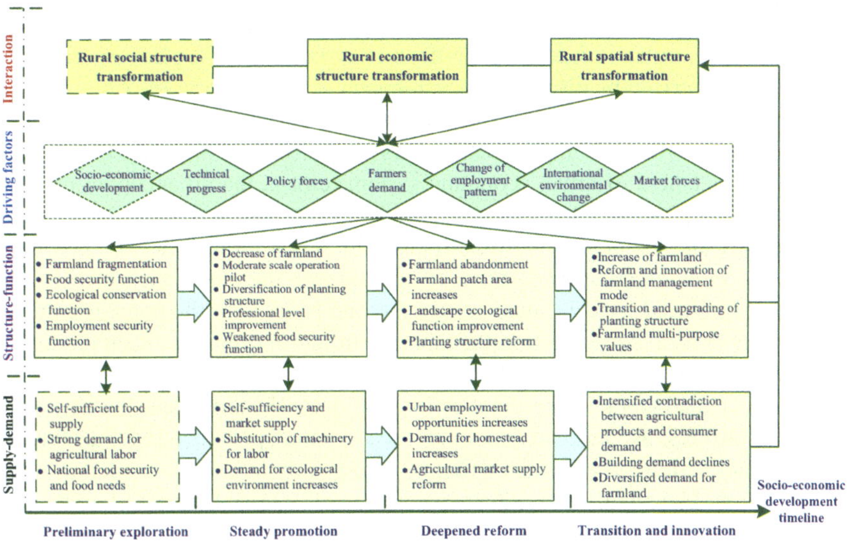

Fig. 2.1 Theoretical analysis framework of FUT in China (Ma et al. 2020)

farmland functional morphologies (Song et al. 2012); the supply function of farmland is constantly changing with the change in human needs, which leads to the "structure-function" transformation of farmland.

Based on the socio-economic development timeline, a theoretical analysis framework of FUT in China was established since its reform and initiation of opening-up policies from the perspective of "structure-function" and "supply-demand", which will be helpful to comprehensively analyze the driving mechanism of FUT (Fig. 2.1).

(1) Farmland use changed from preliminary exploration to steady promotion. In the early stage of reform and development, the Chinese socio-economic development level was relatively low and the level of international import and export trade remained undeveloped, and income from farming was the main source. To guarantee national food security, grain crops are the primary kind of crop planted. In this phase, due to the low-level of agricultural mechanization, the farming modes were dominated by small and refined household production, resulting in the consumption of a large proportion of the agricultural labor force. However, since few bio-chemicals were applied, the environmental load on farmland was small, and the biodiversity was relatively high. During this phase, farmland fulfilled the prominent functions of food security, ecological conservation, and employment security. By the end of the 1980s, with the gradual improvement in the land system, rural policies, technical conditions and economic development level, the food consumption of rural households gradually changed from self-sufficiency to the combination of self-sufficiency and market security. As moderate scale operation of land was tried in rural areas in developed regions and the suburbs of metropolis, the scale of farmland use

showed a steady expansion trend. With the gradual improvement of the professional level, utilization scale and efficiency of farmland, the planting structure was characterized by diversified development.

(2) Farmland use changed from steady promotion to deepened reform. In general, the transitions of rural employment modes promoted FUT. Since the 1990s, with the advancement of urban economic system reform, the urbanization process has gradually accelerated, accompanied by changes to be small-scale peasant economy and the non-agricultural transfer of agricultural labor. The new jobs offered in cities attracted farmers to migrate into the city, which has significantly increased farmers' income. On the one hand, with the non-agricultural transfer of rural surplus labor, the labor input into farming has been gradually reduced. Farmland plots with large farming radius, inconvenient transportation, and poor quality faced marginalization and even abandonment. On the other hand, farmers, as "rational economic men" would spontaneously transfer the farmland with good location conditions and quality to achieve moderate scale operation. Meanwhile, the average patch area of farmland increased, with concurrent improvement to the farmland economic and ecological functions. The objective needs of farmers promoted FUT. At this stage, the phenomenon of family splitting and house building was quite common. In addition, the increase in the income of farmers further stimulated the demand for housing, which accelerated the arrival of the construction boom. As a consequence, as villages were expanded rapidly, a large amount of farmland was occupied, resulting in disorderly development and a serious waste of land resources (Long et al. 2010). Therefore, farmland area rapidly declined during this phase. Changes in the international trade environment also influenced the FUT. After China's accession to the WTO in 2001, foreign low-priced agricultural products began to substitute China's land-intensive products. To strengthen its comparative advantage in agricultural products, China increased the input and output of related agricultural products, thereby bringing the deepened reform of farmland in terms of operation scale and planting structure (Zhang et al. 2018).

(3) Farmland use changed from deepened reform to transition and innovation. To some extent, socio-economic progress promoted FUT. With the continuous advancement of urbanization, the rural population and living space were gradually accumulating in cities and towns, of rural population density was decreasing, and the demand for rural housing construction was also declining (Ge et al. 2018b). As people's awareness of farmland protection has been strengthened gradually, farmland area has been slowly increasing. In addition, as the urban-rural dual structure is gradually broken and the level of agricultural mechanization is improved, moderate scale operation has gradually emerged in various forms, and farmland productivity has increased significantly, remarkably enhancing the socio-economic functions of farmland. Nonetheless, due to the increased amount of chemical fertilizers and pesticides applied, the ecological function of farmland declined during this phase. Policy orientation has facilitated FUT. In the new era, some issues such as unreasonable agricultural

production structure, an unclear property rights system, and difficulty in mobilizing farmers' enthusiasm for agricultural production have become prominent obstacles that restrict rural development and the continued growth of farmers' income. The Chinese government implements and introduces relevant policies and systems, which play the role of macroeconomic regulation and control, innovatively develop the management and transfer models of farmland, promote the transition and upgrading of farmland use and planting structure, and facilitate multi-functional coordinated development of farmland. Thereby, we can adopt the above measures to improve the efficiency of agricultural supply systems, enhance China's international competitiveness in the production of agricultural products, and adjust the development direction of FUT. In addition, the market mechanisms have also promoted FUT. Against the backdrop of trade globalization and consumer-oriented markets, instead of taking grains as the key link, farmers prefer to grow crops with higher economic added-value and stronger competitive advantages (Montgomery 2008). Moreover, as people are demanding more diversified materials of higher quality, the multi-functional value of farmland has been gradually explored. Farmland use tends to pursue diversified coordinated development and high-quality innovative transition.

FUT and rural transformation development and restructuring mutually affect one another. Rural transformation development and restructuring stimulate FUT, but to a certain extent, FUT also affects the transformation of rural social, economic and spatial structures (Tu and Long 2017). For example, in moderate scale farmland operation, the large-scale production of advantageous agricultural products leads to intensive processing of agricultural products and relevant non-agricultural employment, resulting in the socio-economic structure transformation of the rural area. Nevertheless, the marginalization and abandonment of farmland may lead to the gradual decay of a rural agricultural production function, and the rural area will face a transformation development or collapse.

2.3 A Perspective of Intensive Use Level

The intensive process of farmland use is an important manifestation of FUT in China, which mainly presents the change of input and output of per unit area and the farmland use mode. Driven by the pursuit of better returns and the limited farmland area, the organization mode of farmland use tends to increase the input of farmland per unit area, becoming the core driving force to promote the intensive use of farmland. The intensive use of farmland has changed the characteristics of farmland use morphologies and the input-output pattern, which has become an important feature of FUT in China's traditional farming areas.

The intensive process of farmland use is closely related to grain production and rural transformation development. By analyzing the impact of intensive use of farmland on food security and the coupling relationship between intensive use of farmland

and rural transformation development, it can provide better scheme for analyzing FUT from the perspective of intensive use of farmland. Based on this, this section tries to analyze the process and effect of FUT driven by farmland intensive use from the perspectives of intensive use of farmland and evolution of food security patterns.

2.3.1 *Grain Production Security Patterns Driven by Variations in Farmland Use*

2.3.1.1 Evaluation of Farmland Intensive Use Level

Agricultural production is the most profound industry in which transforming the natural environment. The grain production ability per unit farmland area is not only determined by the original physical conditions of farmland, but also by the human investments, such as man-made materials, technology and capital. Currently, the latter is more important than the former. To a large extent, the farmland use level (FUL) may be used to replace the role of farmland quality for measuring grain production.

Considering the diversity of the factors affecting farmland use, to assess the FUL completely and systematically needs a scientific assessment indicator system which takes farmland use characteristics into account (Li et al. 2008). We chose nine indicators reflecting the FUL to establish an assessment system at the provincial level (Table 2.2). Analytic Hierarchy Process (AHP) was used to establish the FUL indicator system with a target layer, a rule layer and an indicator layer. The target layer is the farmland-use level, and the rule layer involves three factors: (i) investment intensity; (ii) management level; and (iii) production effect. In addition, there are nine indicators in the indicator layer. Then we used the AHP and Delphi to evaluate the weight to the rule layer and the indicator layer.

Table 2.2 Indicator system for farmland use assessment (Long and Zou 2010)

Rule layer factors weight	Indicator layer factors weight	Definition
Investment intensity (0.329)	Power Investment (0.361)	Gross power of farming mechanism per ha
	Labor Investment (0.329)	Gross farming labors per ha
	Fertilizer Investment (0.310)	Gross fertilizer utilization per ha
Managing level (0.293)	Multi-cropping index (0.436)	Dividing the crop area by the area of farmland
	Irrigation index (0.328)	Dividing the irrigated farmland area by the area of farmland
	Grain-farmland index (0.236)	The proportion of grain-crop area in the total crop area
Production effect (0.378)	Grain yield (0.357) Farming output value per ha (0.386) Grain output per capita (0.257)	Grain yield per ha

Data are from the China's statistical yearbooks and all provincial statistical yearbooks. To describe the 31 provinces for the whole country, excluding Taiwan Province, more consistently, the territory is divided into six regions according to geographic locations and similar physical conditions, i.e., South China, Central China, North China, The Northeast, The Southwest and The Northwest. To assess the FUL, section data for 1978, 1985, 1995 and 2004 are selected. Data for farming output values are converted as constant prices in 1990. However, some data are lacking, i.e. data for 1985 in Jiangxi and Anhui provinces, data for 1978 in Xinjiang Autonomous Region, data for 1978 and 1985 in Guizhou Province. Considering the integrality of the data and analysis, data for 1985 in Jiangxi, Anhui and Guizhou provinces are substituted by data for 1987 in the same three provinces. Data for 1978 in Guizhou and Xinjiang are treated as deficient.

(1) **Data normalizing**

Since the socio-economic data for the various indicators in Table 2.2 are in different dimensions, to compare these data, they need to be transformed into common units by normalizing all measures, according to Eq. (2.1):

$$X'_{ij} = \frac{X_{ij} - X_{i.\min}}{X_{i.\max} - X_{i.\min}} \tag{2.1}$$

where X'_{ij} is the standardized value of the indicator; ij means the No. i indicator in the No. j rule layer; X_{ij} is the value of the indicator ij; $X_{i.\max}$ is the maximum value of indicator ij of all provinces; $X_{i.\min}$ is the minimum value of the indicator ij of all provinces.

To compare the FUL among all the provinces, the same type data from different provinces are normalized by the same extremum. These guarantees the final FUL of different provinces are comparable.

(2) **Farmland Use Level (FUL)**

We used the weight and the normalized values to calculate the FUL of each province in each year according to Eqs. (2.2) and (2.3):

$$F_j = \sum_{i=1}^{3} X'_{ij} \times W_{ij} \tag{2.2}$$

$$F = \sum_{j=1}^{3} F_j \times W_j \tag{2.3}$$

where F_j is the score of rule layer j; X'_{ij} is the standardized value of indicator; W_{ij} is the weight for indicator ij; F is the FUL; W_j is the weight for rule layer j.

(3) **FUL Changes**

To reflect the FUL changes, we chose the simple measurement to calculate the direct change between comparative years. It reflects the change of human investment in farmland use, to provide more intuitive information explaining the relationship between FUL changes and Farmland-grain elasticity coefficient (FGEC), according to Eq. (2.4).

$$\Delta F_{ij} = (F^j - F^i) \times 100\% \tag{2.4}$$

where ΔF_{ij} is the FUL change of a specific province from year i to year j; F^i and F^j are the specific province's FUL in year i and year j, respectively.

(4) **Farmland-Grain Elasticity Coefficient (FGEC)**

To analyze the relationship between the farmland quantity change and variation of grain output or the impact of farmland change on grain production, the FGEC index was established according to Eq. (2.5):

$$C_j = \frac{(Q_{i+1} - Q_i)/Q_i}{(S_{i+1} - S_i)/S_i} \tag{2.5}$$

where j represents a province; Q_i and Q_{i+1} are j province's grain output in year i and year i+1, respectively; S_i and S_{i+1} are j province's farmland area in year i and year i+1, respectively; C_j is the FGEC, which means the relationship between the farmland change and variation of grain output of j province from year i to year i+1. Before we analyze this index, it is necessary to understand the trend in farmland change, in order to make clear the meaning of the index.

When $C_j > 0$, this indicates that grain output shows the same change in the trend as that of available farmland area. Grain production is remarkably influenced by the changes in farmland area. When the area of farmland increases, the higher the value of C_j, the less the influence of farmland change on grain production and the more the influence of farmland quality change or the changes of FUL on grain production. When farmland area decreases, the higher the value of C_j, the more serious is the reduction in grain production, and the more negative is the influence of farmland area decrease on grain production.

When $C_j < 0$, this indicates that grain output shows the opposite change in the trend as that of available farmland area. Grain production is remarkably influenced by non-farmland factors. The higher of the absolute value of C_j, the more remarkable the influence of non-farmland area factors on grain production: if farmland increases, it means that the decrease in FUL has more negative influence on grain production than the positive influence from an increase in farmland; if the farmland decreases, it means that the FUL improvements have a more positive influence on grain production than the negative influence of farmland decrease. Thus, when $C_j < 0$ and the higher of the absolute value of C_j, the more remarkable the influence of FUL or farmland quality change on grain production than that of farmland area change.

2.3.1.2 Spatio-Temporal Pattern of FUL

(1) **FUL of 1978**

The industrial catch-up strategy adopted by the Chinese central government when the PRC was founded, decreased capital investments in agriculture in order to guarantee capital accumulation in heavy industry and in the chemical industry. This strategy resulted in a weak capital accumulation in agricultural production. Capital and technical investments in agriculture have been at a standstill for a long time, and this resulted in a lag in agricultural productivity before 1978. Poor production conditions led to agricultural production strongly determined by natural conditions, and the FULs of all provinces were generally low. Since better water and heat conditions in southern provinces of China can sustain higher rates of farmland multiple cropping and higher grain yields, i.e. better output values, than those in the northern provinces, the southern provinces had higher FULs than those in the north. With increasing latitude, the FUL index decreases from over 0.3 in the southern provinces to below 0.2 in the northern provinces (Fig. 2.2). In general, all provinces had similar levels of human investment in farmland use, and the differences in FUL for the different provinces were determined to a great extent by natural conditions. Therefore, the spatial patterns of China's FUL showed a descending trend from south to north.

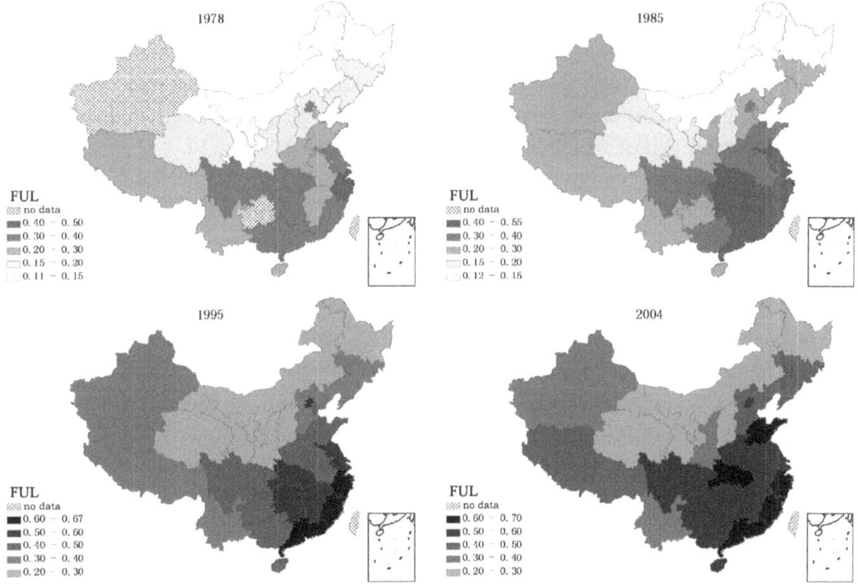

Fig. 2.2 Spatio-temporal pattern of China's FUL between 1978 and 2004 (Long and Zou 2010)

(2) **FUL of 1985**

Since the socio-economic reform policy implemented in 1978, the household responsibility system has been gradually established in rural areas, and the rural workers have become strongly motivated. The downturn in the national industrial catch-up strategy has resulted in more technology and capital being invested in farm production. All of these changes have promoted rapid development in the agricultural sector. The FUL at the national level has risen universally (Fig. 2.2). In the area of South China, Central China and North China, the low topography facilitated extensive mechanization extending, and the coastal advantages obvious in the reform and open-door policy made the economies for these regions develop rapidly. This led to the establishment of a good economic basis for increased capital and technological investments in farm production. It promoted the swift rise of FUL in these areas. FUL of South- and Central China generally exceeded 0.4, and that of North China exceeded 0.3. However, limited by the low farming population and weak economic development, farming in the northwestern area and Heilongjiang province developed slowly, resulting in most of their FULs being under 0.2. Constrained by an upland topography, it is difficult for the southwestern area of China to increase farm mechanization, and together with extremely weak economic development, the FUL for this area increased less than that in 1978, with the exception of Sichuan province which has a relatively well-developed economy.

(3) **FUL of 1995**

With the successful implementation of market-economy system, China's economy has been developing at high speed. Great progress has been made in socio-economic and technology development, farm mechanization has been extended, and fertilizer-use has increased. All of which have led to large increases in the FUL (Fig. 2.2). The eastern coastal provinces have developed at the fastest rate, their FULs exceeding 0.6. In Central China, Northern China and the Northeast, increased mechanization, extensive fertilizer investments, the establishment of irrigation, and an increasing farming population with no shift to industry, have all promoted the effective increase of the FUL. The progressive development of the Southwestern region has guaranteed an increase in human investment, and FULs exceeded 0.4. This has made the region other growth motor for FUL. In the Loess Plateau, Qinghai-Tibet Plateau and northwestern China, because of slow economic development and inferior physical conditions, the development of agriculture has been generally impeded and the FUL shows less improvement, most of which in these areas was between only 0.2 and 0.3.

(4) **FUL of 2004**

During the period 1995 to 2004, there were obvious improvements in each province's economy. However, the impacts of this economic development on the FUL differed. In some regions with a rapid speed of development, e.g. Beijing and Tianjin, the Yangtze River Delta and the southeast coastal provinces, i.e. regions where rapid non-agricultural conversion of farmland and transfer of farm labor, together with a decreasing grain output, agricultural structural adjustments mitigated the positive influence of FUL improvements by the persistent increase of machinery and fertilizer investments. In these areas, the FUL increase was no longer possible (Fig. 2.2). The FUL of Beijing, Tianjing and Shanghai municipalities is only 0.5, i.e. almost the same level as that for 1995. The FUL of southeastern provinces increased only slightly, even with the obvious increase in fertilizer investments and grain output value, and it still at 0.6 to 0.7. The FULs of Zhejiang and Jiangsu provinces have changed very little, because improvements in agricultural production have coexisted with rapid shift of farm workers to non-agricultural jobs and an increasing loss of farmland. On the other hand, the inland economy has also developed rapidly over the last decade. Farming investments and multiple cropping have increased, and the non-agricultural shift of farm workers has been halted, all of which has resulted in improved FULs.

In general, at the national level, FUL has increased due to rapid economic development since 1978. However, FUL may not increase persistently with the sustainable development of the Chinese economy. On the one hand, conversion of farmland and transfer of farm workers together with other agricultural structural adjustments have depressed the FULs. On the other hand, the growth potential of farming machinery and fertilizer investments has kept the FUL at a high and stable level.

2.3.1.3 The Variations in FGEC

(1) **Variations in FGEC between 1978 and 1985**

Although large areas of farmland have decreased, driven by economic developments during 1978–1985, this did not cause a general decline in grain production. On the contrary, most provinces increased their grain production, with the exception of Liaoning, Heilongjiang provinces and Shanghai City. As can be seen from Fig. 2.3, the value of most FGECs was negative and their absolute values were almost below 5. The provinces with a high FGEC absolute value are concentrated in the North- and Central China, where there is a rapid increase in grain production.

The establishment of the household responsibility system motivated farm workers, which in turn guaranteed the base for effective farmland use. Farmland decrease versus FUL increase indicates the growth of grain production at the national level (Figs. 2.3 and 2.4). In the three provinces with decreasing grain production, Liaoning Province's farming level was seriously affected by the flood in 1985, which reduced grain reduction by 4.5 million metric tons. However, this was only a 1.41 million metric tons loss compared to that of 1978, and provides robust proof of how the increase in FUL strengthened grain productivity. There was only a 0.02% decrease from the level of 0.144 in Heilongjiang Province's FUL from 1978 to 1985, but a weak agricultural foundation caused a grain reduction of 0.73 million metric tons. The FUL of Shanghai decreased by 6% and caused a grain reduction of 0.47 million metric tons. Therefore, due to the initial weak agricultural base, it can be assumed that the rise of FUL largely promoted the grain production and guaranteed grain production security under the conditions of general farmland loss.

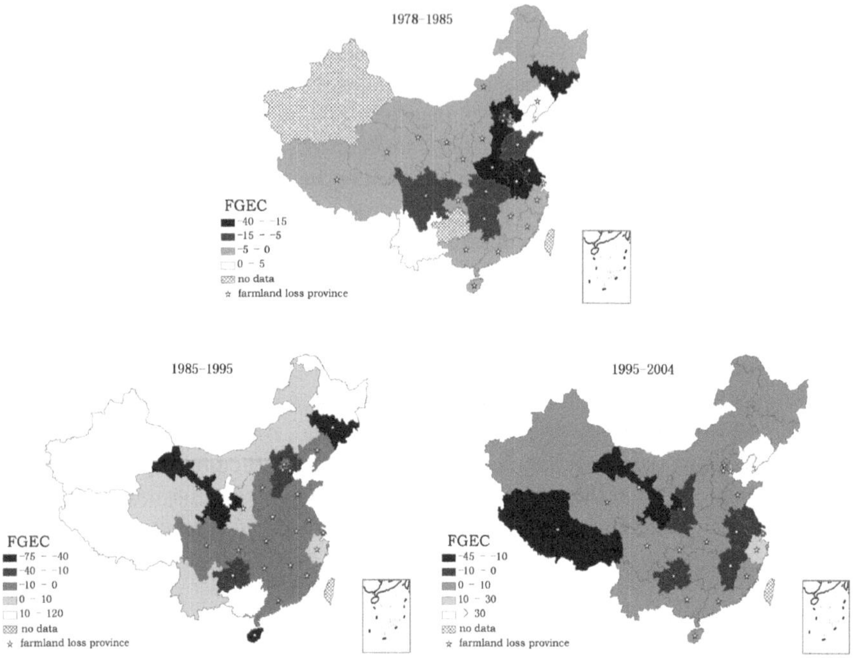

Fig. 2.3 The variations in FGEC between 1978 and 2004 (Long and Zou 2010)

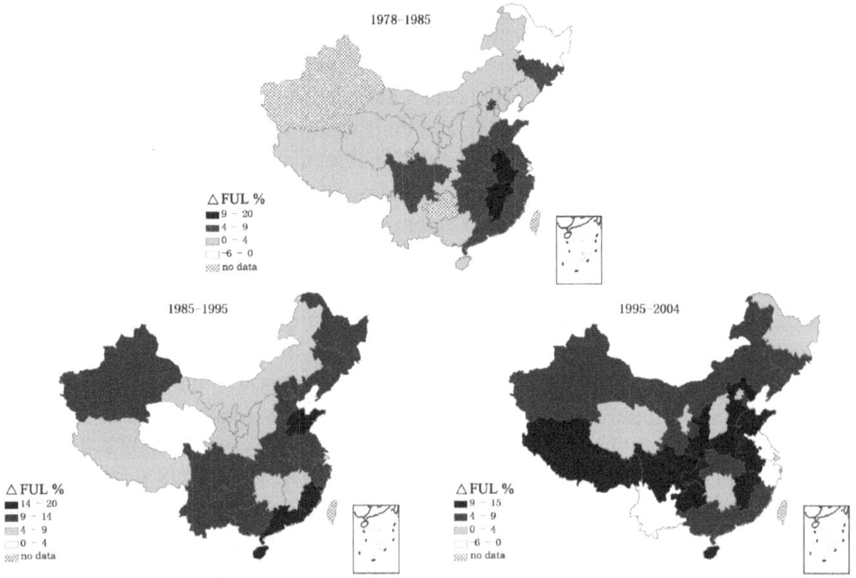

Fig. 2.4 The variations in FUL between 1978 and 2004 (Long and Zou 2010)

(2) **Variations in FGEC between 1985 and 1995**

There is a sharp contrast between farmland protection and arable land reclamation in the western provinces of China and the rapid farmland loss in the eastern provinces during 1985–1995. However, the general growth of grain production in the whole country showed that the FGEC varied over the different regions. FGECs of the western provinces were almost positive while those of the eastern and central provinces were almost negative. Compared to the previous period, the absolute value of FGECs of the western provinces and the northeast increased, and their grain production increased greatly. However, that of the central and eastern provinces changed only slightly and showed a stable growth level for grain production.

The upgraded FUL in the northeast and western provinces and the increase in farmland in the western provinces led to large increases in grain production. However, rapid farmland loss occurred in some developed provinces in the eastern and central regions, but their increasing FUL driven by rapid economic development still guaranteed stable grain production (Figs. 2.3 and 2.4). However, especially in the eastern and coastal areas, the decreasing marginal effects of increasing machinery and fertilizer investments led to a situation where the rapidly increasing FUL did not result in massive growth in grain production, and even led to the reduction of grain production in some provinces. Zhejiang Province is an example. Its farmland decreased by 8.9%, and the FGEC was 1.313 during this period. In 1985, the FUL for Zhejiang was 0.54, while it increased by 0.13, to 0.67 in 1995. This was due to a comparatively high FUL, but massive farmland loss cut down the grain production by 11.7% with low marginal effect of FUL increase.

(3) **Variations in FGEC between 1995 and 2004**

During 1995–2004, it appears that the regions with increased farmland areas have had grain production growth while the ones with decreases in farmland area have had grain production reductions, which resulted in generally positive FGECs (Figs. 2.3 and 2.4). Rapid industrialization and urbanization caused serious farmland loss (Liu et al. 2008) with the implementation of 'grain for green' policy, and large areas of farmland were shifted to other land-use types (Long et al. 2006). The FUL growth in eastern coastal China will threaten grain production security when FUL was relatively high, and in the central provinces a similar situation of stable grain production or even reduction existed when the FUL was increased. The farming structure adjustment also depressed grain production growth. In this decade, grain production growth mainly occurred in those regions with increasing farmland. Therefore, under conditions of high FUL, it is more important to protect farmland areas for grain production security than to increase the FUL. This would also provide a practical scientific basis that would constitute a strict farmland protection objective and strategy in China.

2.3.1.4 Discussion

The serious farmland loss since 1978 has led to great pressure on grain production security, but increasing investments in farmland quality and human incentives have mitigated this pressure to some extent. Farmland protection, in terms of both quantity and quality, is one of the most important measures if national food security is to be guaranteed. FULs at the national level have increased due to the rapid economic development since 1978. The path of this rapid improvement shows a gradient declining from southeast coastal China to inland China with further economic development. However, the FUL may not increase persistently with the sustainable development of economy, because of the conversion of farmland and transfer of farm workers to non-agricultural activities. Agricultural structural adjustments and the unceasing improvements of FUL may not always bring about sustainable and steady growth in grain outputs.

In general, farmland areas and human investments interact with each other to influence grain production. At the beginning of Chinese economic reform, due to the weak agricultural base, improvements in the artificial "quality" of farmlands had great positive effects on maintaining food security. Along with economic development and improvements in the agricultural base, human investments will play only a weak role in increasing grain production and in maintaining food security, without technological breakthroughs in all aspects of agricultural production. Accordingly, the role of farmland areas will become more and more important in maintaining food security. Therefore, considering the law of diminishing marginal utility, in developing countries, the available area of farmland will play a key role in maintaining the security pattern of grain production. This will also provide a practical scientific basis for constituting strict farmland protection objectives and strategies for China. The central government was advised to maintain the total cultivated land area at no less than 1.8 billion mu until 2020, however, in 2006 there were only 1.82 billion mu.

There is a dilemma in relation to farmland protection in China. Recently, some highly productive farmlands were abandoned for two major reasons. On the one hand, the user of the abandoned farmlands left to seek work in the city, found a job, and made a living. Few people are willing to manage farmland with a low economic incentive and/or low profit margin from grain production. On the other hand, many rural workers have left for urban areas, while those who stay on in the rural area are old people, women, sick and disabled men, and children (Long et al. 2010). The latter workforce is unable to manage large farmland areas. Thus, to some extent, preventing further farmland decreases may not ensure food security. We suggest that the central government invest in the construction of a grain production base, improve infrastructure, and develop modern agriculture and scale management, in order to increase the profitability and productivity of farming. In addition, more attention needs to be paid to providing job opportunities for the surplus rural labors. Therefore, both protecting farmland from a transformation to other land-use types, and ensuring its effective management constitute key solutions for maintaining grain production security in China.

2.3.2 *Greenhouse-Led Farmland Transition and Rural Transformation Development*

The FUT in the traditional farming area is closely related to rural transformation development (RTD). The greenhouse-led farmland transition (GFT) provides a new perspective for analyzing the changes of rural production and farmers' livelihood in the traditional farming area, which further create conditions for revealing the evolution of the coupling relationship between the FUT and RTD. Based on the analysis of the internal relationship between GFT and rural development, this section attempts to construct a coupling conceptual model between the FUT and RTD. Taking Qingzhou City as an example, this section analyzes the process of GFT from a county level, which affects rural development. On this basis, Nanxiaowang Village is taken as an example to analyze the coupling process between GFT and RTD. Combined with the afore-mentioned empirical analysis, the main problem and future optimization strategy of the current GFT and RTD were discussed.

2.3.2.1 Coupling Model of GFT and RTD

(1) **Relationship between greenhouse land expansion and farmland transition**

The core concept of the FUT is to reveal the nonlinear change process of farmland morphology (e.g. dominant morphology and recessive morphology) (Long and Qu 2018; Meyfroidt et al. 2018). The transition process of farmland morphology is a concentrated reflection of the evolution of the rural man–land interrelations in the process of urbanization characterized by the expansion of greenhouse land. The comprehensive development mode of modern agriculture characterized by the expansion of greenhouse land and its socio-economic response has profoundly changed the morphology of farmland. Although greenhouse construction has not fundamentally changed the utilization type of farmland, it has significantly changed the cultivation manner and mode of farmland use. The introduction of greenhouse cultivation in the traditional farming areas first changed the function of farmland for grain production and turned to cash crops with higher added value. The pattern of farmers' participation in agricultural production has undergone significant changes, and the degree of marketization of agricultural production has increased significantly. The agglomeration of large-scale greenhouse production not only affects the production mode of farmers, but also changes the agricultural production organization mode of traditional farming area, from a dispersed small-scale farming mode to a clustered community production mode.

Greenhouse construction is mainly used to grow cash crops with higher market value, which changes the original planting mode of grain crops and further strengthens the production function of farmland (Ge et al. 2017). Therefore, the dominant morphological changes of greenhouse-led farmland use mainly include the area and proportion changes of greenhouse land, the changes in use mode, the spatial distribution characteristics, and the spatial structure evolution (Fig. 2.5). The

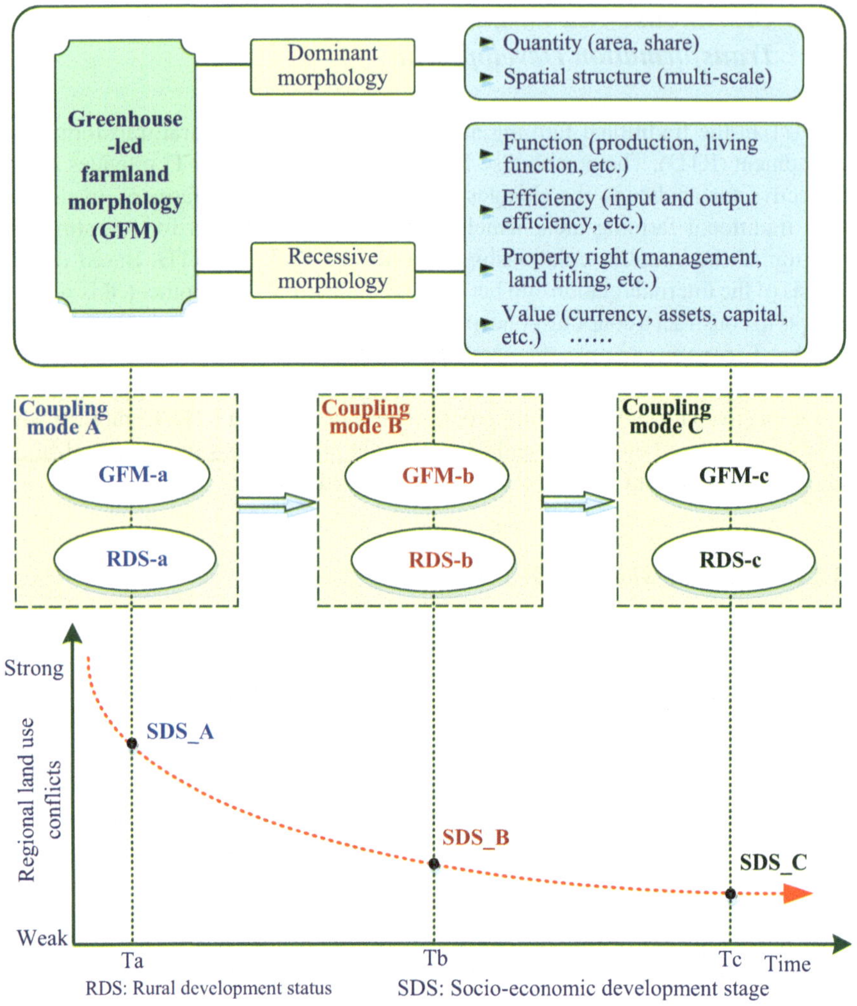

Fig. 2.5 Conceptual framework of the coupling analysis between greenhouse-led farmland transition and rural transformation development (Ge et al. 2019)

recessive morphological changes of greenhouse-led farmland use mainly include (Ge et al. 2018a), for example, farmland functions (production, living and ecological functions) (Zhang et al. 2018), production efficiency (e.g., input and output), property rights realization methods (e.g., organizational modes and rights distribution), and value systems (e.g., asset values and capital value) (Long and Qu 2018). From the morphological changes of the afore-mentioned GFT and RTD, we observe that the production pattern and value system of the farmland after the introduction of greenhouse planting have undergone significant changes. Greenhouses invested large amounts of capital in construction, and greenhouses' production pattern belongs to technical and labor-intensive production mode, which has changed the barriers of low-density input and low-value output of traditional farmland land use and effectively improved the asset value of farmland (Song et al. 2017).

(2) **Expansion of greenhouse land and rural transformation development**

The development process of greenhouse-led rural industry integrates the rural production system into the urban–rural market network. The rural production system promotes the agglomeration of rural development factors under the cross-scale effect of urban–rural factors. The endogenous dynamic mechanism of rural development is continuously strengthened, and the RTD enters a benign ascending channel. The use of farmland in traditional farming area is mainly to produce grain products. The greenhouse-led farmland use has changed the production system and the mode of labor cultivation. The promotion and application of greenhouses have continued to extend around the industrial chain of greenhouse construction, production, and support (Su et al. 2016). In addition, the cash crops produced in greenhouses are mainly aimed at urban market demand, and the rural industrial system is gradually integrated into the urban–rural market network. The production and circulation of cash crops bring more market information to the rural development of the traditional farming area, which increases the intensity of the connection between small farmers and the urban–rural market. In addition, the supply and demand relationship between urban and rural products has made the urban–rural market network connection more closely.

The labor-intensive production of greenhouses creates conditions for local employment of surplus labor, promotes the specialized division of the labor market, and attracts migrant laborers to a certain extent (Caulfield et al. 2019). Furthermore, modern agricultural technology is the main factor in ensuring efficient greenhouse production. Therefore, greenhouse production requires sustained capital and technology inputs. The development of local industries to support greenhouse planting has also improved the local labor force's agricultural skills and provided employment opportunities for young and middle-aged laborers (Chen et al. 2014). The above analysis shows that during the process of URTD, the rural production system characterized by greenhouse production has resulted in important opportunities: the transformation and development of local villages, the strengthening of the docking ability of rural production and the market, and the condensation of the popularity of the local RTD. Greenhouse production has strengthened the endogenous power of rural development by condensing the "popularity". Additionally, the rural production system characterized by greenhouse production promotes the flow of urban–rural development factors (e.g., capital, technology, information), creates conditions for rural development of benign transformation (Su et al. 2016), and promotes the man–land relationship from "constraint" to "coordination".

(3) **Coupling relationship between GFT and RTD**

In the context of the evolution of the rural man–land interrelations, the analysis of the coupling relationship between GFT and rural development is conducive to comprehending the rural transformation (Ge et al. 2018b). During the process of URTD, rural production and living systems undergo significant changes in China's farming area. The geographic division of urban–rural dual division has gradually been replaced by the close relationship between urban–rural networks. The diversification of rural regional function and the alienation of rural man–land relations have presented differentiated stage characteristics. Different rural man–land interrelation transformation stages present different rural man–land relationship morphological characteristics (Ge et al. 2018a), which are closely related to the greenhouse-led farmland use morphologies (GFM). The evolution of rural man–land interrelations is also the power of rural development, and the coordinated evolution of rural man–land interrelations will affect the process of RTD (Ge et al. 2018a; Long et al. 2016). From the above analysis, we observe that the GFT is the specific external appearance of the rural man–land relationship, while the RTD is its specific internalization (Long et al. 2018; Tu et al. 2018). Therefore, the construction of the coupling relationship between GFT and RTD of the rural farming area creates conditions for coordinating rural man–land interrelations and ensuring the orderly transformation of rural areas.

The core concept of GFT is to reveal the nonlinear change process of GFM in the process of regional development, and RTD is a process of describing the changing trend of the regional rural development status (RDS), i.e., rural production status, farmer life and organization patterns, man–land interaction ways, etc. Studies show that the change process of regional land use pattern conflict is the internal mechanism that drives the evolution of land use morphologies. In addition, the changing rural man–land relationship corresponds to the conflict of land use morphologies, which also becomes the key factor in determining the dynamics of rural development (Long and Qu 2018). Therefore, the evolution of rural man–land relationship patterns established a bridge between GFT and RTD through the interaction of response and feedback. Coordination of the coupling relationship between GFT and RTD creates conditions for easing the tension of the man–land interrelations in the traditional farming area and promoting leapfrog development of rural areas (Deng et al. 2015; Long et al. 2018).

During the process of URTD, the coupling relationship between GFM and RDS is an important indicator for judging the regional RTD process. The GFT is characterized by changing the farmland utilization mode and enriching the functions of farmland, changing the production and lifestyle of labor in the traditional farming area, and combining the urban–rural market linkage system to optimize the geographical pattern of man–land interrelations in rural areas (Ge et al. 2018a). Different socio-economic transformation development stages correspond to different characteristics of man–land relationship patterns, and regional GFM and RDS have obvious qualitative characteristics. The evolution process of the coupling relationship between GFM and RDS become an important driving force for reconstructing the rural development (Fig. 2.5). The staged differences in the coupling process between GFT and RTD have also become an important reference for the evolution of man–land interrelations. The coupled evolution process between GFM and RDS in different regions will produce different geographical types (for example, some villages toward a benign development channel, while some villages show little changes, and a few villages experienced recession), which has become an important factor in the rural development diversity of greenhouse cultivation areas.

The evolution process of the coupling relationship between GFM and RDS become an important window to track the change mechanism of rural man–land relationship (Long and Qu 2018; Tu et al. 2018). Under the background of the tension pattern of "more people and less land" in the traditional farming area, the pattern of farmland use is mainly to continue the mode of intensive cultivation to maintain the basic food needs of farmers. In the early stage of URTD (social and economic development stage A in Fig. 2.5), urban–rural relations were separated, and rural production patterns and farmers' livelihoods maintained traditional farming production patterns. At this stage, the coupling relationship between the farmland morphologies and RTD showed a state of mutual inhibition (coupling mode A in Fig. 2.5). The conflict of land use pattern was strong, the possibility of FUT was greater. The GFT has changed the morphologies of farmland utilization, which triggered the core factors of the evolution of the rural man–land relationship and changed the coupling relationship between FUT and RTD.

During the process of URTD, the state of rural development and the morphology of farmland use showed significant changes (social and economic development stage B in Fig. 2.5). The tension pattern between people and land in rural areas was gradually weakened, and the livelihood and production organization of farmers were gradually integrated into the urban–rural network system (Li et al. 2015). As the GFT is strongly constrained by land management policies, there is a risk that the FUT will lag behind the RTD in the evolution of the coupling relationship between GFT and RTD (coupling mode B in Fig. 2.5). In response to the coupling problem between GFT and RTD, the interest appeals of different participants are diversified. Different villages have different ways of dealing with the coupling problems, which will lead to the differentiated evolution path of the coupling relationship between GFT and RTD (coupling mode C in Fig. 2.5), which also leads to the difference of rural development dynamics (Li et al. 2019). Therefore, by regulating the coupling relationship between GFT and RTD, it can provide a basis for the orderly rural transformation development.

2.3.2.2 Process of Greenhouse-Led Farmland Transition in Nanxiaowang Village

Qingzhou City is a major city for producing cash crops such as melons and vegetables in Shandong Province, a national-level ecological construction demonstration zone, and one of the top 100 counties in the country. By the end of 2015, Qingzhou City's vegetable planting area had increased to 41,653 ha: the area of facility vegetable production was greater than 85% of the total area of vegetables planted; more than

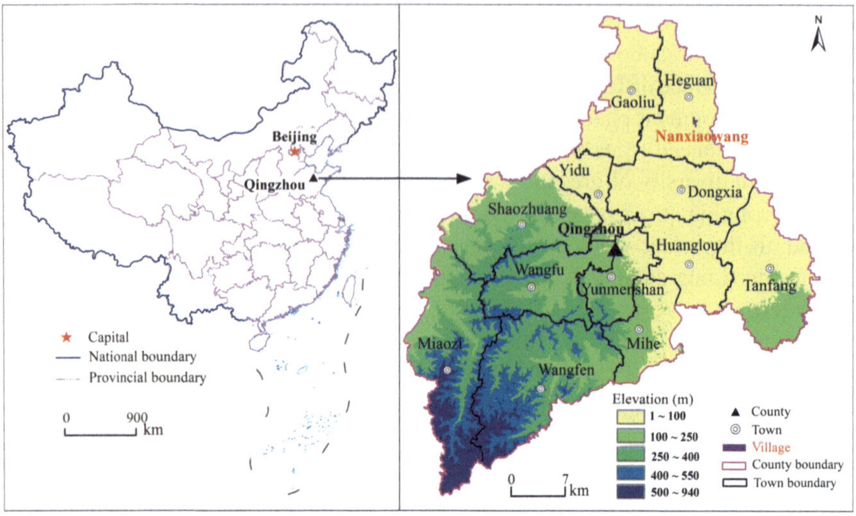

Fig. 2.6 Map of Qingzhou City (Ge et al. 2019)

200,000 vegetable greenhouses of various types were in use, the total output increased to 3.24 million tons, and the output value increased to more than 9 billion RMB yuan.

In addition, Qingzhou City is the largest potted flower production base in northern China. The city's flower and seedling planting area is 13,333 ha, and 146 professional villages produce flowers and seedlings. This industry employs 110,000 people, and the annual transaction volume of flowers and seedlings is 4 billion RMB yuan. Qingzhou City is one of the largest transfer stations in China (Fig. 2.6). This city is also the largest production center and collection and distribution center of potted plants and flowers in the logistics center in China. The development of the vegetable and flower industry in Qingzhou has benefited from the development and promotion of greenhouse technology.

Nanxiaowang Village is in the north of Qingzhou City. Since the 1990s, Nanxiaowang Village has transformed from a typical agricultural village focusing on grain cultivation to a professional village focusing on greenhouse cultivation. In 2008, Nanxiaowang Village took the lead in the establishment of a land share cooperative in Qingzhou City, and relied on land transfer to develop organic agriculture. In 2013, the cooperative signed a "Land Circulation Trust" with China International Trust and Investment Corporation (CITIC). By using the trust financing to develop scale operations, the income of farmers increased significantly. As the core area of greenhouse production in the northern part of Qingzhou City, Nanxiaowang Village has developed a greenhouse production transformation mode from grain production to farmers' independent greenhouse development, and from the self-organized agricultural stock cooperatives to the advanced agricultural production stage based on land trusts. In the context of the continuous evolution of farmland morphologies patterns, the rural development pattern has been continuously updated. Nanxiaowang Village became the representative village of the benign evolution of the coupling relationship between GFT and RDT. The interaction between GFT and RDT jointly promoted the Nanxiaowang to become a fastgrowing star village.

The change of farmland use pattern in Nanxiaowang Village is a visual reflection of the RTD in space. The analysis of the change process of farmland morphology can effectively reflect the changes in rural production mode. In 1995, the type of greenhouse land began to appear in Nanxiaowang Village. Although the area was only 0.64 ha, the emergence of greenhouse land indicates that the village has begun to shift from the traditional grain production mode to the multi-production mixed operation mode under the guidance of market demand. In 2005, the proportion of greenhouse land in Nanxiaowang Village increased to 38.22% of the available farmland, and these greenhouses were all from farmland. In addition, the proportion of rural residential land to village area also increased: increased from 13.65% in 1995 to 15.83% in 2005. The proportion of farmland under the double extrusion of greenhouse land and rural residential land decreased rapidly by 36.79% points (Fig. 2.7). After 2008, Nanxiaowang Village established the Shengfeng Land Share Cooperative (SLSC) to integrate the farmland of the whole village into the cooperative operation. The SLSC conducted high-efficiency agricultural development such as vegetable planting on a large scale.

In 2015, the proportion of greenhouse land to available land increased to 92.39%, and the proportion of farmland has decreased from 61.78% in 2005 to 7.61%. With the increase ratio of greenhouse land in available farmland area, the dominant morphologies of farmland have undergone significant changes in quantity and structure. In addition, as mentioned above, after the large-scale introduction of greenhouse planting in Nanxiaowang Village, the recessive morphologies of farmland have also undergone significant changes. Firstly, the investment in farmland continues to increase, and the increase investment in greenhouse construction and other facilities have significantly upgraded the economic value of farmland. The increase capital has strengthened the capital attribute of farmland and blurred the property

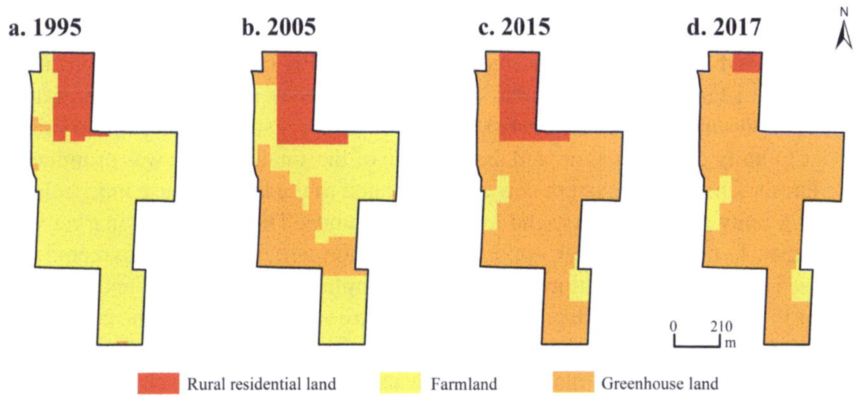

Fig. 2.7 Changes of land use structure of Nanxiaowang Village from 1995 to 2017 (Ge et al. 2019)

rights of farmland owned by the village collective. Coordinate the coupling relationship between GFT and RTD create conditions for optimizing the rural production system and peasant's organizational model. By analyzing the coupling process of the conversion of farmland and greenhouse land, it can be fully reflected that the rural production mode of the village has been gradually transformed from a mixed mode of grain production and modern agriculture to a modern market-oriented agricultural production mode centered on cash crops such as vegetable greenhouses since 2005 (Su et al. 2016). And farmers' production and living mode have been completely changed (Caulfield et al. 2019).

2.3.2.3 Coupling Process of GFT and RTD in Nanxiaowang Village

During the process of URTD, Nanxiaowang Village located in the core area of greenhouse cultivation in the northern part of Qingzhou City. The process of GFT has both common characteristics and unique paths. At different stages of URTD, the coupling state of GFM and RDS presents different stage characteristics. The interaction modes and effects between the changes of GFM and the evolution of RDS at different stages are also significantly different. Summarizing the stage characteristics of the coupling relationship evolution between GFT and RTD in different periods is helpful to clarify the differences between the interaction modes of them, and summarize how the types of differential coupling relationship affect the adjustment of the rural development path. Based on the above analysis, we can optimize the theoretical model of the succession of the coupling relationship between GFT and RTD (Fig. 2.5).

(1) **Traditional grain production stage**

During this period, the proportion of farmland accounted for more than 90% of the available farmland area. The use of farmland was relatively simple, and farmland was mainly used for the production of grain and other foods (e.g., self-sufficient vegetables). The production activities of farmers were closely related to farmland. The rural labor force has been tied to limited land for a long time. The relationship between man and land was tense, and the farmland use morphologies and RDS showed strong inhibition and stability. There were more rural surplus laborers, lower utilization efficiency of farmland, lower input and output efficiency of grain production, and limited food supply (Ge et al. 2018a). The production organization of farmers was a form of family co-production, and the attribute of the small peasant was prominent. The farmer's livelihood system was closely related to the farmland use morphology showing unity and instability, and low-risk resistance. The rural labor market was not active, farmers had fewer opportunities for non-agricultural employment, and accessed to market information was relatively simple. The greenhouse land began to appear in Nanxiaowang Village in 1995. Driven by some rural elites, the cash crops planting began to appear, and the proportion of grain crops began to decline (Li et al. 2019). At this stage, the farmland was mainly used for planting grain crops. The rural production system and the rural development pattern were mainly presented as traditional agricultural villages (Long et al. 2018). The emergence of greenhouse land has changed the morphological characteristics of farmland use. The rural production system has also begun to change.

(2) **Mixed development stage of traditional grain production and greenhouse cultivation**

The remarkable characteristics of Nanxiaowang Village in the transformation from the grain production stage to the modern high-efficiency agricultural production stage from 2000 to 2008 were the diversity of farmland use morphologies, agricultural production methods, and forms of farmer employment. The game process of the expansion of greenhouse land and the remaining cultivation of farmland was accompanied by the expansion of cash crops grown in greenhouses and partial retention of the proportion of grain production (Anderman et al. 2014; Wang et al. 2017). The combination of various agricultural production and management modes was a prominent feature of this stage. And farmers' production methods and livelihood systems also diversify with it.

By 2005, the proportion of greenhouse land increased to 38.22% (Fig. 2.8). At the same time, the rural employment labor market began to flourish, and the rural labor non-agricultural employment opportunities continued to increase. The income of farmers who grew cash crops in the village could fulfill the basic needs of life. The GFT changed the internal and external opportunity of rural development. The cultivation of cash crops promoted the development of the rural market economy, and the emergence of rural elites resulted in new vitality to the development of rural areas (Li et al. 2019). At this time, the production organization system of the village farmers was still dominated by the production of small peasants, and the production behavior of farmers was determined by independent decision-making. Due to the

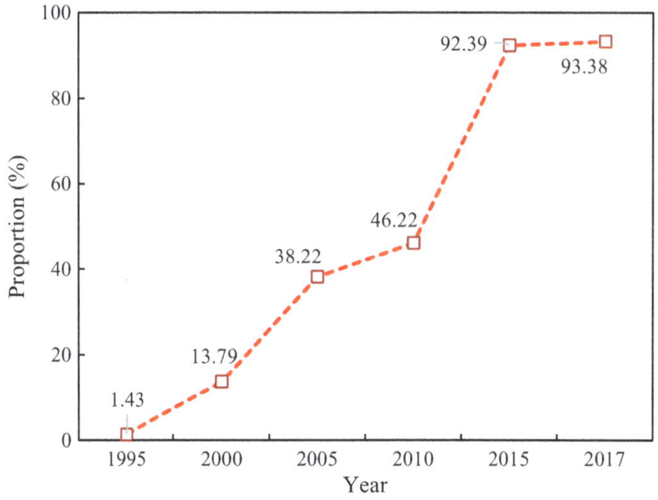

Fig. 2.8 The proportion of greenhouse land in the available farmland in Nanxiaowang Village during 1995–2017 (Ge et al. 2019)

low degree of organization of farmers, farmers had a low-risk resistance to cash crop production. The internal differentiation among farmers has been continuously presented, and part of the demand for labor land circulation to achieve non-agricultural employment has increased. We observe many problems in the coupling relationship between GFT and RTD. Regarding how to effectively solve the problems, Nanxiaowang Village attempted to establish a new form of farmland utilization mode for the land share cooperative system. The village promoted a new stage of transformation and development of rural production and farmer organization mode.

(3) **Modern and efficient agricultural production stage**

Taking the establishment of the SLSC in Nanxiaowang Village in 2008 as a node, the transformation and development of the village entered the stage of modern and efficient agricultural production. To effectively change the restrictive factors of farmland use such as sporadic farmland, backward infrastructure, and low economic benefits, the village established the SLSC in 2008 to reasonably determine the equity distribution plan and dividend standard. Additionally, farmland use morphology ushered in an important period of transformation opportunities. Through the SLSC, the reorganization of the production organization level of the farmers was realized, which led to the formation of the cohesiveness of rural development and strengthened the endogenous impetus of rural development. We observe that the change in the utilization pattern of farmland directly changes the dynamic mechanism of rural development.

After the establishment of the SLSC, large-scale greenhouse cultivation and agricultural production enterprises were gradually cultivated. The proportion of land for greenhouses accounted for more than 90% of the available farmland area, and modern high-efficiency agricultural production began to take shape. The SLSC changed the means of production and management of farmers, and the degree of man–land interrelation in rural gradually decreased. Land as a share capital joins the cooperative, and the asset attributes of the land begin to be observed. The emergence of shareholding cooperation has resulted in a greater degree of differentiation within the farmers. The existence of large growers and agricultural enterprises made the phenomenon of rural employment labor widespread, and the rural employment labor market continually improved (Ge et al. 2018b). There will be more local labor opportunities, and farmers' economic purchasing power was generally stronger. Farmers' household income was higher than in other rural areas during the same period. At this stage, the demand for local housing construction of farmers declined, and the demand for urban public services (e.g., medical, children' schooling) gradually increased. The SLSC became the core element that stimulated the endogenous dynamics of rural development (Long et al. 2016; Long and Tu 2018), revitalizing institutional barriers to industrial development in rural areas and condensing the intellectual resources required in rural development.

(4) Market-oriented comprehensive agricultural development stage

With the enhancement of the modern agricultural production capacity of Nanxiaowang Village, the land share cooperatives broke through the restrictions of the village and unified the operation of more than 100 ha of land in the surrounding villages. Nanxiaowang Village implemented the project of linked change of rural-urban construction land, signed the land transfer contract of the CTTIC trust, and established the Tianlu agricultural comprehensive development company as important nodes; next, after 2016, Nanxiaowang Village gradually entered the market-oriented agricultural comprehensive development stage (Fig. 2.9). The implementation of the project to link urban construction land increase and the rural residential land decrease has saved the farmland resources of the village (Long et al. 2012; Zhang et al. 2018). All the 4.2 ha of farmland savings indicators that have been re-cultivated and flowed into the land share cooperatives, which has strengthened the source of the village collective income and strengthened the role of the village collectives in the village construction and development. In addition, the production and quality of life of farmers have greatly improved, and local public service capabilities have improved.

The SLSC and the CITIC Trust signed the first order of the "Land Transfer Trust" in Shandong Province. The involvement of the land trust institution has created conditions for Nanxiaowang Village to conduct comprehensive agricultural development.

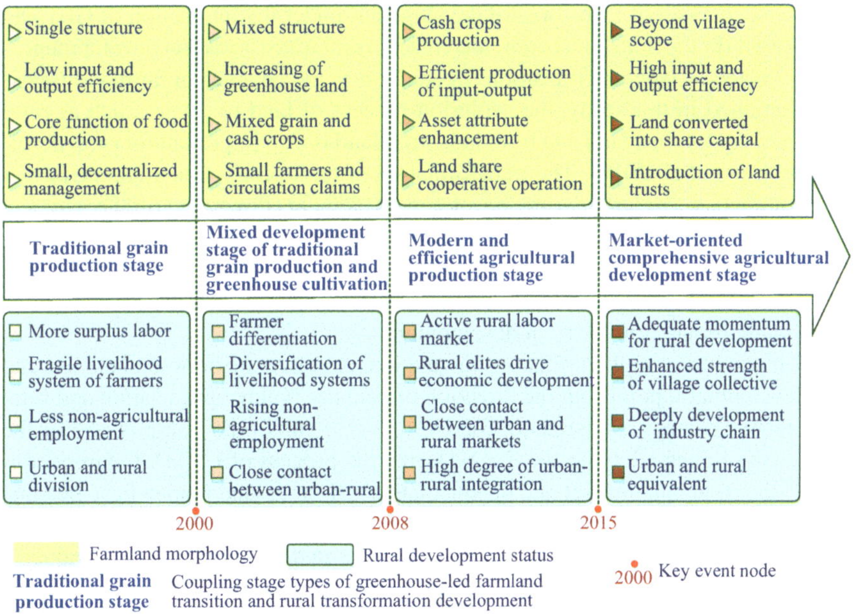

Fig. 2.9 The evolution of coupling stages of greenhouse-led farmland transition and rural transformation development in Nanxiaowang Village (Ge et al. 2019)

Furthermore, it is more convenient for farmers to obtain funds for industrial development from financial institutions by transferring land through trust institutions. On this basis, in 2017, the village and Hangzhou Tianlu Group jointly established Qingzhou Tianlu Agricultural Development Co., Ltd., which started a construction engineering logistics park for deep processing, refrigeration, preservation, e-commerce, cold chain logistics, and agricultural product technology of agricultural products. Therefore, modern agricultural comprehensive development enters a stage of rapid development. At this stage, the identity of the farmers has changed from "farmers" to "modern agricultural industrial workers." The farmers can enjoy multiple benefits such as land rent, floating dividends, and wages, and the farmers' livelihood system is diversified (Long and Tu 2018; Tu et al. 2018). Local employment opportunities are no longer limited to agricultural production. It created conditions for higher levels of talent return. Rural development is full of vitality under the combined effects of internal cohesion rural internal forces and exogenous forces, which have become a typical case for rural vitalization and development (Long et al. 2018).

2.3.2.4 Discussion

This study takes the GFT as the core concern, discusses the evolution process of the coupling relationship between GFT and RTD in China's traditional farming areas. Furthermore, taking Nanxiaowang Village of Qingzhou City as an example to analyze the mechanism of leap-forward development under the interaction of GFT and RTD. Based on the above research, the current rural transformation development is closely related to land use transition. In addition to the changes in farmland transition mentioned in this study, the evolution process of land use types such as rural residential land, forest land and rural industrial land is also important for rural development (Long et al. 2018). The transformation of rural production systems, farmer organization patterns, urban–rural development elements flows, and rural ecological environment provides guarantees for rural sustainable development (Liu 2018; Long et al. 2018; Meyfroidt et al. 2018). An indepth analysis of the coupling mechanism and regional differences between land use and rural development will help to improve the dynamic foundation for rural development.

Optimizing the coupling relationship and interaction mode between land use and rural development, points out the direction of land use policy optimization and rural development strategy adjustment. The evolution of rural land value is one of the core driving forces affecting the RTD. During the process of URTD, the imprisonment of the rural land market and the unequal urban–rural land market have become important reasons for the rural decline (Long et al. 2016). The imperfect development of the rural land market leads to the serious underestimation of the value of rural land, which hinders the accumulation process of rural primitive capital, and also becomes an important reason for the loss of rural production elements, the imbalance of production relations, and the lack of development momentum (Ge et al. 2018a). In addition, the decision-making system of farmers in land use change is the internal mechanism for the improvement of rural development. In turn, it has brought about deep changes in land use. Farmers play a dual role in land use change and rural development, who are both active and passive adaptors. The socio-ecological feedback effect of land use change is an important guarantee for rural sustainable development. The land use system and the rural development system are closely linked through the socio-ecological feedback effect of land use change (Meyfroidt et al. 2018). Therefore, rationally regulating land use policies to cope with the needs of RTD requires comprehensive consideration of factors such as farmers' decision-making system, urban--rural land markets, and social-ecological effects.

During the process of URTD, China's RTD stage varies from place to place, systematically analyzing the regional differences in the coupling relationship between land use transition and RTD provide reference for improving the policies of rural land management and formulating a reasonable rural development strategy (Caulfield et al. 2019; Long and Qu 2018). Taking the GFT as an example, this study attempts to analyze the interaction mechanism between farmland transition and RTD in traditional farming areas through the theoretical model construction and empirical case analysis. However, the geographical differences in rural development in China are significant, and there are also significant differences in the types of land use that dominate rural transformation development (such as rural industrial land transition, rural residential land transition, etc.). How to construct a research framework of coupled analysis between land use transition and rural development with gradient differentiation, and carry out practical research with typical cases, will help to enrich the theoretical and practical research on rural land use transition and RTD.

2.4 A Perspective of Grain Productive Ability

Farmland resources and their utilization are the core of ensuring food security. The grain production ability of farmland is an important manifestation of farmland use morphology. The spatial distribution characteristics and evolution process of farmland use in China can be well reflected by analyzing the matching relationship between farmland resources and grain production. The grain production ability of farmland is also the core of carrying capacity of regional resources and environment. The spatial characteristics and evolution process of grain production ability per unit farmland area intuitively reflect the recessive morphology of farmland use and its transition process. Therefore, the following is an attempt to analyze the recessive morphology of farmland use from the perspective of the spatial matching relationship between farmland resources and grain production and the grain production ability of per unit farmland area, and then analyze the transition process of farmland use from the perspective of grain production ability of farmland.

2.4.1 Spatial Mismatch of Grain Production and Farmland Resources

Grain yield is simultaneously affected by farmland area and other external factors (fertilizers, irrigation, new species, new technology and mechanization). At the beginning of the 21st centaury, as the declining marginal efficiency provided by these external factors with regard to the level of grain production, farmland area again became the main constraint to increase the grain yield. So, retaining certain areas of farmland resources will become a paramount priority in assuring and maintaining China's food security. At present, China's grain production pattern runs counter to the distribution patterns of farmland resources and water-heat conditions. It is urgently important to conduct research on the spatial matching of grain production and farmland resources. Based on the Potential Crop Rotation data, with regards to the situations of irrigation and rain feeding, this section builds a measuring model of potential farmland, and separately measures the spatial matching index of grain yield with actual and potential farmland resources, thus revealing China's grain production pattern.

2.4.1.1 Materials and Methods

The data relating to grain yield, irrigated areas (*IA*) and total cultivated area (*TCA*) of 2,347 counties in China, in 1990, 2000 and 2010, was mainly extracted from "China Statistical Yearbook for Regional Economy" and "China County Statistical Yearbook". The data pertaining to 100 × 100-meter grids of farmland in 1990, 2000 and 2010, was provided by National Data Sharing Infrastructure of Earth System Science (www2.geodata.cn). The data pertaining to 10 × 10-kilometer grids of Potential Crop Rotation Index (PCRI) for the irrigated and rain-fed scenarios in 1990, 2000 and 2010, was provided by Global Change Research Data Publisher & Repository (www.geodoi.ac.cn). It was estimated with GAEZ-model developed by FAO and IIASA, based on DEM data, soil data, meteorological data and arable land. The PCRI includes single, 1.5 times, double and triple crop rotation systems.

The PCRI was resized from 10 km to 100 m grids, in order to calculate the potential farmland combined with actual farmland. Eventually, the actual farmland and potential farmland were calculated at county level.

Potential farmland area denotes the natural potential area of physical farmland. For example, the potential farmland was two times the actual farmland in a double crop system. Therefore, potential farmland is given as (Liu et al. 2014a):

$$y_p = y \times PCRI_i \times i + y \times PCRI_r \times (1 - i) \tag{2.6}$$

$$i = \frac{IA}{TCA} \tag{2.7}$$

where, y_p is the area of potential farmland, y is the area of actual farmland, $PCRI_i$ and $PCRI_r$ is the $PCRI$ under irrigated and rain-fed scenarios, respectively; i is the ratio of irrigated area (IA) to the total cultivated area (TCA).

The model of gravity center fitting is used to reveal the displacement rule of grain yield, farmland and potential farmland and is calculated as follows:

$$x_j = \sum_{i=1}^{n} (T_{ij} \times x_i) / \sum_{i=1}^{n} T_{ij}$$
$$y_j = \sum_{i=1}^{n} (T_{ij} \times y_i) / \sum_{i=1}^{n} T_{ij} \tag{2.8}$$

where, P_j $(x_j\ y_{j)}$ represents the national barycentric coordinates of grain yield (farmland, potential farmland) in year j, P_i $(x_j\ y_{j)}$ stands for the barycentric coordinates of grain yield (farmland, potential farmland) of county i, and T_{ij} stands for the grain yield or area of farmland (potential farmland) of county i in year j.

Spatial mismatch refers to the imbalance between grain yield and farmland area, which is currently a hot topic in China. The spatial mismatch index of grain-to-farmland (SMIGF) is used to measure the spatial relationship of grain-to-farmland and is calculated as follows:

$$SMIGF\_A_i = \left(G_i \Big/ \sum_{i=1}^{n} G_i - F_i \Big/ \sum_{i=1}^{n} F_i \right) \times 100$$
$$SMIGF\_P_i = \left(G_i \Big/ \sum_{i=1}^{n} G_i - PF_i \Big/ \sum_{i=1}^{n} PF_i \right) \times 100 \tag{2.9}$$

$$\sum SMIGF\_A = \sum_{i}^{n} |SMIGF\_A_i|$$
$$\sum SMIGF\_P = \sum_{i}^{n} |SMIGF\_P_i| \tag{2.10}$$

$$\begin{aligned} \Delta SMIGF_i &= SMIGF\_A_i - SMIGF\_P_i \\ &= \left(PF_i / \sum_{i=1}^{n} PF_i - F_i / \sum_{i=1}^{n} F_i \right) \times 100 \end{aligned} \tag{2.11}$$

where, $SMIGF\_A_i$ represents the spatial balance between grain yield and actual farmland area. $SMIGF\_P_i$ is used to measure the spatial mismatch between grain yield and potential farmland area. G_i and F_i are the grain yield and farmland area of county i, respectively, and n is the number of counties. $\Sigma SMIGF\_A$ ($\Sigma SMIGF\_P$) represents the degree of total spatial mismatch between grain yield and farmland area (potential farmland area) in China. Actually, $SMIGF\_A_i$ is the comparison between the location quotient of grain yield and that of the farmland area in county i. When the absolute value of $SMIGF\_A_i$ is less than a certain value, there is a spatial match between grain production and farmland. When the absolute value of $SMIGF\_A_i$ is more than the value and the $SMIGF\_A_i$ is negative, there is a spatial mismatch between grain production and farmland. In addition, the location quotient of farmland area is larger than that of grain yield, which means the use of farmland in county i is inefficient. Conversely, when the absolute value of $SMIGF\_A_i$ is more than the value and the $SMIGF\_A_i$ is positive, the spatial mismatch still exists, but the location quotient of grain yield becomes larger than that of farmland, which in turn means that the farmland in county i is used efficiently. The same is true for $SMIGF\_P_i$. $\Sigma SMIGF\_A$ ($\Sigma SMIGF\_P$) increases with the degree of spatial mismatch between grain production and farmland (potential farmland). $\Delta SMIGF_i$ is the variance between the $SMIGF\_A_i$ and $SMIGF\_P_i$, and is actually the difference between the location quotient of potential farmland and actually farmland. $\Delta SMIGF_i > 0$ implies the great potential of the farmland resources of county i, and vice versa.

The Gini coefficients can be used to detect the equilibrium between grain yield and farmland area, and it is specified as:

$$G = 1 - \sum [a \times (2b - c)] \tag{2.12}$$

where G is the Gini coefficient, a and c are the proportion of farmland area and grain yield of each county, respectively, and b is the cumulative proportion of the grain yield of each county. Usually, the Gini coefficient is between 0 and 1. If the farmland area and grain yield reach a greater degree of equilibrium, the Gini coefficient becomes smaller, and vice versa.

2.4.1.2 The Spatio-Temporal Pattern of Grain Production and Farmland Resources

The distribution of grain production and farmland resources in China are restricted by the "HU Line"[2] (Fig. 2.10). Most of these areas with high grain yield and abundant farmland resources are concentrated in the southeast region of the "HU Line". The concentrating tendency for the spatial distribution of grain yield is becoming more and more obvious. The spatial distribution of the actually and potential farmland are, however, relatively stable. The Potential Crop Rotation Index (PCRI) in China is characterized as high spatial distribution in southern and eastern regions and low spatial distribution in northern and western regions. Thus, in southern and eastern China, potential farmland resources are more common than actual farmland

[2]HU line, is a "geo-demographic demarcation line" discovered by Chinese population geographer HU Huanyong in 1935 (Hu 1935). The imaginary Heihe (in Heilongjiang)-Tengchong (in Yunnan) Line divides the territory of China into two parts: northwest of the line covers 64% of the total area but only 4% of the population; however, southeast of the line covers 36% of the total area but 96% of the population (Long and Li 2012)

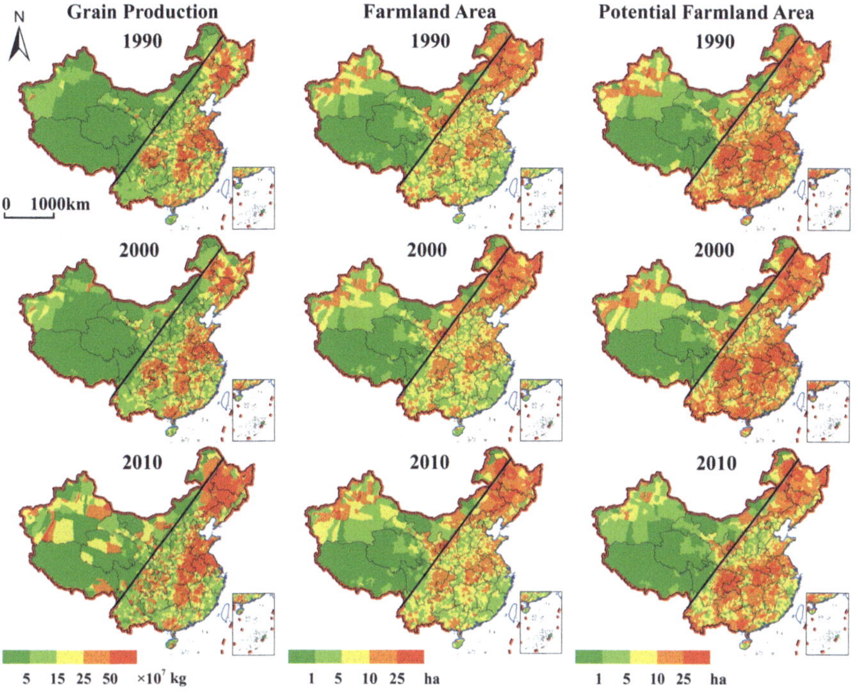

Fig. 2.10 The spatio-temporal pattern of grain production, farmland and potential farmland. The inclined line is called "HU Line" (Li et al. 2017)

resources. When combined with the barycenter model (Fig. 2.11), we found the barycenters of grain yield, farmland area and potential farmland area all shifted to north China. The barycenter of grain yield presents a radical northward shifting. However, that of the farmland area and potential farmland area presents a slight northward shifting. In addition, the barycenter of potential farmland area locates in the south of that of farmland area and grain yield. Therefore, it is necessary to discuss the issue of spatial mismatch between grain yield and potential farmland area.

Grain yield is not only affected by farmland area, but also affected by other factors. The changing rules relating to grain yield and farmland area are reviewed since the founding of the country (Fig. 2.12). At the beginning of the country, when the farmland areas were scarce, grain yield changed positively with the change of increased farmland areas. Since 1960, when farmland areas were relatively sufficient, the input of external factors (fertilizers, irrigation, new species, new technology, mechanization) increased the grain yield. At the beginning of the 21st centaury, the marginal benefits provided by these external factors with regard to the level of grain production began to decrease, so farmland area again became the main constraint to increasing the grain yield. However, due to the implementation of the "Grain for Green" policy in 1998, as well as the increasing phenomenon of rural-to-urban migration, a great number of farmers swarmed into the cities, and the instances of

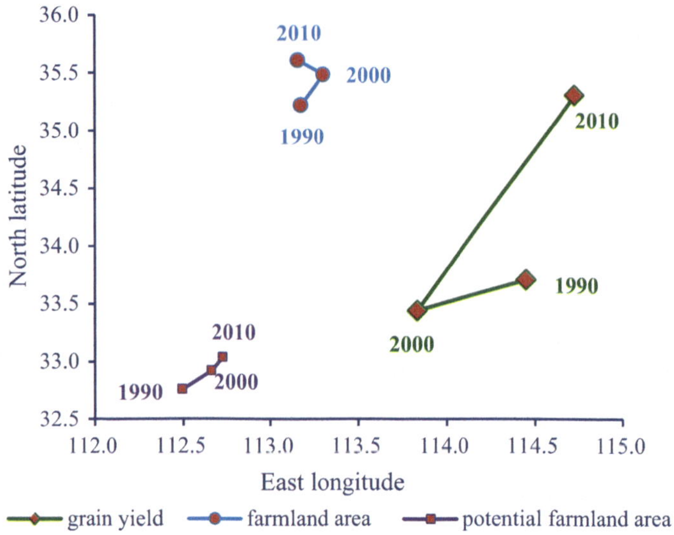

Fig. 2.11 The trajectory of the barycenter of grain yield, farmland area and potential farmland area (Li et al. 2017)

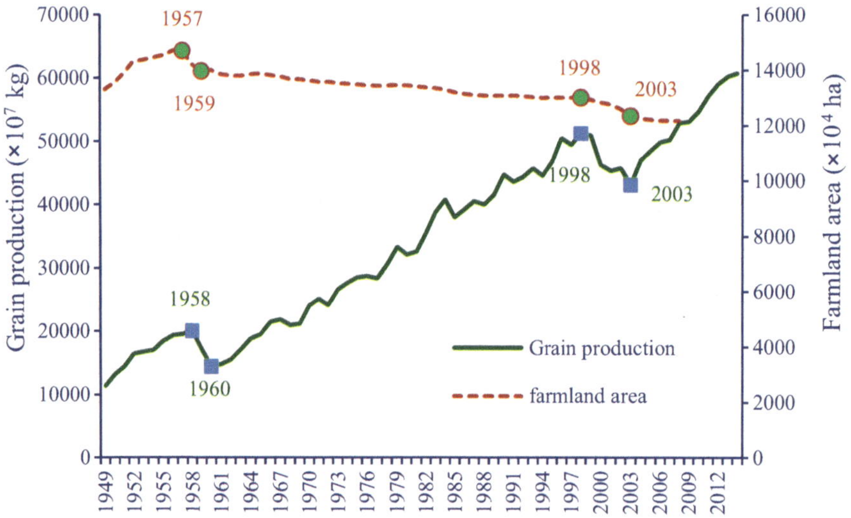

Fig. 2.12 The grain yield and farmland area of China since 1949 (Li et al. 2017)

farmlands being abandoned was staggering. The continuous sharp decrease in the use of farmland area caused a corresponding radical decrease in the grain yield. In 2003 the grain yield had decreased to 0.48 billion tons. In the same year, at the Third Plenary Session of the 16th CPC Central Committee, it was proposed to "carry out

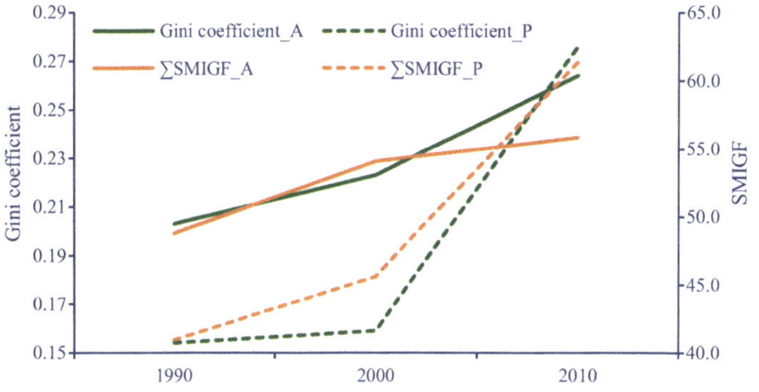

Fig. 2.13 The Gini coefficient and ∑SMIGF of grain production and farmland during 1990–2010 (Li et al. 2017)

the strictest farmland protection policy to ensure the food security of the country". Since then, the rate of decrease in farmland area has decelerated, and the grain yield has continuously increased.

In order to measure the matching degree of grain yield with farmland resources throughout the whole of China, we use the Gini coefficient together with ∑SMIGF. The results of these two expressions possess similar changing characteristics (Fig. 2.13). During the period from 2000 to 2010, both the Gini coefficient and ∑SMIGF presented a trend of sharp increase, and the spatial mismatch between grain production and farmland resources became more serious. Through the comparison between ∑SMIGF_A and ∑SMIGF_P, we found that in the early periods covered by the study, ∑SMIGF_P was always less than ∑SMIGF_A. There was no doubt that, grain production was concentrated in the regions with good hydrothermal conditions, and the farmlands with multi-crop systems were used sufficiently. However, in 2010, ∑SMIGF_P overtook ∑SMIGF_A, and the spatial mismatch between potential farmland resources and grain production was higher than that between actual farmland resources and grain yield. It could be seen that, during this period, the inefficient use and even abandonment of farmland resources with multi-crop systems was very common.

2.4.1.3 The Spatio-Temporal Pattern of SMIGF_A

Based on Jenks Natural Breaks Classification in ArcGIS software, the research used –0.02 and 0.02 as the cutoff threshold values to determine efficient or inefficient uses of farmland, as well as spatially (mis)matched farmland. As shown in Fig. 2.14 the degree of the spatial matching of grain yield and farmland resources was going from bad to worse during this period. As seen from the regions with serious spatial mismatch, the regions with inefficient utilization of farmland was mainly distributed in the Greater Khingan Mountains, northern farming-grazing transitional zones and the northwest area of Xinjiang. In 2010, these inefficient regions extended to the

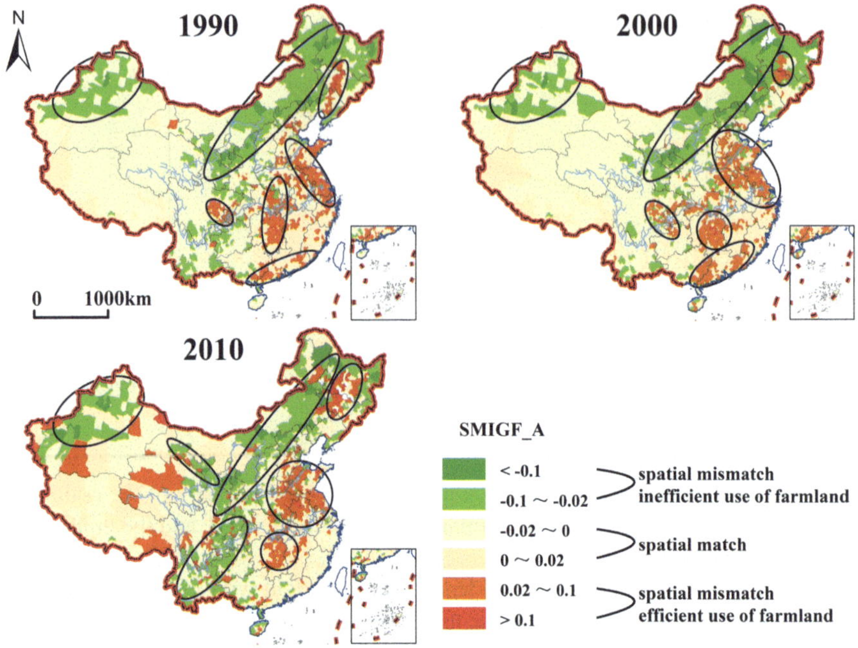

Fig. 2.14 The spatio-temporal pattern of SMIGF_A from 1990 to 2010 (Li et al. 2017)

southwest karst landform area. All of these areas are regions in China with the fragile ecological environments. In the highly efficient farmland utilization regions, or the main grain-producing areas, the spatial distributions showed greater differences. In 1990, the main grain-producing areas were mainly distributed over the Pearl River Delta region, Yangtze River Delta region and the middle reaches of the Yangtze River. In addition, the main grain-producing areas were also scattered in the Sichuan Basin, the Huanghuaihai Plain and northeastern China. In 2000, the barycenter of grain production began to shift to the Huanghuaihai Plain. The remediation of the saline-alkali lands contributed to the increase of grain yield per unit farmland area. The function of grain production in the Huanghuaihai district was strengthened. Meanwhile, due to the ample sunshine and rainfall of the Pearl River Delta region, the grain-producing potential of farmland resources in this area was further tapped, and the function of grain production was enhanced even further. Conversely, the grain-producing function of the middle and lower reaches of the Yangtze River was substantially weakened. Additionally the non-agricultural use and inefficient use of the farmlands persisted, which caused the grain yield to sharply decline.

The decreasing scope of the grain yield was in fact larger than that of farmland areas. The northward shifting of the grain production barycenter began to take shape. In 2010, the pattern of "westward movement and northward expansion" for grain production came into full effect, and the grain-producing function in the Pearl River Delta region substantially deteriorated. In the middle and lower reaches of China's Yangtze

River (known as "a land flowing with milk and honey"), the main grain-producing areas were greatly reduced, and the grain-producing function in the Zhejiang, Hubei and Sichuan Basin was weakened. On the contrary, the grain-producing function in the northeastern and northwestern China was greatly improved. Corresponding with global warming, the farming boundary in the northeastern China extended to the north, while the crop rotation system changed from a one-season pattern to a multi-season pattern (Gao and Liu 2011). Driven by the economic development and the West Development Strategy of China, efforts were intensified to exploit the farmlands in the western regions, and the inputs of agricultural technology and capital were increased rapidly, thus causing the grain yield and even the SMIGF_A to increase.

The dynamic changes in the SMIGF_A are, in essence, direct comparison between the relative changes in the Location Quotients of grain yield and farmland area in different years. The increase of the SMIGF_A indicates that the grain yield allows for a relative higher increase of Location Quotient than farmland area, which in turn shows the improvements in both farmland-utilization efficiency and grain-producing function. Conversely, the Location Quotient has, relatively speaking, reduced, which indicates serious extensive utilization of farmlands and a trend towards a weakening grain-producing function. Overall, the SMIGF_A during the period from 2000 to 2010 changed most radically. The changes in the spatio-temporal pattern during this period tallied with that during the period from 1990 to 2010.

Also during the period from 1990 to 2000, the efficiency of farmland-utilization was an important factor which affected the spatio-temporal pattern of the SMIGF_A. A pattern of southward and northward expansion from the Yangtze River valley emerged (Fig. 2.15).

(1) The regions with decreasing SMIGF_A were mainly situated on the plains in the middle and lower reaches of the Yangtze River, as well as in northeastern and western China. In the 1990s, the plains in the middle and lower reaches of the Yangtze River were the most rapidly urbanized regions in our country. These regions rapidly transformed from a grain-producing function to a function of socio-economic development. Under these circumstances, the move away from the agricultural and food-producing utilization of farmland was inevitable. Driven by both the decrease in farmland area and the increasingly inefficient use of farmland, the decline rate of the Location Quotient for grain yield far exceeded that of farmland area. Different from the decreases of both grain yield and farmland areas, however, the areas showing a decline of SMIGF_A in northeastern and western China also commonly showed an increase of farmland area and a decrease in grain yield. Land remediation and reclamation supplemented the farmland resources, but the increase in farmland area hardly offset the decrease in grain yield accelerated by the inefficient use of farmland. Together, these factors caused the SMIGF_A in these regions to decline. On this basis, the inefficient use of farmland is the main cause of the country's decreased grain yield.

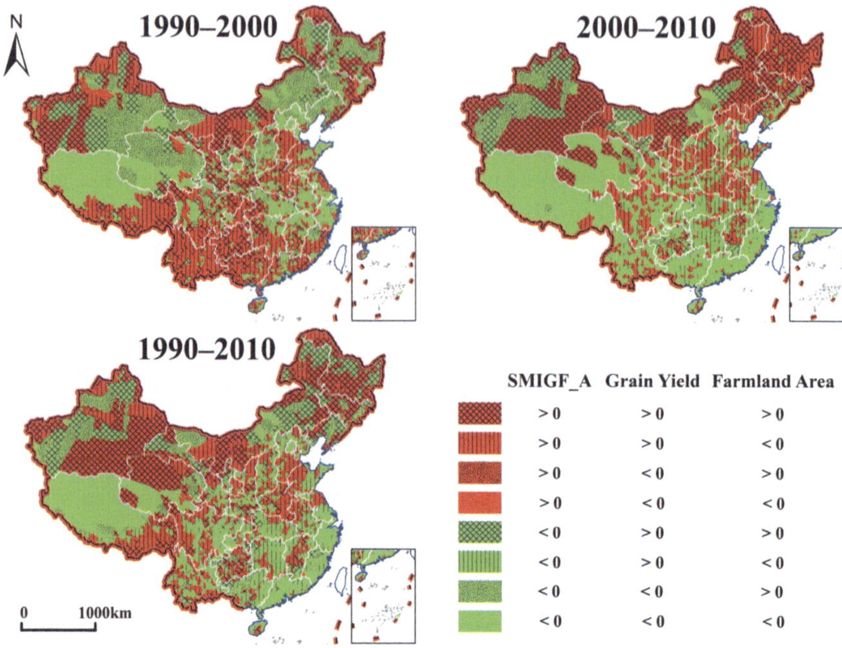

Fig. 2.15 The dynamic change of SMIGF_A, grain yield and farmland area (Li et al. 2017)

(2) In the regions with an increasing SMIGF_A, the grain yield also commonly increased. In the southern, southwestern, western and northeastern China, both grain yield and farmland area increased simultaneously. The rate of increase of the Location Quotient for grain yield was higher than that for farmlands, with the intensive level of the farmlands improved. However, in the Huanghuaihai Plain, the regions with increasing SMIGF_A commonly showed a trend of increasing grain yield and decreasing farmland area. As the region with highest density of population, industries and towns in northern China, in the urbanization process of Huanghuaihai Plain, the non-agricultural use of farmland was an indisputable fact. However, the Huanghuaihai district possesses a natural advantage in terms of grain production, with deep soil layers, fertile soil, ample light and heat resources, and optimum levels of rain and heat over the growing period. There is a good mix and match between light, heat, water and soil in this area. Additionally, the flat landscape makes larger-scale operations easier to achieve. Any negative effect of decreasing farmland area on grain yield can be reduced by improving the efficiency of farmland utilization, which will in turn help achieve the goal of increasing the region's grain yield.

During the period from 2000 to 2010, as the marginal benefits of external factors in the grain production process decreased successively, the farmland area again became the main constraint to increasing the grain yield. Both SMIGF_A and farmlands showed a pattern of reducing in the south and increasing in the north (Fig. 2.15).

(1) The southern and eastern parts of China (Guangdong, Fujian, Zhejiang and the southern area of Jiangsu) presented a phenomenon of decreases in both farmland area and grain yield, leading to a SMIGF_A decrease. These regions' marketization level is very high, and the handicraft, business and electronic industries in many villages are very advanced. The farmers have a high level of consciousness in terms of starting their own businesses, so the regions almost naturally experienced a transformation from agricultural to non-agricultural industries. This transformation made the inefficient use of and even the abandonment of farmlands inevitable.
(2) In most areas within the Yangtze River Valley, SMIGF_A decreased. At the same time, this decrease was accompanied by an increase in grain yield and a decrease in farmland area. Despite the increased grain yield, the Location Quotient for grain yield in this region still decreased, due to the larger and increasing amplitude of grain yield in the northeastern and western areas of China. In addition, the decrease of the Location Quotient for grain yield was larger than that for farmland areas. The grain-producing function decreased.
(3) In the county areas between the Yangtze River Valley and the northern farming-grazing transitional zone, the SMIGF_A showed both increasing and decreasing patterns, while the farmland areas in this region presented a trend of general reduction. The areas with decreasing SMIGF_A tallied with the distribution of city groups.
(4) Large increases in both grain yield and farmland area were evident in three northeastern provinces of China, as well as in Inner-Mongolia and the northwestern region. As the efficiency of grain production improved, these areas became the main regions with newly-increased grain yield. In addition, the increment of the Location Quotient for grain yield was far greater than that of farmlands, resulting in a general increase in the SMIGF_A in these regions.

2.4.1.4 The Spatio-Temporal Pattern of SMIGF_P

The spatio-temporal pattern of SMIGF_P presented a pattern of "northward and southward expansion" and "westward movement and northward expansion" during the two periods of 1990–2000 and 2000–2010, respectively (Fig. 2.16). (1) The inefficient farmland utilization regions are mainly distributed in the northern farming-grazing transitional zones and the southwest karst landform regions, as well as along the middle reaches of the Yangtze River. In 2000, the southwest karst landform areas realized a balance between grain yield and potential farmland resources. In 2010, this balanced pattern was broken. The southwest karst landform areas and the plain area in the middle reaches of the Yangtze River were the main districts experiencing population output. The phenomenon of the inefficient use of farmland resources was very common, so these regions' SMIGF_P showed a negative value. (2) For the main grain-producing areas, the spatial heterogeneity were very obvious. In 1990, the main grain-producing areas were mostly located in the Hunan Dongting Lake area,

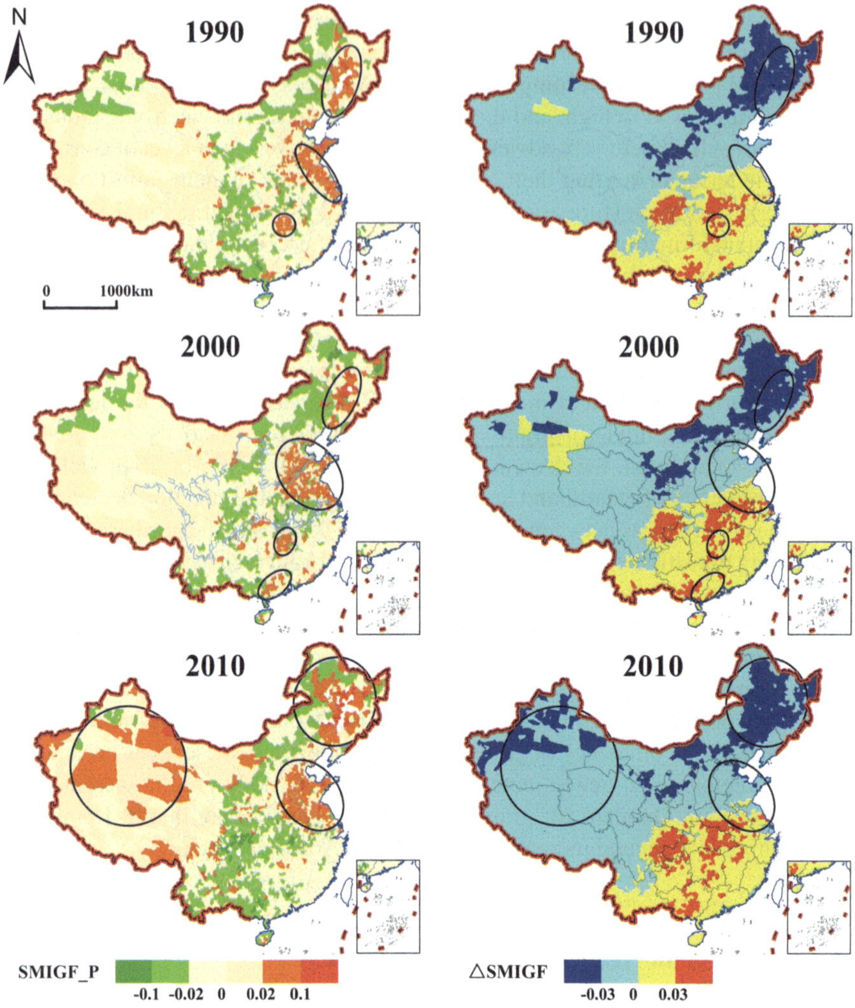

Fig. 2.16 The spatio-temporal pattern of SMIGF_P and ΔSMIGF during the period from 1990 to 2010 (Li et al. 2017)

Jiangsu, Shandong and the Northeastern Plain. In 2000, the main grain-producing areas extended southwards to the border areas of Guangdong and Guangxi, and northwards to the Huanghuaihai Plain. In 2010, the main grain-producing areas showed a significantly large margin of "westward movement and northward expansion". As a result, the main grain-producing function of the border areas of Guangdong and Guangxi and the Dongting Lake area declined significantly, while the regions with a higher grain-producing function expanded rapidly in the northeastern and northwestern areas.

By comparing the spatio-temporal patterns of SMIGF_A (Fig. 2.14) and SMIGF_P (Fig. 2.16), we found that the Huanghuaihai Plain and the northeastern regions became the main grain-producing areas in 2000 and in 2010, respectively, with the full exploration of farmlands' potential. Additionally, in 2010, the western region, with a water shortage, also began to undertake a grain-producing function. Contrarily, the SMIGF_A and SMIGF_P in the regions of the middle sections of the Yangtze River and Sichuan Basin varied greatly. The calculating result of SMIGF_A indicates that, in these two regions, the Location Quotient for grain yield was higher than that of actual farmland areas, and these were the main grain-producing areas in China. While the calculating result of SMIGF_P showed that, in these two regions, the grain yield and potential farmland areas realized a spatial matching, even in 2010, it appeared that the Location Quotient for grain yield was smaller than that for potential farmland areas, with seriously extensive and inefficient use of farmland resources. Finally, the relative grain yield was low, and these regions' function as main grain-producing areas declined. Therefore, we can see that the advantage of multi-crop systems in the middle reaches of the Yangtze River and Sichuan Basin has not yet been fully exploited, and the farmlands' potential has not yet been fully realized.

With the northern edge of the Yangtze River Valley as a boundary, the spatial distribution pattern of ΔSMIGF formed a north-south differentiation rule. The southern district possessed greater potential farmland resources, especially in the middle reaches of the Yangtze River, Sichuan Basin and Guangxi Basin. In contrast, the potential of farmland resources in northern areas was relatively low, especially in the northern farming-grazing transitional zones and in the northern Xinjiang. By comparing the spatial distributions of the grain-farmland mismatching regions with that of ΔSMIGF, we found that in 1990 and 2000, most main grain-producing regions were distributed in the areas with greater potential of farmland resources (ΔSMIGF > 0). In 2010, a serious phenomenon existed, whereby the main grain-producing regions shifted to the north. As a result, in the regions with greater farmland potential in Guangdong, Guangxi, the middle reaches of the Yangtze River and the Sichuan Basin, the grain-producing function deteriorated. Contradictorily, the northeastern area, Huanghuaihai Plain and the western region, all of which had lower farmland potential, became important functional zones for grain production. It was obvious that a great change in the grain-producing pattern of China has occurred. A large number of farmlands in the south with multi-crop systems were abandoned. The grain barycenter shifted to the Huanghuaihai Plain and northeastern areas. Even Xinjiang, with a serious water shortage, gradually became the main grain-producing area of China. This kind of grain-producing pattern runs counter to the distribution patterns of potential farmland resources and water resources, and their spatial mismatch has increased the ecological risks.

2.4.1.5 Discussion

The spatial mismatch of the grain production and farmland resources in China is affected by many factors, including the exter-nal factors (such as urbanization, the rural-to-urban migration, irrigation technology and land remediation technology) and thenatural factors (such as global warming). Since the 1990s, the farmland resources have been occupied as a means to provide more development space for the towns and cities, expecting to experience a rapid urbanization. In the areas where groups of cities were located, this phenomenon of farmland occupation was the most radical. This change of use of farmland is also leading to the decrease of grain-producing function in these areas. Simultaneously, the rural population has undergone a large-scale transfer to urban areas, and the phenomenon of the absence of agricultural management mainstream appeared which has further led to the inefficient use of and even the abandonment of farmland (Chen et al. 2014). This phenomenon is especially obvious in provinces which have experienced large population outputs, such as Sichuan and Hunan. The development of water-saving irrigation technology and the "South-to-North Water Diversion" Project, at least to a certain extent, has eliminated the restriction of water shortages on grain production in northern areas, thus improving the regions' grain-producing efficiency. Land remediation technology has allowed formerly saline-alkali lands and barren desert lands to be exploited as farmland. The saline-alkali land remediation in Huanghuaihai district helped the region to become the main grain-producing area in China at the beginning of the 21st century. In addition, Global warming has led to a northward shifting of the grain-producing boundary in China. The crop system in the north has also been enhanced. In general, the above factors affect the farmland productivity. However, the spatial mismatch between grain production and farmland resources also causes many problems, such as an imbalance in regional structures, ecological risks, agricultural production risks, and concerns over the price of food. The imbalance in regional structures is manifested as the imbalance of the main producing areas and the main selling areas. As the Chinese population migrates to the southeastern region, this region is becoming the main grain-selling area. This phenomenon is in strong spatial contrast to the pattern of the "westward movement and northward expansion" of the main grain-producing areas. At present, the grain production pattern is inconsistent with hydrothermal conditions, which poses significant risks in terms of water and ecology. Due to the dry climate of the Xinjiang district, its irrigation coefficient on grain production reaches 80%, resulting in a high dependence on groundwater, which in turn frequently causes an overuse of groundwater. The northeastern region has enlarged its farmlands area by reclaiming meadow lands and wet-lands, which is rapidly leading to a decline in the ecological function of those wetlands. In addition, due to the overuse of black soil and irrational fertilization practices, the black soil is now facing ecological risks such as the reduction of the organic layer's thickness and soil salinization. Meanwhile, large-scale north-to-south grain diversion increases the transportation cost of grain (Li et al. 2015). Confronted with the reality of low-priced imported grain, the "ceiling effect" of grain production in China is becoming more and more severe, with the grain production price risk increasingly highlighted.

In order to cope with the problems caused by the spatial mismatch between grain production and farmland resources, this study provides some advice in view of actual conditions. As the contribution of farmland area to the improvement of grain yield is strengthened, farmland protection in terms of both quantity and quality will increasingly become the most fundamental measure to ensure the nation's food security. For the sake of matching the spatial pattern of multi-crop rotation indices, with high levels in the south and low levels in the north, the farmland in South China which are of the highest quality should be preferentially categorized as basic farmland under permanent protection (especially in the plains of the middle reaches of the Yangtze River, Sichuan Basin and Guangxi Basin) (Lichtenberg and Ding 2008; Ding 2003). Only by that could we give full play to the advantage of multi-crop rotation planting and efficiently use the local rich hydrothermal resources, so as to improve the efficient use of farmland resources and gradually renew the grain-producing function of South China (Su et al. 2016). Conversely, in the regions with the fragile ecological environment, such as the northeastern area and West China, based on the water, soil and climatic conditions, the crop rotation system should be decided upon scientifically. Drought-tolerant crops should be planted, and the ground water should be recovered by means of leaving lands fallow and rotating crops.

In addition, in view of the inefficient use and even abandonment of farmland caused by population outflow, and given the opportunity presented by migrant workers returning home due to the economic decline, the Chinese government should actively provide the returning farmers with help and support in terms of financing, training in agricultural techniques, and so on. Such policies would make it possible for the migrant workers to again throw themselves into grain production. In addition, the government should encourage the returning farmers to extend the industry chain by taking advantage of their more advanced social resources and market consciousness than the local farmers have at present. It makes sense to build the grain industry chain by integrating "production-processing-logistics distribution", promoting the integrating development of the primary, secondary and tertiary industry, increasing the additional value of the grain industry, and ensuring food security. Such policies would at the same time help with the reemployment and increase the income of migrant workers who are returning home. To ensure the implementation of the above-mentioned regional policy suggestions, it would also be necessary for China to reform and innovate some aspects of its agricultural policies, such as creating a rural land circulation system, grain price forming mechanism, farmers' subsidy policy and a grain collection and storage policy. Such practices, if implemented, would protect the farmers' benefits, ensure the stability of grain prices, promote large-scale land management and commonly defuse the food security issues currently facing China.

Finally, under the background of radical barycenter shift of grain production, it is important to point out that appropriate policies should be issued in different regions, which give fair amount of consideration to cost factor along with benefits. As we all know,external factors are important since they bear cost implications. And the marginal effects of external factors are also affected by natural factors. The hydrothermal conditions in the northern region are congenitally deficient, and the northward shifting of the grain-producing barycenter, at least to a certain extent, increases the overall cost of grain-production. Thus, the specific policy should be released in each region according to their external factors and natural factors, so as to save the additional cost and achieve the maximized efficiency.

2.4.2 Influences of Farmland Use Transition on Grain Production

In the context of URTD, the spatio-temporal pattern of farmland distribution and the population transition jointly drive FUT. Based on the change of per capita farmland area (PCFA), this research constructs a theoretical model for the temporal and spatial transition of farmland. Using 1 km grided PCFA data for statistical analysis, the study found that 71% of farmland covered areas across China experienced FUT during 1990–2010. Trends for PCFA were reversed; that is, there was a gradual decline during 1990–2000, which changed to progressive growth during 2000–2010. In addition, the temporal transition trends with respect to farmland are significant in China; the elevations of 1000 m and 500 m constitute two important demarcation lines in the spatial transition of farmland, representing inflection points for PCFA change trends and growth rates, respectively. Based on the spatio-temporal pattern of PCFA during 1990–2010 of China, we validated the theoretical model of FUT from the view of temporal transition and spatial transition.

We used the coupling relationship between PCFA and per capita grain possession (PCGP) to further analyze the influencing mechanisms of FUT on grain production transformation (GPT). The results indicate that significant GPT can be observed in the FUT process in China, with significant regional differences. Traditional agricultural areas have experienced a gradual transition from "Intensive type" to "Modern type;" there has been a "Declining type" transition for grain production in South China. The transformation trend of grain production type in traditional farming areas indicated that the highly intensified grain production mode has been difficult to adapt to the eco-driven agricultural production system transformation (Smith et al. 2017). In addition, the declining capability of grain production in the south of the Yangtze River is not only the result of regional industrial structure adjustment, but also the self-adjustment of the farmers' livelihood system in the process of socio-economic transition (Zhang et al. 2016).

FUT and GPT are closely related, and they constitute an important part of the URTD of China. Regulating the mechanisms and processes associated with FUT may be possible based on innovations in land management systems and policy (Zhang et al. 2017). Under the background of urbanization, this study revealed the characteristics of PCFT through theoretical and empirical tests, and provided the theoretical basis for optimizing the FUT. Furthermore, this study provided a theoretical foundation for determining the rational production scale in China's GPT process. Because of the coupling relationship between demographic transition (Hussain 2002) and FUT determine the transformation process of agricultural production organization in China (e.g. land circulation, proper operation scale of farmland). Therefore, it is of great practical significance to deeply research the coupling relationship between FUT and GPT in China, and to formulate regional regulation policy for FUT and grain production. During the urbanization process of China, the key thresholds (e.g. the time node and spatial transition characters of FUT) and the reasonable scale of farmer production will be the focus of further research in the future. In addition, the influencing mechanism and modes of FUT to regional land use transition will be the key contents to deepen the empirical and theoretical research of FUT.

2.4.2.1 Farmland Use Transition and Grain Production Transition

FUT is influenced by changes in agricultural labor, which cause transitions in agricultural production patterns. Population and socio-economic transitions jointly drive PCFA from a rapid decline to gradual increase. In the process of FUT, the migration of agricultural surplus labor changes traditional small-scale production patterns. The proportion of part-time peasant households continuously increases, accompanied by a constant increase in labor-saving inputs into grain production (Oseni and Winters 2009). The "Green Revolution" of agricultural production (Koohafkan et al. 2012), characterized by technological change, has driven peasant households to continuously increase their capital input per unit area to produce grain. As a result, FUT makes it possible to realize professional and large-scale grain production. The spatial transition of farmland gradually creates a spatial pattern that optimizes grain production. In high-elevation areas, farmland with poor tillage conditions and low marginal utility is the first to withdraw from grain production (Su et al. 2017). As land use patterns gradually change (such as the forest transition), grain production activities begin to concentrate step-by-step in low-elevation areas with good water and soil conditions. The spatio-temporal transition of farmland further facilitates GPT in the spatio-temporal pattern and production mode.

FUT provides the operational path and time window to regulate GPT. In the process of FUT, PCFA transitions from a rapid decline to a gradual increase. The PCFA transition take exactly the time window needed to adjust grain production. In this case, effectively identifying the key node of FUT is of vital significance to promote large-scale grain production. This gradually increases PCFA and transforms the small-scale production pattern. Rational land management policy regulation can accelerate the FUT process. In contrast, missing this window period may postpone the FUT process or weaken its effect. In this stage, GPT related systems and policy designs, including land management system, household registration system, and social food security system, greatly impact GPT's effect and progress. In addition, with respect to the design of farmland guarantee strategies targeting urban agglomeration districts with continuously reduced PCFA), the FUT process can be optimized through a planning strategy that rationally withdraws marginal farmland and that designs fallow crop rotation for areas with excessive farmland use (Smith 2013).

2.4.2.2 Research Framework of the Coupling Relationship Between Per Capita Farmland Transition and Per Capita Grain Possession

(1) **Per capita farmland area**

To tackle the problem of low statistical data precision, and to overcome the difficulty encountered in spatial econometrics, this research introduces high-precision grid data capturing farmland and population density, and applies GIS-based statistical analysis techniques to verify the theoretical model for FUT. The data for farmland along 100 m grids for 1990, 2000, and 2010 were obtained from the National Data Sharing Infrastructure of Earth System Science (www2.geodata.cn). Mainland China's population density on a 100 m grid was downloaded from the WorldPop program (Gaughan et al. 2016); the data is produced by a random forest regression model using China township population census data. DEM data at a resolution of 90 m were obtained from the China Natural Resources Data Center (http://www.data.ac.cn). County-level socio-economic statistical data (including grain yield and population number) were obtained from China's County Statistical Yearbook (NBSC 2011). Basic geographic data for this study included administrative boundaries and administrative centers derived from National Basic Geographic Information Center (http://ngcc.sbsm.gov. cn/). To characterize FUT across the country and its influences on grain production, China was segmented into 11 districts, based on agricultural production characteristics and physic-geographical environment (Fig. 2.17).

The calculation process to generate a national 1 km grid resolution of PCFA mainly consists of data calculations and checks. The national 100 m gridded population density data did not cover the Tibet Autonomous Region and Taiwan Province; as such, unless otherwise specified, the calculation processes in this research uniformly excluded the Tibet Autonomous Region and Taiwan Province. However, the data available for this study covered more than 99% of China' farmland area, population quantity, and grain yield, reflecting the country's basic conditions.

To support grid data quality, artificial visual inspection and zonal statistical tests were first conducted using farmland data and population density data, respectively. The results demonstrated that data precision exceeded 90%, meeting the demands of the study. To eliminate the geometric error of the two types of data in the spatial dimension, an ArcGIS spatial aggregate tool was used to generate national 1 km resolution farmland data and population density data; a grid calculation tool was used to obtain 1 km resolution PCFA data. A triple standard deviation test was performed on PCFA; for data that failed to pass the test, a 3 × 3 low pass filtering tool was adopted for smoothing processing.

PCFA (ha/person) represents per capita farmland area at each pixel, which is calculated using:

$$PCFA = \frac{Farmarea}{Popdensity} \tag{2.13}$$

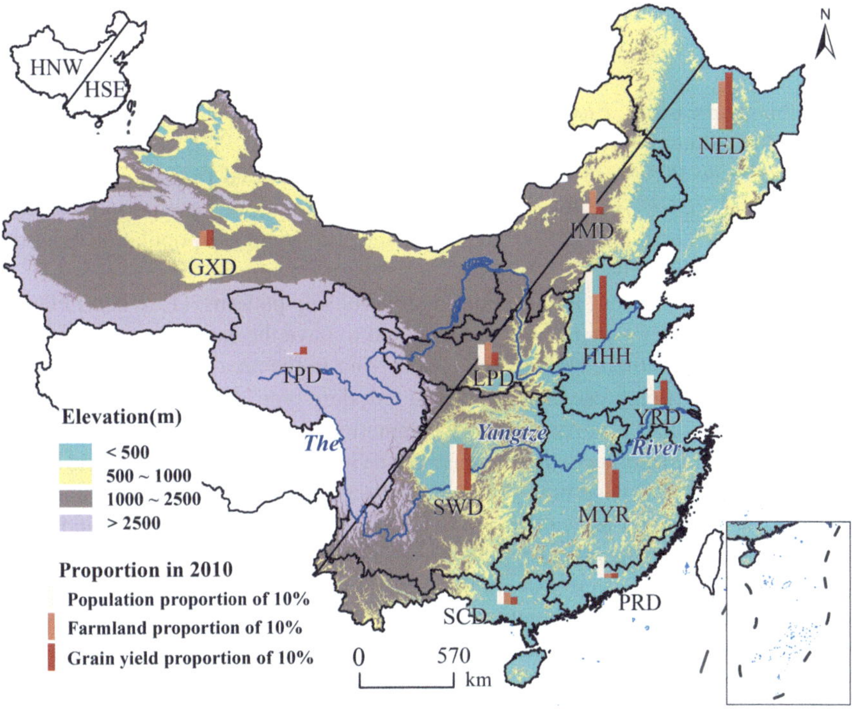

Fig. 2.17 Map of research districts (Ge et al. 2018a). *Note* HSE, HNW, NED, HHH, YRD, MYR, PRD, SCD, SWD, LPD, IMD, GXD and TPD represent the southeast district of Hu line, the northwest district of Hu line, the northeast district, the Huang-Huai-Hai plain, the Yangtze River Delta, the middle-lower reaches of Yangtze River, the Pearl River Delta, the South China district, the Southwest district, the Loess Plateau district, the Inner Mongolia district, the Gansu-Xinjiang district, and the Tibetan Plateau district, respectively

In this expression, Farmarea represents farmland area in each 1 km × 1 km grid, and Popdensity represents the population (number of people) in each grid. PCFT indicates the change in PCFA in each grid, which is calculated as Eq. (2.14).

$$PCFT = \frac{PCFA_{t2} - PCFA_{t1}}{PCFA_{t1}} \tag{2.14}$$

$PCFA_{t2}$ and $PCFA_{t1}$ represent the grid cell PCFAs at the base period and end period, respectively.

(2) **Population-Farmland-Grain Yield Coefficient (PFGC)**

To discuss the relationship between FUT and GPT, this resaerch constructs PFGC to reveal the coupling relationship between PCFA and PCGP. PFGC is calculated as follows:

$$PFGC_x = \frac{(\overline{PCFA_{x,t2}} - \overline{PCFA_{x,t1}}) / \overline{PCFA_{x,t1}}}{(PCGP_{x,t2} - PCGP_{x,t1}) / PCGP_{x,t1}} = \frac{\overline{PCFT_x}}{(PCGP_{x,t2} - PCGP_{x,t1}) / PCGP_{x,t1}} \tag{2.15}$$

$$PCGP_x = \frac{GP_x}{POP_x} \tag{2.16}$$

In this expression, $PFGC_x$ represents the coupling relationship pattern of population-farmland-grain yield of county x; $\overline{PCFA_{x,t1}}$ and $\overline{PCFA_{x,t2}}$ represent the mean grid-scale PCFAs of county x at base period and end period, respectively; $PCGP_{x,t1}$ and $PCGP_{x,t2}$ represent the PCGPs of county x at base period and end period, respectively; GP_x and POP_x represent the grain yield and population quantity of county x, respectively.

In theory, there are four PFGC coupling relationship patterns: (1) When PFGC > 0 and PCFA > 0, PCFT and PCGP are assumed to have the same change trend and increase at the same time. When FUT is occurring in the area where this pattern is seen, grain production presents a significant "Modern type" transition trend. Where there is no FUT yet, it is considered a "Reclamation type" grain production pattern, that is, more farmland is needed to increase grain yield. (2) When PFGC > 0 and PCFT < 0, PCFA and PCGP have both declined, and regional grain production guarantee capability has declined as well. (3) When PFGC < 0 and PCFT > 0, it indicates that, notwithstanding the increase of PCFA, PCGP has declined, and that the decline trend of grain production has also gradually become significant. We define the above two coupling patterns, i.e., (2) and (3), as the "Declining type" transition patterns of grain production. (4) When PFGC < 0 and PCFT < 0, notwithstanding the decrease of PCFA, PCGP has increased, and the "Intensive type" transition of grain production is significant.

2.4.2.3 Macro Patterns of Population, Farmland, and Grain Production of China

The distribution of China's population and farmland resources are influenced by the natural environment, with significant regional differences. China has abundant total agricultural resources, but there is insufficient per capita possession. The PCFA is only 0.1 ha, equivalent to only 12% the level in the United States. The mean farmland area per household is 0.38 ha, below the global average as well as that of India, Japan, and South Korea (FAO 2016). The proposed Hu line is of vital significance for uncovering the macro distribution of China's population (Hu 1935; Wu and Guo 1994); it also serves as the demarcation line between humid and semi-moist monsoon climate; the arid and semi-arid climate; and the macro demarcation line between low and medium-elevation terrain and high-elevation terrain in China. This water and soil environment has profoundly influenced the distribution of population, farmland, and grain yield in China. In 2010, the HSE generated 93.79, 86.63, and 93.06% of the population, farmland area, and grain yield of China (Fig. 2.18).

In the context of URTD, the mainland population continuously migrated from inland provinces to coastal regions from 1990 to 2010. The proportions of PRD and YRD populations (compared to China's total population) rose from 5.15 and 9.78% in 1990 to 7.85 and 10.85% in 2010, respectively. However, China's population distribution patterns were generally relatively stable, with no fundamental

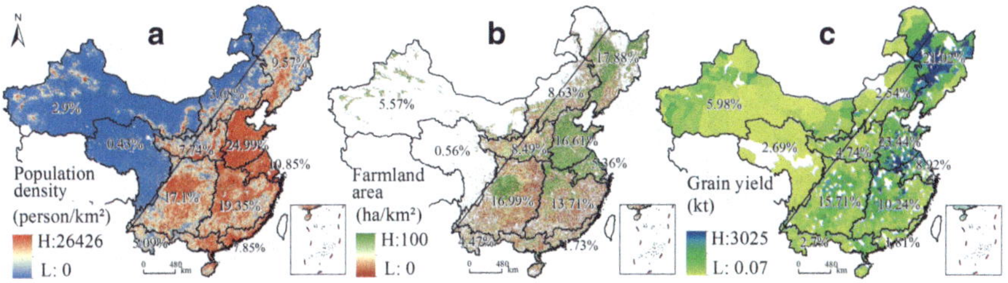

Fig. 2.18 The spatial pattern of population (a), farmland (b), and grain yield (c) by districts in 2010 (Ge et al. 2018a)

transformations. During this period, the center of China's newly-developed farmland gradually shifted from Northeast China to Northwest China (Zuo et al. 2014). From 1990–2000, the newly-increased farmland area of NED amounted to 2.3 Mha; its proportion of China's total farmland area rose from 16.31% in 1990 to 17.67% in 2000. From 2000–2010, the newly-increased farmland area of GXD amounted to 0.83 Mha; its proportion of China's total farmland area rose from 4.95% in 2000 to 5.57% in 2010.

Meanwhile, the SWD, MYR, and PRD experienced a continuous decrease in farmland area, and the center of farmland in China continuously shifted northward. As the center of farmland shifted northward (Li et al. 2017), China's center of grain yield also shifted northward. Bounded by the Qinling Mountains-Huaihe River line, the areas in north China witnessed a continuous increase in grain yield, while south China experienced a continuous decrease of grain yield. In the macro context, the evolution in population and farmland spatio-temporal patterns initiated China's original drive towards FUT.

(1) Overall pattern of PCFA

PCFA is small in east China and large in west China; areas distributed along the Hu line create a high-value sector. The Hu line is the macro demarcation line for terrain and population distribution in China, constituting an important geographical boundary marking regional differences in PCFA. The PCFA for HNW is about four times the PCFA of HSE (Fig. 2.19a, b). In HSE, areas with a PCFA less than 0.3 ha accounts for more than 65% of the total farmland coverage of this district; in HNW, these values represent only about 24% of the total. In east China, the population distribution is highly concentrated, and areas suitable for cultivation have had a long development history. There is also a spatial overlap of farmland resources and population; that is, areas with abundant farmland resources usually have a large population. This results in a limited PCFA.

In brief, the self-sufficient smallholder production pattern and the agricultural culture of "intensive cultivation" are both closely related to PCFA distribution (Wu and Guo 1994; Ge et al. 2018a). Northeast China has a relatively short development history, and effective agricultural development did not begin until the large-scale inflow of population during modern times. Combined with large farmland resources reserves (Liu et al. 2010), the PCFA of Northeast China is very abundant. The northwest district of the Hu line, located in inland China, is sparsely populated, and the

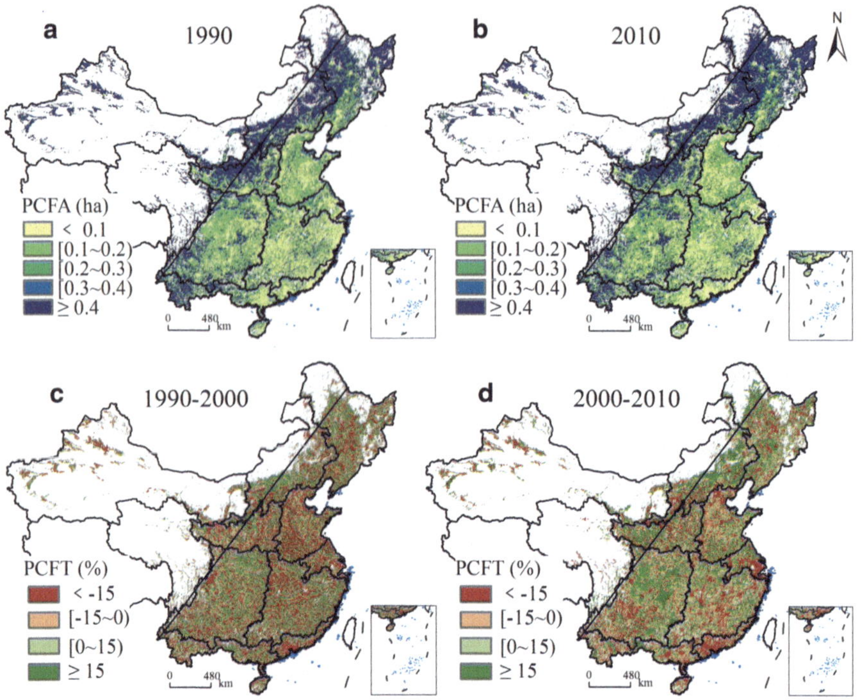

Fig. 2.19 Spatio-temporal pattern of PCFA and PCFT during 1990–2010 (Ge et al. 2018a)

areas suitable for cultivation have a relatively short development history. The large-scale agricultural development of Xinjiang (in this district) started after the founding of new China; given the great increase in farmland resources during “migration and reclamation” processes (Gu et al. 2017), this district’s PCFA is more abundant than in east China.

Since the 1990s, the spatial distribution of PCFA experienced significant changes in China because of rapid urbanization. During 1990–2000, most of China saw a general reduction of PCFA (Fig. 2.19c), which was closely related to both the rapid increase the population and the broad occupation of farmland as rural residential land (Long et al. 2012). During this period, the total population of China grew at an average rate of 10.68%, increasing from 1.143 billion to 1.267 billion. In this stage, many people born before the implementation of the family planning policy reached marriage age (Hussain 2002). This led to an increasingly urgent demand for farmland that could provide housing in rural areas (Long et al. 2012).

Studies have suggested that a clear coupling relationship between rural residential land and farmland area was a significant driver of the overall reduction of PCFA in this stage (Long and Li 2012). In traditional agricultural areas, this trend was even more significant. Taking the HHH as an example, during this period, its PCFA declined by 18.63% (Table 2.3). A significant amount of high-quality farmland was occupied

Table 2.3 Statistical summary of the proportions of population, farmland area, and grain yield and PCFAs by district during 1990–2010 (Ge et al. 2018a)

	Proportion of population quantity (%)			Proportion of farmland area (%)			Proportion of grain yield (%)			PCFA (ha)		
	1990	2000	2010	1990	2000	2010	1990	2000	2010	1990	2000	2010
HSE	94.21	93.97	93.79	87.62	87.18	86.63	95.18	94.43	93.06	0.34	0.33	0.36
HNW	5.79	6.03	6.21	12.38	12.82	13.37	4.82	5.57	6.94	1.38	1.43	1.44
NED	10.23	9.79	9.57	16.31	17.67	17.88	15.07	15.49	21.02	0.81	0.89	0.94
HHH	25.19	25.04	24.99	17.29	16.76	16.61	21.88	22.86	23.44	0.16	0.13	0.13
YRD	9.78	9.83	10.85	5.88	5.6	5.36	9.32	9.48	8.92	0.18	0.16	0.16
MYR	21	20.07	19.35	14.16	13.83	13.71	14.46	12.64	10.24	0.16	0.15	0.15
PRD	5.15	6.77	7.85	2.01	1.88	1.73	3.17	3.14	1.81	0.13	0.11	0.1
SCD	5.08	4.99	5.09	4.58	4.48	4.47	4	3.9	2.7	0.34	0.32	0.34
SWD	19.26	18.59	17.1	17.35	17.05	16.99	18.77	18.37	15.71	0.24	0.24	0.26
LPD	7.82	7.89	7.74	8.72	8.61	8.49	4.03	4.02	4.74	0.42	0.4	0.41
IMD	3.8	3.69	3.68	8.32	8.63	8.63	1.95	1.98	2.54	1.14	1.2	1.34
GXD	2.51	2.73	2.9	4.83	4.95	5.57	4.44	4.86	5.98	1.61	1.53	1.58
TPD	0.41	0.4	0.43	0.54	0.55	0.56	2.69	3.03	2.69	1.78	1.79	1.73

Note HSE, HNW, NED, HHH, YRD, MYR, PRD, SCD, SWD, LPD, IMD, GXD and TPD represent the southeast district of Hu line, the northwest district of Hu line, the northeast district, the Huang-Huai-Hai plain, the Yangtze River Delta, the middle-lower reaches of Yangtze River, the Pearl River Delta, the South China district, the Southwest district, the Loess Plateau district, the Inner Mongolia district, the Gansu-Xinjiang district, and the Tibetan Plateau district, respectively in an unordered and low-efficiency manner, resulting inwasted farmland resources (Shi et al. 2013). During 2000–2010, the distribution of PCFA underwent relatively significant changes in China; that is, the core areas experiencing a reduced PCFA began to gradually shift from the rural areas to peripheral urban areas.

During urbanization, the PCFA rapidly increased in the areas with massive population outflow. For example, the PCFA of HSE increased by 11.51% in this period. In traditional agricultural areas, the proportion of part-time peasant households significantly increased (Cai 2010). As surplus agricultural labor migrated away, the demand for rural residential land began to decline, and PCFA transitioned from a rapid decline to a gradual increase. In areas with rapid urbanization, the proportion of farmland occupied for urban construction increased rapidly, resulting in the rapid reduction of PCFA (Cartier 2001; Huang et al. 2017). For example, the PCFA of the PRD decreased by 14.38% in this period; its grain production capacity declined rapidly as well.

2.4.2.4 Coupling Relationship Between Per Capita Farmland Transition and Per Capita Grain Possession

In the context of FUT, there are significant regional differences in the coupling of PCFT and PCGP in China, and the coupling relationship between FUT and GPT has consistently evolved. During 1990–2000, the areas with PFGC > 0 and PCFT > 0

Table 2.4 Zonal proportions of county numbers of different coupling types between PCFT and PCGP during 1990–2010 (Ge et al. 2018a)

	PFGC 1990–2000			PFGC 2000–2010		
	Reclamation type (%)	Declining type (%)	Intensive type (%)	Modern type (%)	Declining type (%)	Intensive type (%)
NED	36.75	31.33	31.93	64.46	13.25	22.29
HHH	5.65	33.63	60.71	44.64	19.94	35.42
YRD	12.50	35.71	51.79	50.89	33.93	15.18
MYR	16.18	60.29	23.53	22.43	55.88	21.69
PRD	23.53	54.41	22.06	0	100	0
SCD	24.73	31.18	44.09	11.83	81.72	6.45
SWD	43.26	25.06	31.68	29.08	54.85	16.08
LPD	13.04	52.17	34.78	46.38	15.94	37.68
IMD	23.76	45.54	30.69	42.57	33.66	23.76
GXD	30.00	31.43	38.57	32.14	28.57	39.29
TPD	17.39	53.26	29.35	25.00	60.87	14.13

Note NED, HHH, YRD, MYR, PRD, SCD, SWD, LPD, IMD, GXD and TPD represent the northeast district, the Huang-Huai-Hai plain, the Yangtze River Delta, the middle-lower reaches of Yangtze River, the Pearl River Delta, the South China district, the Southwest district, the Loess Plateau district, the Inner Mongolia district, the Gansu-Xinjiang district, and the Tibetan Plateau district, respectively

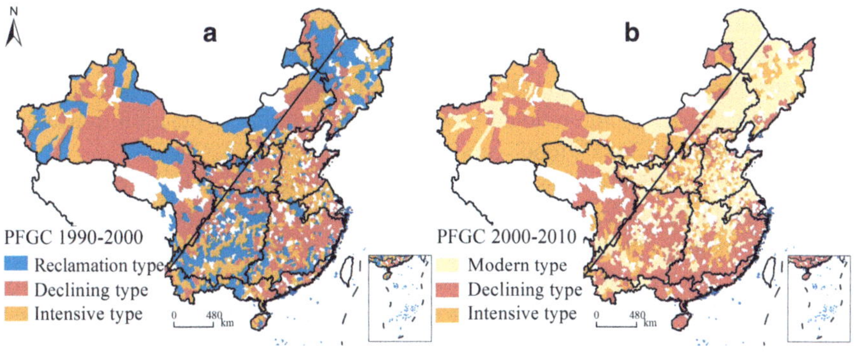

Fig. 2.20 Coupling relationship between PCFT and PCGP (Ge et al. 2018a)

were mainly distributed in the west of NED and part of SWD (Table 2.4); before FUT, the increase of PCFA in these areas was mainly due to new farmland reclamation, that is, they are of the "Reclamation type" (Gu et al. 2017). The areas with PFGC > 0 and PCFT < 0 and with "Intensive type" coupling relationship were mainly distributed in the HHH, the northern YRD, and NED (Fig. 2.20a). These areas mainly relied on increasing input per unit farmland area to increase grain yield and PCGP (Wang et al. 2017; Ge et al. 2018a). The areas with PFGC < 0 and PCFT < 0 and with "Declining type" coupling relationship were mainly distributed in the PRD and the Jiaodong Peninsula (Table 2.4). These areas saw a continuous decline in grain production because of their earlier socio-economic transition.

During 2000–2010, in the context of FUT, the areas with PFGC > 0 and PCFT < 0 and with "Modern type" coupling relationship were mainly distributed in the HHH, NED, and other major grain producing areas (Fig. 2.20b). Those with "Intensive type" coupling relationship were mainly distributed in GXD (Wang et al. 2017). Those with "Declining type" coupling relationship were broadly distributed in regions south of the Yangtze River, as evidenced by the rapid decline of grain production in South China (Li et al. 2017). Among the traditional agricultural areas of China (Table 2.4), the HHH, the Dongting-Jianghan Plain of MYR, and NED experienced a gradual transition from "Intensive type" to "Modern type." This was accompanied by a continuous increase of PCGP and a significant trend of GPT.

2.4.2.5 Discussion

The process of FUT determines the rational path of GPT in China. By analyzing the spatio-temporal evolution process of the coupling relationship of PFGC, this section investigates the coupling relationship between PCFA and PCGP at a county level in the process of FUT. GPT is closely related to FUT, and the FUT process can be optimized to further regulate China's GPT process. The theoretical model of PCFA transition and the optimal FUT regulation path described above provide beneficial references about the GPT of China. The predictions and judgments about FUT time provide a theoretical foundation to optimize the land management system (especially farmland management policy) in China. The information also supports scientific predictions about opportunities to adjust grain production policy (Deng et al. 2006; Wang et al. 2017). The FUT and GPT trend and process are closely related to the population transition (Chen et al. 2014), within the larger socio-economic transition. Scientifically mastering the rational process of FUT during population transitions helps further ensure China's food security.

FUT provides a scientific basis for determining the rational scale of grain production in China. In the process of GPT in China, an efficient peasant household production scale constitutes the core of stabilizing the grain production system (Deng et al. 2006). This guarantees food security and maintains stable social transition. Many scholars have investigated the optimal peasant household production scale of China from different perspectives (Li et al. 2017). Government sectors are also constantly promoting "new-type agriculture managing bodies," encouraging "land circulation," and launching other grain production policies to address FUT (Su et al. 2016). There are many disputes about the appropriate peasant household production scale in China (Ge et al. 2018a); the process for scientifically determining China's grain production organization system is a core need to master the GPT.

The theoretical and empirical studies of this study, based on FUT and GPT, suggest that FUT exhibits significant regional differences, and the spatio-temporal evolution law of the coupling characteristics between the two provides a scientific basis for mastering the GPT of China. It also offers important references to determine a rational grain production scale in China. Determining an effective grain production scale is closely associated with the FUT process; a coordinated push of the two support the GPT process (Long et al. 2016). When farmland resources are concentrated into the hands of a few people too quickly, it increases the number of landless peasants, and may give rise to serious socio-economic problems (Hussain 2002; Cai 2010). When the process of farmland circulation lags behind the FUT process, it reduces agricultural production efficiency and wastes farmland resources. Thus, the GPT of China relates closely to both food security and socio-economic transition; scientifically mastering the trend and time nodes of FUT will provide a basis for scientifically guiding GPT in China.

2.5 A Perspective of Agricultural Labor Changes

The transformation process of rural agricultural labor during the process of urbanization has become an important basis for insight into the evolution of rural man-land relationship. The evolution of the quantity and structure of agricultural labor becomes the core driving force of the change of farmland use morphology. Farmland use morphology is closely related to the transformation process of rural labor, which provides effective means and methods for deepening the study of FUT. The aging of rural labor is an important feature of the evolution of rural agricultural labor structure, which changes the input-output pattern and organizational farming patterns of farmland use, and further changes the dominant and recessive morphologies of farmland use. The quantity of rural labor is an important basis for the use pattern and organizational pattern of farmland. The urban-rural migration of agricultural labor changed the quantity and production pattern of rural labor. Therefore, the internal relationship between agricultural labor quantity and grain yield can better reflect the internal process of transition of farmland use morphology driven by the change of agricultural labor quantity. Next, this section will study the internal relationship between the aging of rural labor and the transition of farmland use, and the coupling relationship between the quantity of agricultural labor and grain yield, trying to explore the internal logic of FUT from the perspective of agricultural labor.

2.5.1 *Effects of Rural Aging on Farmland Use Transition*

From the perspective of labor employing quantity in the planting industry, we explore the spatio-temporal coupling relationship between rural aging and land use transitions and further expands the research on the relationship between agricultural input and FUT. The research results increase the understanding of to the spatial differentiation characteristics of rural aging and land use transitions in China's traditional farming area from a macro perspective, which is of great significance to the macro-control and precise decision-making of regional agricultural production as well as, the more effective regulation of rural human-land interrelations. Land use transitions are influenced by impetus, such as population, economy, technology, institution, culture and location, but are subjected to the availability of data; thus, we discuss the spatio-temporal evolution of rural aging and labor employing quantity from the macro level. On the household level, the expansion of grain crops and cash crops were both significantly affected by agricultural labor (Su et al. 2016). In the future, a probe into the impetus and mechanism of farmers' land use decision-making behavior in the context of rural aging from the micro level and analysis of the relationship between rural aging and land use transitions in different landform conditions to compensate for the deficiency of mechanism analysis at the macro scale are necessary. Based on the relationship between rural aging and agricultural planting structure, its impact on the development of agriculture and corresponding coping strategies deserve further discussion.

2.5.1.1 Theoretical Analysis of Land Use Transitions and Rural Aging

Population is an important factor affecting agricultural production and its changes dominate the direction of agricultural development and land use transitions. In the process of urbanization, the migration of labor from rural to urban areas and the responses of behavioral main stream bring about the change of regional land use morphologies and promote the land use transitions in the rural. On the one hand, the aging of rural population promotes the restructuring of production factors, which is the internal driving force of land use transitions. The transfer of rural labor accelerates the aging of rural population, resulting in the reduction of rural agricultural labor supply and the rise of labor costs, and puts forward new demands for the transitions of rural agricultural labor factors. The decrease in quantity and quality of rural labors, and the rise of labor price will induce substitution between input factors, prompting farmers to change the input of agricultural capital and technology, thus leading to the transition of farmland which is labor substitution effects. On the other hand, the change of agricultural production factors (aging of labor force) requires the corresponding land use morphologies to adapt to it. The changes of labor input in the process of land use transitions affect agricultural production in two aspects of labor quantity and quality. As a rational economic man, in order to obtain economic benefits, he will adapt to such changes by adjusting the planting structure of cash crop and grain crop, which is land substitution effect.

When agricultural economists study the impact of aging on agricultural production, they often emphasize the importance of managers' physical strength and human capital in agricultural production. Age and labor ability had a significant inverted "U"-type relationship. First, with the increase in age, the agricultural production and management workers gradually increased their labor ability. After reaching a peak, the physical strength and human capital of the operators gradually decreased.

How does rural aging affect farmers' land use decisions and promote FUT? In the research of relevant scholars, this process is subdivided into two processes: rural aging and its impact. The increase in life expectancy and the decline in fertility makes the age structure of the rural population enter the platform of aging, and the population mobility aggravates the aging of the rural population and the agricultural working time of elderly individuals. Under the combined effect of the three factors, the aging degree of the rural population is increasing. In the context of changing population pattern, rural aging affects agricultural production in the aspects of dominant and recessive morphologies of farmland use. The dominant influence is mainly through the supply of agricultural labor (quantity, structure), the structure of farmland use (farmland abandonment, planting structure), and the agricultural output structure (output value, output). The recessive effects of rural aging are presented through large-scale production (land transfer, agricultural mechanization) and functions (economic, ecological, social and cultural functions) (Fig. 2.21).

The deepening of regional rural aging and population mobility have reduced the supply of agricultural labor, and led to an increase in labor costs. To adapt to the current shortage of household labor, the aging of laborers, and the weakening of labor ability, farmers will make rational production decisions according to the price changes of agricultural products and market demand. They will choose to plant crop types that are easy to cultivate, use less labor, reduce the planting area of crops, plant crops with higher labor productivity to maximize the economic benefits (Su et al. 2017; Caulfield et al. 2019). From the perspective of FUT, this adaptation can be called farmland use recessive transition caused by agricultural labor supply (Ge et al. 2018a; Long and Qu 2018). Aging population will encourage farmers to increase the input of machinery to make up for the shortage of labor force. The impact of rural aging on planting structure is significantly different in terrain. The plains are flat, and the farmland is concentrated, contiguous, and suitable for mechanical cultivation. In

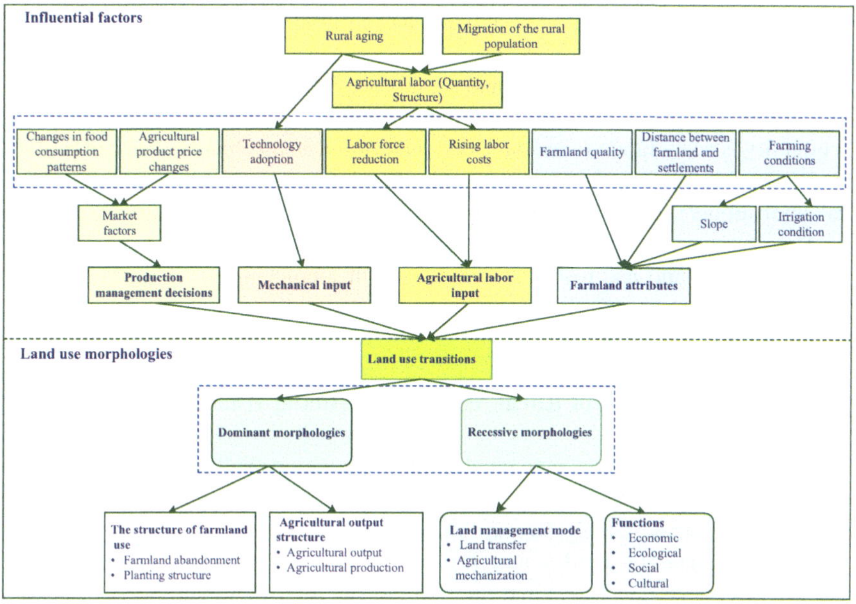

Fig. 2.21 Analysis framework of farmland use transitions influenced by rural aging (Liao et al. 2019)

the case of insufficient labor, the households increase the proportion of food crops with high degree of mechanization. In mountainous and hilly regions, due to the topographic relief and farmland dispersedly, it is difficult to implement agricultural mechanization. Therefore, rational and economic farmers plant fewer food crops and increase the economic crops area.

The adjustment of farmland use decision (including farmland abandonment) is more manifested in the transition of land use dominant morphology caused by the change in agricultural labor supply (Shao et al. 2015; Liu 2018; Zhang et al. 2018). Due to the changes in the price of agricultural input and output, the profit of farming is reduced to zero or is even negative, that is, farming beyond the margin of zero, no matter how the farmers adjust the proportion of investment, is always beyond this margin. The aggravation of the aging results in a shortage in the agricultural-age workforce, and the cost of farming increased, rational farmers have no incentive to cultivate, thus, the farmland is abandoned (Macdonald et al. 2000). The adjustment of farmland use decision is the interpretation of human social attributes in the process of interaction between the two major elements of "human" and "land". Intensive agricultural production in the plains reduces the possibility of farmland abandoned. Mountainous and hilly regions, have poor physical conditions (high slope and poor irrigation condition), farmland is relatively poor, and the number and quality of agricultural laborers decrease due to aging. When agricultural output is insufficient to compensate for the marginal wages of agricultural labor and lack of agricultural successors, the possibility of marginalization of farmland land in these areas greatly increase.

Rural aging mainly affects FUT and the adjustment of agricultural planting structure based on the quantity and quality of labor supplies. The amount of labor employing quantity is a key factor to measure this interpretation mechanism and can be used to verify how the change in dominant agricultural production morphology (agricultural labor supply) due to the rural aging affects the recessive morphology of farmland use. Therefore, we mainly discuss the impact of rural aging on the recessive morphology of farmland use from the perspective of labor employing quantity in the planting industry.

2.5.1.2 Spatial Pattern of Rural Aging in the Huang-Huai-Hai Region

The Huang-Huai-Hai region is located south of the Great Wall, north of the Huaihe river, east of the Taihang mountains, and western of Henan mountains and encompasses five provinces, namely, Henan, Anhui, Jiangsu, Shandong, and Hebei, and two municipalities, namely, Beijing and Tianjin. The Huang-Huai-Hai region accounts for 4.3% of China's land area, and the share of the total farmland, population, and grain yield is 20.6%, 22.5%, and 30.8%, respectively. The Huang-Huai- Hai region is a main grain-producing area of China and ranks first among the nine traditional Chinese farming areas (Long et al. 2018). The rural industrial structure of this region has long been dominated by agriculture, especially planting (Guo et al. 1990). Additionally, its grain output provides a major guarantee for China's food security; in 2010, the output of oil in this region was 8.87 million tons and that of cotton was 2.11 million tons, accounting for 35.47 and 27.47% of the national total, respectively.

Aging is a combination of the uneven distribution of the elderly population in the space and gradual change in population age structure. Due to regional differences in natural environment, economic environment, population environment and other factors affecting aging, the degree of aging varies by region. Many indicators are used to measure the aging degree, including the elderly population coefficient, aging index, and the median age. At present, scholars mainly use the proportion of the elderly population (the proportion of the population aged over 60 or 65 years) as the indicator (Goll 2010; Długosz 2011). This indicator can directly reflect the aging phenomenon of the regional population. Therefore, this study selects the proportion of the population aged 65 years and over in the total population as a measurement index of the aging degree. We use the growth rate of aging to measure the spatial difference of rural aging and its dynamic process of evolution over time.

$$TA_i\left(65^+\right) = \left[\frac{p_i^{t+n}\left(65^+\right) - p_i^t\left(65^+\right)}{p_i^t\left(65^+\right)}\right] \times 100\% \tag{2.17}$$

where TA_i (65^+) represent the growth rate of elderly population over 65 years old. The larger the value, the faster the regional aging speed; while p_i^t (65^+) refer to the population of elderly over 65 years old; $p_i^{t+n}(65^+)$ represent the elderly population over 65 years old after n years.

The United Nations classification criteria means where the proportion of people aged 65 years and over exceeds 7.0% belongs to the elderly-aged areas. Combined with the actual situation of rural aging in the Huang-Huai-Hai region, rural aging is divided into five stages: young type [0–5%), adult type [5–7%), initial senile type [7–10%), middle senile type [10–15%) and later senile type ($\geq$ 15%). Based on this classification, the spatial distribution map of regional rural aging was obtained (Fig. 2.22). Comparative analysis demonstrated that the rural aging in the Huang-Huai-Hai region has the following characteristics:

(1) From 2000 to 2010, the degree of rural aging in the region gradually deepened, and the trend of aging in rural areas was consistent with the national level but higher than the national level. In general, the degree of rural aging in the region changed significantly and is in the transitional stage from the initial senile type to the middle senile type, and later senile type. In 2000, the regional rural aging population was 7.99% (the whole of China was 7.50%) and increased to 10.15% in 2010 (the whole of China was 10.10%).
(2) From the perspective of the number distribution of county units, the number of county units with the young type, adult type, and later senile type is small, and the number of county units in the initial senile stage and middle senile stage is large. From 2000 to 2010, the number of young type county units remains

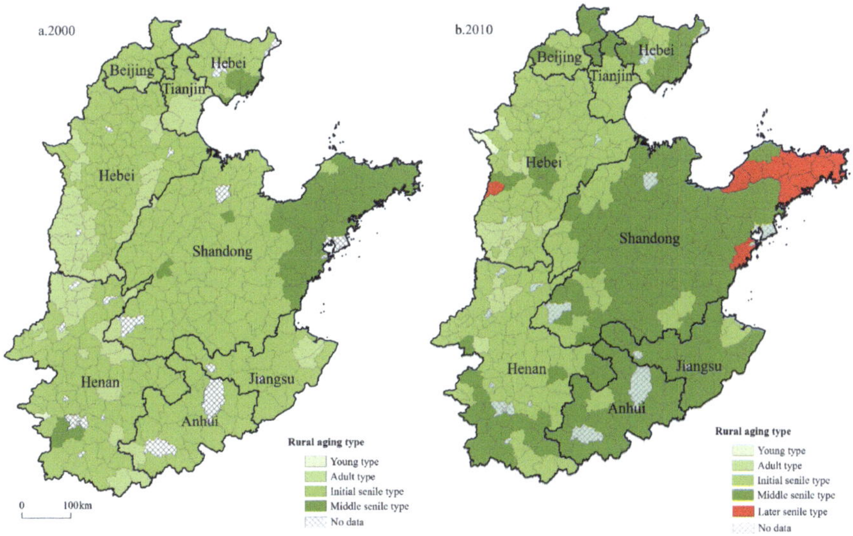

Fig. 2.22 The aging degree of rural population in the Huang-Huai-Hai region in 2000 and 2010 (Liao et al. 2019)

unchanged, and the number of adult type counties decreased from 58 to 17; the number of initial senile type county units decreased from 263 to 171, which accounted for 49% of the total number of counties; the number of middle senile type county units increased from 29 to 151, which accounted for 43% of the total number of counties. In the past decade, the rural aging of 176 counties has changed significantly (Fig. 2.23). There were 39 counties transformed from the adult type to the initial senile type and mainly distributed in the border areas of Hebei Province and Henan Province; 127 counties transformed from the initial senile type to the middle senile type, and were mainly distributed in Shandong Province, central of Hebei Province, and most of Anhui Province.

(3) Rural aging has generally maintained a low growth rate. According to formula (2.17), the growth rate of regional rural aging was calculated and then divided into four grades according to the natural breakpoint classification method. The analysis found that the negative growth zone ($R < 0$) is mainly located in the cities of Jiaozuo, Hebi, Handan, and Xingtai in Hebei Province. In 211 counties,

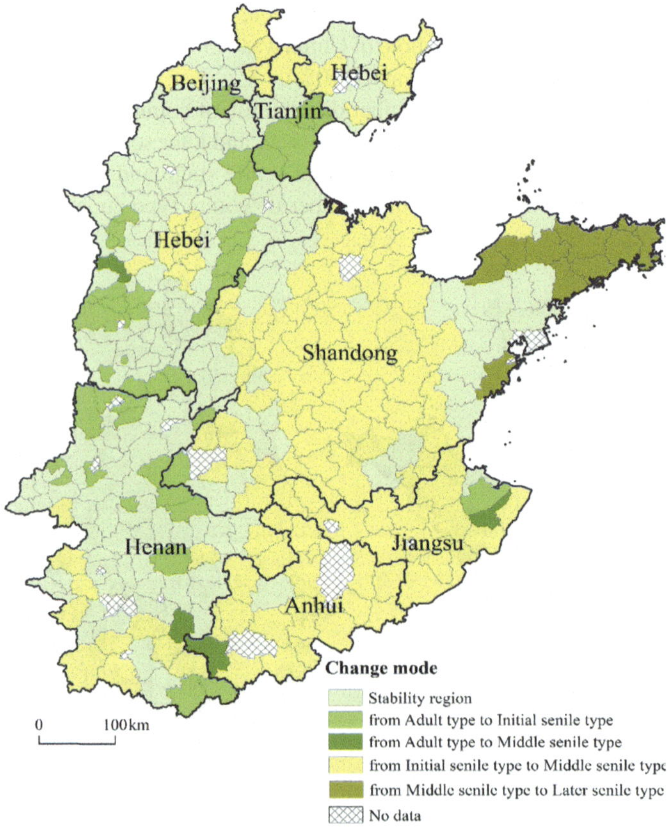

Fig. 2.23 Population aging change mode during 2000–2010 (Liao et al. 2019)

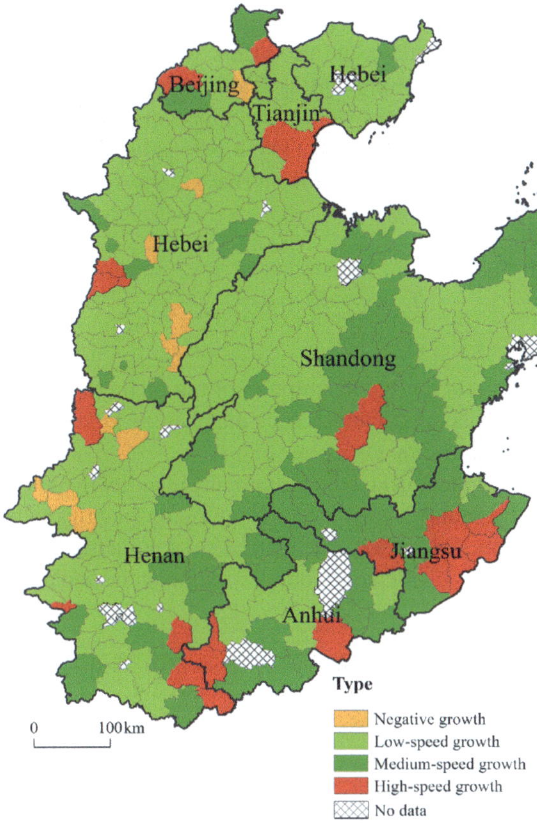

Fig. 2.24 Spatial pattern of population aging during 2000–2010 (Liao et al. 2019)

rural aging maintained a low speed ($0 \leq R < 26\%$), accounting for 60.11% of the total number of county units, distributed in Hebei, Henan, and western Shandong. The medium-speed growth ($26\% \leq R < 46\%$) counties accounted for 29.06%. There were 27 counties in high-speed growth ($R \geq 46\%$), and the spatial distribution is relatively scattered (Fig. 2.24).

2.5.1.3 Spatial Pattern of Labor Employing Quantity in the Planting Industry

Rural aging mainly affects the supply of labor from the aspects of quantity and quality of agricultural labor and then affects the agricultural production. On the one hand, the deepening of aging and population mobility have reduced the proportion of agricultural employees; on the other hand, the elderly laborers are significantly inferior to the middle-aged laborers due to physical function decline. The labor costs required to be consumed in different crop production processes differ, and the labor quantity refers to the number of days the producers work directly in production process. The quantitative adjustment of planting scale can be effectively reflected by labor consumption, which can effectively reflect the impact of rural aging on planting decision adjustment. The unit labor of crops in two municipalities and five provinces of the Huang-Huai-Hai region are compiled from the "China Agricultural Products Cost-benefit Data Compilation" (Table 2.5).

$$T = \sum_{i=1}^{5} S_i \times k_i \tag{2.18}$$

In the formula, T is the total labor consumption of regional crops; i is crop type (grain, oil plants, cotton, tobacco, vegetable); S_i is the planting area of the i crop; K_i is the unit labor consumption of crop i.

The number of counties with reduced labor employing quantity in each province is increasing. According to Formula (2.18), the labor employing quantity in the planting industry in 2000 and 2010 are calculated. The natural breakpoint method is adopted, the labor employing quantity is divided into four sections, and the two municipalities

Table 2.5 The changes of labor employing quantity in crop during 2000–2010 (unit: day/mu) (Liao et al. 2019)

Province	Year	Grain	Oil plants	Cotton	Tobacco	Vegetables
Beijing	2000	8.8	\	\	\	31.0
	2010	6.6	\	\	\	30.9
Tianjin	2000	8.0	9.0	21.7	\	47.5
	2010	6.6	8.0	15.2	\	42.1
Hebei	2000	11.8	13.8	21.7	43	47.1
	2010	6.0	10.3	23.9	30.5	36.4
Jiangsu	2000	9.6	10.8	32.7	\	44.1
	2010	5.8	9.2	28.8	\	30.4
Anhui	2000	11.7	11.4	27.5	32.2	53.4
	2010	6.2	8.8	22.2	25.2	39.0
Henan	2000	12.1	11.4	28.5	41.9	55.1
	2010	7.1	6.6	24.7	31.7	38.1
Shandong	2000	11.2	15.3	29.2	41.4	48.1
	2010	6.5	10.6	23.9	30.1	27.7
National average	2000	12.2	14.2	29.1	40.1	45.3
	2010	6.6	9.0	20.1	30.5	31.9

Note Grain crops mainly include rice, wheat and corn; Oil crop include rapeseed and peanut two kinds of main oil crops; 1 mu = 1/15 ha

Table 2.6 Comparative analysis of the labor employing quantity in planting industry in each province (unit: %) (Liao et al. 2019)

Labor employing quantity in planting industry		(0, 669) days	[669, 1404) days	[1404, 2426) days	[2426, 4880) days
Beijing	2000	66.67	33.33	0	0
	2010	66.67	22.22	11.11	0
Tianjin	2000	0	50	33.33	16.67
	2010	0	83.33	16.67	0
Hebei	2000	25.22	55.65	17.39	1.74
	2010	51.30	41.74	6.09	0.87
Jiangsu	2000	22.22	38.89	11.11	27.78
	2010	16.67	44.44	38.89	0
Anhui	2000	5.26	15.79	63.16	15.79
	2010	15.79	73.68	10.53	0
Henan	2000	10.26	20.51	35.89	33.33
	2010	50	15.38	16.67	17.95
Shandong	2000	5.71	30.48	47.62	16.19
	2010	60	23.81	4.76	11.42

and five provinces in the Huang-Huai-Hai region are counted in the four categories (Table 2.6).

By comparison, it is found that the number of counties in Tianjin with [669, 1404) days increased, and the proportion of counties within the region increased from 50% in 2000 to 83.33% in 2010. The number of county units in the four types of sections in Shandong and Henan provinces has been adjusted, and the labor employing quantity is decreasing. The proportion of county units in Shandong Province with (0, 669) days increased from 5.71 to 60.00% from 2000 to 2010, and the reasons for this are the improvement of the level of agricultural mechanization and the adjustment in planting structure. In the past decade, 58% of the counties in the region have reduced the planting area of economic crops such as oil and cotton, which need to input more time. Compared with other provinces, the amount of labor employing quantity in the planting industry in Beijing was lower in 2000 and 2010.

From 2000 to 2010, labor employing quantity in the planting industry in the Huang-Huai-Hai region was generally reduced. Due to the overall impact of natural conditions and socio-economic conditions, there was a significant spatial difference in the labor employing quantity of the planting industry (Fig. 2.25). In the past decade, labor employing quantity decreased in 311 counties, accounting for 88.86% of the sample in the region. The rapid reduction area ($\leq -26\%$) was mostly distributed in the west of Shandong hilly region, the Taihang and Yanshan mountainous region, most of the agricultural areas of the Hebei-Shandong-Henan Low-lying Plain, and the eastern part of the Huang-Huai-Hai Plain agricultural region. The areas with

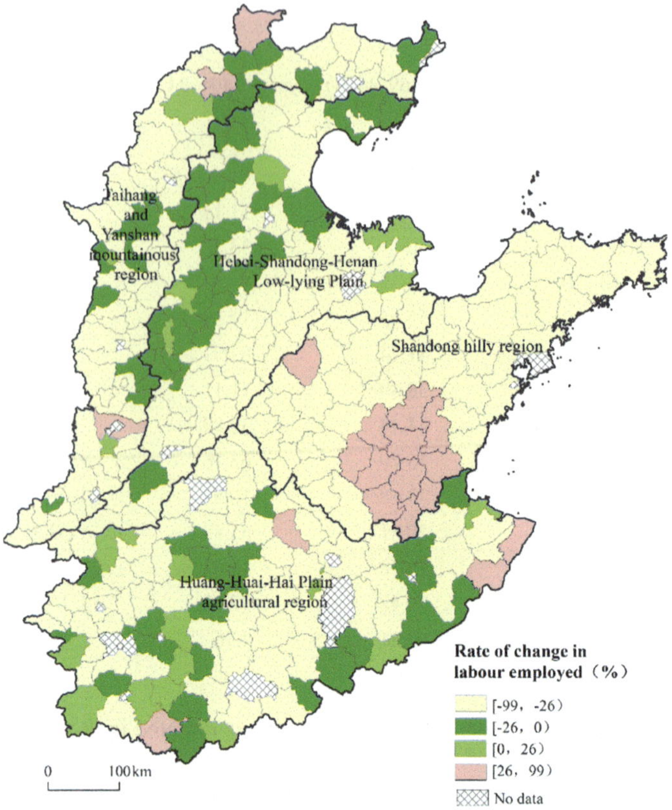

Fig. 2.25 The spatial pattern of labor employing quantity change (Liao et al. 2019)

increased labor employing quantity were mainly distributed in the central part of Shandong Province, and in the west and east of Nanyang City in Henan Province. There is a similar spatial distribution between the labor employing quantity reduction area (change rate < 0) and the increase in grain output (change rate of grain ≥ 0) (Fig. 2.26).

From 2000 to 2010, the grain output in the Huang-Huai-Hai region increased, and the increased counties accounted for 84.62% of the effective samples. The areas where the labor employing quantity was reduced and the output of grain crops was increased were mainly distributed in the south western part of the Taihang and Yanshan mountainous region, Hebei-Shandong-Henan Low-lying Plain, the central part of the Huang-Huai-Hai Plain agricultural region, and the southern part of the Shandong hilly region. The labor employing quantity of grain crop is less than that of cash crops, in the context of the deepening of rural aging and the shortage of effective rural labor; farmers prefer to increase the planting area of crops with less labor employing quantity, and the increase in the area planted with grain crops has led to increased grain production. In addition, the increase in grain production is also

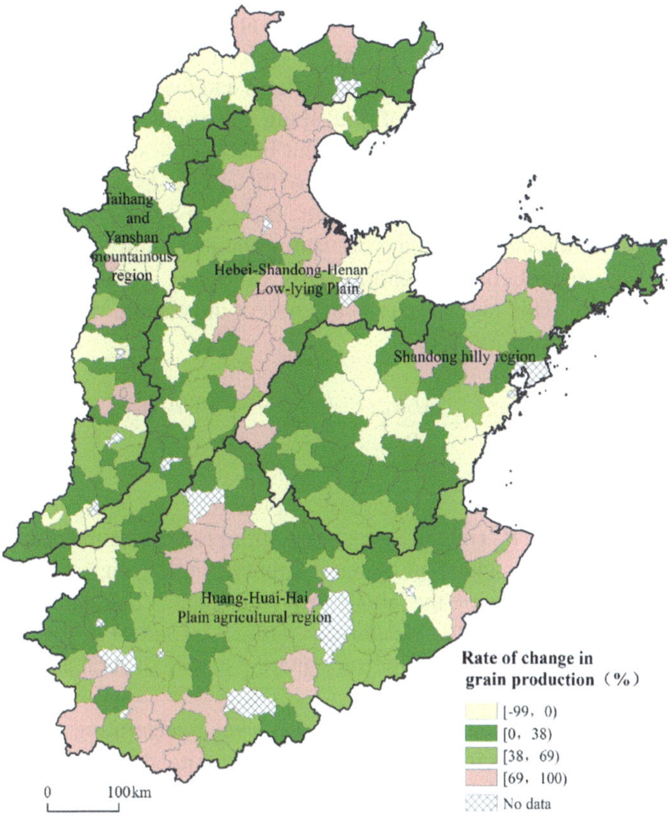

Fig. 2.26 The spatial pattern of grain yield change (2000–2010) (Liao et al. 2019)

related to the input of grain production materials such as grain crops, pesticides, fertilizers, agricultural machinery as well as grain prices, grain subsidies, and land use policies.

2.5.1.4 Coupling Characteristics of Rural Aging and Labor Employing Quantity in the Planting Industry

The quantity and quality of labor affect agricultural production. Therefore, when examining the impact of rural aging on agricultural production, it is also necessary to determine the interaction between rural aging and agricultural production changes. Labor employing quantity in the planting industry is acritical variable to reflect the time input of laborers in the agricultural production process. Using labor employing quantity to quantify land use transitions is a critical part of land use recessive morphology, which can reflect the changes in regional land use morphology to a certain extent. The elasticity coefficient is the ratio of the growth rate of two indicators related to each other in a certain period of time, and a measure of the dependence of the growth of one variable on the growth of another (Ma et al. 2019a). To further understand the relationship between the aging level and the change in labor employing quantity in the dynamic process of rural aging, this study attempts to draw on the calculation method of the elasticity coefficient in economics, by establishing rural aging—labor employing quantity elasticity coefficient *(ALEC)* to reflect the sensitivity of aging to the labor employing quantity. Based on *ALEC*, this study analyzes the direction of aging and the change in planting labor quantity, and then explains the temporal and spatial coupling characteristics of aging and the change in labor employing quantity. This method can better reflect the change in two variables in space and time than other methods.

$$ALEC_{ij} = \frac{ALR_{ij}}{APR_{ij}} = \frac{(AL_{ij} - AL_{i0})/AL_{i0}}{(AP_{ij} - AP_{i0})/AP_{i0}} \tag{2.19}$$

where $ALEC_{ij}$ refers to the *ALEC* of county i in the year of j; ALR_{ij} (change rate of amount of labor) and APR_{ij} (rate of aging population) represents the change rates of labor employing quantity and rural aging of county i in the year of j, respectively. Similarly, AL_{ij} and AP_{ij} refer to the labor employing quantity and the degree of rural aging of county i in the year of j, respectively; while AL_{i0} and AP_{i0} refer to the labor employing quantity in the planting industry and the degree of rural aging of county i in the base year.

Under the existing agricultural productivity level, accelerating the transfer of agricultural labor and reducing agricultural labor are critical means to improve the productivity of agricultural labor, but the transfer of agricultural labor inevitably increases the speed of the rural aging. Based on the analysis of the changes in $ALEC_{ij}$, this study divides the coupling types between the changes in rural aging and labor employing quantity in the planting industry into six types: growth, lagged, extensive, intensive, shrink and imbalance (Table 2.7). The specific description is as follows: (1) If $ALR_{ij} > 0$ and $APR_{ij} > 0$, the increase in the rural aging has played a positive role in labor employing quantity. When $ALEC_{ij} > 1$, it indicates that the growth rate of labor employing quantity is faster than the degree of rural aging, which is the growth type. When $0 < ALEC_{ij} < 1$, the growth rate of labor employing quantity is lower than the degree of rural aging, which is the lagged type. (2) If $ALR_{ij} > 0$ and $APR_{ij} < 0$, t the decrease in rural aging is accompanied by the increase in labor employing quantity, and the decrease in rural aging has a "reverse" effect on the labor employing quantity, which belongs to the extensive type. (3) If $ALR_{ij} < 0$ and $APR_{ij} > 0$, the degree of rural aging is increasing while the labor employing quantity is decreasing, and the increase in rural aging has a "reverse" inhibition effect on the change in labor employing quantity, which is the intensive type. (4) If $ALR_{ij} < 0$ and $APR_{ij} < 0$, the degree of rural aging and the labor employing quantity both decrease. When $ALEC_{ij} > 1$, the reduction in labor employing quantity is faster than the rural aging, which is the shrink type; when $0 < ALEC_{ij} < 1$, the labor employing quantity is declining slower than the degree of rural aging, which is the imbalance type.

Table 2.7 Coupling relation types between the changes in rural aging and labor employing quantity (Liao et al. 2019)

Type	ALR_{ij}	APR_{ij}	$ALEC_{ij}$	The characteristics of ALEC
I	$ALR_{ij} > 0$	$APR_{ij} > 0$	$ALEC_{ij} > 1$	Labor employing quantity is growing faster than the degree of rural aging; Growth type
II	$ALR_{ij} > 0$	$APR_{ij} > 0$	$0 < ALEC_{ij} < 1$	Labor employing quantity is growing slower than the degree of rural aging; Lagged type
III	$ALR_{ij} > 0$	$APR_{ij} < 0$	$ALEC_{ij} < 0$	Labor employing quantity grows, but the degree of rural aging decreases; Extensive type
IV	$ALR_{ij} < 0$	$APR_{ij} > 0$	$ALEC_{ij} < 0$	Labor employing quantity declines, but the degree of rural aging increases; Intensive type
V	$ALR_{ij} < 0$	$APR_{ij} < 0$	$ALEC_{ij} > 1$	Labor employing quantity is declining faster than the degree of rural aging; Shrink type
VI	$ALR_{ij} < 0$	$APR_{ij} < 0$	$0 < ALEC_{ij} < 1$	Labor employing quantity is declining slower than the degree of rural aging; Imbalance type

At present, China is more inclined toward the intensive coupling relationship model in which the population is aging and the labor employing quantity in the planting industry is decreasing. The continuous non-agricultural transfer of agricultural labor force has caused the aging of agricultural production to go mainstream, but the improvement in agricultural production infrastructure such as the agricultural mechanization level, construction of farmland water conservancy construction, and water conservancy irrigation, have continuously improved the agricultural production conditions. The improvement in the external conditions of agricultural production offsets the impact of the agricultural labor and the decline of human capital on agricultural production caused by rural aging and releases laborers from the constraints of physical labor, which to some extent compensate for the aging of agricultural labor force, reducing the labor employing quantity in the planting industry. The extensive type (labor employing quantity increase, but the degree of rural aging decreases) shows that the proportion of the elderly population to the total population is reduced, and the degree of non-agricultural transfer is low. In other words, agricultural production continues to rely on the massive input and expansion of labor factors to achieve agricultural economic growth, which is not conducive to improvement in agricultural production efficiency and the transformation of traditional agriculture (Ge et al. 2018b). The growth type and lagged type coupling modes contribute less to the improvement in agricultural production efficiency than that of the intensive type. In addition, the shrink type and imbalance type coupling modes may accelerate the transfer of agricultural labor, which is helpful for developing modern agriculture with intensive technology and high efficiency.

Based on the rural ALEC, according to the two types of rural aging growth zones and reduction zones, the coupling relationship between rural aging and labor employing quantity at the county level in the Huang-Huai-Hai region are further analyzed.

The spatial coupling characteristics of rural aging and labor employing quantity in the Huang-Huai-Hai region are obvious, the labor employing quantity in the planting industry in most areas decreases with the increase in the rural aging. The proportion of labor employing quantity in the Taihang and Yanshan mountainous region, Hebei-Shandong-Henan Low-lying Plain, Shandong hilly region and Huang-Huai-Hai Plain agricultural region decreased with the increase in rural aging level and was 87%, 92%, 79%, and 83%, respectively. This finding shows that the deepening of the rural aging results in the agricultural labor shortage, and the age structure of agricultural laborers is aggravated. Farmers, who are rational economic individuals, will adjust their planting strategies, reduce the planting area of crops with large labor employing quantity, and expand the planting area of labor-saving crops. The areas where the rural aging and labor employing quantity increased at the same time were mainly concentrated in the middle the of Shandong hilly region. The growth rate of the economic crops planting area in Shandong hilly region is obviously faster than that of grain crops, and the unit labor of economic crops is significantly higher than that of grain crops. Therefore, labor employing quantity in the planting industry is increasing.

The results show that: (1) during the research period, there were 337 counties in the Huang-Huai-Hai region with rural aging increasing (Fig. 2.27), and 39 counties with increased labor employing quantity (ALR > 0 and ALEC > 0), including 21 growth type counties (ALR > 0 and ALEC > 1), mainly distributed in the middle of the Shandong hilly region, the southern part of the Huang-Huai-Hai Plain agricultural region, and the north western part of the Taihang and Yanshan mountainous region. There were 18 lagged type (ALR > 0 and 0 < ALEC < 1) counties scattered in the Huang-Huai-Hai Plain agricultural region and Hebei-Shandong-Henan Low-lying

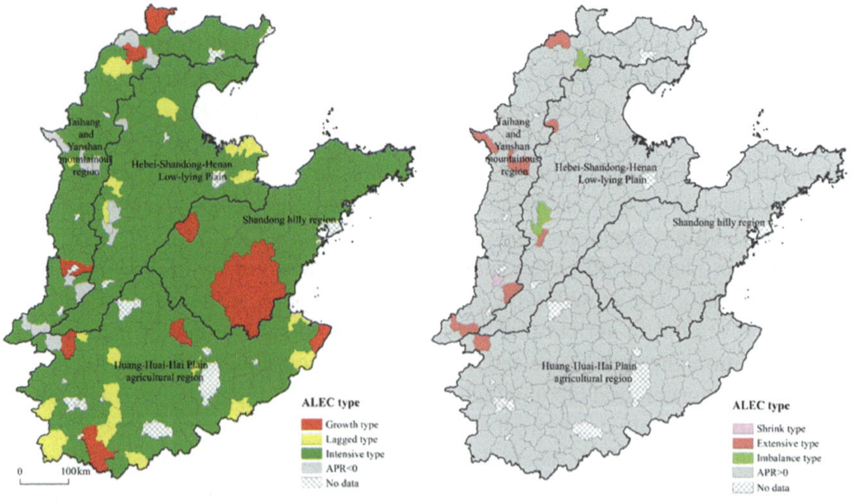

Fig. 2.27 Spatio-temporal pattern of ALEC types at county level in the Huang-Huai-Hai region during 2000–2010 (Liao et al. 2019)

Table 2.8 The change of ALEC at county level in the Huang-Huai-Hai region during 2000–2010 (Liao et al. 2019)

APR	Type	Coupling type	*ALR*	*ALEC*	County number (%)
APR > 0	I	Growth type	*ALR > 0*	*ALEC > 1*	21 (5.98)
	II	Lagged type	*ALR > 0*	*0 < ALEC < 1*	18 (5.13)
	IV	Intensive type	*ALR < 0*	*ALEC < 0*	298 (84.90)
APR < 0	III	Extensive type	*ALR > 0*	*ALEC < 0*	10 (2.85)
	V	Shrink type	ALR < 0	ALEC > 1	1 (0.28)
	VI	Imbalance type	*ALR < 0*	*0 < ALEC < 1*	3 (0.85)

Note Data outside the brackets refers to the number of counties belonging to a certain type, percentage indicates the proportion of the counties belongs to certain type in the total counties

Plain. There were 298 counties belonging to intensive type (ALR < 0 and ALEC < 0) (Table 2.8), mainly distributed in the central region of the Hebei-Shandong-Henan Low-lying Plain and the Huang-Huai-Hai Plain agricultural region. (2) In the 14 counties with reduced rural aging, 10 counties had a decrease in labor employing quantity (ALR > 0 and ALEC < 0), belonging to the extensive type, and mainly distributed in the Hebei- Shandong-Henan Low-lying Plain and Taihang and Yanshan mountainous region. One had a labor employing quantity declining faster than the degree of rural aging and belonged to shrink type (ALR < 0 and ALEC > 1) was distributed in the southern part of Taihang and Yanshan mountainous region. Three counties belonged to the imbalance type (ALR < 0 and 0 < ALEC < 1), and were distributed in the Hebei-Shandong-Henan Low-lying Plain.

2.5.1.5 Discussion

From the perspective of labor employing quantity in the planting industry, this study explores the spatio-temporal coupling relationship between rural aging and land use transitions and further expands the research on the relationship between agricultural input and agricultural land use transitions. The research results increase the understanding of to the spatial differentiation characteristics of rural aging and land use transitions in China's traditional farming area from a macro perspective, which is of great significance to the macro-control and precise decision making of regional agricultural production as well as, the more effective regulation of rural man-land interrelations. Land use transitions are influenced by impetus, such as population, economy, technology, institution, culture and location (Song 2017), but are subjected to the availability of data; thus, this study only discusses the spatio-temporal evolution of rural aging and labor employing quantity from the macro level. On the household level, the expansion of grain crops and cash crops were both significantly affected by agricultural labor. In the future, a probe into the impetus and mechanism of farmers' land use decision-making behaviour in the context of rural aging from the micro level and analysis of the relationship between rural aging and land use transitions in different landform conditions to compensate for the deficiency of mechanism analysis at the macro scale are necessary. Based on the relationship between rural aging and agricultural planting structure, its impact on the development of agriculture and corresponding coping strategies deserve further discussion.

Land is not only an important production factor for sustaining socio-economic development, but also plays an important role in promoting rural industrial transformation and upgrade, social security and ecological environment conservation. Land use reflects the evolution of the relationship between "human" and "land" in rural areas of China, as well as the current situation and problems of China's agricultural and rural development. In the process of urbanization, significant changes have taken place in the relationship between urban and rural areas, such as arable land loss, land abandonment, population aging, emergence of hollowing villages and so on (Liu and Li 2017), which have a profound impacts on land use and the strategic adjustment of land use policy. In order to reverse the severe situation of China's land development and protection in the process of urbanization, the Chinese government has formulated a series of land use policies and strategic guiding measures to deal with these problems and challenges, including requisition-compensation balance of arable land, rural land consolidation and so on (Long et al. 2012; Liu et al. 2014b). These policies emphasize the combination in the aspects of quantity, quality, ecology of farmland, but neglect the leading role of human in the process of land use transitions. In view of the facts that land use and agricultural production are directly or indirectly related to different stakeholders, land use transitions in the context of rural aging should be promoted by formulating comprehensive strategies from the perspectives of farmers, governments and other stakeholders, and seek for different modes. On the institutional level, it is necessary for the government to establish a large-scale farmland management system and an improved land transfer system to guide the orderly transfer of rural land contractual management rights.

2.5.2 Coupling Analysis of Grain Yields and Agricultural Labor Changes

2.5.2.1 The Analysis of Coupled Relationship Between Grain Yields and Agricultural Labor

Agricultural labor is indispensable for grain production. Under classical dual economic theory, Lewis (Lewis 1989) and Ranis-Fei models (Ranis and John 1969) have been used to map the process of labor transformation from traditional agricultural to modern production and have revealed the necessary conditions and effects of agricultural labor transfer on grain production. Thus, existing research demonstrates that the relationship between grain yields and changes in agricultural labor is affected by supply of the latter (Lipton 1980; Oseni et al. 2009). Therefore, changes in agricultural labor can be used to characterize supply and reveal the coupled relationship between changes in grain yields and labor. In China, since the reform and opening-up policies were enacted in 1978, a number of characteristics defined changes in agricultural labor including regional increases and decreases. Building on previous analyses, three coupled models are developed between grain yields and changes in agricultural labor.

The coupled relationship between grain yield and changes in agricultural labor conforms to a Lewis-Fei-Ranis model given the context of an agricultural labor increasing zone (Ranis and John 1969). Thus, the coupled relationship between grain yields and agricultural labor changes basically conforms with the reverse of dual economic structure transformation (Fig. 2.28a). In other words, increasing agricultural labor over the period between T_1 and T_2 has enhanced mankind's ability to develop nature. Grain yields have therefore constantly increased as the scale of regional agricultural production has expanded as the result of increasing labor. However, at the same time, the marginal diminishing effect of grain production has also become increasingly obvious. Based on this model, agricultural labor productivity

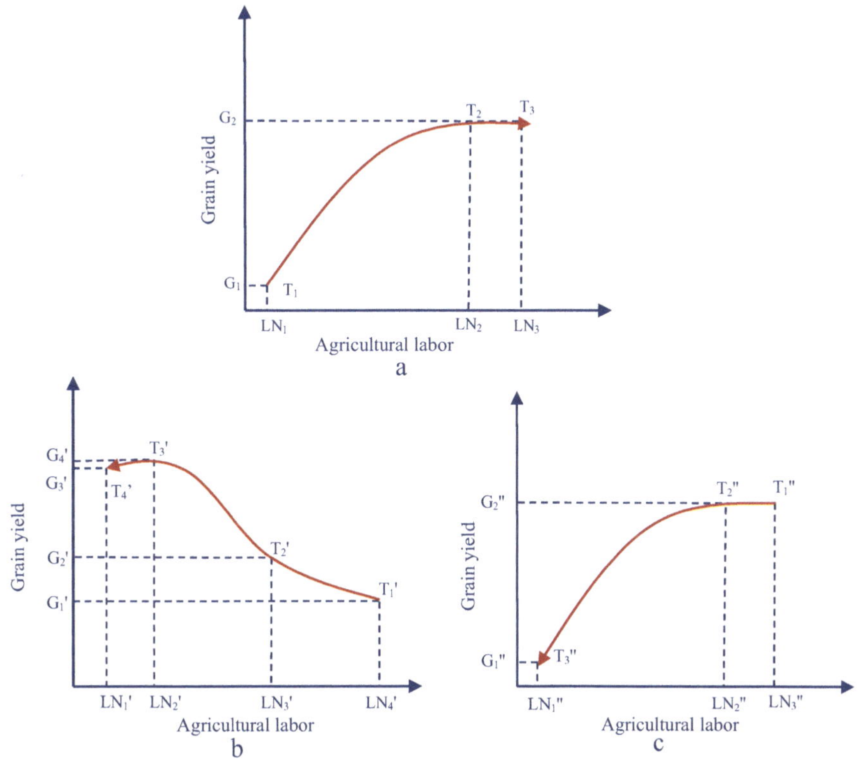

Fig. 2.28 Conceptual models illustrating the coupled relationship between grain yields and changes in agricultural labor (Ge et al. 2018b). *Note* Coupled models in agricultural labor increasing zones (**a**), traditional agricultural zones (**b**), rapid agricultural transformation zones (**c**)

reaches a maximum and the effects of natural resource constraints become prominent at the point T_2. Subsequent to T_2, even if the volume of the available agricultural labor continues to increase, grain yields will stall and the relationship between food supply and demand will lead to severe social crises, necessitating urgent reforms of existing production relationships. As a result of this process, the coupled relationship between grain yields and agricultural labor changes enter a transitional period.

There are two coupled modes-of-relationship that concern grain yield and decreases in agricultural labor, one of which dominates in traditional agricultural zones (TAZs), while the other is prevalent in rapid agricultural transformation zones (RATZs). Different models of coupled relationships involve different processes, as illustrated in Fig. 2.28b which high lights the coupled relationship between grain yields and decreases in agricultural labor in a TAZ. During the early stage of this process (T_1), agricultural labor includes a high degree of surplus but at a low level of productivity, directly limiting grain yields. At this time, per capita grain occupancy remains low and is integrally linked to poverty in rural areas; the relationship between grain yields and agricultural labor is considered in an 'antagonism' stage at this point. In contrast, during the period between T_1' and T_2', urban areas come to the forefront of development and a 'Lewis' urban-rural dual structure is apparent. Affected by the costs of opportunity and comparative benefits, agricultural labor begins to transfer and non-agricultural employment is seen coupled with increasing non-farm incomes.

This transition leads to structured employment and a gradual improvement in quality of life. Coupled with these developments, people also being able to obtain advanced technology as well as more funds which results in improvements in agricultural labor productivity and further increases in grain yields. All this has the effect of alleviating the so called 'man-land' and 'man-grain' interrelations characteristic of TAZs (Rozelle et al. 1999). Over the course of the period between T_2' and T_3', grain yields reach their highest point, G_3', as agricultural labor decreases and productivity increases. Due to rapid urbanization and industrialization, labor productivity, agricultural production technology, and management systems are all continuously innovated, while the scale and specialization of agricultural production is continuously promoted. At the same time, marginal declines in the effects of agricultural production also come to the forefront; after point T_3', for example, grain yields will decline because of the problem of rural 'hollowing' as the income gap widens between the urban and rural population. This process is accompanied by the widespread marginalization of agricultural land, which leads directly to widespread land abandonment (Chan 2010).

The coupled relationship between grain yields and decreases in agricultural labor in RATZs (Fig. 2.28c) conform to a dual economic structure model with respect to the relationship between the labor transfer and variations in agricultural output (Rains and John 1969). However, both the economic basis and locations of RATZs are an improvement compared to those of TAZs, ensuring agricultural transformation. In other words, the more non-agricultural employment opportunities there are, the lower the opportunity cost to abandon grain production, which reduces the social security function of grain. In cases where declines in agricultural labor do not lead to decreases in grain yield between T_1'' and T_2'', inputs of other production factors compensate for the impact of decreases in agricultural labor. During the process of regional economic transformation in China, agricultural production was lower in efficiency than other kinds of production; this was then gradually replaced by other industries, and regional grain self-protection capacity declined.

The three coupled relationships that exist between grain yields and changes in agricultural labor can occur simultaneously in different areas because of differences in levels of regional development and location conditions. It is also noteworthy that these three coupled models can also appear continuously in the same region over different time periods; for example, agricultural labor first increased and then decreased as the result of regional agricultural transformation, before eventually being transferred. Thus, one of the aims is to reconstruct the couplings between grain yield and agricultural labor changes which characterize differences within the same and different time periods.

2.5.2.2 The Grain-Labor Elasticity Coefficient (GLEC)

The GLEC can be defined as the ratio between the rate of grain yield change divided by the rate of agricultural labor change over a given time period (Liu et al. 2010), The GLEC is calculated as follows:

$$GLEC_i = \frac{GYR_i}{ALNR_i} = \frac{(GY_{it2} - GY_{it1})/GY_{it1}}{(ALN_{it2} - ALN_{it1})/ALN_{it1}} \tag{2.20}$$

In this expression, i denotes the county number, while GYR_i and $ALNR_i$ refer to the rate of grain yield and the rate of agricultural labor number change respectively in a county, i. Thus, GY_{it2} and GY_{itl} refer to the grain yields of county i during the time periods t_2 and t_1, while ALN_{it2} and ALN_{itl} denote agricultural labor in county i during t_2 and t_1, respectively. Similarly, GLEC denotes the elasticity coefficient between grain yield and agricultural labor, which indicates a coupling relationship between grain yield changes and agricultural labor changes over the period between t_1 and t_2.

Thus, by analyzing changes in GLEC, the coupled relationship between grain yields and agricultural labor is revealed. In other words, if ALNR is greater than zero and GLEC is less than zero, an increase in agricultural labor and a decrease in grain yields is implied; increases in the former will have a reverse effect on the latter. Conversely, if ALNR and GLEC are both greater than zero, agricultural labor and grain yields vary in the same direction, and can explain the positive effect of increases in the former on the latter. In a case where both indexes are less than zero, agricultural labor decreases while grain output increases and the former has a positive effect on the latter. While if ALNR is less than zero and GLEC is greater than zero, decreases in agricultural labor will exert an inverse effect on increasing grain yield. The data include agricultural labor statistics, the number of rural employees, and total grain yields for the years 1991, 2000, and 2010. Data for 1991 were provided by the China Natural Resources Data Center (http://www.data.ac.cn), while those for 2000 and 2010 come from the China County (City) Social and Economic Statistical Yearbook. Thus, with the exception of counties where statistical data is absent, this research includes a total of 2,006 effective statistical units. The basic geographical data we used are derived from the Resource and Environment Science Data Center of the Chinese Academy of Sciences (http://www.resdc.cn/). Because administrative divisions at the county level across China were continuously adjusted between 1991 and 2010, those used in this study are adjusted against standard divisions of 2010.

2.5.2.3 Spatio-Temporal Coupling Between Grain Yields and Changes in Agricultural Labor at the County Level

(1) **Grain yields and changes in agricultural labor at the county level**

Characteristics of spatial coupling between grain yields and agricultural labor in China are obvious; in places where there is more agricultural labor, grain yields are higher. In addition, the 'Hu Huanyong Line' (the 'Hu Line') determines the nature of this spatial pattern (Fig. 2.29). To the southeast of the 'Hu Line', agricultural labor and grain yields are higher than to the northwest, especially within river alluvial plains, while in contrast, to the northwest of this line, the proportion of agricultural labor is generally low and grain yields are very limited. These characteristic spatial couplings between grain yield and agricultural labor across China are inseparable from physical factors, including water, soil, air temperature, and heat, as well as the spatial distribution pattern of population (Fig. 2.29).

(2) **GLEC spatial characteristics between 1991 and 2000**

Between 1991 and 2000, agricultural labor decreased in 1078 counties mainly located to the southeast of the 'Hu Line' (Fig. 2.30a). Of these, 363 were characterized by decreasing grain yields (i.e., ALNR less than zero and GLEC greater than, or equal to, zero) and are mainly distributed in Zhejiang Province and on the eastern Shandong

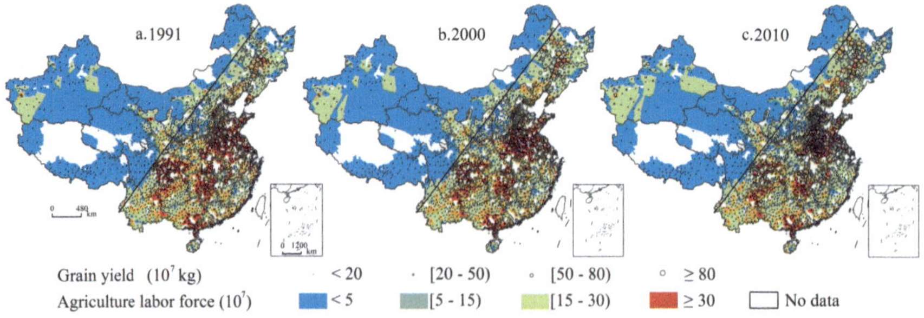

Fig. 2.29 Maps showing the spatio-temporal pattern in the coupled relationship between grain yields and agricultural labor at the county level over the period between 1991 and 2010 (Ge et al. 2018b)

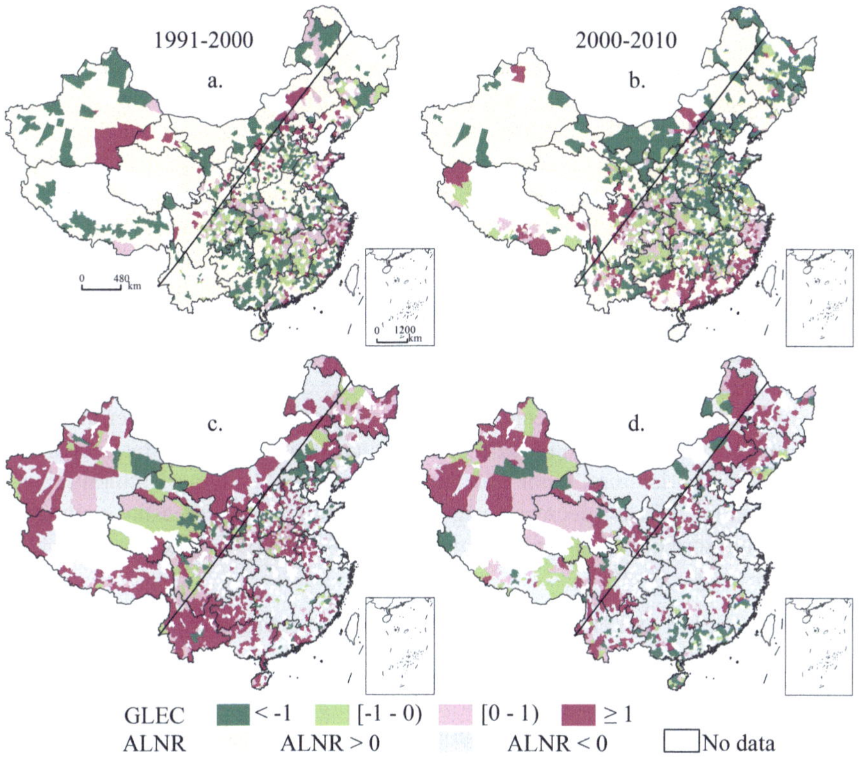

Fig. 2.30 Maps showing spatio-temporal patterns of GLEC at the county level between 1991 and 2010 (Ge et al. 2018b)

Peninsula. A total of 715 counties were characterized by increasing grain yields (i.e., ALNR and GLEC less than zero), mostly concentrated in areas including the Huang-Huai-Hai Plain and Sichuan Basin.

At the same time, 712 counties exhibited increases in both agricultural labor and grain yields (i.e., ALNR greater than zero and GLEC greater than, or equal to, zero), mostly located in the central and western parts of Xinjiang Uygur Autonomous Region (Xinjiang), on the Yunnan-Guizhou Plateau, in the Henan-Shandong-Anhui junction area, and in the west of Sichuan Province (Fig. 2.30d). A further 228 counties are characterized by increasing agricultural labor and decreasing grain yields (i.e., ALNR greater than zero and GLEC less than zero), mainly scattered in northern Qinghai and western Liaoning provinces. Indeed, over the time period of this survey, counties characterized by 'increasing grain yields and agricultural labor', 'decreasing grain yields and agricultural labor', 'increasing grain yields and decreasing agricultural labor', and 'decreasing grain yields and increasing agricultural labor' comprise 35.54%, 18.09%, 35.64%, and 10.72%, respectively, of the total number.

(3) **GLEC spatial characteristics between 2000 and 2010**

Between 2000 and 2010, results show that 1385 counties experienced a decrease in agricultural labor, mainly in eastern China (Fig. 2.30b). Of these, 482 saw a decrease in grain yield (i.e., ALNR less than zero and GLEC greater than, or equal to, zero) and are mainly scattered in the southeastern coastal provinces and Guangxi Province, while 903 counties accounting for 45% of the total experienced an increase in grain yields (i.e., ALNR and GLEC less than zero), principally located on the Huang-Huai-Hai Plain, in the Hetao area, and on the northeastern plain. At the same time, 432 counties chiefly concentrated in Xinjiang and eastern Inner Mongolia experienced an increase in both agricultural labor and grain yields (i.e., ALNR greater than zero and GLEC greater than, or equal to, zero) (Fig. 2.30d); of these, just 189 counties mostly distributed in eastern Tibet and in Hainan Province experienced an increase in both agricultural labor as well as a decrease in grain yields (i.e., ALNR greater than zero and GLEC less than zero). Counties at this time that are characterized by 'increasing grain yield and agricultural labor', 'decreasing grain yield and agricultural labor', 'increasing grain yield and decreasing agricultural labor', and 'decreasing grain yield and increasing agricultural labor' accounted for 21.54%, 24.03%, 45.01%, and 9.42% of the totals, respectively.

(4) **The evolution and classification of the coupled relationship between grain yields and changes in agricultural labor**

Changes in GLEC show that the effects of agricultural labor changes on grain yields both decline and exhibit significant regional differences. Results show that between 1991 and 2010, the average GLEC value rose from 3.81 to 3.99 in counties characterized by increasing agricultural labor. This result indicates that the effect of increasing agricultural labor on grain yields is still present in less developed areas; however, between 2000 and 2010, this county type decreased to just 307 with grain yields accounting for 41.22% of the total in 2000, before falling further to 25.81% in 2010. Over the same period, the average GLEC values of counties characterized by decreasing agricultural labor increased from –2.56 to –1.76, which shows that the effect of decreases in this factor on enhancing grain production also declined. Grain yields in this county type accounted for 58.78% of total yields in 2000, increasing to 74.19% in 2010, becoming the major coupling between these two factors. Against the background of an overall decline in the availability of agricultural labor, average GLEC across China rose from –0.25 between 1991 and 2000 to –0.16 between 2000 and 2010. This increase is indicative of the positive effect of agricultural labor decreases on declines in grain production. As a result, the role of agricultural labor in promoting grain production has gradually declined in China while the role of non-labor-related factors continues to strengthen. Spatial patterns within the coupled relationship between grain yields and agricultural labor also reveal a number of obvious regional differences, with the 'Hu Line' acting as a large-scale GLEC boundary. Counties that exhibit a coupled relationship in the opposite direction are located mainly to the southeast of the 'Hu Line' (exception of developed areas in the southeastern coastal region of China), where agricultural labor is decreasing and grain yields are simultaneously increasing. Similarly, counties that exhibit a coupled relationship in the same trend direction are mostly concentrated to the northwest of the 'Hu Line', characterized by increases in both grain yields and agricultural labor (Heerink et al. 2007).

Because of agricultural transformation and development, different coupled relationships co-exist simultaneously between grain yields and changes in agricultural labor in different regions. This phenomenon is one distinguishing feature of Chinese agricultural transformation and development that is different to other countries. Indeed, coupled relationships between increasing agricultural labor and grain yields are mainly seen in under-developed pastoral regions and in the agro-pastoral ecotones of western China, while coupled relationships between increasing grain yields and decreasing agricultural labor are mainly concentrated in traditional farming areas which account for the highest total grain yields countrywide. Coupled relationships between decreasing grain yields and agricultural labor are seen in areas that have experienced rapid transitions in agricultural development, especially in developed southeastern coastal China, a region characterized by much more employment in non-agricultural sectors as well as a gradual withdrawal from grain production. Because these coupled relationships also reveal regional differences in agricultural transformation and development processes, it is necessary to formulate differentiated policies to address regional grain production security.

2.5.2.4 Discussion

This coupled relationship between grain yields and changes in agricultural labor exhibits synchronous diversification. Thus, based on theoretical analyses and empirical research, the results of this study highlight three coupled relationships between grain yields and agricultural labor in the pastoral and agro-pastoral ecotone, in traditional agricultural areas, and in the rapid agricultural transition zone. These three relationships have co-existed with one another over a long time period, while the 'Hu Line' represents a significant boundary delineating the coupled relationship between grain yields and changes in agricultural labor. Traditional agricultural areas characterized by this coupled relationship are mainly located to the southeast of the 'Hu Line' where agricultural labor decreases and grain yields increase. However, decreases in both grain yields and agricultural labor mainly characterize the rapid agricultural transition zone of southeastern coastal China, while a coupled relationship within pastoral and agro-pastoral ecotones is also seen to the northwest of the 'Hu Line' where both grain yields and agricultural labor increase.

The coupled curve between grain yields and changes in the proportion of agricultural labor exhibits stepped fluctuations but has been continually strengthened over time. Indeed, this coupled relationship between grain yields and changes in the proportion of agricultural labor has mainly been positive over time, indicating that the transformation of agricultural labor has had a positive effect on promoting grain yields albeit with different spatial patterns. Compared with other countries, the agricultural labor market, land ownership, and man-land interrelations in China exhibit specific characteristics as regional socio-economic development levels as well as the social security function of grain differ markedly. The coupled relationship between grain yields and agricultural labor also exhibits significant regional differences.

Chinese food security and the transfer of agricultural labor are important current research topics. Based on the GLEC and ATLEC values, this study describes the coupled relationship between grain yields and changes and proportions in agricultural labor. Analyses show that the overall effect of agricultural labor transfer is declining. At the same time, grain production organization mode, new types of agricultural management bodies, land resource managerial systems, and other new factors of grain production (Long et al. 2016) as well as their impacts on grain production are increasing. Given a market economic background, the factors influencing changes in grain production are also increasing, but because they interact with one another, it is difficult to quantitatively measure their degree of influence on agricultural labor. Thus, based on detailed natural environment and socio-economic data, quantitative description of the impacts of agricultural labor on grain yield changes will be a future research focus.

The coupled relationship between grain yields and changes in agricultural labor not only reflects processes of regional agricultural transformation, but also tracks the path and nature of regional spatial transformation and development. At present, nearly half of the Chinese population still lives in rural areas, and thus their survival and development remain closely associated with grain production. The social security function of grain determines the relationships between fair, efficient, current, and long-term social security and ecological protection as part of the processes of rural transformation and development. A sustainable rural development system (Long et al. 2016), which guarantees food security and promotes benign rural advancement, is thus an important factor in formulating future agricultural and rural policies in China.

2.6 A Perspective of Function Evolution

2.6.1 Farmland Function and Farmland Use Transition

Meeting the demands of subsistence and development, which is regarded as the essence of farmland function, is the endogenous dynamic and ultimate goal of farmland development and utilization. Farmland management with both commodity and non-commodity outputs results in a multifunctional complex of material production and non-material production. The demand for farmland tends to be diversified and high-end for the advancement of society and the economy, which reflects the transformation from emphasizing traditional economic value to focusing on the multiple values of the economy, society, and ecology. Based on grain or other cash crop production, the economic function brings financial gains to rural households and promotes national economic development. Generally, social function indirectly manifests as providing basic employment for farmers, maintaining social stability, and ensuring food security. Undoubtedly, farmland plays an important role in ensuring regional and national food security because of its status of major production base of agricultural products. Therefore, the function of ensuring food security was not taken into consideration, while the function of guaranteeing employment was emphasized in this research. Ecological function mainly embodies as preserving biological diversity, regulating the climate, purifying the environment, maintaining the resilience of field ecosystems, and enhancing the landscape esthetic. Farmland, the functions of which are constrained by multiple factors, such as human activity, individual concept and consciousness, regional cultural traditions, socio-economic systems and policy, and farmland ecosystem carrying capacity, is the coupling of the natural ecological subsystem and the artificial sub-system. Furthermore, a FF system is a synthesis formed by interrelated and interacting sub-functions, rather than the simple superposition of each single function (Matson et al. 1997).

2.6.2 *Methods*

2.6.2.1 Establishing an Evaluation Index System

Considering data accessibility and the difficulty of quantification, an evaluation index system covering economic, social, and ecological functions was established by following scientific, systematic, and hierarchical principles (Table 2.9).

(1) In order to evaluate economic function scientifically, the yield, output value, and potential crop yield of farmland were taken into consideration. The output value of crop farming (Fleskens et al. 2009) and grain yield per farmland area (Tan 2014) comprehensively reflect the output level affected by natural conditions and varying degrees of input production factors. Based on farmland spatial distribution, soil conditions, and elevation, the GAEZ (Global Agro-Ecological Zones) model, which incorporates natural factors, pest, and disease damage, was adopted to calculate potential farmland crop yield (Evans and Fischer 1999). (2) When selecting the indices for social function evaluation, emphasis was placed on farmland carrying capacity of rural labor employment and the effects of agricultural mechanization on farmers' employment. As a rule, the larger the per capita farmland area is, the stronger its employment absorption capacity. Small-scale and dispersive management with lower earnings give rise to the transfer of rural labor from traditional agriculture to non-agricultural industries. The proportion of workers who engaged in agriculture, forestry, animal husbandry, and fisheries can directly reflect the reliance of famers on farmland. The total power of agricultural machinery per farmland area is a firsthand measurement of agricultural mechanization. The increase of labor-saving inputs may lower the dependence of agricultural production and management on rural labor, thus weakening the employment carrying capacity of farmland. Ecological pressures due to yield-increasing inputs and population growth, which were depicted by

Table 2.9 The evaluation index system for farmland function (Zhang et al. 2018)

Criteria	Index	Unit	Direction	Weight(3)
Economic function index	Grain yield per farmland area	ton/hm^2	+	0.29
	Output value of crop farming per farmland area	yuan/hm^2	+	0.28
	Potential crop yield of farmland	kg/hm^2	+	0.43
Social function index	Per capita farmland area	hm^2/person	+	0.53
	Proportion of the employees who engaged in agriculture, forestry, animal husbandry, and fisheries in the total rural employees	%	+	0.21
	Total power of agriculture machinery per farmland area	kW·h/hm^2	–	0.26
Ecological function index	The equivalent quantity of fertilizer input per farmland area exceeds the upper limit of the safety standard of fertilizer input per farmland area(1)	kg/hm^2	–	0.17
	Population per farmland area	person/hm^2	–	0.25
	Percentage of farmland in ecological land(2)	%	+	0.32
	Function index of landscape aesthetics	–	+	0.26

Note (1) The upper limit of the international safety standard of fertilizer input per farmland area is 225 kg/hm^2; (2) Ecological lands mentioned here include farmland, forest land, grassland, water bodies, and unused land; and (3) A compound method of entropy weighting and multiple correlation coefficient weighting was used to determine indices weight (Luo and Cai 2016)

the excess level of fertilizer and population per farmland area, were taken into consideration when selecting indicators for measuring ecological function. The percentage of farmland in ecological lands represents the importance of farmland to eco-environmental security, due to the large share of farmland in HHHP. In the process of evaluating landscape aesthetic function, the concentration of farmland and accessibility were emphasized, and the relief amplitude was ignored due to the flat terrain and smaller regional differences. It is generally accepted that centralized and contiguous farmland has greater aesthetic value. Farmland is more attractive for urban residents compared to rural residents. However, distance and accessibility determines where urban dwellers will go.

2.6.2.2 Calculating the Function Index

The first step was to standardize the evaluation index using the maximum difference normalization method. Then, each sub-function index was calculated by the formulas below:

$$F(econ) = \sum_{i=1}^{n} \mathrm{w}(econ)_i * f(econ)_i$$

or

$$F(soci) = \sum_{i=1}^{n} \mathrm{w}(soci)_i * f(soci)_i$$

or

$$F(ecol) = \sum_{i=1}^{n} \mathrm{w}(ecol)_i * f(ecol)_i \tag{2.21}$$

where F (econ), F (soci), and F (ecol) represent economic, social, and ecological function indices, respectively. Similarly, $\mathrm{w(econ)}_i$, $\mathrm{w(soci)}_i$ and $\mathrm{w(ecol)}_i$ are the weights of index i for each sub-function, respectively; and $\mathrm{f(econ)}_i$, $\mathrm{f(soci)}_i$ and $\mathrm{f(ecol)}_i$ are the respective standardized index values. F (econ), F (soci), and F (ecol) range from 0 to 1, and the larger the value, the stronger the corresponding sub-function is.

2.6.2.3 The Coupling Pattern of Farmland Function Evolution

There are relationships of relevance, influence, and constraints among economic, social, and ecological functions, which may have positive or negative effects. Although a multi-factor comprehensive evaluation method may reflect the comprehensive function of farmland, it is unable to directly describe the coupling and coordination degree among the sub-functions. Hence, a coupling and coordination model was used to quantitatively depict the strength of the interaction between the three sub-functions and reveal the state of synergetic evolvement. The formula for the coupling degree model is as follows:

$$C = \left[(F(econ) * F(soci) * F(ecol) / \left(\frac{F(econ) + F(soci) + F(ecol)}{3} \right)^3 \right]^{1/3} \tag{2.22}$$

where C is the coupling degree, which varies between 0 and 1. When $C = 0$, this means the relationship between three sub-functions is disordered. On the contrary, when $C = 1$, the coupling degree of each sub-function reaches the maximum, which means that they are fully coupled.

In order to further explore the degree of coordinated development, a coordination model was used, and the formula is below:

$$D = C * T^{1/2}, T = \alpha * F(econ) + \beta * F(soci) + \gamma * F(ecol) \quad (2.23)$$

where C is the coupling degree and D is the coordination degree; T is the comprehensive evaluation index of farmland functions. A combined method of entropy weighting and multiple correlation coefficients weighting was used to calculate weights. The values of α, β, and γ were 0.21, 0.40, and 0.39, respectively.

2.6.2.4 The Selection of Driving Factors and Quantitative Recognition Method

(1) **Driving factors**

Farmland function evolution (FFE) is the result of the interactions and influence between regional indigenous factors and external driving factors. The state of FFE is determined by the power of driving forces. Topography, climate, hydrology, and farmland resources are basic natural factors that provide support for FFE. Urbanization, industrialization, agriculture modernization, and regional development policies are external factors that give impetus to the FFE by boosting the flow of urban-rural elements and industrial interaction (Fig. 2.31).

Regional indigenous factors: (1) Natural farmland conditions, which are regarded as the prerequisite of farmland use, fundamentally affect agricultural production and resource supply capacity, as well as the direction of FFE. In spite of the fact that the change in natural factors is weak and not very obvious over the short term, farmland productivity varies widely because of the large span from the south to the north and clear regional differences of precipitation in HHHP. Accordingly, elevation, average annual precipitation and distance from major rivers were selected as the major natural driving factors. (2) Transportation accessibility is an effective indicator to estimate location advantage. Hence, distances to national, provincial, and county roads per area were used to provide a comprehensive estimation of regional transportation accessibility.

External driving factors: (1) It is universally acknowledged that establishing a multidimensional evaluation index system of urbanization covering the economy, society, and space (land) is a difficult task at the county level due to the overwhelming obstacle of collecting socio-economic statistical data. Considering the powerful effects of urbanization on surrounding regions, we first established an evaluation index system for urbanization at the level of the prefecture (Table 2.10). In order to avoid repetition in indices, indices related to industrialization were not involved in the economic urbanization evaluation index system. Next, prefecture cities were classified into five levels according to their comprehensive influence (Fig. 2.32).

Then, the scores associated with their influence received from corresponding core cities were calculated using a linear attenuation model that was based on the influence sphere of core cities designated by the Weighted Voronoi Method. (2)

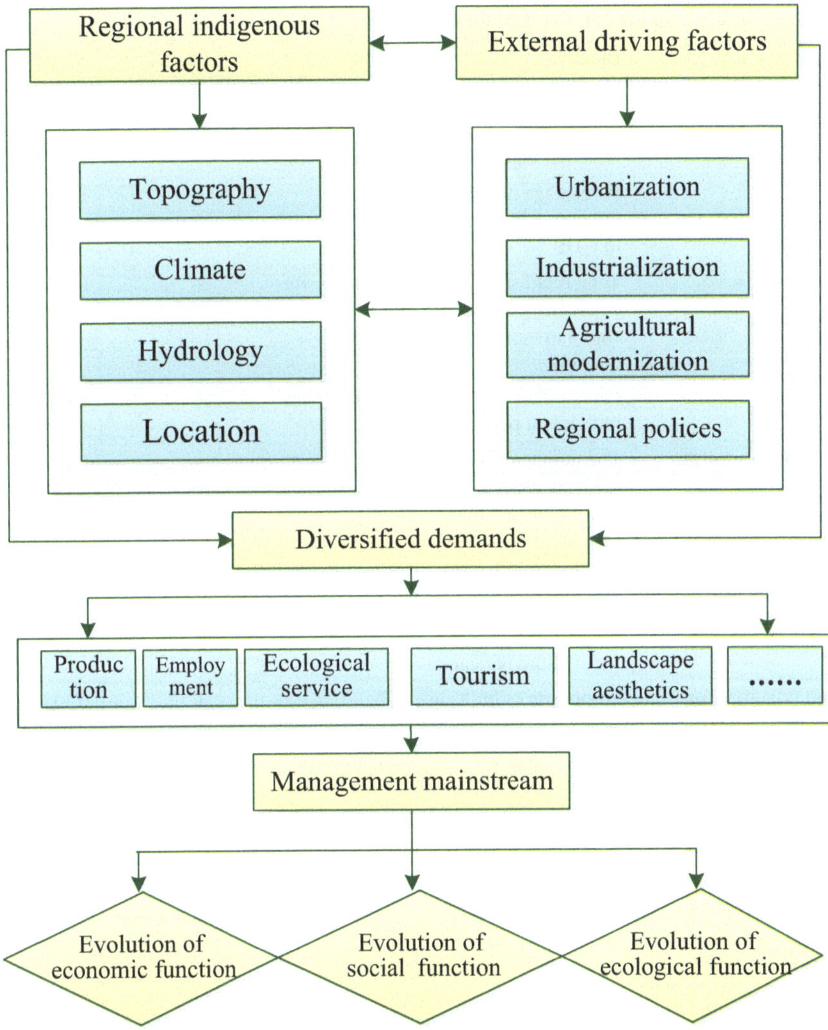

Fig. 2.31 The conceptual framework for the dynamic mechanisms of FFE (Zhang et al. 2018)

The number of industrial enterprises per unit land area and percentage of industrial added value in GDP at the county level were used to indicate the industrialization level. (3) The increase in yield-increasing and labor-saving inputs triggered by agricultural modernization boosted farmland productivity and the de-agriculturalization transfer. Simultaneously, excessive fertilizer inputs can cause severe agriculture non-point source pollution and threaten farmland ecosystems. (4) Regional development policies including farmland protection, balancing urban and rural construction land, grain purchasing and storage, water-saving agriculture and environmental agriculture, exerted limitation or guidance on FFE.

Table 2.10 The evaluation index system for urbanization level (Zhang et al. 2018)

Criterion	Index	Unit	Direction	Weight*
Economic urbanization	GDP per land area in municipal districts	10,000 yuan/km^2	+	0.12
	GDP per capita	yuan/person	+	0.11
	Percentage of tertiary industry added value in GDP	%	+	0.07
	Total retail sales of consumer goods per capita in municipal districts	yuan/person	+	0.03
	Fixed assets investment per capita	yuan/person	+	0.05
Social urbanization	The number of beds in medical and health institutions per million people	–	+	0.08
	The number of buses per million people	–	+	0.04
	Local fiscal expenditure per capita	yuan/person	+	0.10
Land urbanization	Proportion of built-up area	%	+	0.13
	Built-up area per capita	m^2/person	+	0.13
	Urban road area per capita in municipal districts	m^2/person	+	0.12

Note *Combined method of entropy weighting and multiple correlation coefficients was used to determine indices weights

Farmland management mainstream: as the management mainstream of farmland, the rural population has the dual status of producer and customer, which can satisfy the multiple demands of urban and rural residents. The proportion of the rural population and the percentage of the population with an educational level above junior middle school in the total population was adopted to show the peculiarities of farmland management from the perspective of quantity and quality.

(2) **Spatial econometric regression analysis**

In light of the relevance of the sub-function indices for different regions, the use of a spatial econometric model to explore influence factors can make up for the disadvantages of a traditional econometric model, because it explores the spatial correlation among explanatory variables or random error terms. There are many kinds of spatial econometric models. Spatial constant coefficient models, including a Spatial Lag Model (SLM) and a Spatial Error Model (SEM), were chosen for this study (Fang et al. 2013).

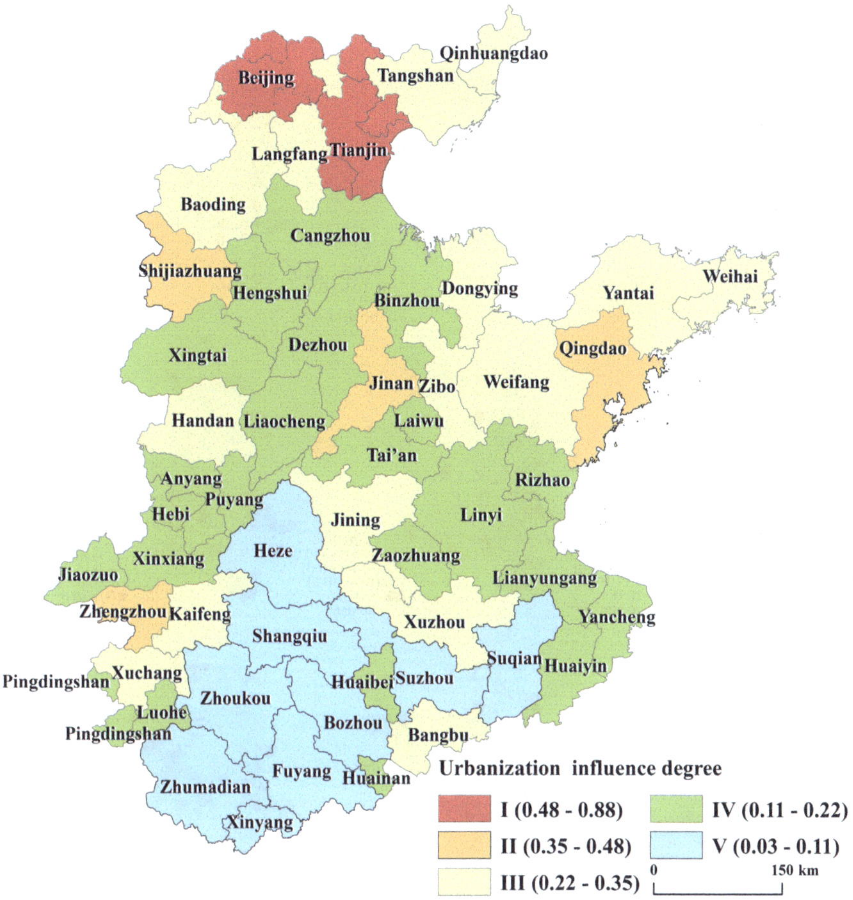

Fig. 2.32 Urbanization influence classification for cities (Zhang et al. 2018)

2.6.3 Spatio-Temporal Characteristics

2.6.3.1 Spatio-Temporal Patterns of Farmland Function Evolution at the County Level

Spatio-temporal patterns of economic function evolution: economic function in HHHP generally weakened during 1990–2010 and showed prominent spatial differentiation between the areas of growth and decline. Economic function grew overall and the rate of change was normally distributed. The regions of growth were mainly concentrated in the middle of Hebei Province, the east of Henan Province, and the west of Shandong Province, while decreasing areas were mostly distributed in and around the Beijng-Tianjin metropolitan area, the Yantai-Weihai urban area, and Jinan (Figs. 2.33a and 2.34a). Skewness and Kurtosis clearly increased due to the sharp

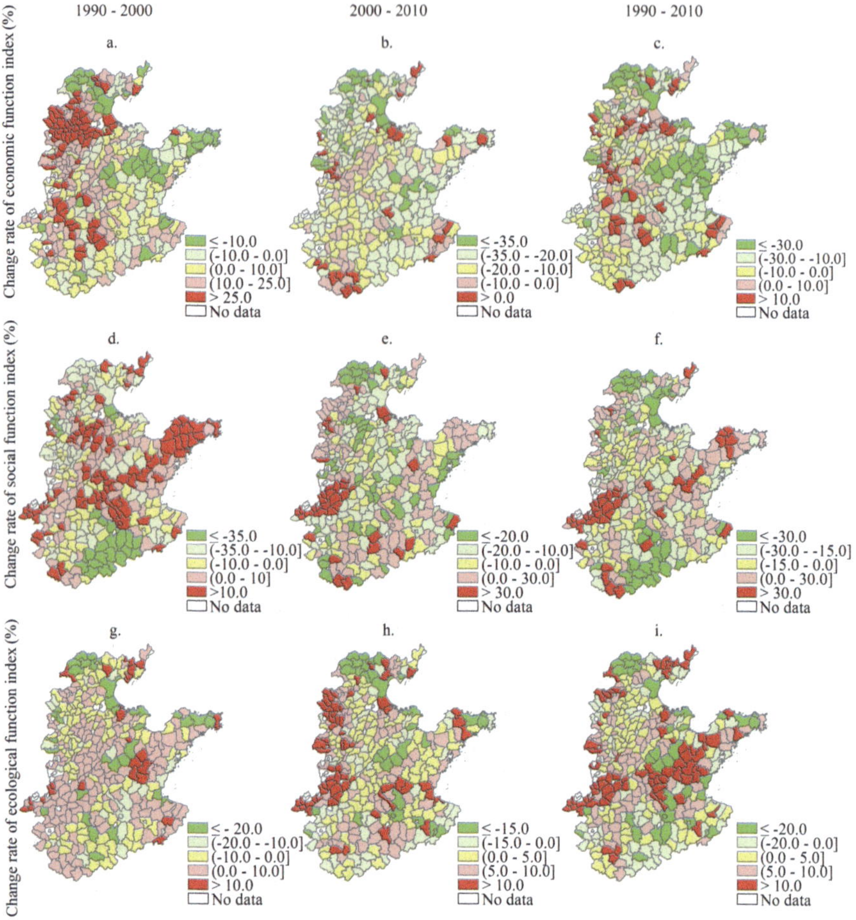

Fig. 2.33 The spatial patterns of the change rate of FF in HHHP during 1990–2010 (Zhang et al. 2018)

increases of economic function in some counties during 2000–2010 (Figs. 2.33b and 2.34b). Compared to the last period, the number of growing counties decreased with only 27 counties growing, scattered around the southern parts of Hebei and Henan Provinces. The growth areas of economic function were mainly located in the middle and south of Hebei Province and the east of Henan Province during 1990–2010 (Figs. 2.33c and 2.34c). The last 20 years has witnessed a continuous decline in areas surrounding Beijing and Tianjin, Yantai and Weihai, and Jinan, which were deeply affected by urbanization and industrialization. Meanwhile, agriculture production efficiency was low in northern Anhui Province because of the larger proportion of traditional agriculture and the smaller fraction of high efficiency facility agriculture and leisure agriculture.

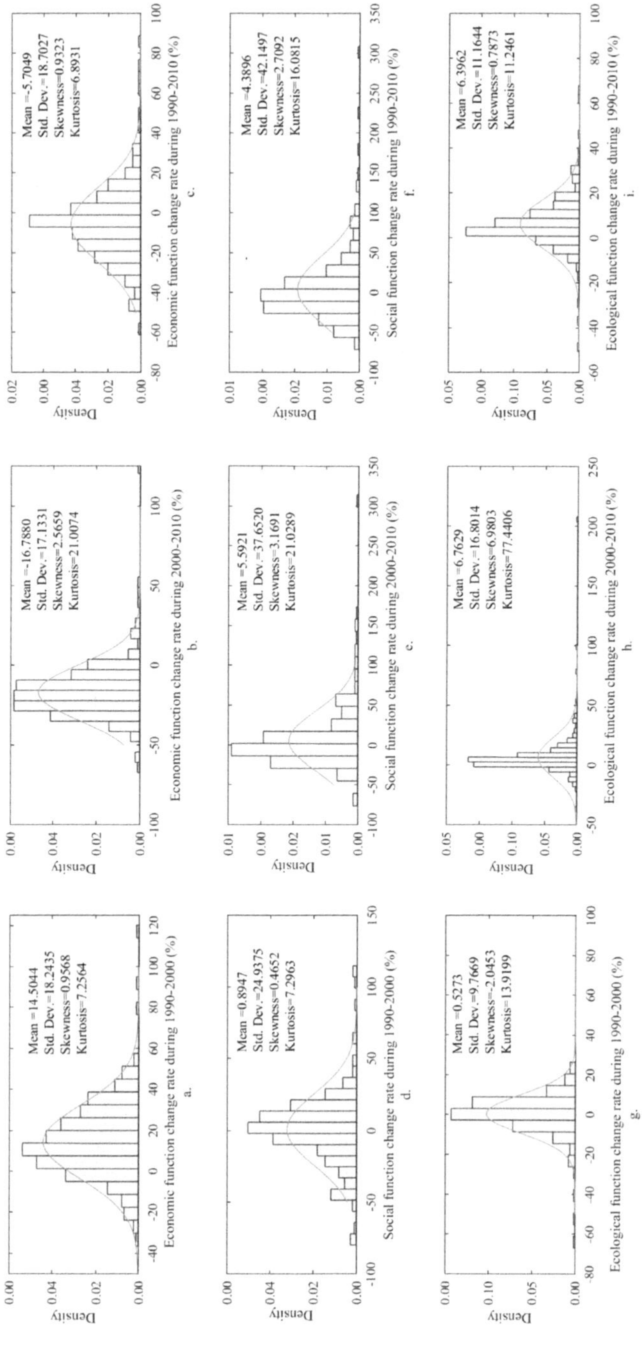

Fig. 2.34 The density graphic of the change rate of FF in HHHP (Zhang et al. 2018)

Spatio-temporal patterns of farmland social function evolution: 51.18% of the counties experienced a clear decrease in social function, which was mainly distributed around the north of Anhui Province and the Beijing-Tianjin metropolitan area (Figs. 2.33d and 2.34d). The number of counties that grew was 83, which were mostly concentrated in the west of Cangzhou, Yantai, Weihai, Jining, Linyi, and Weifang. The number of growing counties dropped by 10% during 2000–2010, while traditional growth areas, including Weifang, Yantai, and the north part of Zhengzhou, still maintained a rather strong growth trend. Some regions with extremely high growth rates contributed to the significant peak and fat tail of the density map. The spatial patterns of social function change during 1990–2010 (Fig. 2.33f) was roughly the same as that during 2000–2010; meanwhile, a small set of cities including Dezhou, Cangzhou, and Hengshui declined somewhat during 2000–2010. The social function index change rate for 89.12% of the counties ranged from −35 to 35%, and Skewness and Kurtosis fell compared to 2000–2010 (Fig. 2.34f). The developed rural areas surrounding Beijing and Tianjin were dominated by industry and commerce. The rapid development of rural industry and tourism, coupling with scarce farmland resources and the high proportion of non-agricultural employment, contributed to the continuous decline of employment carrying capacity. Rural areas with high population density, high population growth rate, and lagged non-agricultural industries underwent reductions in their social function indices due to urbanization and industrialization, compared to high baseline index values. Where there were low levels of agricultural industrialization or imperfect industrialization, rural populations in Xinxiang, Jiaozuo, Hebi, Anyang, Zhoukou, and Zhumadian in Henan Province were still highly dependent on farmland. Yantai, Weihai, and the adjacent hilly areas in the east of the Shandong Peninsula were characterized by large farming areas per capita and high proportions of cash crop. Relatively-developed modern agriculture, urban agriculture, and environmental agriculture jointly affected the employment choices of the rural population in these areas to a great degree.

Spatio-temporal patterns of ecological function evolution: ecological function evolution was characterized by a larger fluctuation and clear differences in spatial patterns between the different periods. Of all the counties, 47.35% experienced ecological function declines, especially around Beijing-Tianjin, the east of Shandong province, and the north of Anhui Province (Figs. 2.33g and 2.34g). Growth areas were mainly concentrated in the north and south of Henan Province, and the east and west of Shandong Province. The sharp increase in farmland area of most counties in Yantai, Weihai, and Qingdao led to farmland playing a significant role in environmental conservation. Abundant farmland and lower fertilizer and pesticide inputs jointly contributed to the increase in ecological function. Ecological function rates of change were between −50 and 50% during 2000–2010 (Figs. 2.33h and 2.34h), while the Kurtosis reached 77.44 due to outliers, which caused the high peak and fat tail on the density map. During this period, the growth areas were mainly distributed in the west of Hebei Province, the north of Henan Province, and the north of Anhui Province. During 1990–2010, the core areas of growth centered around the north of Henan Province and the middle of Shandong Province, and areas around Beijing, Tianjin, and Jinan were the primary decreasing areas.

2.6.3.2 The Spatio-Temporal Characteristics of the Coupling Effect Among Economic, Social, and Ecological Function

The "economy-society-ecology" composite system of farmland is open, dynamic, and complex, and if one component is out of order it would result in the decline of system coordination and cause deviation from the given targets. (1) Economic function was closely associated with natural property and human management activities. Productivity can be modified by the input of various elements and changes in utilization. As a basic carrier and spatial limit to urbanization and industrialization, land resources de-agriculturalization triggered by the increasing objective demand, which furtherly causing the decline of employment carrying capacity and ecological protection function. Driven by the dual goals of food security and farmer income and production, pesticides, fertilizers, and agricultural machinery certainly will increase if agricultural technology is improved. Moderate pesticide, fertilizer, and other yield-increasing inputs will either enhance the productivity of farmland or cause farmland ecosystem degradation. Increasing labor-saving inputs directly enhance productivity and free more farmland labor, as well eliminate traditional, labor intensive agriculture and weaken the labor absorption capacity. (2) Farming is the last source of peasant employment and it is farmers who realize the economic value of farmland. Modern agricultural practices, such as urban and ecological agriculture, lead to localized employment and directly motivate efficient utilization and intensive management. Meanwhile, if farmland carried excessive population, it would initiate agricultural involution, which is against with the promotion of agriculture productivity and rural economic development. Similarly, the dual effects of population growth and farmland decline jointly exert pressure on existing farmland. (3) A favourable farmland environment is the basic condition and support for maximizing its economic and social function. On the contrary, environmental damage such as soil salinization, soil erosion, heavy metal pollution, and agricultural non-point source pollution will constrain production and the social function.

The synergistic evolution of population, land use, and industry inevitably coincides with changing interactions among farmland economic, social, and ecological sub-systems (Fig. 2.35). The results from the coupling degree model showed that there was clear spatial variation in coupling degree (Fig. 2.36a, b, c).

(1) Areas with a coupling degree greater than 0.6 were in high numbers, widely distributed, and concentrated. Limited by urbanization and industrialization, traditional agriculture contributed less to national economic growth and rural household income. The nearly equivalent de-agriculturalization rates of land, population, and industry made a smaller difference among farmland sub-function indices for most counties across different periods, thus creating the high coupling degree. (2) Areas with a low coupling index were mainly concentrated around Beijing, Tianjin, Jinan, Yantai, and Weihai. Compared to 1990, the degree of coupling in 2000 and 2010 was relatively lower. The decline of primary industry and the rise of industry and commerce jointly triggered the transition in employment structure. Farmland sub-functions diverged as farmland transformed to

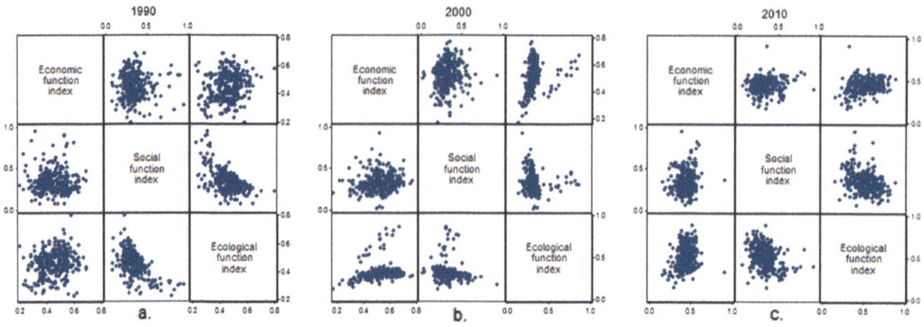

Fig. 2.35 The matrix scatter diagram of farmland economic, social, and ecological functions indices in HHHP in 1990, 2000 and 2010 (Zhang et al. 2018)

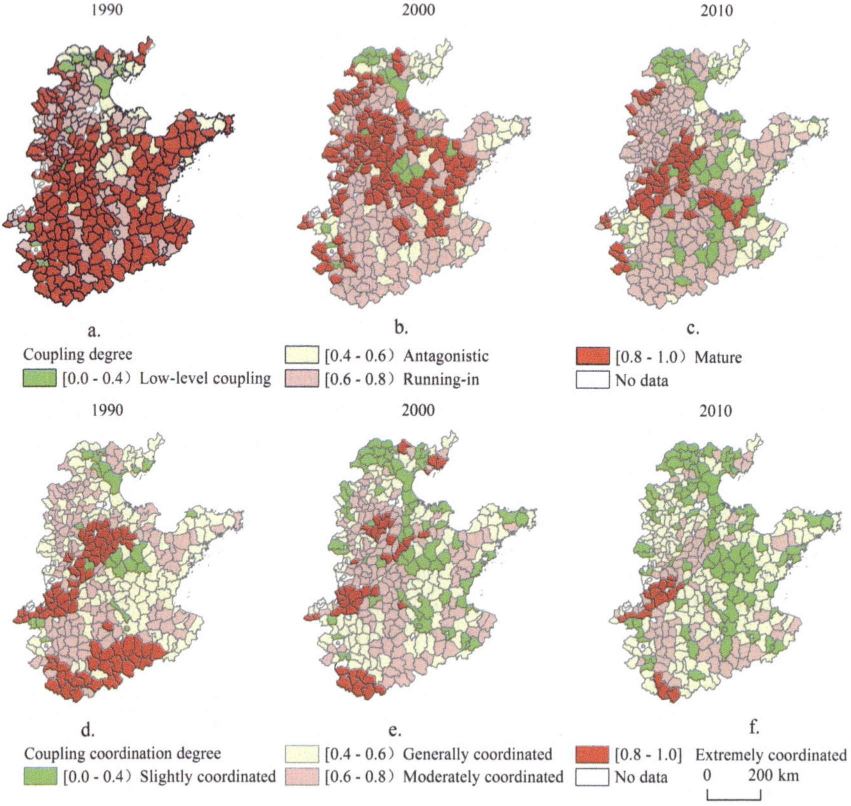

Fig. 2.36 The spatial distribution of the coupling and coordination degrees for farmland economic, social, and ecological functions in HHHP in 1990, 2000, and 2010 (Zhang et al. 2018)

developed land, which caused the coupling degree to decrease. (3) From the perspective of temporal evolution, the coupling degree of the sub-functions had a general downward trend, especially in the areas surrounding Beijing and Tianjin. Areas with high coupling values shrank as a result of the unbalanced development of farmland sub-functions.

In order to conduct further research on the coupling and coordination relationships between the three sub-functions mentioned above, a coupling coordination model was adopted to comprehensively analyze the coordination level. The results showed that: (1) Compared to coupling degree, areas with corresponding value intervals shrank dramatically and the low value zones expanded. (2) Counties with a high coupling coordination degree were mainly located in Dezhou, Hengshui, the south of Henan Province, the north of Anhui Province, and other undeveloped areas dominated by traditional agriculture. In general, most counties in the Shandong peninsula underwent sustainable economic, societal, and ecological development through the transformation and upgrade of traditional agriculture, forming an advanced coupling morphology rather than underdeveloped areas with high dependence on farmland for socio-economic development. (3) When it comes to dynamic evolution, initial areas with high values gradually shrank and areas with low values expanded around the original center. The farmland sub-functions of economy, society, and ecology gradually were differentiated by de-agriculturalization of population, land and industry. Abundant farmland resources and agricultural modernization enhanced economic function. The migration of surplus rural labor and the decline in people engaging in agriculture due to urbanization brought the farmland socio-economic function index down, and resulted in the decline of coupling coordination degree.

2.6.4 Dynamic Mechanisms

The economic, social, and ecological function indices, along with the explanatory variables, were imported into GeoDa and analyzed using Ordinary Least Squares (OLS), SLM, and SEM to explore the driving factors of FFE. Model selection was based on the results of a spatial correlation test. If Moran's I and LR-LM Lag were significant, the LR-LM Error was not significant, and the LM-Lag was larger than the LM-Error, SLM was chosen; otherwise, SEM was chosen. Log L likelihood (Log L), Likelihood Ratio (LR), the Akaike information criterion (AIC), and the Schwartz Criterion (SC) are usually taken into consideration in the process of model selection. The larger the Log L is, and the smaller the AIC and SC are, the better the fit (Liu and Long 2016). These indicators were also used to compare OLS, SLM, and SEM, and the largest Log L represents the most appropriate choice. According to the results (Table 2.11), SEM was more suitable for explaining economic and ecological function evolution, and SLM was more appropriate for social function evolution.

The spatial econometric analysis results at the county level indicated that regional indigenous factors and external driving factors jointly determined the direction and

Table 2.11 Results of spatial correlation tests for FFE in HHHP (Zhang et al. 2018)

Spatial dependence test	Economic function	Social function	Ecological function
Moran's I (error)	9.8575***	6.7193***	9.2617***
Lagrange Multiplier (lag)	38.4304***	28.1586***	19.1501***
Robust LM (lag)	1.7955*	5.3022**	2.3325
Lagrange Multiplier (error)	54.1750***	23.0057***	47.2461***
Robust LM (error)	17.5401***	0.1493	30.4286***
Lagrange Multiplier (SARMA)	55.9706***	28.3079***	49.5786***

Note *: Significant correlation at the 10% level; **: Significant correlation at the 5% level; ***: Significant correlation at the 1% level

extent of FFE. The specific dynamic mechanisms of economic, social, and ecological function evolution are as follows (Table 2.12).

Economic function had a clear positive correlation with traffic accessibility, the number of industrial enterprises per land area, percentage of industrial added value in GDP, and urbanization influence index at a significance level of 1%, as well as average annual precipitation at a significance level of 5%. Additionally, it had a negative correlation with elevation at a significance level of 10%, and there was no significant correlation with distance from major rivers, proportion of rural population, and education level. (1) Compared to the plains areas, hilly areas were characterized by scarce and scattered farmland, high levels of difficulty for automated cultivation, and expensive irrigation costs, which severely restricted agricultural production and management and reduced the efficiency of farmland utilization. For instance, Fangshan and Shunyi counties, which are located around Beijing, developed a prosperous valley economy, and planting became a minor choice of farmers. Specialized tourism and other non-agricultural industries relied on the location advantage of metropolitan suburbs and specialized resources of mountain areas, becoming the main source of farmers' income. (2) Theoretically,precipitation is correlated positively with farmland production capacity, but the average annual precipitation showed a decreasing trend from south to north, and inland from the coast. Due to the export-oriented economy, coastal areas had higher levels of urbanization and industrialization, which caused agriculture to weaken in the national and rural household economies. (3) With the implementation of high standard basic farmland construction, distance from major rivers was no longer a constraint for field irrigation due to the improvement

Table 2.12 The spatial regression analysis results for FFE and influential factors in HHHP (Zhang et al. 2018)

Factors	Economic function			Social function			Ecological function		
	OLS	SLM	SEM	OLS	SLM	SEM	OLS	SLM	SEM
Elevation	– 0.13**	– 0.18***	– 0.31*	0.13	– 0.06	– 0.05	– 0.49***	– 0.42***	– 0.58***
Distance from major rivers	0.04	0.03	– 0.23	– 0.06	– 0.01	0.11	0.25***	0.12**	0.06**
Precipitation	0.07**	0.05**	0.27**	0.07	0.05	0.30	0.006**	0.01**	0.20**
Traffic accessibility	– 0.13***	– 0.13***	– 0.13***	– 0.05**	– 0.06**	– 0.06**	–	–	–
The number of industrial enterprises per land area	– 0.14**	– 0.07*	– 0.13***	– 0.19**	– 0.13**	0.08***	– 0.07**	– 0.10**	– 0.08***
Percentage of industrial added value in GDP	– 0.22**	– 0.14*	– 0.12***	– 0.09***	– 0.11***	– 0.13***	– 0.24***	– 0.21**	– 0.20***
Proportion of rural population	0.15	0.14	0.17	0.20***	0.19***	0.20***	– 0.04**	– 0.03**	– 0.02**
Education level	0.002	0.001	0.001	– 0.003***	– 0.002***	– 0.002***	– 0.001	– 0.01	– 0.01
Urbanization influence index	– 0.15***	– 0.119***	– 0.11***	– 0.076**	– 0.06**	– 0.06***	– 0.07***	– 0.06**	– 0.09**
W-Y		0.71***			0.67***			0.61***	
Lambda			0.88***			0.81***			0.84***
R^2	0.19	0.33	0.41	0.23	0.34	0.36	0.42	0.49	0.55
Log-L	210.86	233.29	245.89	258.16	275.32	277.62	241.30	255.81	267.81
AIC	– 401.71	– 444.58	– 471.78	– 496.32	– 528.64	– 535.23	– 462.61	– 489.62	– 515.62
SC	– 365.16	– 404.37	– 435.22	– 459.76	– 488.43	– 498.67	– 426.05	– 449.40	– 479.06

Note *: Significant correlation at the 10% level; **: Significant correlation at the 5% level; ***: Significant correlation at the 1% level in irrigation facilities. (4) Favourable location nearby regional central cities was a prerequisite for economic growth. The boom in non-agricultural industries was motivated by the transfer of capital, information, technology and, talents from metropolitan regions to the surrounding areas. (5) Impacts of industrialization imposed on economic function were summarized into two points. On the one hand, the rapid increase in the proportion of industrial outputs in the GDP reduced the proportion of agriculture outputs. On the other hand, so much high quality farmland was occupied that the economic output declined. (6) The increasing frequency of urban and rural interactions, the accelerated flow of economic elements, and the rapid development of non-agricultural industries were attributed to changes in economic aggregation and structure, and manifested as an increase in urbanization. The increasing demand of residential land, public services, and infrastructure induced by population growth in urban areas, resulted in the loss of farmland and a decline in economic function. It should be noted that economic function of areas oriented around traditional agriculture were enhanced because of the promotion of agricultural industrialization and modernization, but across the entire HHHP it generally declined.

Social function evolution did not have a significant correlation with natural factors, but it was negatively correlated with traffic accessibility, the number of industrial enterprises per land area, and the urbanization influence index at a significance level of 5%, while percentage of industrial added value in GDP and education level were correlated at a significance level of 1%. Additionally, social function had a strong positive correlation with proportion of rural population at a significance level of 1%. (1) Well-equipped transportation networks in favour of rural-urban migration, promoted part-time farming and employment de-agriculturalization of surplus rural labor. (2) In the process of industrialization and urbanization, working in cites turns to be the best choice for rural middle-aged labor driven by increasing industry and demand. Moreover, the increasing number of land-deprived residents who were compelled to leave agriculture became the primary reason for the weakening of farmland social function. (3) Farmland management was still the main way of maintaining minimum subsistence levels and resisting social risks due to the large and densely distributed rural population in HHHP. Human resources, whose employment capacity was restrained by education level, were the most dynamic factor in rural development. Well-educated farmers may more easily adapt to multiple employment patterns and reduce their dependence on farmland.

Ecological function evolution had a strong negative correlation with elevation, number of industrial enterprises per land area, and percentage of industrial added value in GDP at a significance level of 1%, and proportion of rural population and the urbanization influence index were significant at the 5% level, but there was a positive correlation with distance from main rivers and average annual precipitation at a significance level of 5%. (1) Scarce farmland, together with land development in hilly areas resulted in the decline of the proportion of ecological farmland. (2) Extensive economic growth, excessive resource exploitation, and poor management jointly caused disorderly development and soil and water pollution, which further led to the sharp drop in total farmland and the degeneration of soil quality. High quality farmland remained occupied but inferior farmland underwent environment degeneration, in spite of the implementation of farmland protection policies. Meanwhile, farmland ecosystems were further damaged by excessive fertilizer and pesticide inputs to increase grain yield, driven by growing demand. (3) Shrinking areas and an increasing rural population imposed more pressure on farmland.

2.6.5 Discussion

All sub-functions of farmland gradually declined due to urbanization and industrialization in the HHHP, while the value they possess was continuously improved in virtue of resource scarcity. Hence, a comprehensive understanding of FFE and the declining trend of some sub-functions is critical to achieving modern agriculture and upgrading traditional agriculture, as well as promoting endogenous development and increasing the efficiency of agricultural production.

Multifunctional farmland use and management should be aimed at people's diverse demand at given socio-economic development context. Dominant function transfer from economic and social oriented at a low socio-economic level to ecological and cultural oriented, while other derivative functions appear in company with the flourished non-agricultural industries. The decline of economic function is mainly attributed to construction land encroachment and the decreased status of agriculture in both national and household economy. More importantly, countermeasures such as improving basic mechanisms of land supply and planning, landscape design, and ecological design should be implemented to enhance farmland ecological function.

In view of the existing obvious regional difference of FFE, distinguished and diversified multifunctional farmland use and management policies are needed to be proposed for optimizing the layout of farmland functions. Over emphasizing the requisition-compensation balance will result in superficial and shoddy balance of farmland area. In order to internalize the cost of farmland loss and enhance farmland ecological protection, it is necessary to develop ecological agriculture from the perspective of regional eco-environmental security. As for underdeveloped rural areas, advocating modern agriculture is an appropriate choice for promoting economic benefits and enhancing the capacity of labor absorption.

Limited by the complexity of farmland system, there is still not a generalized and perfect FF classification system. A widely accepted classification system, which can provide the theoretical basis for FF assessment, urgently needs to be established. Current research on FFE mainly focuses on the macro scale rather than micro scale, such as the village and household. Therefore, studies of FFE from the micro perspective will be the emphasis of future research.

2.7 Farmland Use Transition in China: An Integrated Perspective

As the Chinese economy is currently undergoing a transition from high-speed to medium-high speed development, China's economic development has entered a transition period featuring structural adjustment, layout optimization, and quality and efficiency improvement. Due to the increased pressure on economic growth and the need for comprehensive reform, there are urgent requirements for scientific research into rural development and land use, and it has also become imperative to adjust land use structure and function (Liu et al. 2018). As one of the major factors influencing rural development, Farmland use transition (FUT) is an important indicator reflecting rural changes, and in-depth research has been performed from the perspectives of the conceptual connotation, method innovation, and influencing factors (Lambin and Meyfroidt 2010; Long and Qu 2018).

As the largest developing country in the world, China has the largest population in the world. By the end of 2017, the area of farmland in China was 134.9 million hectares, ranking third in the world, but the area per capita is only one-third of the world average, that is, China feeds nearly 20% of the global population with 7% of the world's farmland. Certainly, faced with the problem of the shortage of farmland resources, how to rationally develop, utilize and manage farmland land resources has become the focus and challenge of management decision-makers (Wu et al. 2018). However, under the backdrop of socio-economic transition, the question of differences in the degree of farmland transition in different regions arises, as well as how and if people's demand for farmland function changed in different periods, and what factors might drive the demand. Additionally, methods to improve farmland utilization efficiency in the future should be understood. Answering these questions may provide an approach to address the root.

Focusing on the above research gaps, this study establishes an evaluation index system for quantifying FUT in China in the two dimensions of dominant and recessive transitions. Then, K-means clustering was used to quantitatively divide the phases of FUT in China over the past four decades, and the phase characteristics and driving factors of FUT were summarized. Finally, some policy suggestions were proposed to promote the optimal regulation and management of farmland resources, targeting providing a scientific basis for ensuring national food security, promoting efficient resource allocation, alleviating the tense man-land interrelations, achieving rural revitalization, and enhancing human well-being.

2.7.1 Data and Methodology

2.7.1.1 Data Sources

To highlight the FUT characteristics over a long time, the relevant data of FUT phases are selected from the national level data from 1978 to 2017. At the beginning of reform and opening up, some indicators from prefecture-level data are difficult to obtain. Therefore, the relevant data on the spatio-temporal evolution of China's FUT are selected from the city-level data of 1990, 2000, 2010 and 2017. Socio-economic data comes from China Economic and Social Big Data Research Platform (http://data.cnki.net), the Statistical yearbook of Chinese cities, and China's Land and Resources Bulletin. In the process of data handling, some missing data were interpolated by using the provincial statistical yearbooks and the national economic and social statistical bulletins of the sub-districts. Finally, 322 valid statistical units were obtained. To eliminate the impact of price factors on the analysis of economic functions of farmland, the economic data in this study were translated into 1990 comparable prices.

2.7.1.2 Methods

(1) **Evaluation system**

Following the principles of comprehensiveness, representativeness and accessibility, and based on the method of combining literature and expert knowledge, the indicators were selected from the two dimensions of dominant and recessive morphology transition of farmland. A comprehensive evaluation index system consisting of five factors layers and twenty-two individual indicators concerning the FUT was initially constructed. Based on the preliminary establishment of the index system, sixteen indices available for data were selected. Linear regression was used to perform the collinearity test for all indices, and indices with a variance inflation factor value greater than 10 were eliminated. Pearson correlation analysis was further conducted for a single index of the same factor layer in pairs, retaining only one index with a significant correlation, and 11 quantifiable indices were finally obtained (Table 2.13).

The dominant farmland transition was characterized by the land reclamation index and farmland change rate, indicating the transition of the structure and quantity of farmland in one phase. The recessive morphology of farmland has multiple attributes, including the quality of farmland, property rights, operation mode, input and output, and efficiency and benefit (Qu and Long 2018). It is difficult to quantify the property right and operation mode of farmland, however, all those transition changes will have an impact on the input and output of farmland. Therefore, this study mainly evaluates the recessive FUT from the input, output, and function of farmland (Li and Li 2019). According to the multi-functional theory of farmland, the recessive transition of farmland is divided into social, economic and ecological functions, and further quantitative analysis is performed.

Table 2.13 Evaluation index system of FUT (Ma et al. 2020)

Decision layer	Target layer	Factor layer	Indicator	Indicator interpretation	Attributes	Weight*
Index of FUT	Dominant transition (Structure) 0.36	Farmland structure	Land reclamation index	farmland area/Total land area	+	0.47
		Amount of farmland	Farmland change rate	$(T_2\text{-}T_1)/T_1^{**}$	+	0.53
	Recessive transition (Function) 0.64	Social function	Total power of agricultural machinery per unit area	total power of agricultural machinery/farmland area	+	0.50
			Per capita farmland area	farmland area/Agricultural practitioners	+	0.34
			Labor transfer index	Non-agricultural population/Total population	+	0.15
		Economic function	Farmland productivity	Added value of primary industry/farmland area	+	0.30
			Farmland grain productivity	grain yield/farmland area	+	0.44
			Agricultural output contribution	Added value of primary industry/GDP	+	0.25
		Ecological function	Agricultural environmental impact index	Consumption of chemical fertilizers/farmland area	–	0.45
			Multiple cropping index	Total sown area/farmland area	–	0.34
			Farmland ecosystem diversity index***	Characterizing the degree of biodiversity of ecosystems	+	0.21

Note *A method combining entropy weighting and multiple correlation coefficient weighting was used to determine indices weights (Luo and Cai 2016). **, T_1 represents the area of farmland at the beginning of the study, and T_2 represents the area of farmland in the next period. ***, Farmland ecosystem diversity index (FED), $FED = -\Sigma b_i \ln b_i$, b_i is the ratio of the planting area of each variety of crops to the total planting area of crops

(2) **Evaluation model**

The degree of FUT is evaluated by using the multi-index method. The dominant transition index (*DTI*), social transition index (*STI*), economic transition index (*ETI*), ecological transition index (*ECTI*) and comprehensive transition index (*CTI*) of farmland are calculated as follows:

$$DTI = \frac{sep_d^-}{sep_d^+ + sep_d^-} \tag{2.24}$$

$$STI = \frac{sep_s^-}{sep_s^+ + sep_s^-} \tag{2.25}$$

$$ETI = \frac{sep_e^-}{sep_e^+ + sep_e^-} \tag{2.26}$$

$$ECTI = \frac{sep_{ec}^-}{sep_{ec}^+ + sep_{ec}^-} \tag{2.27}$$

$$CTI = w_d \times DTI + w_s \times STI + w_e \times ETI + w_{ec} \times ECTI \tag{2.28}$$

where, sep_d^+, sep_s^+, sep_e^+, sep_{ec}^+ and sep_d^-, sep_s^-, sep_e^-, and sep_{ec}^- represent, respectively the Euclidean distances of the solutions with the optimal solution and the worst solution, which are obtained by the TOPSIS model. The specific formula refers to Ma et al. (2019). The w_d, w_s, w_e, w_{ec} represent, respectively, the weights of dominant, social, economic, and ecological transition. The higher the values of *DTI*, *STI*, *ETI*, *ECTI*, and *CTI* are, the higher the degree of FUT is, and the interpretation and characteristics of each indicator are as follows (Table 2.14).

(3) **Quantitative method for dividing FUT phases**

Based on the evaluation index system of FUT in Table 1 and the SPSS software analysis platform, K-means clustering is used to divide the phases of FUT. The K-means clustering algorithm is a common clustering analysis method, and has the advantage of being able to test the pre-given number of categories to select the most reasonable number of categories. The main steps are as follows: First, select K objects in the data space as the initial center, where each object represents a cluster center. Second, for the data objects in the sample, according to their Euclidean distance from the cluster centers, they are divided into classes corresponding to the nearest cluster centers (the most similar) according to the nearest distance criterion. Then, update the clustering center by taking the mean of all objects in each category as the clustering center of this class, and calculating the value of the target function. Finally, determine whether the values of the cluster center and the objective function change. If they do not change, output the results; if they change, return to step two.

Table 2.14 Classification and characteristics of FUT (Ma et al. 2020)

Type	Explanation	Feature
Dominant transition index (*DTI*)	Used to describe the quantity and structural changes of farmland	The higher the degree of dominant transition, the more the number of farmland increases in this phase; and vice versa
Social transition index (*STI*)	Used to describe the social function transition of farmland	The higher the degree of social transition, the more social function of farmland increases in this phase; and vice versa
Economic transition index (*ETI*)	Used to describe the economic function transition of farmland	The higher the degree of economic transition, the more economic function of farmland increases in this phase; and vice versa
Ecological transition index (*ECTI*)	Used to describe the ecological function transition of farmland	The higher the degree of ecological transition, the stronger the environmental impact of farmland is in this phase, the higher the possibility of polluted and degraded farmland occurs, and the more biodiversity of the farmland ecosystem occurs; and vice versa
Comprehensive transition index (*CTI*)	Used to describe the comprehensive degree of dominant transition and recessive transition of farmland	The higher the degree of comprehensive transition, the more reasonable and positive the "structure-function" of farmland use is in this phase; and vice versa

2.7.2 *Spatio-Temporal Characteristics*

2.7.2.1 Spatio-Temporal Patterns of Dominant Transition of Farmland

In the past three decades, the farmland area in China has shown a trend of first rapid decrease followed by a slow increase (Fig. 2.37). In the first period, there were 144 regions with reduced farmland area, accounting for 45% of the total. In the second period, there were 161 regions with reduced farmland area, accounting for 50.31%, and in the third period, there were 83 regions with reduced farmland area, accounting for 25.94%. During the entire study period (1990–2017), there were 58 regions with a reduced farmland area, accounting for 18.13%, which indicated that the farmland area in most prefecture-level cities has experienced a process of decreasing first and then increasing slowly.

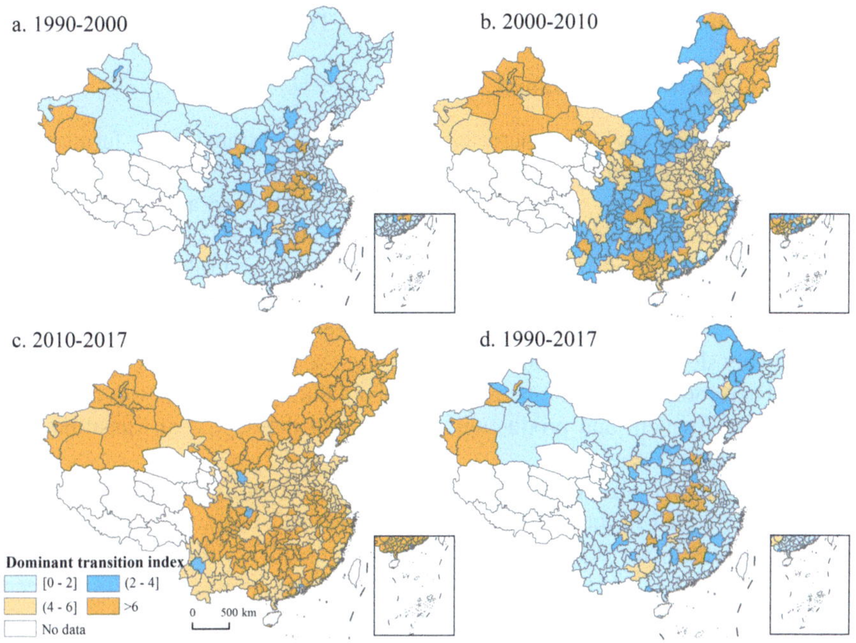

Fig. 2.37 Spatio-temporal patterns of dominant transition of farmland in China during 1990–2017 (Ma et al. 2020)

The spatio-temporal evolution characteristics of the dominant farmland transition in different periods were analyzed and it is found that in the first period, the *DTI* of farmland was generally low, and 86.02% of the research units their transition indices were between 0–2. The regions with relatively high transition index were mainly located in central China, including Henan, Hunan, and other provinces and cities. In the second period, the dominant farmland transition was more active, and the *DTI* of farmland in the eastern coastal areas and Xinjiang were higher than 4. The central regions along the "Hu Line" were relatively low in transition, and the *DTI* was between 2–4, accounting for 33.23%, and the farmland area in this type of region decreased significantly. The third period was the most active phase of dominant transition of farmland. The regions with DTI of farmland between 6–10 accounted for 46.89%; and 51.24% regions were higher than 10, mainly distributed along the Yangtze River and northern China, including Xinjiang, Inner Mongolia, and Heilongjiang provinces. The farmland area and land reclamation index showed a slow upward trend during this period.

In general, the *DTI* of farmland in China was relatively small during 1990–2017, and the farmland area and land reclamation index showed an overall upward trend. The transformation hotspots were mainly concentrated in Central China, including Henan, Hunan provinces and some cities of other provinces.

2.7.2.2 Spatio-Temporal Patterns of Recessive Transition of Farmland

(1) Social transition of farmland

The spatio-temporal evolution characteristics of the social transition of farmland in different periods were analyzed and found that (Fig. 2.38).

In the first period, the *STI* of farmland was generally low, and 92.86% of the research units their social transition indices were between 0–3. The regions with relatively high transition index were mainly located in Central China, including Nanyang, Zhoukou, Fuyang, and Liu'an City. During this period, China's urbanization level was still underdeveloped, the transfer of the agricultural labor force was relatively slow, and farmland transfer and large-scale management were in their infancy. Therefore, the social function of farmland during this phase was still dominated by employment and production security, and the degree of transition was relatively small.

In the second period, the regions with active social transition of farmland were mainly concentrated in the southeast coastal areas of China, such as Yunnan and Guizhou. Among them, there were 97 prefecture-level cities with an *STI* between 3–6, accounting for 30.12%; there were 173 prefecture-level cities with an *STI* between 6–10, accounting for 53.73%; and 45 prefecture-level cities with an *STI* higher than 10, accounting for 13.98%.

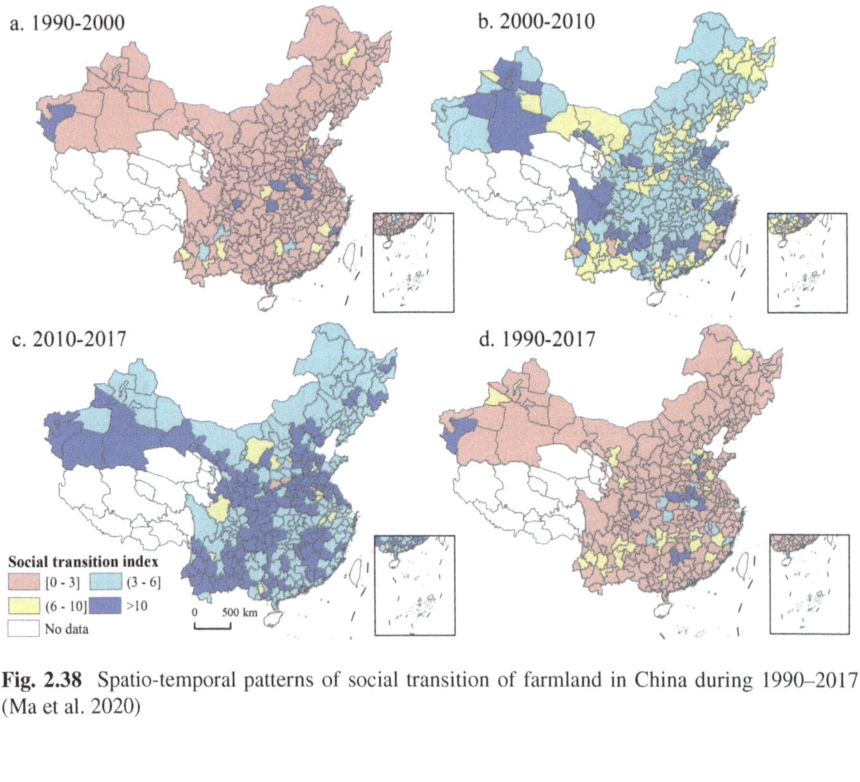

Fig. 2.38 Spatio-temporal patterns of social transition of farmland in China during 1990–2017 (Ma et al. 2020)

The third period was the most active phase of the social transition of farmland. The STI values of farmland in 40.99% of the regions were higher than 10, mainly distributed on both sides of the "Hu Line", including Hebei, Henan, Shandong, Hunan, Jiangxi, Yunnan, Chongqing, Sichuan, and Xinjiang. Further analysis of the driving factors found that on the one hand, with the socio-economic development and scientific and technological progress, the level of agricultural mechanization improved, which has a substitution effect on the agricultural labor force. On the other hand, with the increase in urban employment, high-paying employment opportunities attract non-agricultural transfer of agricultural labor. During this period, the social function of farmland gradually shifted from agricultural production and employment functions to diversification.

(2) **Economic transition of farmland**

Compared with the social transition of farmland, the economic transition of farmland in China occurs earlier and more significantly (Fig. 2.39).

In the first period, the regions with a high degree of economic transition of farmland were mainly distributed in the eastern coastal areas and Guangxi, Yunnan, and Guizhou provinces and some cities in other provinces. Among them, there were 13 regions with farmland *ETI* values above 10, accounting for 4.04%, and 189 regions with *ETI* values between 3–6, accounting for 58.70%. By analyzing the main driving factors, we found that those regions have good resource endowment and the economic

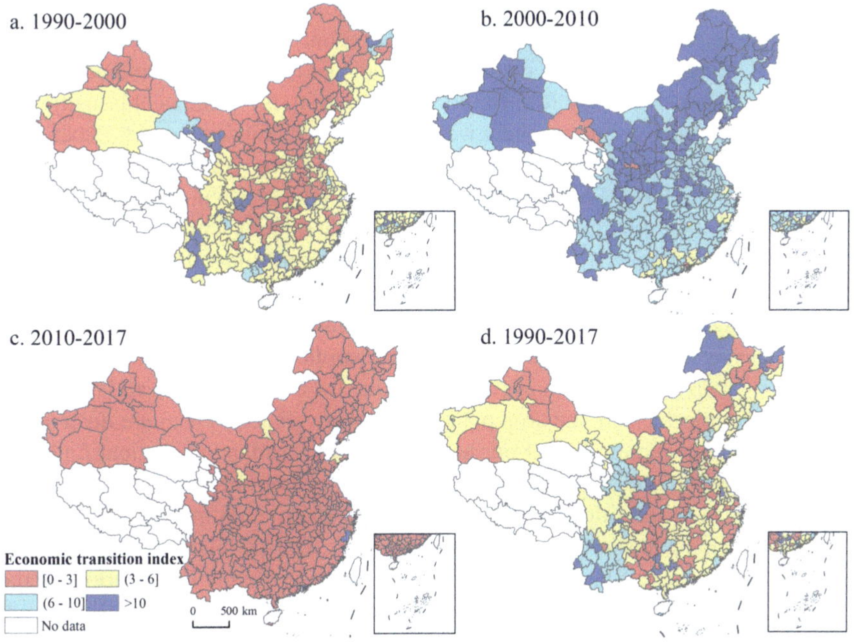

Fig. 2.39 Spatio-temporal patterns of economic transition of farmland in China during 1990–2017 (Ma et al. 2020)

benefit of farmland increased rapidly. The second period is the most active stage of the economic transition of farmland. There were 190 prefecture-level cities with an *ETI* between 6–10, accounting for 59.01%, mainly distributed in the south eastern part of China. One hundred and nine prefecture-level cities with a farmland *ETI* higher than 10, account for 33.85%, mainly distributed in the prefecture-level cities on both sides of the "Hu Line", including Heilongjiang, Inner Mongolia, Shanxi, Shaanxi, Hubei and Yunnan provinces. During this period, China's socio-economic development was rapid, the industrial structure transition was obvious, and under the influence of economic globalization and international trade, the economic function of farmland was gradually transformed from simple food supply to multiple forms of economic production functions, including the production of grain, economic crops planting, and flower and seedling management. The third period was a stable development period after the economic function transition of farmland in China. The *ETI* was generally low, and the regions with an *ETI* of less than 3 accounted for 97.20%. In this period, other derivative functions of farmland have received more attention.

In the whole phase, the economic transition of farmland was not obvious, but the structure and connotation of the economic function of farmland have undergone an essential transition.

(3) **Ecological transition of farmland**

The frequency of the transition of the ecological function of farmland was consistent, showing a trend characteristic of slow to rapid transition (Fig. 2.40). The higher the *ECTI* of farmland, the greater the impact on the farmland ecosystem.

In the first period, due to the relatively small amount of chemical fertilizer applied, the crop planting structure was relatively simple, and the trend of farmland ecological transition was relatively low. In the second period, the ecological transition of farmland was relatively active. There were 163 research units with an *ECTI* between 3–6, accounting for 50.62%, mainly distributed in eastern coastal China, including Beijing, Tianjin, Hebei, Shandong, Guangxi, Guangdong, Zhejiang, Shanghai, and some cities in other provinces. There were 124 research units with an *ECTI* between 6–10, accounting for 38.51%, mainly distributed in Inner Mongolia, Shaanxi, Hubei, Hunan, Sichuan, and Yunnan provinces. During this phase, the increase in fertilizer application significantly impacted the transition of farmland ecological function. Statistical analysis showed that the average fertilizer application in China increased from 0.46 t/ha in 2000 to 0.67 t/ha in 2010. In addition, due to the influence and regulation of the market mechanism, the crop planting structure tended to be more diversified, which also had a certain impact on the transition of farmland ecological function. In the third period, the ecological function of farmland tended to be stable, and the regions with high transition degrees were mainly distributed in Yunnan, Shaanxi, Gansu, Xinjiang, and some cities in other provinces. During this phase, the ecological security function of farmland ecosystem has been gradually taken seriously, the amount of fertilizer utilization and multiple planting index were gradually reduced, and the implementation of the project of *returning farmland to forest and grassland* has also played a certain role in promoting the ecological function transition.

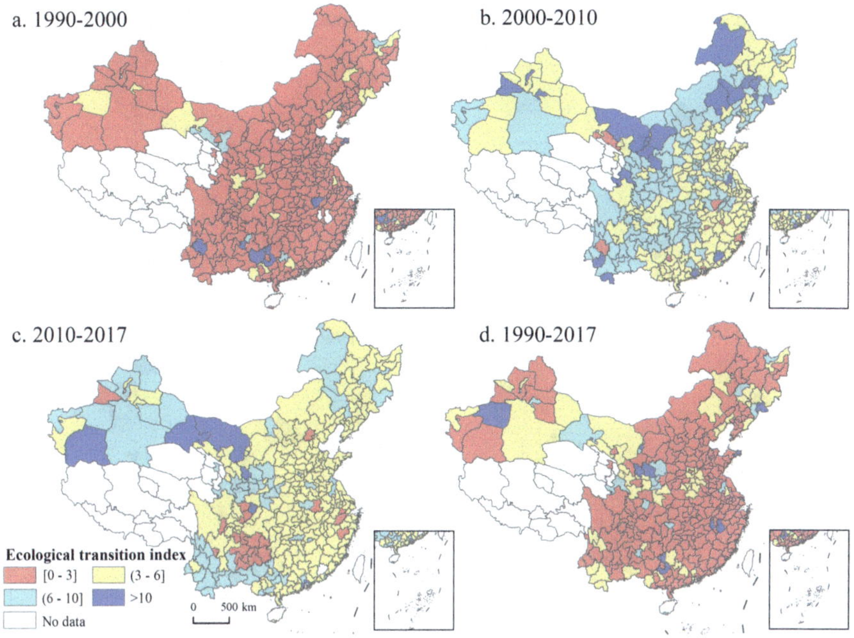

Fig. 2.40 Spatio-temporal patterns of ecological transition of farmland in China during 1990–2017 (Ma et al. 2020)

2.7.2.3 Spatio-Temporal Evolution of Comprehensive Transition of Farmland

In general, the comprehensive transition of China's farmland has experienced several periods from a slow transition to a rapid transition, and to gradual stabilization (Fig. 2.41).

In the first period, the *CTI* of farmland was relatively low. There were 277 research units with a *CTI* between 0–3, accounting for 86.02%, and only 15 prefecture-level cities had a *CTI* of farmland higher than 10, mainly distributed in Henan Province. During this phase, the economic transition of farmland was active, and the transition of the social and ecological functions of farmland was not obvious. In the second period, affected by the international trade environment and rapid urbanization, the comprehensive transition of farmland was relatively active. There were 72 prefecture-level cities with a *CTI* between 3–6, accounting for 22.36%, mainly distributed in Beijing, Tianjin, Shanghai, Chongqing, Guangdong, Fujian, Jiangsu and other relatively developed provinces and cities. There were 226 regions with a *CTI* between 6–10, accounting for 70.19%. The *CTI* values of farmland in some regions in Guangxi and Shaanxi were higher than 10. During this period, the "structure-function" of China's farmland has undergone a relatively intense transition. From 2010 to 2017, there were 36 regions with a *CTI* of higher than 10, accounting for 11.18%, mainly

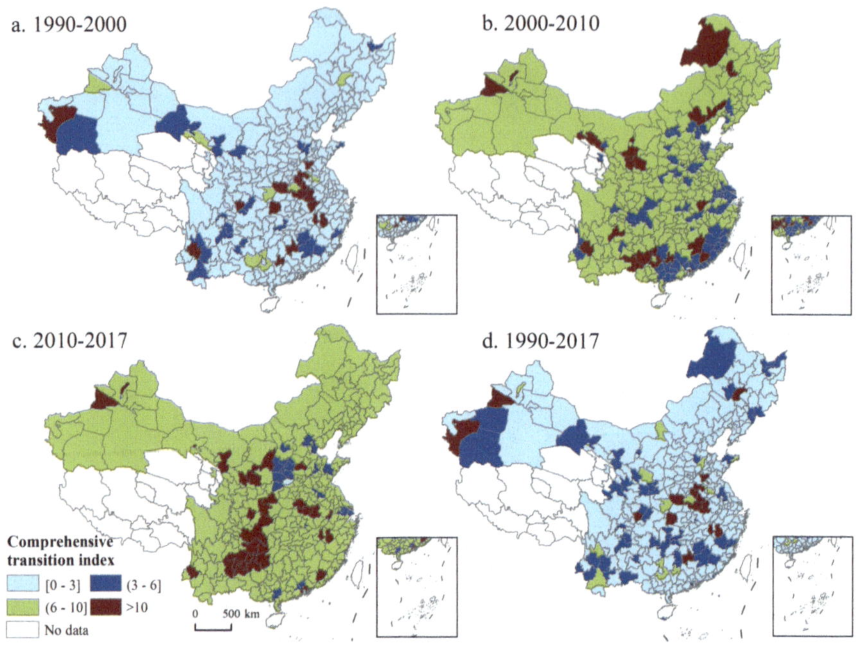

Fig. 2.41 Spatio-temporal patterns of comprehensive transition of farmland in China during 1990–2017 (Ma et al. 2020)

distributed in Guizhou, Chongqing, Henan, and some cities in other provinces. There were 265 prefecture-level cities with a *CTI* between 6–10, accounting for 82.30%. During this period, with the change in policy orientation and people's demand for farmland, the function of farmland utilization gradually shifted to multifunctional, among which the most obvious transitions were the social and ecological farmland functions.

2.7.3 Phase Division

2.7.3.1 Phase Division of FUT in China from the Perspective of "Structure-Function"

In this study, K-means clustering is used to divide the phases of FUT. The cluster analysis itself cannot remove the correlation between the indicators, while principal component analysis is a good supplement, and the principal component scores are consistent in size and there is no correlation, that can be directly used as the input for cluster analysis. Therefore, this study combines principal component analysis with the K-means clustering method to divide the phases of FUT.

Based on 11 secondary indicators in Table 2.13, principal component analysis was used for factor analysis. When the three common factors were extracted, KMO = 0.716 > 0.70, indicating that the factor analysis was effective. Table 2.15 shows that the eigenvalues of the first three principal components are greater than 1, and their cumulative contribution rate is 91.898%. That is, the three common factors integrate the information of the original 11 variable indicators, and the total variance of 91.898% can be explained by three potential factors.

As shown in Table 2.16, we can find that the common variance of most variables are above 0.7, and most of them exceed 0.9, indicating that the three common factors have good explanatory power and can reflect the information of the original variables well. When the factor loading is greater than 0.5, it indicates that the common factor has a good explanatory power for the corresponding index. The first common factor (Fac_1) mainly corresponds to five indices, X_1, X_3, X_5, X_6, and X_9. The second common factor (Fac_2) mainly corresponds to one index of X_{11}. The third common factor (Fac_3) mainly corresponds to X_7. According to the values of the three common factors mentioned above, the score values of the common factors are calculated. After testing, the score values of the three common factors are completely linearly independent, that is, the information represented does not overlap. Therefore, K-means clustering analysis can be performed based on the scores of these three common factors over the years, and the phases of China's FUT from 1978–2017 are divided. Through the comprehensive comparison test of different phase divisions, we found that the four-phase classification can ensure that the three common factor indicators pass the F-test, and their statistics are 202.466, 58.268, and 9.821, for Fac.1, Fac.2, and Fac.3, respectively. All the significance levels were 0 and passed the significance level test (Table 2.17).

The four decades from 1978 to 2017 were divided into four clusters in Table 2.18. In general, the four types of cluster classifications in each year have obvious and continuous clustering attribution. Considering the continuity and proximity of farmland in each phase, the cluster types of some clustering classification years with discrete characteristics were merged into the neighboring years before or after them. For example, the cluster of 2010 belongs to cluster four, but the neighboring years all belong to cluster two, thus 2010 is merged into the phase of cluster two. Finally, four phases of China's FUT in the past four decades were obtained: the first phase (1978–1984), the second phase (1985–1995), the third phase (1996–2002), and the fourth phase (2003–2017).

2.7.3.2 Evolution Characteristics of FUT Phases in China

Based on the quantitative division of China's FUT phase, combined with the relevant policies and systems of rural land use and internal and external environmental changes since the reform and opening up, this study summarizes the regular characteristics, leading functions and the impacts on rural development of each phase of FUT (Fig. 2.42).

Table 2.15 Common factor variance (Ma et al. 2020)

Component	Initial eigenvalue			Extract the sum of squares of loads			Sum of squares of rotating loads		
	Total	Percentage of variance	Cumulative percentage	Total	Percentage of variance	Cumulative percentage	Total	Percentage of variance	Cumulative percentage
X_1	6.515	59.224	59.224	6.515	59.224	59.224	6.221	56.559	56.559
X_2	2.556	23.232	82.456	2.556	23.232	82.456	2.67	24.271	80.83
X_3	1.039	9.442	91.898	1.039	9.442	91.898	1.217	11.067	91.898
X_4	0.729	6.625	98.522						
X_5	0.071	0.649	99.172						
X_6	0.04	0.361	99.533						
X_7	0.028	0.255	99.787						
X_8	0.013	0.12	99.908						
X_9	0.006	0.057	99.965						
X_{10}	0.002	0.023	99.987						
X_{11}	0.001	0.013	100						

Table 2.16 The rotated factor loading matrix (Ma et al. 2020)

Index	Variable	Common variance	Factor loading		
			Fac.1	Fac.2	Fac.3
Land reclamation index	X_1	0.991	0.916	−0.378	0.096
Change rate of farmland	X_2	0.600	0.107	−0.463	−0.612
Total power of agricultural machinery per unit area	X_3	0.974	0.950	0.229	0.140
Per capita farmland area	X_4	0.985	0.313	−0.886	0.319
Labor transfer index	X_5	0.986	0.977	0.151	0.091
Farmland productivity	X_6	0.940	0.918	0.267	0.160
Farmland grain productivity	X_7	0.815	−0.55	−0.223	0.680
Agricultural output contribution	X_8	0.958	−0.973	−0.105	0.021
Agricultural environmental impact index	X_9	0.980	0.934	0.309	−0.112
Multiple cropping index	X_{10}	0.982	−0.861	0.490	−0.014
Farmland ecosystem diversity index	X_{11}	0.896	−0.033	0.933	0.154

Note The extraction method is the principal component analysis; the rotation method is Kaiser standardized maximum variance method; the rotation converges after six iterations

Table 2.17 Analysis of variance (Ma et al. 2020)

Common factor	Clustering		F	Significance
	Mean square error	df		
Fac.1	0.761	3	202.466	0
Fac.2	0.799	3	58.268	0
Fac.3	0.266	3	9.821	0

(1) **Preliminary exploration phase (1978–1984)**

In the early 1980s, China began to implement the household contract responsibility system, which separated rural land into collective ownership and household contractual management rights. This realized the great innovation in China's rural land system, solved the problem of farmers' food and clothing in a short period of time, and improved the agricultural production efficiency. However, due to the introduction of the concept of "market", a large amount of farmland in rural areas has been converted into construction land and production land such as forest land, orchards, and fisheries. From 1981 to 1985, the annual average reduction of farmland area reached 4.87×10^5 hm^2. In 1985, the reduction of farmland area was as high as 1×10^6 hm^2.

During this phase, the household contract responsibility system was gradually established, which greatly stimulated the enthusiasm of farmers' production, and the

Table 2.18 Four clusters of year distribution (Ma et al. 2020)

Year	Clustering	Distance	Year	Clustering	Distance	Year	Clustering	Distance
1978	1	0.199	1992	3	0.112	2006	2	0.229
1979	1	0.195	1993	3	0.145	2007	2	0.205
1980	1	0.116	1994	3	0.161	2008	2	0.206
1981	1	0.025	1995	3	0.174	2009	2	0.231
1982	1	0.101	1996	4	0.547	2010	4	0.237
1983	1	0.166	1997	4	0.123	2011	2	0.290
1984	1	0.231	1998	4	0.102	2012	2	0.255
1985	3	0.224	1999	4	0.086	2013	2	0.283
1986	3	0.168	2000	4	0.076	2014	2	0.238
1987	3	0.128	2001	4	0.087	2015	2	0.204
1988	3	0.108	2002	4	0.223	2016	2	0.171
1989	3	0.062	2003	2	0.264	2017	2	0.172
1990	3	0.091	2004	2	0.230			
1991	3	0.070	2005	2	0.236			

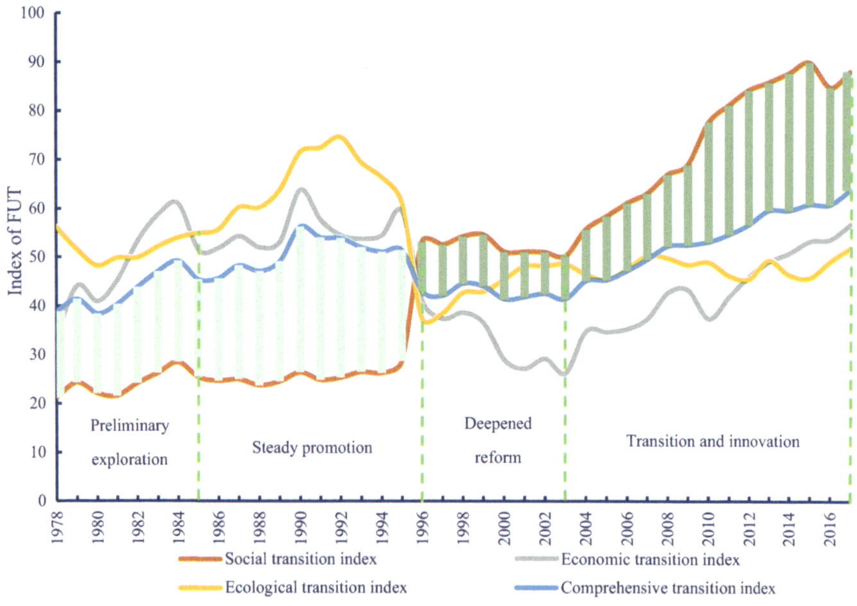

Fig. 2.42 Phase division of FUT in China from 1978 to 2017 (Ma et al. 2020)

economic and employment security functions of farmland gradually began to appear. However, due to farmland irrational use, the contradiction between population and land began to appear.

(2) **Steady promotion phase (1985–1995)**

Since 1985, with the adjustment of agricultural planting structure and the steady development of the rural economy, rural household food consumption has changed from basic self-sufficiency to the combination of self-sufficiency and market security. In the same year, agricultural green production began to move toward legalization, relevant departments paid more attention to quality and safety while also paying attention to the supply of grain. The ecological security function of farmland ecosystem gradually received more attention. In 1986, Central Document No. 1 proposed the establishment of a rural reform pilot zone, which attempted to undertake moderate-scale land management in rural areas and large urban suburbs in developed areas, and to a certain extent, improved the level, scale and production efficiency of agricultural specialization.

During this phase, the employment and production security functions of farmland were slightly weakened, economic function rose in waves rises with time, and the ecological functions such as the purification ability of farmland ecosystems and landscape ecological effects were steadily increasing.

(3) **Deepened reform and deepening phase (1996–2002)**

With the acceleration of urbanization and the advancement of science and technology, new opportunities and challenges have been brought to the use of farmland. On the one hand, with the rapid development of the secondary and tertiary industries in cities, a large quantity of urban jobs has been created. In addition, the state encourages the transfer of the rural labor force, and the rural surplus labor flows to cities, China's "migrant labor tide" has begun to emerge. On the other hand, China has adjusted the development strategy of industry and agriculture, and strengthened the back-feeding of the industry for agriculture. The industrialization of agriculture has been deepened, and the degree of specialization, scale, and efficiency of the agricultural planting industry have been greatly improved, which further promotes the transfer of farmland and moderate scale management. In addition, since China joined the WTO in 2001, agriculture has suffered a large impact. The contribution of agriculture to the GDP has declined, the contribution of agricultural output has declined from 19.60% in 1995 to 12.30% in 2003, and the grain output per unit area of farmland has declined from 4.89 tons/ha in 1995 to 3.49 tons/ha in 2003.

Furthermore, coordinating the development of man and nature has become a national strategy. The quality and safety of agricultural products and the governance of rural human settlements have gradually received attention. Agricultural subsidies and ecological compensation policies have been introduced one after another, which provided policy guarantees for the rational use of farmland. During this phase, the economic function of farmland gradually declined, and the social, ecological, and employment security functions gradually increased.

(4) **Transition and innovation phase (2003–2017)**

Recently, the urban-rural dual pattern has been gradually broken in China, and the land fragmentation and the pattern of small-scale farmers' production and management under the family contracted production pattern have increasingly become prominent obstacles restricting the development of agricultural modernization and the continuous increase in farmers' income. To alleviate the contradiction between population and land, and improve farmland efficiency, the Chinese government has proposed a series of policy systems.

Central Document No. 1 both in 2013 and 2014, proposed to encourage the development of various forms of moderate scale operations. In 2015, in order to improve the efficiency of the agricultural supply system, promote the structural adjustment of agriculture and the transformation and upgrade of the crop industry, the Central Rural Work Conference proposed strengthening the structural reform of the agricultural supply side. In 2016, the General Office of the CPC Central Committee and the General Office of the State Council issued the *"Opinions on improving the measures for the division of rural land ownership contracting rights and management rights"*, which advances the reform of the rural land property right system to promote the separation of collective ownership, farmer's contractual rights and management rights. The reform of the property rights system is of great significance for the liberation of "people" and "land" and to meet the development requirements of a moderate scale (Huang et al. 2019; Zhong et al. 2018). After a series of institutional reforms and innovations, the function of farmland has also changed, and the social, economic, and comprehensive transition indices have steadily increased, and ecological functions have increased in volatility.

In addition, with the diversification and high quality of people's material needs, many farmland functions have been gradually exploited or manifested, and the cultural leisure and entertainment functions of farmland have become increasingly prominent (Zhang et al. 2018). During this phase, the use of farmland tends to be high-quality transformation and innovation, and the functions of farmland tend to be diversified and coordinated.

2.7.4 *Discussion and Implications*

2.7.4.1 What Are the Key Factors Contributing to FUT?

China's land use structure is closely related to economic development, and there are obvious regional differences. Therefore, not all regions will experience the same phase of FUT, and some regions may stay in a phase for a long time, while others may move rapidly between different phases depending on the combined effects of many complex factors. On the one hand, with the development of social economy, people's demand for products changes, thus stimulating the transformation of the farmland supply; on the other hand, the negative ecological environment changes caused by human activities will also have an impact on farmland use (Lambin and Meyfroidt 2010).

The transformation of farmland in China is affected by multiple factors, so there are obvious regional differences in geographical space. The key factors affecting the transformation of farmland can be divided into two categories. First, the differences in natural conditions and geographic locations. For instance, the transformation of farmland use in suburbs of large cities is faster and deeper, while the transformation of farmland in remote and backward rural areas is even slower (Long and Qu 2018). Due to the differences in resource endowment, topography and geographical location conditions often affect the crop planting structure, farmland operation scale, and the degree of farmland utilization, thereby making differences in the speed and depth of FUT.

Second, the impact of socio-economic development on rural population, industry and economy. Economic globalization and agricultural policy affect the transformation of farmland use at a macro level. With most agricultural frontiers now closed or closing around the world, one of our most significant challenges is to manage land for multiple goals and the needs of different groups (Radel et al. 2019). In order to cope with the challenges brought by globalization, government have introduced relevant agricultural policies to achieve the purpose of adjusting the agricultural industrial structure and highlighting the comparative advantages of agricultural products in the competition, thus facilitating the recessive transformation of farmland utilization. Urbanization and population migration affect the transformation of farmland use at a micro level. Due to geographical inequality in wealth and in the cost of living, people tend to migrate to areas where living is more comfortable. Migration may have a greater impact on national and local economies and land use (Radel et al. 2019). In terms of receiving land systems, due to the sharp increase in population, people's demand for residential and commercial space increases, which drives the transformation of farmland to construction land and public service facility land, and the cost of grain production increases with the increase in land rent, so the agricultural production function of this type of region is gradually degradation. In terms of sending land systems, migration can lead to a contraction of farming activities and associated forest expansion. In addition, the increase in population during the process of urbanization and industrial structure changes are also important factors that contribute to the FUT, which are mainly manifested by excessive plundering of resources, increasing pressure on existing agricultural land, and threatening the ecological environment of farmland (Enaruvbe et al. 2019).

2.7.4.2 Policy Suggestions and Enlightenment on Sustainable Use of Farmland

(1) Promote coordinated development of farmland space and it's functional forms

From a phase where development scale and economic growth were overemphasized, today's China has entered to a new one where green and eco-friendly development is appreciated and encouraged. In the new era, the aims of comprehensive land consolidation in China are to enhance land use efficiency and benefit, ensure the sustainable use of land resources, and improve the ecological landscape environment. A series of measures, such as consolidation, exploitation, reclamation, remediation and protection, are taken to improve human living and production conditions, conserve ecological space, and ultimately promote the harmonious development of man and nature (Long et al. 2019). Land consolidation should take the moderate scale of ecological land as the premise, and ensure the reasonable quantity structure and space allocation between ecological land and farmland is an important foundation for realizing the protection of farmland in the new era, also an important means to promote the construction of ecological civilization. Regional land use should be based on the sustainable carrying capacity of resources and environment. By calculating the minimum farmland scale, the optimal ecological land scale and the maximum construction land scale of each region, the coordinated development of agricultural production space, ecological protection space and construction land development space can be guaranteed (Zhong et al. 2018). In the future, the utilization of farmland and the optimization of territorial space should be combined to optimize and adjust land use structure. Through land consolidation, increasing farmland area, improving farmland quantity and quality, repairing the ecological environment, and improving the field infrastructure to increase land productivity (Tan 2017) can be realized. Meanwhile, the multi-functions of farmland in food production, landscape shaping, biodiversity protection, ecological security, leisure, and tourism should be enhanced relying on the favorable landscape conditions created by land consolidation, which are also helpful for improving the competitiveness of rural development.

(2) Strengthening zoning planning and farmland governance and exerting the comparative advantages

China has a vast territory, with significant regional differences including the resource endowment of farmland in different areas. By judging the new opportunities and challenges faced by farmland use in different regions, designing the scientific path of farmland development management is vital from the aspects of institutional reform, policy implementation and planning (Ma et al. 2019b; Tu et al. 2018).

For the regions with appropriate resource endowments, such as plain farming areas. First, the government can improve the efficiency of farmland use by encouraging the circulation of agricultural land and large-scale operation, and properly disposing the farmland that is temporarily unsuitable for large-scale operations. Second, pay attention to the environmental effects in the process of using farmland. On the one hand, it is necessary to guide farmers to use chemical fertilizers and pesticides scientifically; on the other hand, through the establishment and improvement of the farmland quality monitoring platform and evaluation mechanism, the quality of farmland, the level of agricultural ecological environment pollution and the stability of regional agricultural systems are measured, to as to realize the ecological protection and dynamic monitoring of farmland resources. For the regions with relatively harsh resource endowments, such as plateau, hilly and mountainous areas that are dominated by the ecological conservation function of farmland, the protection and inheritance of traditional agricultural civilization and national culture should be focused on. In addition, based on the resource and environmental carrying capacity, moderately developing modern agriculture with local characteristics, while further improving the level of agricultural mechanization and modernization, are effective measures for relieving the social pressure of modern agricultural scale management (Liu et al. 2014b). As for the abandoned farmland, we can incorporate it into the management system of farmland management, to promote the rehabilitation and fertility of inferior farmland.

(3) **Enhancing the awareness of farmland protection and promoting the construction of ecological civilization**

Agricultural production can lead to land degradation, water quality degradation and loss of ecosystem function and biodiversity in surrounding watersheds. In the past four decades, China's chemical fertilizer application has increased by nearly six fold, and while the fertilizer utilization rate is only 32%, far below the world average of 55%, it has led to serious environmental pollution issues. Meanwhile, problems such as low efficiency of agricultural water use and outdated facilities have become shortcomings that constrain China's agricultural production. In response to the above questions, this study advances policy recommendations for improvement.

First, encouraging innovation and research and development of related technologies, so as to improve the utilization rate of fertilizers and agricultural infrastructure conditions, and reduce the damage of agricultural production to the ecological environment. Second, the government should play a macro-control role in the economic and industrial structure, and guide farmers to undertake agricultural production through scientific methods and technical means, such as through the adjustment of crop planting structure and other policy measures to promote structural changes of agricultural production and enhance the technical efficiency of farmland use (Cao et al. 2018). In addition, strengthening the publicity and education of resource protection, so as to enhance the public awareness of the rational development and effective use of farmland resource is important (Wu et al. 2017).

2.8 Conclusions

Farmland is the key resource for sustaining rural development and agricultural production. Farmland use morphology is closely related to rural development status. Studying the process and mechanism of FUT in China under rapid urbanization is an important part of analyzing the internal mechanism of rural transformation development. Starting from the internal relationship between rural transformation development and FUT, this chapter establishes the research focuses and core perspectives of FUT. The changes in both dominant and recessive morphologies of farmland use are closely related to rural transformation development. By analyzing the evolution mechanism of farmland use morphology, it is beneficial to analyze the internal logic of FUT.

This chapter establishes a theoretical analysis framework for FUT in China from the perspectives of "structure-function" and "supply-demand", and analyzes the process and mechanism of FUT from the perspectives of intensive use level, grain production ability, changes of agricultural labor and function evolution of farmland. This chapter mainly tries to analyze the driving mechanism of FUT from the perspective of driving the evolution of the recessive morphology of farmland use, and provides a comprehensive analysis scheme for understanding the transition process of farmland use under the background of URTD. By analyzing the transition of farmland use from multiple perspectives, it reveals the interactive process of FUT and rural transformation development more deeply, and provides a decision reference for optimizing rural man-land interrelations and promoting the transformation development in traditional farming areas.

It is found that deepening the research on FUT provides the basis for optimizing the policy of farmland use and protection. The rapid loss of farmland resources since the reform and opening up has increased the risk of food security. Economic development cannot bring about the continuous improvement of the intensive use level of farmland, nor can it necessarily bring about the continuous and stable growth of grain yield. The change of the traditional farmland use (such as from planting grain crops to planting cash crops in greenhouses) is conducive to promoting the intensive use level of farmland and promoting the rapid development of rural areas. The spatio-temporal distribution characteristics and evolution process of farmland resource are important factors that determine the grain production ability and its spatial pattern. The interactive process of FUT and rural population transition has become an important driving force to promote the evolution of the rural man-land interrelations. The aging of rural population and the change of agricultural labor are important points to analyze FUT. The evolution process of farmland function is a comprehensive evaluation of the change of farmland use morphology, and its evolution process and mechanism analysis provide support for deepening the research on FUT.

Presently, China's land use structure is closely related to economic development, and there are obvious regional differences. First of all, the farmland use in China has experienced several phases from slow transition, rapid transition to steady transition and innovation transition. Second, the dominant transition phase, the social transition and ecological transition of farmland, mainly occurred during 2000–2010 while the economic transition of farmland occurred earlier, from 1990 to 2000. Spatially, the economic transition of farmland gradually shifted from the southeast coastal area to the inland. Third, from 1978 to 2017, China's FUT can be divided into four phases: the preliminary exploration and development phase (1978–1984), the steady development and promotion phase (1985–1995), the market-oriented reform and deepening phase (1996–2002), and the reform, innovation and transition phase (2003–2017). Finally, with the diversification and high quality of people's physical requirements, the cultural leisure and entertainment functions of farmland have become increasingly prominent.

In view of the current challenges of farmland, this chapter puts forward a series of policy recommendations, which include deepening the reform of the land system, combining the utilization of farmland with the optimization of land use space, promoting land consolidation, strengthening the zoning planning and governance of farmland, and promoting the construction of ecological civilization. Function evolution is the core of FUT. The function of farmland will transform with changes in socio-economic development, human demand, and policy instruments. In the new era, the use of farmland tends to be a high-quality transformation and innovation, and the functions of farmland tend to be more diversified and coordinated. Therefore, in the future, we should focus on the transformation of farmland utilization caused by the change in human demand and new policy instruments, as well as the impact of farmland transformation on farmers' livelihoods, ecological environment and agricultural structure change around the world.

References

Anderman TL, Remans R, Wood SA et al (2014) Synergies and tradeoffs between cash crop production and food security: a case study in rural Ghana. Food Secur 6:541–554

Assunção JJ, Braido LH (2007) Testing household-specific explanations for the inverse productivity relationship. Am J Agr Econ 89(4):980–990

Bardhan PK (1973) Size, productivity, and returns to scale: An analysis of farm-level data in Indian agriculture. J Polit Econ 81(6):1370–1386

Barrett CB, Bellemare MF, Hou J (2010) Reconsidering conventional explanations of the inverse productivity–size relationship. World Dev 38(1):88–97

Bhalla SS, Roy P (1988) Mis-specification in farm productivity analysis: the role of land quality. Oxford Econ Pap 40(1):55–73

Cai F (2010) 人口转变、人口红利与刘易斯转折点 (Demographic transition, demographic dividend, and Lewis turning point in China). China Econ J 3:107–119

Calogero C, Savastano S, Zezza A (2013) Fact or artifact: the impact of measurement errors on the farm size-productivity relationship. J Dev Econ 103:254–261

Cao Y, Dallimer M, Stringer LC et al (2018) Land expropriation compensation among multiple stakeholders in a mining area: explaining "skeleton house" compensation. Land Use Pol 74:97–110

Carter MR (1984) Identification of the inverse relationship between farm size and productivity: an empirical analysis of peasant agricultural production. Oxford Econ Pap 36(1):131–145

Cartier C (2001) 'Zone Fever', the arable land debate, and real estate speculation: China's evolving land use regime and its geographical contradictions. J Contemp China 10:445–469
Caulfield M, Bouniol J, Fonte SJ et al (2019) How rural out-migrations drive changes to farm and land management: a case study from the rural Andes. Land Use Pol 81:594–603
Chan K (2010) A China paradox: migrant labor shortage amidst rural labor supply abundance. Eurasian Geogr Econ 51:513–530
Chen R, Ye C, Cai Y et al (2014) The impact of rural out-migration on land use transition in China: past, present and trend. Land Use Pol 40:101–110
Deng X, Huang J, Rozelle S et al (2006) Cultivated land conversion and potential agricultural productivity in China. Land Use Pol 23:372–384
Deng X, Huang J, Rozelle S et al (2015) Impact of urbanization on cultivated land changes in China. Land Use Pol 45:1–7
Ding C (2003) Land use policy reform in China: assessment and prospects. Land Use Pol 20:109–120
Długosz Z (2011) Population ageing in Europe. Procedia Soc Behav Sci 19:47–55
Enaruvbe GO, Keculah KM, Atedhor GO et al (2019) Armed conflict and mining induced land-use transition in northern Nimba county. Glob Ecol Conserv, Liberia. https://doi.org/10.1016/j.gecco.2019.e00597
Evans LT, Fischer RA (1999) Yield potential: its definition, measurement, and significance. Crop Sci 6:1544–1551
Fang Y, Xie W, Lin Z (2013) 信息技术对服务业创新影响的空间计量分析 (A spatial econometric analysis of impact of ICT on service innovationc). Acta Geographica Sinica 68:1119–1130
FAO (2016) The state of food and agriculture. Food & Agriculture Organization of the UN (FAO), Washington
Fleskens L, Duarte F, Eicher I (2009) A conceptual framework for the assessment of multiple functions of agro-ecosystems: a case study of Trás-os-Montes olive groves. J Rural Stud 25:141–155
Gao J, Liu Y (2011) Climate warming and land use change in Heilongjiang Province, Northeast China. Appl Geogr 31:476–482
Gao Q, Liu T, Kong X (2013) 家庭农场的制度解析:特征、发生机制与效应 (Institutional analysis of family farm: characteristics, mechanism and effect). Economist 6:48–56
Gaughan AE, Stevens FR, Huang Z et al (2016) Spatiotemporal patterns of population in mainland China, 1990 to 2010. Sci Data 3(1):160005
Ge D, Long H, Ma L et al (2017) Analysis framework of China's grain production system: a spatial resilience perspective. Sustainability 9(12):1–21
Ge D, Long H, Zhang Y et al (2018a) Farmland transition and its influences on grain production in China. Land Use Pol 70:94–105
Ge D, Long H, Zhang Y et al (2018b) Analysis of the coupled relationship between grain yields and agricultural labor changes in China. J Geogr Sci 28:93–108
Ge D, Wang Z, Tu S et al (2019) Coupling analysis of greenhouse-led farmland transition and rural transformation development in China's traditional farming area: a case of Qingzhou city. Land Use Pol 86:113–125
Ge Q, Zhao M, Zheng J (2000) 20世纪中国土地利用变化研究 (Land use change of China during the 20th century). Acta Geographica Sinica 55(6):698–706
Goll M (2010) Ageing in the European Union: where exactly? Rural areas are losing the young generation quicker than urban areas. In: Statistics in focus population & social conditions, statistical office of the European communities (Eurostat), Luxembourg-Kirchberg, pp 1–15
Gu X, Li M, Xu D (2017) 中国可持续发展遥感监测报告 (Report on remote sensing monitoring of China sustainable development). Social Scientific Academic Press, Beijing
Guo H, Lu X, Lu L (1990) 黄淮海地区乡村发展研究 (A research on the rural development in Huang-Huai-Hai region). Geogr Res 9(2):130–137
Heerink N, Qu F, Kuiper M et al (2007) Policy reforms, rice production and sustainable land use in China: a macro—micro analysis. Agr Syst 3:784–800

Heltberg R (1998) Rural market imperfections and the farm size-productivity relationship: evidence from Pakistan. World Dev 26(10):1807–1826
Henderson H (2015) Considering technical and allocative efficiency in the inverse farm size-productivity relationship. J Agr Econ 66(2):442–469
Hu H (1935) 中国人口之分布——附统计表与密度图 (The distribution of population in China with statistics and maps). Acta Geographica Sinica 2(2):33–74
Huang A, Xu Y, Sun P et al (2019) Land use/land cover changes and its impact on ecosystem services in ecologically fragile zone: a case study of Zhangjiakou City, Hebei Province, China. Ecol Indic 104:604–614
Huang Z, He C, Zhu S (2017) Do China's economic development zones improve land use efficiency? The effects of selection, factor accumulation and agglomeration. Landscape Urban Plan 162:145–156
Hussain A (2002) Demographic transition in China and its implications. World Dev 30:1823–1834
Jiang G, Zhang R, Ma W et al (2017) Cultivated land productivity potential improvement in land consolidation schemes in Shenyang, China: assessment and policy implications. Land Use Pol 68:80–88
Koohafkan P, Altieri MA, Gimenez EH (2012) Green agriculture: foundations for biodiverse, resilient and productive agriculture systems. Int J Agr Sustain 10:61–75
Lamb RL (2003) Inverse productivity: land quality, labor markets, and measurement error. J Dev Econ 71(1):71–95
Lambin EF, Meyfroidt P (2010) Land use transitions: socio-ecological feedback versus socio-economic change. Land Use Pol 27:108–118
Lewis (1989) Dual economy. Beijing Economic College Press, Beijing
Li S, Li X (2019) The mechanism of farmland marginalization in Chinese mountainous areas: evidence from cost and return changes. J Geogr Sci 29:531–548
Li T, Long H, Zhang Y et al (2017) Analysis of the spatial mismatch of grain production and farmland resources in China based on the potential crop rotation system. Land Use Pol 60:26–36
Li X, Zhu H, Tan M et al (2008) 土地利用集约度的测度方法 (Measurement of land use intensity). Prog in Geogr 27(6):12–17
Li Y, Hans W, Liu Y (2019) Why some rural areas decline while some others not: an overview of rural evolution in the world. J Rural Stud 68:135–143
Li Y, Li Y, Westlund H et al (2015) Urban–rural transformation in relation to cultivated land conversion in China: Implications for optimizing land use and balanced regional development. Land Use Pol 47:218–224
Liao L, Long H, Gao X et al (2019) Effects of land use transitions and rural aging on agricultural production in China's farming area: a perspective from changing labor employing quantity in the planting industry. Land Use Pol 88:104152
Lichtenberg E, Ding C (2008) Assessing farmland protection policy in China. Land Use Pol 25:59–68
Lipton M (1980) Migration from rural areas of poor countries: the impact on rural productivity and income distribution. World Dev 8:1–24
Liu J, Zhang Z, Xu X et al (2010) Spatial patterns and driving forces of land use change in China during the early 21st century. J Geogr Sci 20:483–494
Liu L, Xu X, Liu J et al (2014a) 1990–2010 年中国耕地变化对粮食生产潜力的影响 (Impact of farmland changes on production potential in China during recent two decades). Acta Geog Sin 69:1767–1778
Liu Y (2018) Introduction to land use and rural sustainability in China. Land Use Pol 74:1–4
Liu Y, Fang F, Li Y (2014b) Key issues of land use in China and implications for policy making. Land Use Pol 40:6–12
Liu Y, Li J, Yang Y (2018) Strategic adjustment of land use policy under the economic transformation. Land Use Pol 74:5–14
Liu Y, Li Y (2017) Revitalize the world's countryside. Nature 548(7667):275–277

Liu Y, Long H (2016) Land use transitions and their dynamic mechanism: the case of the Huang-Huai-Hai plain. J Geogr Sci 26:515–530
Liu Y, Wang L, Long H (2008) Spatio-temporal analysis of land-use conversion in the eastern coastal China during 1996–2005. J Geogr Sci 18:274–282
Long H (2015) 论土地利用转型与土地资源管理 (Land use transition and land management). Geographic Res 34(9):1607–1618
Long H, Ge D, Zhang Y et al (2018) Changing man-land interrelations in China's farming area under urbanization and its implications for food security. J Environ Manage 209:440–451
Long H, Heilig GK, Wang J et al (2006) Land use and soil erosion in the upper reaches of the Yangtze River: some socio-economic considerations on China's Grain-for-Green Programme. Land Degrad Dev 17:589–603
Long H, Li T (2012) The coupling characteristics and mechanism of farmland and rural housing land transition in China. J Geogr Sci 22:548–562
Long H, Li Y, Liu Y et al (2012) Accelerated restructuring in rural China fueled by 'increasing vs. decreasing balance' land-use policy for dealing with hollowed villages. Land Use Pol 29:11–22
Long H, Liu Y, Li X et al (2010) Building new countryside in China: a geographical perspective. Land Use Pol 27:457–470
Long H, Qu Y (2018) Land use transitions and land management: a mutual feedback perspective. Land Use Pol 74:111–120
Long H, Tu S (2018) 土地利用转型与乡村振兴 (Land use transition and rural vitalization). China Land Sci 32(7):1–6
Long H, Tu S, Ge D et al (2016) The allocation and management of critical resources in rural China under restructuring: problems and prospects. J Rural Stud 47:392–412
Long H, Zou J (2010) Grain production driven by variations in farmland use in China: an analysis of security patterns. J Resour Ecol 1(1):60–67
Long H, Zhang Y, Tu S (2019) Rural vitalization in China: a perspective of land consolidation. J Geogr Sci 29(4):517–530
Lu X, Shi Y, Chen C et al (2017) Monitoring cropland transition and its impact on ecosystem services value in developed regions of China: a case study of Jiangsu province. Land Use Pol 69:25–40
Luo C, Cai Y (2016) 湖北省农产品主产区耕地资源功能的时空演变 (The stage characteristics and spatial heterogeneity of cultivated land resource function evolution in agricultural producing areas of Hubei Province). Econ Geogr 36(3):153–161
Ma L, Long H, Chen K (2019a) Green growth efficiency of Chinese cities and its spatio-temporal pattern. Resour Conserv Recy 146:441–451
Ma L, Long H, Tu S et al (2020) Farmland transition in China and its policy implications. Land Use Pol 92:104470
Ma L, Long H, Zhang Y (2019b) Agricultural labor changes and agricultural economic development in China and their implications for rural vitalization. J Geogr Sci 29:163–179
MacDonald D, Crabtree JR, Wiesinger G et al (2000) Agricultural abandonment in mountain areas of Europe: environmental consequences and policy response. J Environ Manage 59:47–69
Matson PA, Parton WJ, Power AG et al (1997) Agricultural intensification and ecosystem properties. Science 277:504–509
Meyfroidt P, Chowdhury RR, De Bremond A et al (2018) Middle-range theories of land system change. Global Environ Change 53:52–67
Montgomery MR (2008) The urban transformation of the developing world. Science 5864:761–764
NBSC (2011) 中国县(市)社会经济统计年鉴 (China county statistical yearbook). China Statistics Press, Beijing
Oseni G, Winters P (2009) Rural nonfarm activities and agricultural crop production in Nigeria. Agr Econ 40:189–201
Qu Y, Long H (2016) 城市土地利用隐性形态空间分异及其影响因素——以中国289个地级以上城市为例 (Spatial differentiation of the recessive morphology of urban land use and its influential factors: a case study of 289 prefecture-level cities in China). Econ Geogr 36(10):1–8

Qu Y, Long H (2017) 基于开发利用与产出视角的区域土地利用隐性形态综合研究——以黄淮海地区为例 (The integrated research on regional land use recessive morphology from the perspectives of exploitation and output: the case of the Huang-Huai-Hai Region). Geogr Res 36(1):61–73
Qu Y, Long H (2018) 中国耕地利用隐性形态转型的多学科综合研究框架 (A framework of multi-disciplinary comprehensive research on recessive farmland transition in China). Acta Geographica Sinica 73(7):1226–1241
Radel C, Jokisch BD, Schmook B et al (2019) Migration as a feature of land system transitions. Curr Opin Env Sust 38:103–110
Ranis G, Fei JC (1969) A theory of economic development. American Econ Rev 4:533–565
Rozelle S, Taylor JE, DeBrauw A (1999) Migration, remittances, and agricultural productivity in China. American Econ Rev 89:287–291
Sen AK (1962) An aspect of Indian agriculture. Econ Weekly 14:243–246
Shao J, Zhang S, Li X (2015) Farmland marginalization in the mountainous areas: characteristics, influencing factors and policy implications. J Geogr Sci 25:701–722
Shi W, Shi W, Tao F et al (2013) Changes in quantity and quality of cropland and the implications for grain production in the Huang-Huai-Hai plain of China. Food Secur 5:69–82
Smith A, Snapp S, Chikowo R et al (2017) Measuring sustainable intensification in smallholder agroecosystems: a review. Glob Food Secur 12:127–138
Smith P (2013) Delivering food security without increasing pressure on land. Glob Food Secur 2:18–23
Song B, Robinson GM, Zhou Z (2017) Agricultural transformation and ecosystem services: a case study from Shaanxi Province, China. Habitat Int 69:114–125
Song X (2017) 论土地利用转型的研究框架 (Discussion on land use transition research framework). Acta Geographica Sinica 72(3):471–487
Song X, Ouyang Z, Li Y et al (2012) Cultivated land use change in China, 1999–2007: policy development perspectives. J Geogr Sci 22:1061–1078
Su S, Zhou X, Wan C et al (2016) Land use changes to cash crop plantations: crop types, multilevel determinants and policy implications. Land Use Pol 50:379–389
Su S, Sun Y, Lei C et al (2017) Reorienting paradoxical land use policies towards coherence: a self-adaptive ensemble learning geo-simulation of tea expansion under different scenarios in subtropical China. Land Use Pol 67:415–425
Tan M (2014) The transition of farmland production functions in metropolitan areas in China. Sustainability 6:4028–4041
Tan M (2017) Uneven growth of urban clusters in megaregions and its policy implications for new urbanization in China. Land Use Pol 66:72–79
Tu S, Long H (2017) Rural restructuring in China: theory, approaches and research prospect. J Geogr Sci 27:1169–1184
Tu S, Long H, Zhang Y et al (2018) Rural restructuring at village level under rapid urbanization in metropolitan suburbs of China and its implications for innovations in land use policy. Habitat Int 77:143–152
Tuan Y (1971) Geography, phenomenology, and the study of human nature. Canadian Geogr 15(3):181–192
Wang X, Huang J, Rozelle S (2017) Off-farm employment and agricultural specialization in China. China Econ Rev 42:155–165
Wu C, Guo H (1994) 中国土地利用 (Land use in China). Science Press, Beijing
Wu K (2013) 家庭农场的理论分析 (Theoretical analysis of family farm). Econ Rev 6:65–69
Wu Y, Hui E, Zhao P et al (2018) land use policy for urbanization in China. Habitat Int 77:40–42
Zhang Q, Gao W, Su S et al (2017) Biophysical and socioeconomic determinants of tea expansion: apportioning their relative importance for sustainable land use policy. Land Use Pol 68:438–447
Zhang S, Liu S, Zhang C et al (2010) 土地流转与农业现代化 (Land transfer and agricultural modernization). Manage World 7:66–85

Zhang Y, Li X, Song W et al (2016) Land abandonment under rural restructuring in China explained from a cost-benefit perspective. J Rural Stud 47:524–532
Zhang Y, Long H, Ma L et al (2018) Farmland function evolution in the Huang-Huai-Hai plain: processes, patterns and mechanisms. J Geogr Sci 28:759–777
Zhong T, Qian Z, Huang X et al (2018) Impact of the top-down quota-oriented farmland preservation planning on the change of urban land-use intensity in China. Habitat Int 77:71–79
Zuo L, Zhang Z, Zhao X et al (2014) Multitemporal analysis of cropland transition in a climate-sensitive area: a case study of the arid and semiarid region of northwest China. Reg Environ Change 14:75–89

Chapter 3
Rural Housing Land Transition in China

Abstract Rural housing land evolves with socio-economic development, and reveals a need for understanding and predicting human impacts on rural land use. Analysis of rural housing land is helpful for investigating the rural man-land interrelations. This chapter aims to theorize and verify rural housing land transition, as well as reveal the mechanism from a comprehensive perspective. In examining the linkages between economic growth and rural housing land change, theoretical and empirical studies have indicated that rural housing in every region will undergo specific stages—the proportion of rural housing land in the increase of total construction land will decline gradually with the development of the local economy, and the end of the transition corresponds to a new equilibrium between rural housing and other construction activities. In order to deal with the lack of long-term statistical data of land use in China, horizontal comparison and transect research methods were adopted to associate land-use change with socio-economic development based on China's obvious regional differentiation of socio-economic development. Demographic changes open a window for investigating rural housing land transition. The changing regional patterns of China's rural population and settlements showed that a positive pattern of co-evolution of population and residential land in rural areas has not appeared, and land utilization tends to be extensive, especially in northern and eastern coastal provinces. Rapid industrialization and urbanization in China have resulted in a severe phenomenon of 'village hollowing' caused by the dual-track structure of socio-economic development separating rural and urban areas.

3.1 Introduction

The far-reaching transition launched by Deng Xiaoping in the past four decades has transformed rural China from a traditional centrally planned economy to a market-based economy, which has impacted the social, economic and environmental landscape, with key trends including mass migration from rural to urban areas and increasing living standard. All these changes can be reflected in the utilization of rural housing land due to the existing human-land conflicts.

H. Long, *Land Use Transitions and Rural Restructuring in China*,
https://doi.org/10.1007/978-981-15-4924-3_3

Rural housing land plays a significant role in rural transformation development, which is second in importance to farmland. It evolves with socio-economic development, and reveals a need for understanding and predicting human impacts on rural land use. As a perfect entry, rural housing land is helpful for investigating the rural man-land interrelations. While researches about rural housing land transition remain under-theorized, and a more nuanced understanding of how rural housing land responds to the changes in employment, industry structure, and increasingly modernized society is necessary. Population outflow functions as the initial engine leading to changes in rural housing land use. As the statistics from the World Bank revealed, rural depopulation is becoming increasingly severe between 1960 and 2015 in China, with the proportion of rural population dropping 47%. Massive rural laborers flood into cities and towns influenced by the rising opportunity cost of farming, the number of which reached 247 million in 2016 in China. Almost 85% of the rural households in China have at least one family member who gets employed in the non-agricultural sector. Since rural communities are blighted by depopulation and the abandonment of buildings and land, a unique phenomenon of 'village-hollowing' came into being in China. Significant changes in demographic structures, employment opportunities, community organization, lifestyles and standard of living, accessibility and rural culture have taken place in rural China, which resulted in dramatic transformations of rural housing land use. There are two categories of rural housing land transition. One is the changes of dominant morphology of rural housing land, which means the quantity, structure and spatial pattern of rural housing land; another is the changes of recessive morphology, which includes the quality, property rights and management mode of rural housing land. Thus, rural housing land transition is a complex process, therefore, a sketch of the complexity and the multidimensional driving forces are crucial for regulating rural land use.

This chapter has five parts, in the first part, an overall picture of rural housing land transition in parallel with rural transformation development was provided. In the second part, a theoretical framework for rural housing land transition in China was proposed and verified. And the pattern and causes of population and settlements changes in China and their effects on China's rural development were examined in the third section. Then, the evolution of the hollowed village was analyzed. The final part attempts to explore the spatio-temporal patterns and underlying driving forces based on the multi-scale analysis.

3.2 Background

Since the reform and opening up in 1978, China has experienced rapid urbanization and industrialization in terms of rural outmigration and great changes in man-land interrelations. The transformation of rural settlements has become one of the most significant signs of China's rural development. Rural housing land transition not only can be understood from quantity, spatial morphology and function, but also links with the transformation of employment structure, lifestyle, and household structure.

Urbanization and industrialization function as the engines of the growth of farmers' living standards, which drives the increase of the capacity and desire of farmers to improve their housing conditions, and further causes the accelerated extension of rural housing land. Thus, it is common that one household has two or more housing sites in China's rural areas. The change in the peasant lifestyle is mainly reflected in the miniaturization of the family size. The original large family of several generations has been replaced by nuclear families, further generating a large demand for rural housing land. Due to the absence of long-term planning and proper control, the above phenomena have led to the disorderly spread of homesteads. At the same time, the massive expansion of urban land use will inevitably encroach on its surrounding rural areas. Affected by the endogenous development and the exogenous thrust of cities, both extension and shrinkage of rural housing land emerged in different regions. But, generally speaking, the area of rural housing land experienced an growth trend across the whole country, which has increased from 248 million mu (1 mu = 0.067 ha) in 2000 to 250 million mu in 2008 with the annual growth rate of 0.07%. China has a huge number of villages, taking account of a large proportion of construction land. According to statistics from the Ministry of Housing and Urban-Rural Development of China, there were 2.647 million natural villages[1] and 572,000 administrative villages in China at the end of 2007. From 1978 to 2008, the number of rural households in China increased from 174 million to 257 million, while the rural permanent resident population during the period decreased to 713 million in 2009.

So far, the population loss in rural areas has not resulted in the reduction of rural housing land but increase. Rural hollowing is parallel with the extension caused by newly-built houses. The coexistence of the unreasonable behaviors of building new houses without demolishing the old one has caused a colossal waste of rural land and severe damage to cultivated land and has become the primary problem facing the rural transformation development. As the institutional structure of China's urban-rural dual structure has not yet been fundamentally broken, the conflict of interest between urban and rural areas and the pressure on the employment of migrant workers have not yet been alleviated, and the lagging situation of rural land system reform has not been completely improved, China's rural hollowing will be intensified continuously. Therefore, rural land consolidation has become an important foundation for rural transformation development (Liu 2007).

At present, village remediation and new rural construction targeting at increasing cultivated land and maintaining requisition-compensation balance, as well as improving the comprehensive agricultural production capacity, have been carried out in full swing everywhere. But in the implementation process, due to the long-term relative disadvantaged position of agriculture, rural areas, and farmers, the social welfare and security functions of homesteads have been intentionally or unintentionally ignored.

[1]Natural village is different from administrative village in rural China. Generally, people with kinship live within one small-area region, therefore, they share common surnames in most cases. While administrative village consists of several natural villages, which has a wider geographical scope.

The value-added distribution of rural land values is clearly biased towards urban areas, and the village renovation goals remain at a superficial level of increasing cultivated land and improving traffic conditions. Besides, the optimization of village and township systems, the allocation of public resources, the cultivation of rural industries, the improvement of farmers' housing conditions, and the guarantee of farmers' employment also should be concerned (Long et al. 2012). Currently, it is an urgent need to build a strategic system for protecting people's livelihoods that relies on comprehensive management to optimize rural production, living and ecological space, so as to achieve a smooth transformation of rural housing land use.

Rural outmigration is a process that must be experienced in the process of urbanization. The so-called "semi-urbanization" of the amphibious state of migrant workers and the decentralized rural industrialization and urbanization policies of "leave the land but not leave the countryside" and "enter the factory but not into the city" have shaped a "semi-urbanized area" with mixed urban and rural land use, which will inevitably evolve into a completely urbanized state. This objective law determines that it is the general trend to promote the intensive use and transformation of rural housing land. However, the size of per capita rural housing land has continued to increase, and the rural population transfer has not been linked to the reduction of rural settlements, which brings new opportunities for tapping the potential of rural housing land and coordinating the allocation of urban and rural land.

China is undergoing a major economic and social transformation. Based on the system of restricting rural housing land use and the planned economic system, the economic and social conditions on which it depends are gradually losing. Thus, the current rural housing land circulation system is increasingly incompatible with the increasingly perfect market economy system. Accordingly, various problems have emerged, which mainly manifests in that it seriously distorted the market-oriented allocation of rural land and seriously affected the process of rural transformation development: (1) Restricting the transfer of rural surplus labor and hindering the development of rural economy; (2) Narrowing the farmers' financing channels and it is not conducive to the goal of building new countryside-in the aspect of developing production; (3) Leading to the chaos of the free circulation of rural housing land and causing the waste of rural land resources, and (4) Strengthening the dual structure of urban and rural sectors and hindering the coordinated development of urban and rural areas. Therefore, whether the transformation of rural housing land can be successfully implemented will directly affect the rural development and the balanced development of urban and rural areas in China.

3.3 Rural Housing Land Transition in China: Theory and Verification

3.3.1 Background

Currently, issues related to land-use and land-cover change (LUCC) have attracted interest among a wide variety of researchers, ranging from those who are modeling the spatial and temporal patterns of land conversion, to those who try to understand the causes and consequences of land-use changes (Irwin and Geoghegan 2001). To some extent, LUCC is still a complex issue regarding its process, dynamic and driving forces (Lambin et al. 2003). Therefore, a single research approach does not suffice for a complete analysis of LUCC. Instead, a combination of multiple approaches is necessary for LUCC research (Lambin et al. 2003). More attention has been paid to LUCC in developing countries, especially in China (Haberl et al. 2001). While there are numerous studies analyzing the loss of arable land in eastern and coastal China due to urban sprawl, the transformation of arable land into rural housing land has found much less attention.

In 2000, according to the statistical data from the Ministry of Land and Resources of China (MLRC), rural housing land in China amounted to 16.5 Mha, which accounted for 67.3% of China's total construction land (this includes rural housing land and urban and other build-up lands). Why do villagers in China's most densely populated and productive agricultural regions use so much of the scarce farmland to construct housing? And why has the Chinese government, which has legislated to conserve arable land so as to ensure national food security, been unable to control house-building activities in the country side? These open questions indicate that studying rural housing land will make a big contribution to the LUCC research in China, as well as providing some basis for stipulating policies in favor of rural land resources control. The aims of this section are to develop a theoretical approach in an attempt to account for a specific land-use change process, i.e. rural housing land transition in China, to explore a new integrated approach, and analyze the transition of rural housing land in the Transect of the Yangtse River (TYR) by presenting empirical evidence, and to practically apply the outcomes to land management issues.

3.3.2 A Theoretical Basis and Hypothesis

3.3.2.1 Theoretical Basis

The most fundamental obstacle to progress in understanding and predicting human impacts on terrestrial ecosystems lies in the lack of a comprehensive and integrative theory of human-environment relationships, which can be applied to explain empirical observations and predict new results. The integrative characteristics of LUCC research would require an understanding and modeling that incorporates the principles from such theories. Although our understanding of LUCC has improved since early studies on deforestation by Myers (1980) and Mather (1990), it does appear that theoretical elaboration is underdeveloped. While it seems that LUCC research still remains under-theorized, applications making use of structural elements that specifically attempt to examine and explain these issues are rare. The discipline, in principle, should be most able to meet the demands of geographical research (Grainger 1995a). There is tremendous scope for this kind of study, which needs knowledge concerning both physical geography and human geography.

Land use morphology is the overall pattern of actual land cover in a country or region at a given time, comprising the main land use types. It will vary with socio-economic development. Land use transition refers to the changes in land-use morphology over time, and it usually corresponds to a particular socio-economic development stage (Grainger 1995a). Land use transition, therefore, includes both temporal and spatial dimensions. Massey (1999) explored the possibility that there may be commonalities between physical geography and human geography in emerging ways of conceptualizing space, time and space-time. At present, the theorization of forest transition in both temporal and spatial dimensions is relatively mature (Grainger 1995b).

Before 1978, traditional central planning economic policy had been carried out in China. At that time, the rural population in China had no choice but to work in collective farms (with weak incentives for work), and all members shared the output equally. There was no obvious difference in the degree of development in the regional rural economy in China, despite the existing enormous regional diversity in climate, terrain, natural and human resources. However, China has been experiencing rapid transitions since Deng Xiaoping launched economic reforms in 1978. The traditional central planning economy was replaced by a market-based economy (a more laissez-faire/neo-liberal approach) and the primarily agricultural economy is being transformed into an urban, industrial economy. The early 1980s saw high growth in primary sector output in the wake of de-collectivization and the privatization of land-use rights under the "household responsibility system", which to a great extent stimulated rural economic growth at the early stages of China's transition. Coastal China, with the advantages of location, was the initial area of reform and opening, which also has been one of the most developed areas. An important dimension of regional inequality in China has been the uneven rate of growth between the eastern, central and western provinces (Kanbur and Zhang 1999).

There is an obvious gradient of regional economic growth along the Yangtse River, which can be considered a miniature of regional disparity in China's economic development, and also a geographical as well as a historical continuum, representing important development stages in China. Walker (2001) hypothesizes a two-stage sequence in the linkages between urban and rural sectors. This relationship, in turn, is mainly conditioned by the degree of development in the regional economy. The change process of rural housing links closely with urban and rural sectors and is affected by the degree of development in the regional economy. Accordingly, a

Transect along the Yangtse River makes possible the study of rural housing land transition in China.

3.3.2.2 Understanding the Change Process of Rural Housing in China

In developing a broad conceptual framework for understanding the change of rural housing land in China, we need to pay analytical attention both to endogenous growth factors within a specific period and also to the corresponding management measures. In China, the growth of rural housing land had been very slow before 1978. There were two reasons. On the one hand, it was the Chinese tradition to have three or four generations living in the same house (so-called "Si Shi Tong Tang"), and owning one more house or yearning for capacious house had been considered as the source of capitalism (so-called "capitalism's tail") and should have been eliminated before 1978. On the other hand, low income gained from working in collective farming made improving farmers' housing conditions almost impossible.

With the implementation of the market-based economy after 1978, many farmers were becoming affluent and they began to prefer multi-functional, more comfortable or spacious houses. The rural household model of "Si Shi Tong Tang" is no longer popular. Many young couples prefer to have their own housing. China still has a huge amount of rural population. According to the fifth national population census conducted in 2000, China's rural population was 827.8 million, or 63.9% of the total population. Rural housing has been expanding sharply. The social and demographic aspirations of families and the reconfiguration of rural households' economic activities are major stimuli of the rural house-building craze. Farmers prefer to build their houses in the valleys (instead of the hillside), and they prefer a good location and settle close to roads and other available infrastructure, which contributed to the loss of agricultural land. In order to conserve arable land, provincial regulations for rural housing were defined (Table 3.1), which are still unable to effectively control housing construction in the rural area.

With the progress of urbanization, however, some village houses are vacant in China's developed rural areas (especially in eastern China), either because their owners have two or more houses or because they are rural-urban migrators and live permanently in urban areas, which resulted in a serious waste of land resources. The vacant rural housing land of rural-urban migrants in the suburb is usually changed to urban and other build-up lands. A spot of vacant rural housing land of rural-urban migrants in the rural areas is often transferred to other farmers who may need it. If in a rural area there is a large amount of vacant rural housing land, it will be changed into cultivated land through carrying out a land consolidation project, which is financed by the Central Government of China.

Table 3.1 Some existing provincial regulations for rural housing in China (Long et al. 2007)

Province	Maximum area (m^2)	Explanatory note
Beijing (1991)	167	In suburb or other areas with high population density and little cultivated land
	200	In other areas
Tianjin (1992)	133	In suburb
	167	In other areas where cultivated land per capita (CLpc) is less than 667 m^2
	200	In other areas where CLpc exceeds 667 m^2
Chongqing (1991)	20–25*	In suburb
	25–30*	In other areas
Heilongjiang (1987)	250	In suburb
	350	In other areas
Jilin (1994)	270	In suburb
	330	In other areas
Liaoning (1987)	300	A household with 4 or fewer people in the villages with 1333 m^2 or more CLpc
	400	A household with 5 or more people in the villages with 1333 m^2 or more CLpc
	200	A household with 4 or fewer people in the villages with CLpc less than 1333 m^2
	266	A household with 5 or more people in the villages with CLpc less than 1333 m^2
Hebei (1987)	133	In the villages with 667 m^2 or less CLpc
	167	In the villages with CLpc above 667 m^2
	233–467	In sandbar or mountainous regions
Henan (1991)	133	In suburb or in plain with CLpc less than 667 m^2
	167	In plain with 667 m^2 or more CLpc
	200	In hilly or mountainous regions
Shandong (1992)	133–167	In suburb
	133–200	In plain
	267	In hilly or mountainous regions
Jiangsu (1989)	133	In a suburb or in villages with CLpc less than 667 m^2
	133–200	In the villages with CLpc between 667 m^2 and 1333 m^2
	200–267	In the villages with 1333 m^2 or more CLpc or in hilly or mountainous regions
Hubei (1987)	140	If cultivated land may be occupied by housing construction activities
	200	If not

(continued)

Table 3.1 (continued)

Province	Maximum area (m^2)	Explanatory note
Anhui (1989)	160	In suburb
	220	In northern plain
	160	If cultivated land may be occupied by housing construction activities in hilly or mountainous regions
	300	If not
Jiangxi (1989)	130–180	If rural housing will be built on used housing land or idle land
	100–130	If cultivated land may be occupied by housing construction activities
Zhejiang (1994)	125	If cultivated land may be occupied by housing construction activities
	140	If not (the two numbers may be lowered according to the scale of household and local conditions)
Fujian (1989)	20*	Usually
	120	A household with more than 6 people
Guangdong (1991)	80	In plain
	120	In hilly regions
	150	In mountainous regions
Guangxi (1988)	22*	If cultivated land may be occupied by housing construction activities
	30*	If not
	160	Even a household with 8 or more people if cultivated land may be occupied by housing construction activities
	250	Even a household with 8 or more people if not
Shanxi (1989)	200	Usually
	133	In the villages with CLpc less than 667 m^2
	267	In the villages with 2667 m^2 or more CLpc
Xinjiang (1989)		The maximum area is defined by County Government according to local conditions
Yunnan (1994)	20* (100)	In suburb (even a household with 5 or more people)
	30* (150)	In hilly or mountainous regions (even a household with 5 or more people)

Notes Data are from corresponding provincial ordinances implementing Land Administration Law of the People's Republic of China, and the digits in brackets under different province names are the particular years from which the ordinances are valid; The numbers with "*" mean maximum area per capita, and others without "*" mean maximum area per household

3.3.2.3 Hypothesis

The relationship between land-use and social change is a dynamic process, which should be viewed from a long-term perspective both in historical and spatial dimensions. At present, both rural development and urban development in China are in a transition period. On the one hand, we can observe the transformation of a traditional agricultural society into a modern industrial and urban society; on the other hand, the economy is changing from a traditional planned economy to a modern market system. However, there are obvious regional discrepancies in the process of social and economic transformation. As is typical for a large and rapidly developing country, regional development is highly unbalanced. Different regions have diverse developing advantages but are also confronted with different development tasks. They will be at different stages of development and, therefore, will be characterized by different types of land-use patterns.

Land-use changes, while restricted by physical conditions, are mainly driven by socio-economic factors. They can be mainly characterized by the changes in cultivated land and construction land, which are tightly inter-related with human production activities. The expansion of construction land usually occurred at the cost of cultivated land. Usually, with socio-economic development, construction land will increase. Since China was founded in 1949 and experienced several campaigns that hampered economic development, the total amount of construction land has not changed much in recent decades. Therefore, we chose to analyze only the increase of construction land. The proportion of rural housing land in the increase of total construction land at the regional level will have obvious differences, which are associated with the rural population ratio, socio-economic developmental level and ecological conditions. There is a high correlation between the amount of rural housing land and the rural population in China. The underlying trends of population growth and socio-economic development are probably the most significant factors leading to rural housing land transition in China. We assume that the proportion of rural housing land in the total construction land will decrease with the process of urbanization, which will lead to the rural-urban migration and the change from rural housing land to urban and other build-up land and, in the later phase of urbanization, the proportion will reach nearly a fixed value. Currently, the speed of urbanization in developed (eastern) areas of China is fast, and the share of rural housing in the increase of total construction land will decline; however, this share will increase in less-developed (western and central) areas of China, because of the low urbanization rate and dominating rural development. Accordingly, we assume that rural housing land in every region will undergo specific stages—the proportion of rural housing in the increase of total construction land will decline gradually with the development of the local economy, and the end of the transition corresponds to a new equilibrium between rural housing and other construction activities (Fig. 3.1).

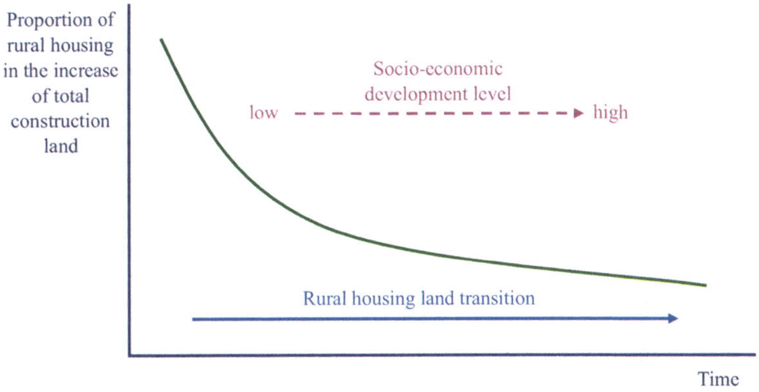

Fig. 3.1 Theoretical hypothesis on rural housing land transition trend (Adapted from Long 2006)

3.3.3 Materials and Methods

3.3.3.1 Study Area

Since land-use change is majorly triggered by socio-economic factors, the socio-economic developmental level was considered the most important factor to define the study area for the research of rural housing land transition. Transect of the Yangtse River, described as Fig. 1.7 in Chap. 1 was selected as our study area, has an underlying gradient both in physical/environmental factors as well as in socio-economic factors (Fig. 3.2).

3.3.3.2 Data Source and Processing

The analysis of land-use changes in TYR is based on two single land-cover (LC) maps at a 1:100,000 scale. These two maps were derived from two sets of historical Land-sat TM images, which were taken in 1986/1987 and 1999/2000, respectively, by the Institute of Geographic Sciences and Natural Resources Research and Institute of Remote Sensing Applications, the Chinese Academy of Sciences (Liu et al. 2003). There are six major LC types (cultivated land, forested land, grassland, water body, unused land and rural and urban settlements), and 25 sub-categories in the original LC datasets. For our in-depth analysis of land use changes in TYR, we reclassified the original LC types into eight common classes—paddy fields, dry land, forested land, grassland, water body, urban and other build-up lands, rural housing land and unused land. In addition, statistical land-use data in counties from 1987 to 2001 were used for analyzing land use transition. The data from 1987 to 1995 are from the corresponding yearbooks of former China's State Land Administration Bureau, and those from 1996 to 2001 are from MLRC. The socio-economic statistical data utilized in the analyzing process are from China's State Statistical Bureau.

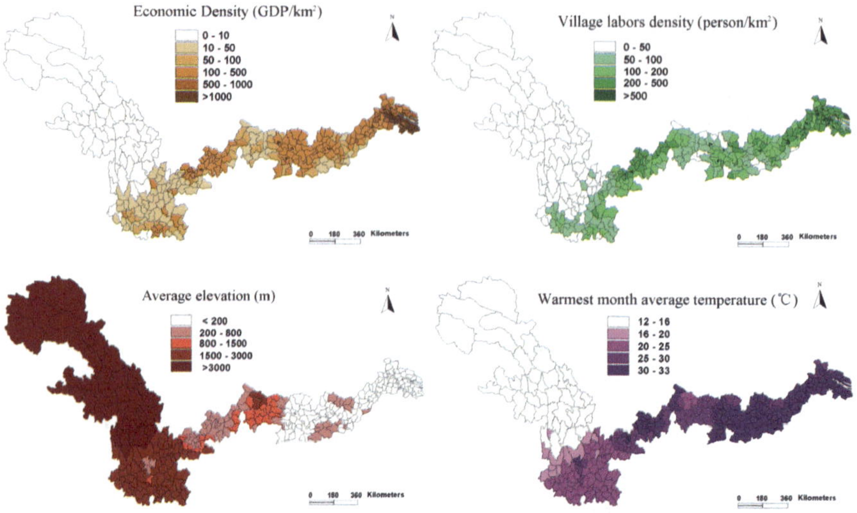

Fig. 3.2 Existing gradients of socio-economic factors and environmental factors in the study area (Long et al. 2007) (Economic Density was calculated by dividing the GDP of each county by the total area of the county, and the unit of GDP is 10^4 RMB¥, exchange rate US$ to RMB¥: 1–8.3; Data of socio-economic factors are from China's State Statistical Bureau, and that of environmental factors are from International Institute for Applied Systems Analysis.)

3.3.3.3 Methods

A land-use change matrix is obtained by using two reclassified LC maps representing the internal variations of LC in TYR between 1986/1987 and 1999/2000. We use the ESRI's ArcGIS spatial analyst module to identify the land-use changes between the two periods. For each LC category i in the matrix, the change between the periods is calculated by the following equation:

$$CH_i = (p_{i.} - p_{.i})/p_{.i} \times 100 \tag{3.1}$$

where

CH_i Is the change of LC in row *i* relative to the previous compared year;
$p_{i.}$ Is the row total of area for the category *i*;
$p_{.i}$ Is the column total of area for the category *i*.

To explore the internal conversions between different LCs, which took place in two compared periods, we treat the change (decrease or increase) of an LC in a given year relative to the compared year as a result of several "loss or gain" conversions. Thus, for any type of "conversion loss to" or "conversion gain from", the percentage taken by this type in the total "loss or gain" conversion of an LC is calculated as:

$$\begin{cases} P_{loss(i),j} = (p_{j,i} - p_{i,j})/(p_{i.} - p_{.i}) \times 100 \quad i \neq j \\ P_{gain(i),j} = (p_{i,j} - p_{j,i})/(p_{i.} - p_{.i}) \times 100 \quad i \neq j \end{cases} \tag{3.2}$$

where

$P_{loss(i),j}$ Is the percentage taken by type *j* in the total "conversion loss" of category row *i*;
$P_{gain(i),j}$ Is the percentage taken by type *j* in the total "conversion gain" of category row *i*;
$p_{i,j}$, $p_{j,i}$ Is the individual entry in the matrix

In order to regionalize the regional types of rural housing land changes in the study area, an Aggregation Index (AI) is calculated to reflect the characteristics of the spatial pattern of changed rural housing land in every prefecture. We take it as our base indicator of regionalization. The value of AI is between 0 and 100 and it is calculated by the following equation[2]:

$$AI = \left[\frac{g_{ii}}{\max - g_{ii}}\right](100) \tag{3.3}$$

where

g_{ii} Is the number of like adjacencies (joins) between pixels of patch type (class) i based on the single-count method;
$\max - g_{ii}$ Is the maximum number of like adjacencies (joins) between pixels of patch type (class) i based on the single-count method

Although the lack of data concerning rural housing is one of the major reasons hampering rural housing studies, and the lack of explanatory theories may be another factor. Land use transitions are often studied with long time-series statistics. Although global environmental data have grown substantially in recent decades (Mather 2005), time-series data for China are unreliable, because the statistical system was disrupted several times since 1949 by political and economic campaigns. However, China has a vast territory with obvious regional differentiation in the level of socio-economic development, which makes it is possible to apply a spatial comparative research method to study land use transition. This method is using the differentiation in regional development to compensate for data deficiencies in long time series, reflecting the change process of rural housing land.[3]

[2]McGarigal, K., S. A. Cushman, M. C. Neel, and E. Ene. 2002. *FRAGSTATS: Spatial Pattern Analysis Program for Categorical Maps.*

[3]We assume that regions with lower levels of development represent the situation in earlier time periods; while more developed regions would represent more recent trends in land use transition.

3.3.4 Results

3.3.4.1 Land-Use Changes in the Study Area

From Table 3.2 it can be seen that LC has shifted greatly over the period from 1986/1987 to 1999/2000 in TYR. In this period, paddy fields and dry land decreased by 2.8% and 0.9%, respectively. In contrast, rural housing land and urban and other build-up land increased by 15.9% and 45.7%, respectively, in 1999/2000. The changing extent of other land-use types is relatively small. In order to explore the internal conversions between different LCs, we calculated the percentages taken by corresponding types in loss or gain conversions by formula (3.2) utilizing the data in Table 3.2. For example, the percentage taken by paddy fields (PF) in the gain conversions of rural housing (RH) equals "(142.7 − 22.1)/(1082.9 − 934.7)", i.e. 81.4%. Table 3.3 illustrates the percentages taken by corresponding land-use classes in such loss or gain conversions in TYR from 1986/1987 to 1999/2000. From Table 3.3, it is clear that the decrease of paddy fields in 1999/2000 was mainly caused by the increase of urban and other build-up lands (36.7%) and rural housing land (35.4%), and the shrinking of dry land was also mainly caused by these two with 52.6% and 21.1%, respectively. However, in terms of spread of rural housing land and urban and other build-up land in 1999/2000, almost all of these changes occurred at the expense of paddy fields and dry land, which account for 81.4% and 12.8% for rural housing land, and 66.6% and 25.3% for urban and other build-up land (Table 3.3). To a large extent, land-use change in TYR from 1986/1987 to 1999/2000 was characterized by a serious conversion of paddy fields and dry land to rural housing land and urban and other build-up lands.

3.3.4.2 Regional Types of Rural Housing Land Change

The rural housing land takes up nearly 65% of the construction land in the study area (Table 3.2). Rural housing land was identified from other construction lands according to its special image characteristics, which include obvious geometrical shape, clear boundary and diversified configuration. Information about changes in rural housing land was derived from two reclassified maps from 1986/1987 and 1999/2000. The results demonstrated that changes in rural housing land were mostly gained from other land-use types (mainly in paddy fields), and only small areas of rural housing land were changed into other types of land (Table 3.2). The patches of new rural housing land are small and scattered. In order to reveal the changed spatial pattern and to identify regional types of change, the map data (vector format) were converted into a raster format with a spatial resolution 1000 × 1000 m using ESRI's ArcGIS spatial analyst module. AIs of 42 prefectures and five counties (with the large area) in the source area of the Yangtse River were calculated by the formula (3) and arranged in the sequence from the lower reaches to the upper reaches (Fig. 3.3).

Table 3.2 Land use change matrix of each compared LULC in 1986/1987 and 1999/2000, and changes in 1999/2000 (Long et al. 2007)

LULC in 1999/2000 (10^3 ha)	LULC in 1986/1987 (10^3 ha)								Total	Changes in 1999/2000	
	PF	DL	FL	GL	WB	UB	RH	UL		10^3 ha	%*
PF	**11596.1**	154.0	46.9	25.4	51.4	5.9	22.1	1.3	11903.2	−340.2	−2.8
DL	182.6	**9021.2**	402.8	300.1	12.4	5.2	12.9	1.1	9938.1	−90.1	−0.9
FL	63.5	441.1	**28065.2**	3040.2	36.4	1.4	2.5	566.4	32216.7	−123.2	−0.4
GL	25.7	299.7	3207.7	**29513.1**	166.3	0.7	2.8	3395.9	36611.9	137.5	0.4
WB	99.7	23.2	35.0	101.7	**4461.1**	0.7	2.0	313.6	5037.1	−40.1	−0.8
UB	130.7	52.6	7.9	2.6	4.0	**394.3**	3.9	1.1	597.0	187.3	45.7
RH	142.7	31.8	10.3	4.7	2.2	0.9	**886.6**	3.7	1082.9	148.2	15.9
UL	2.4	4.7	564.2	3486.6	343.5	0.6	1.9	**10478.7**	14882.6	120.7	0.8
Total	12243.4	10028.2	32339.9	36474.4	5077.2	409.7	934.7	14761.9	112269.4		

Note PF = paddy fields, DL = dry land, FL = forest land, GL = grass land, WB = Water body, UB = urban and other build-up land, RH = rural housing land, UL = unused land. * The percentages of changes were calculated by formula (3.1), for example, that of PF equals "(11903.2 − 12243.4)/12243.4", i.e. −2.8%

Table 3.3 Internal conversions between LCs from 1986/1987 to 1999/2000 and the percentages are taken by corresponding types in such loss or gain conversions (Long et al. 2007)

LC	Type (1)	Percent (%)	Type (2)	Percent (%)	Type (3)	Percent (%)	Type (4)	Percent (%)
Paddy fields (PF)$^+$	UB	36.7	RH	35.4	WB	14.2	DL	8.4
Dry land (DL)$^-$	UB	52.6	FL	42.5	PF	−31.7	RH	21.1
Forested land (FL)$^-$	GL	136.0*	DL	−31.1	PF	−13.5	RH	6.3
Grass land (GL)$^+$	FL	121.8*	UL	−65.9	WB	46.9		
Water body (WB)$^-$	GL	160.8*	PF	−120.5	UL	74.6		
Urban and other build-up land (UB)$^+$	PF	66.6	DL	25.3	FL	3.4		
Rural housing land (RH)$^+$	PF	81.4	DL	12.8	FL	5.2		
Unused land (UL)$^+$	GL	75.1	WB	24.8	DL	3.0		

Note −Conversion loss to, +conversion gain from, * "conversion gain" occurred even the net change was "conversion loss" (and vice versa)

Through analyzing the changes in the AI and considering the continuity in the territory, five regional types, from the lower reaches to the upper reaches, were defined as follows: Shanghai-Chaohu, Tongling-Yichang, Enshi-Chongqing, Luzhou-Diqing and Ganzi-Yushu. Figure 3.4 is a gridded map illustrating the regional types of rural housing land change in the study area. It can be seen from Table 3.4 that there are obvious differences among these five regions in both physical and socio-economic aspects.

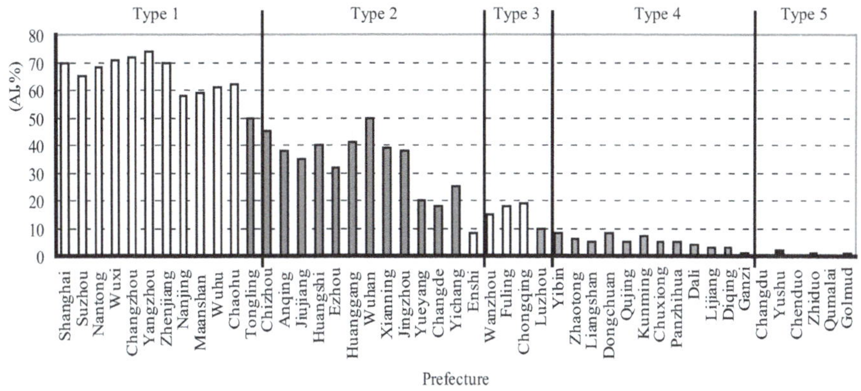

Fig. 3.3 Aggregation Index (AI) of every prefecture in the study area (Long et al. 2007)

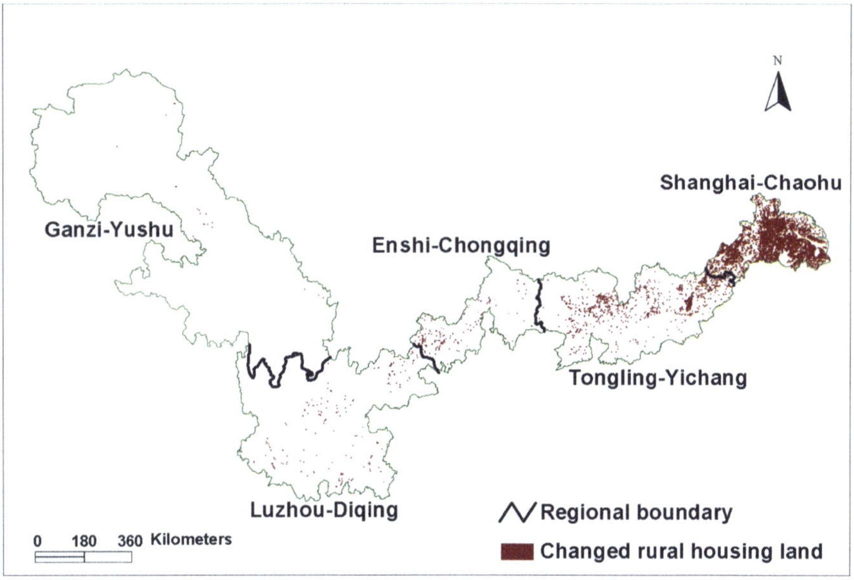

Fig. 3.4 Regional types of rural housing land change in the study area (Long et al. 2007)

3.3.4.3 Analysis of Rural Housing Land Transition

It can be seen from Fig. 3.5 that the proportion of rural housing in the increase of total construction land declines gradually from the upper reaches to the lower reaches, i.e. from Ganzi-Yushu to Luzhou-Diqing, Enshi-Chongqing, Tongling-Yichang and to Shanghai-Chaohu. The proportion in developed areas is lower than that in developing areas, even less-developed areas, because of rapid urbanization. However, the change of the curves in Fig. 3.5 is not only affected by a single factor of urbanization but also

Table 3.4 Comparison of selected physical and socio-economic indicators in different regional types (Long et al. 2007)

Region	The average value of AI	Number of included counties	Total area (thousand km^2)	Annual average precipitation (mm)	Coldest month average temperature (°C)	Average elevation (m)	Population density (persons/km^2)	Rural population density (persons/km^2)	Economic Density (10^4 RMB¥/km^2)	Possessing grain per capita (kg)
Shanghai-Chaohu (Type 1)	66.36	59	68	1120	−2.5	12.7	726.8	605.3	838.72	514.6
Tongling-Yichang (Type 2)	36.23	87	173	1333	−1.2	179.1	329.8	289.7	269.13	449.2
Enshi-Chongqing (Type 3)	15	35	90	1142	0.4	742.4	320.6	279.2	176.27	459.7
Luzhou-Diqing (Type 4)	5.75	95	267	1021	−1.9	1843.7	134.6	119.6	48.01	382.2
Ganzi-Yushu (Type 5)	0.72	36	522	452	−18.4	4260.7	3.6	2.9	1.03	265.8

Source See Fig. 3.2

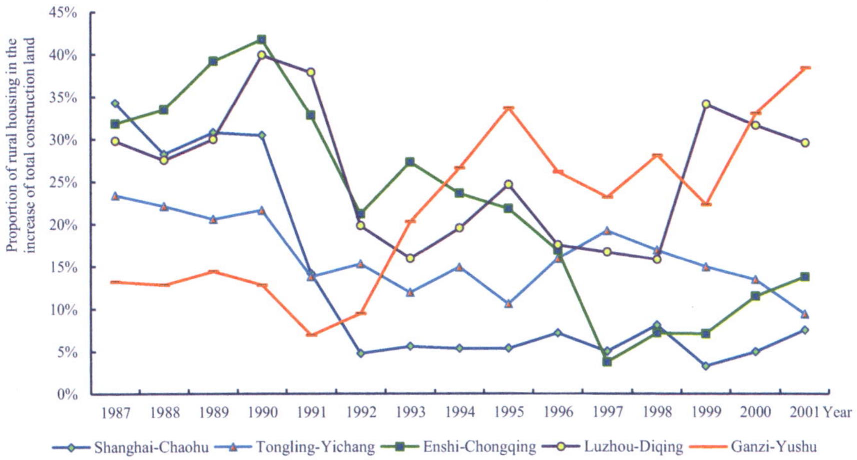

Fig. 3.5 Change curves of the annual proportion of rural housing in the increase of total construction land in different regions (Long et al. 2007)

influenced by physical factors, farmer's income, concerned policy and other factors. The decline of farmer's income will naturally lower the proportion of rural housing in the increase of total construction land; e.g., there is an obvious descending trend in every curve in 1991 and 1992 (Fig. 3.5), which is related to the continuing decline in the price of agricultural products from 1990 to 1993 in China.

In contrast with the other four regions, the region of Ganzi-Yushu is characterized by the lowest economic developmental level, where economic density is only 10.3 thousand RMB¥ per km^2, a vast territory with a sparse population where population density is merely 3.6 people/km^2 and the most adverse circumstances with average elevation above 4200 m and coldest month average temperature of −18.4 °C (Table 3.4). Before 1990, the annual share of rural housing in the increase of total construction land is relatively low, because local people were mainly herdsmen who usually roved around as nomads. However, the share of rural housing in the increase of total construction land has increased by a big margin since the policy of "Si Pei Tao" (which means to form a complete set with four kinds of measures, i.e. surrounding pasture with railings and implementing rotated herding, building covered pen for livestock, cutting grass to store as forage and building settlements for nomad) was carried out since 1990. Currently, the annual share is still rising and has reached 38.4% in 2001, although there was a bit of decline because of widespread suffering from big snow-disasters in this region both in 1995 and in 1999 (http://www.nj.qhei.gov.cn/gzdt/ysz.shtml).

The economic developmental level in the region of Luzhou-Diqing is higher than that of Ganzi-Yushu, but lower than that of the three eastern regions. Its local socio-economic development mainly focuses on agriculture. The average elevation in most part of this region is above 1500 m (Fig. 3.2), and most of the area belongs to a special region, such as mountainous or hilly areas, frontier regions and regions with ethnic minorities, in which the area limitations of rural housing are permitted to be more flexible. The annual proportion of rural housing in the increase in total construction land has been at a high level, although it has fluctuated sometimes, and reaches 29.5% in 2001. However, it can be seen from the curve in Fig. 3.5 that the increase in the proportion in this region is lower than that in Ganzi-Yushu.

In the region of Enshi-Chongqing, local economic development is at a medium level in TYR (Fig. 3.2 and Table 3.4), but recently the speed of economic development has accelerated, which caused the share of rural housing in the increase of total construction land to decline. It declined especially around 1997 because Chongqing was set-up as a municipality directly under the Central Government in 1997, including Qianjiang prefecture and three cities in the rank of the prefecture, i.e. Chongqing, Wanxian and Fuling, in Sichuan province formerly. Increased urban and other build-up lands far exceeded that of rural housing land around 1997, which caused the share of rural housing in the increase of total construction land between 1997 and 1999 less than 10%. Only by 2001 has the share reinstated to the so-called normal level with 13.7%.

Influenced by the developed coastal areas, the economic developmental level in the region of Tongling-Yichang is higher than that in Enshi-Chongqing. The annual proportion of rural housing in the increase in total construction land in this region also has an overall declining trend, from 23.4% in 1987 to 9.3% in 2001.

The region of Shanghai-Chaohu is characterized by a smooth relief, favorable climate, dense population and the highest economic developmental level in TYR (Fig. 3.2 and Table 3.4). The annual share of rural housing in the increase of total construction land in this region has maintained the level of 5% or so, after a rapid decline between 1990 and 1992. This demonstrates a trend of equilibrium between rural housing and other construction activities.

In general, the change curves in Fig. 3.5 illustrate the comparative role between the urban region and rural area in the aspect of regional socio-economic development. It is found that every region is in a different stage of economic growth, respectively, through calculating the value of purchasing power parity GDP per capita (PPPGDPpc) of every region (Table 3.5).

By calculating the average proportion of rural housing in the increase of total construction land in every region between 1992 and 2001,[4] the change trendline shows that the average proportion decreases gradually from the upper reaches to the lower reaches in TYR. The power change trendline can also be regarded as the curve of regional land use transition, indicating the rural housing land transition phases for every region (Fig. 3.6).

[4]To some extent, analyzing the proportion of rural housing in the increase of total construction land after 1992 will be more helpful to reflect the underlying laws of the proportion change, because only after implementing the policy of "Si Pei Tao" since 1990 were settlements for nomad in the region of Ganzi-Yushu begun to be built; moreover, there is an obvious descending trend of the proportion in every region (Fig. 3.5) in 1991 and 1992 because of the declining price of agricultural products.

Table 3.5 Corresponding economic growth stage of every region in TYR divided by PPPGDPpc (US$ of 1998) (Long et al. 2007)

Region	The values of PPPGDPpc	Economic growth stages classified by Li and Hou (2001) according to the values of PPPGDPpc
Shanghai-Chaohu	5891	The intermediate phase of industralization (5350–8590)
Tongling-Yichang	4166	The primary phase of industralization (3010–5350)
Enshi-Chongqing	2807*	The phase of producing primary products (1700–3010)
Luzhou-Diqing	1821**	
Ganzi-Yushu	1443	Originality phase (<1700)***

Notes (1) The values of PPPGDPpc (US$ of 1998) of every region were calculated by dividing GDP per capita in a region by the coefficient of 1.959 according to the study of Li and Hou (2001), and the data of GDP per capita are from China's State Statistical Bureau. (2) Six economic growth stages were classified by Li and Hou (2001) according to the values of PPPGDPpc (US$ of 1998) from 1700 to 22730, i.e. phase of producing primary products, primary phase of industrialization, intermediate phase of industrialization, senior phase of industrialization, primary phase of developed economy, and senior phase of developed economy. The author called: * is the final phase of producing primary products; ** is an early phase of producing primary products, and *** economic growth stage whose value lowers than 1700 is the originality phase

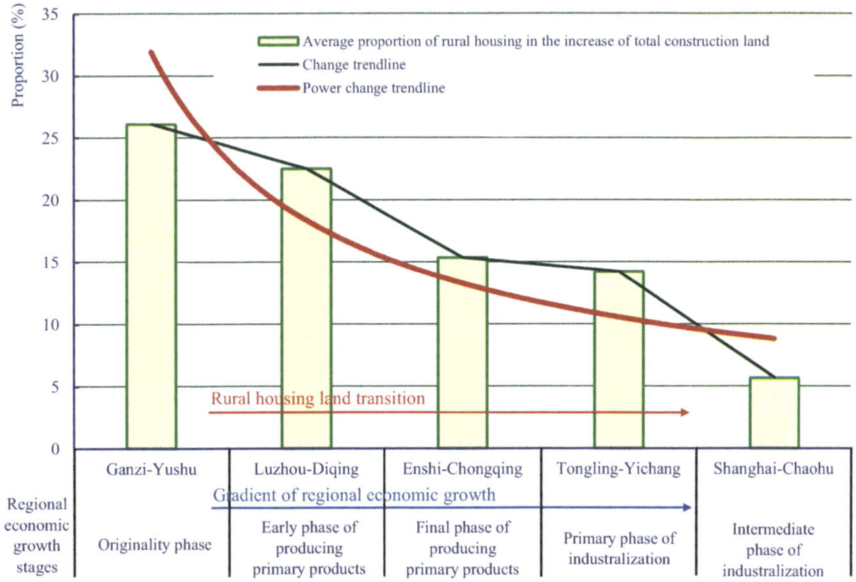

Fig. 3.6 Corresponding rural housing land transition phase and economic growth stage of every region in the study area (Long et al. 2007)

3.3.5 Discussion and Conclusions

According to our research, the concerning management measures for rural housing land in TYR are suggested as follows:

(1) Rural housing land in the regions of Ganzi-Yushu, and Luzhou-Diqing will most likely increase in the next decades. At present, the rural housing land per capita in Yushu prefecture is 126.6 m^2, which is lower than that in Hainan prefecture (with 374.4 m^2 per capita), one of the relatively developed areas in the same Qinghai province. However, the nomads in the two prefectures have only begun to build settlements. The present settlement area is still small and will increase with the development of the local economy. Accordingly, the concerned local management departments should pay more attention to guiding and carrying out scientific village planning, so as to promote the construction of the village's infrastructure and optimize villager's living conditions

(2) The proportion of rural housing in the increase of total construction land in the two regions of Enshi-Chongqing, and Tongling-Yichang is declining gradually and tending towards an equilibrium between rural housing and other construction activities. Although the rural housing in these regions can fundamentally meet the needs of local people, the desire to improve their living conditions is getting stronger with the development of the local economy. Farmers prefer to build their houses in the valleys (instead of the hillside), and they prefer to settle close to roads and other available infrastructure. The houses usually have a small-scale household (in which only three persons or so live), and larger courtyards. There is a short cycle of replacing an old house with a new one. As a result, there is more idle rural housing land, vacant rural housing, with unoccupied land in villages, and often one household owns two or more houses. Hence, local management departments should actively promote the consolidation of rural built-up land, set down and implement rational plans of consolidating villages according to the general land use planning and local developing strategy, in order to enhance the intensity of land-use in town and village and save land for further needs. Meanwhile, local governments should register idle rural housing land, vacant rural housing, unoccupied land in villages, and cases where one household owns two or more houses according to local conditions, in order to estimate the actual potential of existing build-up land. When houses are reconstructed, extended or newly built, farmers should make full use of unoccupied land in villages, old housing land, uncultivated slope land and derelict land. Compensation schemes should be implemented by the local government so as to let villagers who have vacant rural housing or two or more empty houses return them to the collective organization.

(3) There is already a trend of equilibrium between rural housing and other construction activities in the region of Shanghai-Chaohu. Idle and vacant rural housing land is more prominent in this region than in other two regions of Enshi-Chongqing and Tongling-Yichang because of its fast urbanization step,

and because more peasants are working in the city. There are also more rural-urban migrants. In order to enhance the utilization of existing rural housing land, it is necessary to find an effective way to shift the utilization model of rural housing land. Therefore, some encouraging and restraining mechanism (in the form of the policy) should be implemented to utilize rural housing land intensively, e.g., improving legal and managerial system, perfecting the benefit allocation system and standardizing existing market system about shifting the utilization model of rural housing land.

Although China is rapidly transforming itself from an agricultural into an industrial nation, sustainable rural development will be an important theme in developing China for a long time. In order to protect the cultivated land, on the one hand, we should control the expanding of cities and should pay more attention to curbing excessive rural housing and consolidating rural housing land on the other hand. Land use planning and management are one of the most important measures for economic development. For the purpose of strengthening the management of land resources and enhancing their capability for future economic development, the author argues that there are some problems in the current rural housing land managerial system in China, such as there are considerable differences in the regulations for rural housing among provinces, even between the provinces with similar physical and socio-economic conditions (Table 3.1). This situation will do harm to the intensive utilization of land resources and the implementation of national measures concerning land resources. Usually, a region has its particular rural housing land transition phase corresponding to its socio-economic development level, which needs particular management measures. So, the Central Government is suggested to set down uniform regulations for rural housing according to regional socio-economic developmental level, physical conditions and rural housing land transition phase.

The lack of long-term statistical data on land use in China is a barrier to the research of land use transition. However, China's vast territory and obvious regional differentiation in the level of socio-economic development make it possible to use a spatially comparative research method. By taking the differentiation in regional development we can compensate data deficiencies in a long temporal series, for studying land use transitions. The concept of Transects is a valuable research tool for land-use research. It can be considered a new integrated approach because it allows us to combine land-use change with socio-economic development.

In the five regional types of rural housing land change defined according to the distribution characteristics of an aggregation index, research results indicated that each region is in a different phase of rural housing land transition, which corresponds to a particular socio-economic developmental level. This confirms above hypothesis that the development of rural housing in every region will undergo specific stages—the proportion of rural housing in the increase of total construction land will decline gradually with the development of the local economy, and the end of the transition corresponds to a new equilibrium between rural housing and other construction activities. Finally, some managerial countermeasures and policy implications were obtained by applying this research to land management issues. The author argues that there are some problems in the current rural housing land managerial system in China, and the Central Government needs to set down uniform regulations for rural housing according to regional socio-economic developmental level, physical conditions and rural housing land transition phase.

3.4 A Perspective of Population Change

3.4.1 Background

It is well known that China is a nation with strong rural roots. Despite rampant urbanization, 52% of its population still lived in rural areas in 2009, and the national economy has been built on agricultural foundations since ancient times. However, China has transformed rapidly since Deng Xiaoping launched the economic reforms of 1978. The traditional centralized economic system changed to being market-based and the economy, which was once primarily agricultural, has become increasingly urban and industrial (Long and Woods 2011; Long et al. 2011). Whereas during the 1980s and 1990s the restructuring of rural China was largely driven by internal national dynamics and reforms, and arguably had a limited impact outside China, the outcomes of the current policies for rural development will be of global significance because rural China is increasingly integrated into global social and economic networks (Long and Woods 2011). Global integration has also brought convergence between the dynamics driving change in rural China, and those experienced in other parts of the world. Increased political, economic and cultural openness has exposed rural localities in China to globalization processes—such as global trade and economic competition, international migration, cultural assimilation, and political issues such as climate change and food security—that are already contributing to the reconstitution of rural places elsewhere (Woods 2007).

Recently, the restructuring process in rural China has been influenced to a great extent by such national macroeconomic development strategies as implementing a household responsibility system, developing township and village enterprises (TVEs) and building a new countryside. These have resulted in tremendous interconnected changes in the rural population, lifestyles, employment structure, industrial structure and community organization (Li et al. 2010; Long et al. 2010, 2012; Putterman 1997; Unger and Chan 1999). For example, rural population change also directly influences the utilization and management of rural land. Since rural communities are blighted by depopulation and the abandonment of buildings and land, a unique phenomenon of 'village-hollowing' came into being in China (Long et al. 2012). This part aims to examine the pattern and causes of population and settlement change in China and its effects on China's rural development.

3.4.2 Materials and Methods

China has two different statistical methods of measuring rural population: registered and resident populations. The rural registered population is based on household registration (hukou) and on the administrative classification of rural areas from which the sampling frame for the rural household survey is derived. On the other hand, rural resident persons have lived in their locality for more than six months in the year preceding the census/survey. The rural settlement, also called rural housing/residential land, comprises land utilized by rural residents for dwelling and living, that is, land for building houses and other structures or affiliated facilities (Long et al. 2007).

In this study, national rural registered population data were collected from the China Population Statistics Yearbook (1993–2006) and the China Population and Employment Statistics Yearbook (2007–2010), while national rural resident population data were from the China Population Statistics Yearbook (2001) and the China Statistics Yearbook (2001–2010) (Li et al. 2010). Provincial rural resident population data and rural registered population data were from the China Population Statistics Yearbook (2001–2006) and the China Population and Employment Statistics Yearbook (2007–2009) (Li et al. 2010). We focused our study on the period of 2000–2008 as provincial rural resident population data have only been published since 2001. National and provincial land use data (1996–2008) were from the Department of Land and Resources of each province (autonomous region or municipality). The ESRI's ArcGIS 9.2 spatial analysis module was used to analyze the databases and to identify spatial relationships.

3.4.3 Results

3.4.3.1 Population Change in China's Countryside

Figure 3.7 shows that since economic reform and an open-door policy were adopted in 1978, the rural resident population and the rural registered population entered a transitional period of rapid decline in 1996 and 2001 respectively. To put it more precisely, the rural registered population has gone through a rapid decline since 2001 following a period of rapid growth from 1978 to 1991 and slower growth between 1992 and 2000. The rural resident population shows rapid growth from 1978 to 1995 and a rapid decline since 1996. The deviation between the rural registered population and the rural resident population peaked at 170 million in 2009, most of whom flocked into cities and, thus, a floating population came into being. Meanwhile, improved economic conditions in rural regions and diversification of employment away from agriculture have reduced the traditional dependency on the 'big family'. Traditional models of the 'coalition family' and the 'stem family', with diverse family members forming a single intergenerational household, are being replaced by smaller nuclear

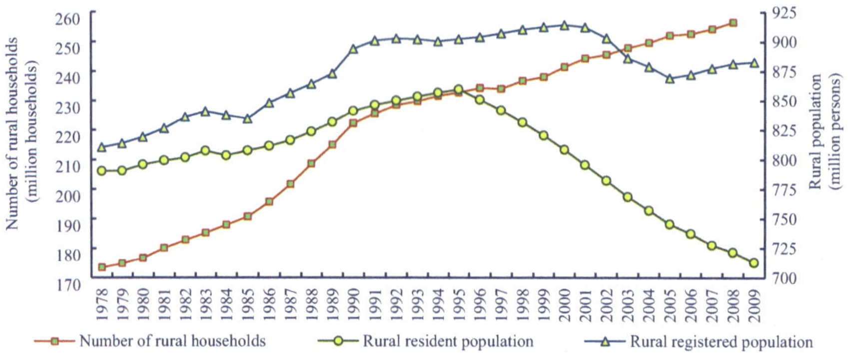

Fig. 3.7 Change of China's rural population and households during 1978–2009 (Long et al. 2013)

family households (Long et al. 2012). The number of rural households in China increased from 173.5 million in 1978 to 256.6 million in 2008 (Fig. 3.7).

Figure 3.8 shows that the areas with the rapid decline of the rural registered population were mainly located in the economically prosperous eastern coastal provinces which have experienced rapid urbanization and industrialization, while demographical change in the northern provinces was relatively slow. However, there was an obvious increase of rural registered population in central and southwest rural China where there are low levels of economic development. The rural resident population of the eastern coastal provinces also declined rapidly, as well as that of the major grain-producing areas with huge migrant and surplus rural labor. There is a universal decline trend of the rural resident population in China except for a minor increase in Guizhou, Tibet, Xinjiang and Shanghai (Fig. 3.8).

3.4.3.2 Settlement Change in China's Countryside

China's rural residential land has been growing at an annual rate of 0.12%, representing an average annual increase of 20 thousand ha since 1996 (Fig. 3.9). However, there is an obvious provincial difference, with a decline in the central provinces due to the emigration of individuals and families from rural areas and the reclamation of their settlements for rural purposes while a significant increase took place in the surrounding provinces (Fig. 3.10).

Figure 3.11 shows that China's rural residential land per capita in terms of rural resident population continued to increase by 20.76% from 1996 to 2008, and it also increased by 3.60% in terms of rural registered population, which exceeded the national standard limit of 150 m^2 (MCC 1993) by 55.80% and 25.79%, respectively.

In 2008, the spatial pattern of China's rural residential land per capita, in terms of rural registered population, showed that the value is high in the north and east and low in the south and west, and the value in 24 provinces exceeded 150 m^2. However, in terms of the rural resident population, it showed an obvious pattern of high-values

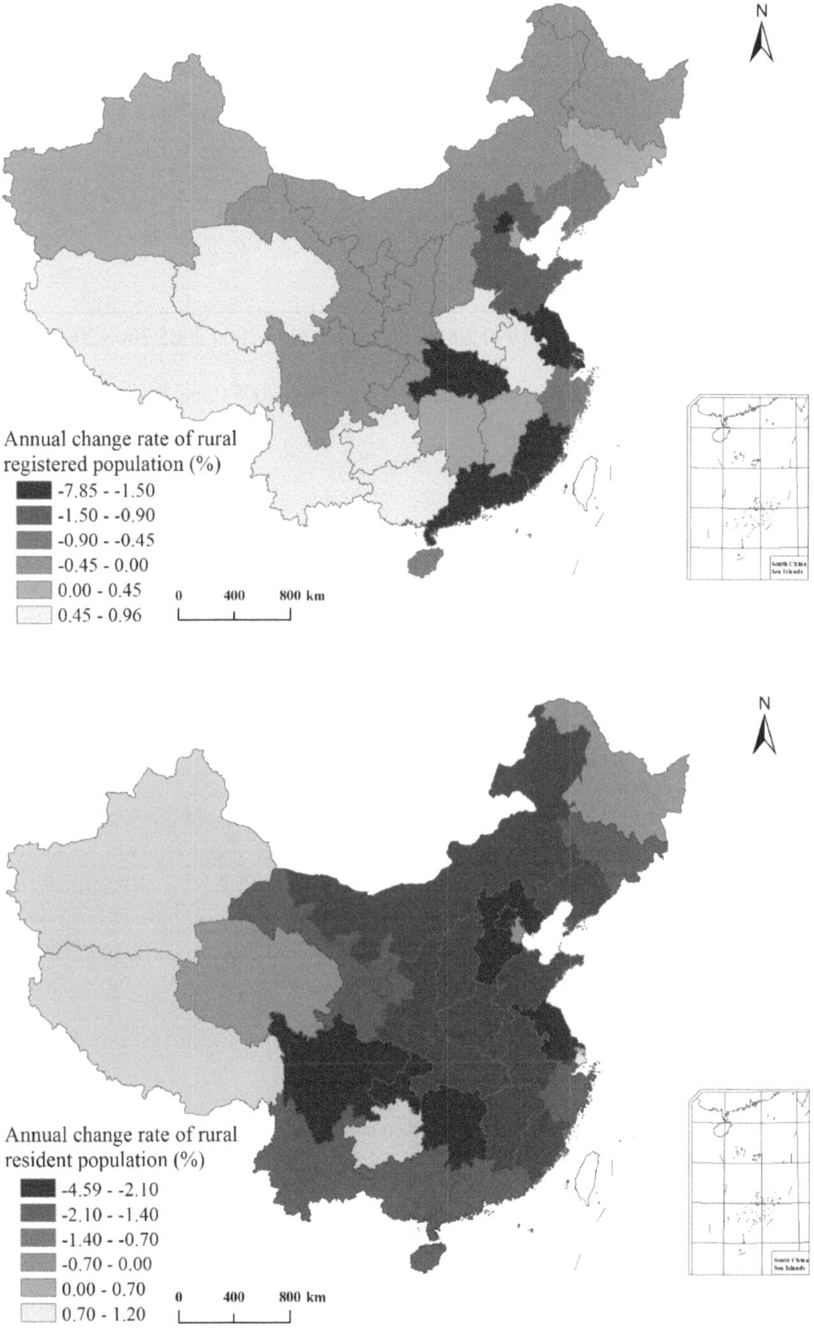

Fig. 3.8 Spatial pattern of China's rural population change during 2000–2008 (Long et al. 2013)

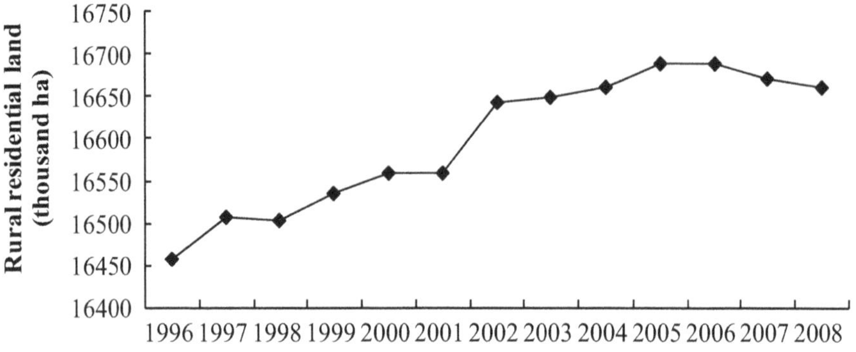

Fig. 3.9 Change in rural residential land in China during 1996–2008 (Long et al. 2013)

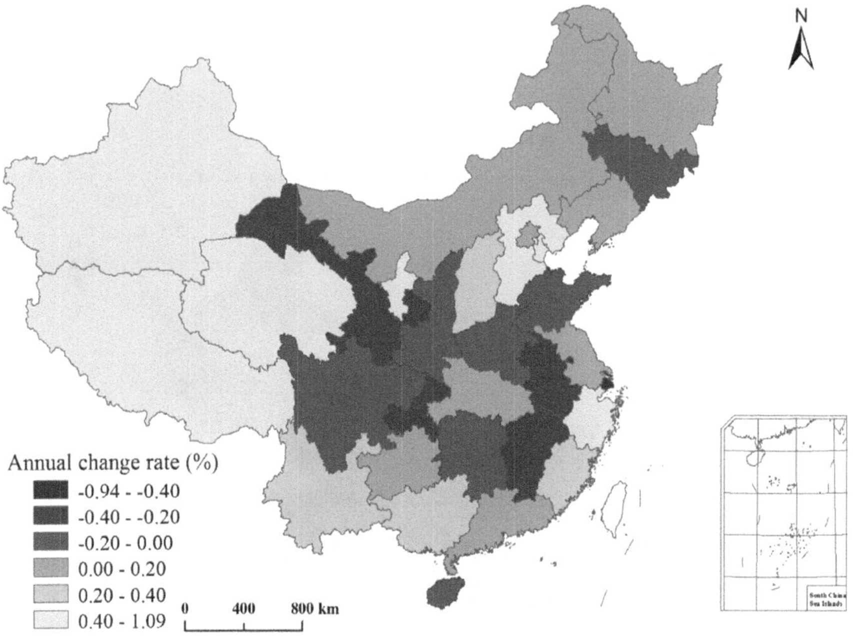

Fig. 3.10 Spatial pattern of rural residential land change in China during 2000–2008 (Long et al. 2013)

in the north and low-values in the south, and there were 27 provinces with the value exceeding the national standard (Fig. 3.12).

During the period 2000–2008, the rural residential land per capita in terms of the rural registered population took on a complex pattern, but there was obvious growth in eastern coastal China; however, in terms of the rural resident population, all the provinces except one experienced an increasing trend (Fig. 3.13).

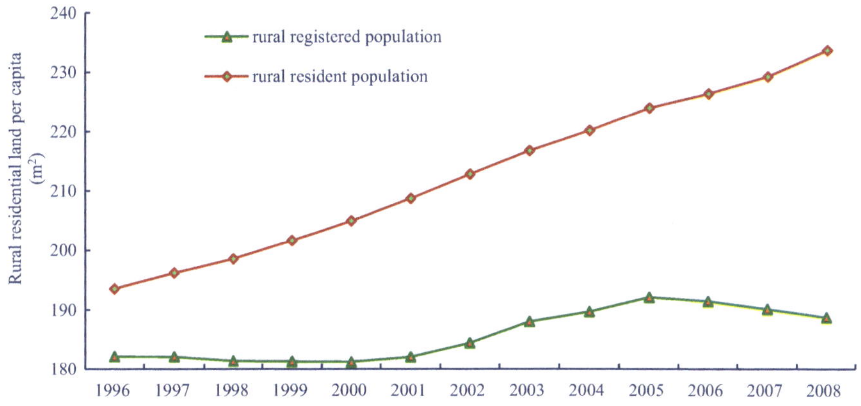

Fig. 3.11 Change of rural residential land per capita in China during 1996–2008 (Long et al. 2013)

3.4.3.3 Causes

To a great extent, the patterns of population and settlement change in China have been shaped by the 'dual-track' structure of rural-urban development, including the contribution of economic, institutional-managerial, socio-cultural and environmental factors (Long et al. 2012). The rural population migrating to urban areas in search of highly paid jobs effectively became 'floating'. The post-1978 economic liberation together with the growth of TVEs increased real per capita incomes and enabled investment in new more spacious and comfortable houses after abandoning the old ones. However, in the context of institutional and managerial factors, the absence of rules and regulations aimed at dealing with vacant or abandoned housing means that local governments are limited in their capacity to act to address the problem of irrational expansion of rural settlement. The 'dual-track' policy of rural-urban development in China dictates that the millions of rural migrants to cities are not granted permanent household registration in cities and are not eligible for many social welfare entitlements and employment opportunities that are reserved for permanent urban residents, which encourages many rural migrants to work and live in the cities but keep vacant rural housing to give the opportunity of returning home (Li et al. 2010). In some cases, villagers are forced to emigrate and resettle in response to both natural and human induced local environmental conditions.

The factors mentioned above combine to drive the rural population and settlement change in China. And the process is embedded in the wider context of economic restructuring and rural-to-urban migration in China. More importantly, it will evolve differently over time and may have different effects in different places (Long et al. 2007, 2012).

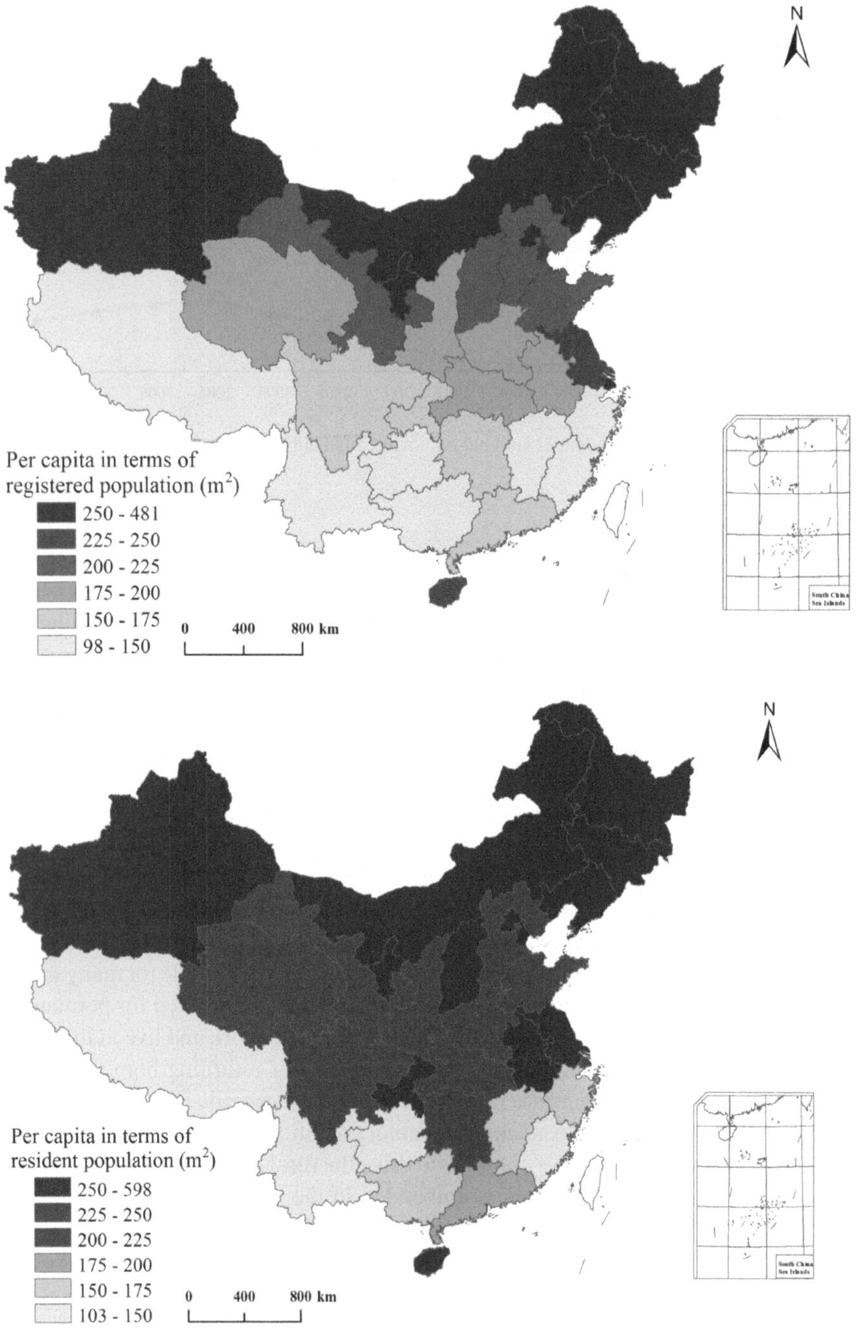

Fig. 3.12 Spatial pattern of rural residential land per capita in China in 2008 (Long et al. 2013)

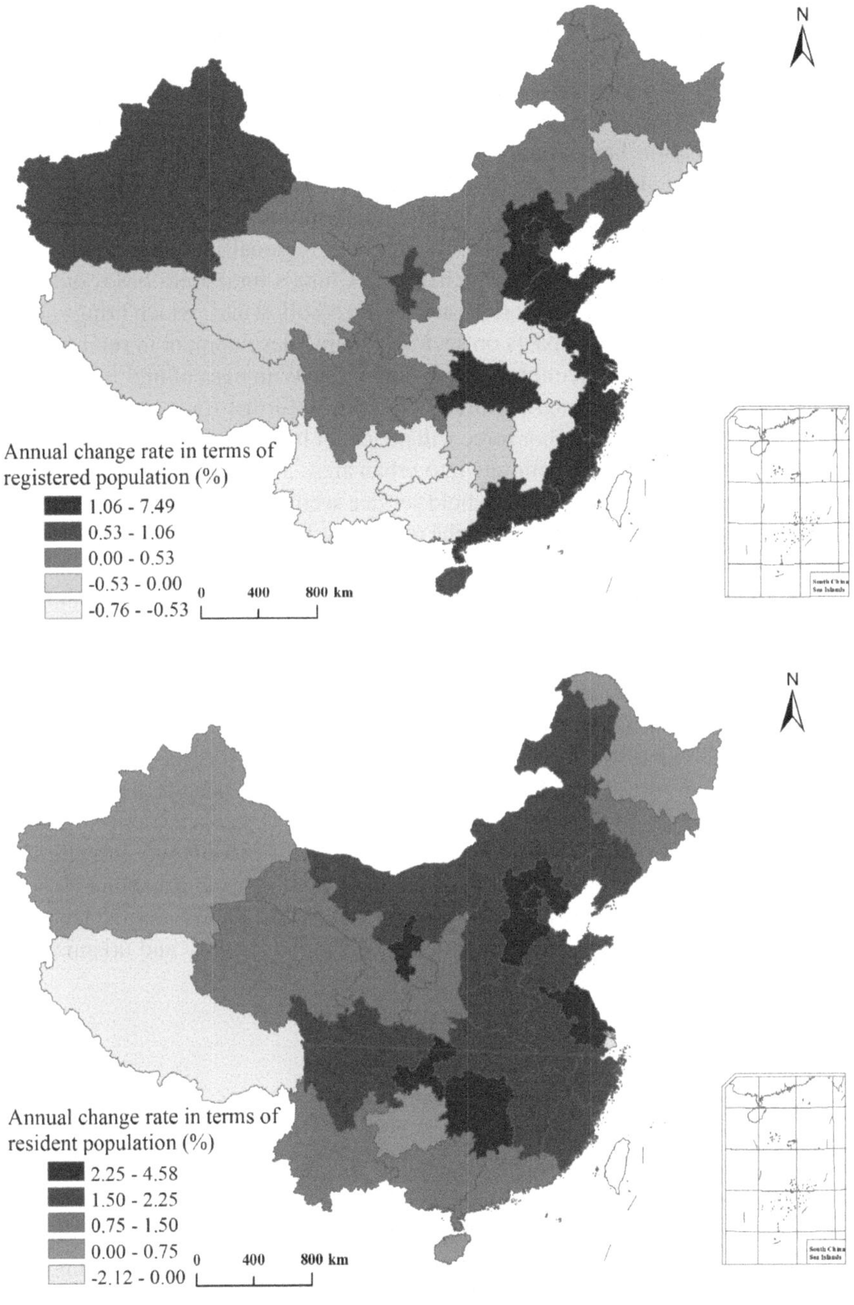

Fig. 3.13 Spatial pattern of rural residential land per capita change in China during 2000–2008 (Long et al. 2013)

3.4.3.4 Effects of the Change in China's Rural Development

Rural industrialization and modernization are closely interrelated in the context of migration, employment, land use and environments. Reform-induced industrialization and urbanization have rapidly altered the physical and human landscapes in China's rural areas, as evidenced by the substantial rate of rural housing development, rural poverty, rural-to-urban migration, agricultural to non-agricultural land conversion, widening rural-urban income gap, and regional inequalities (Li et al. 2010). The population and settlement change in China's rural areas has resulted in the phenomena of 'floating population' and 'village-hollowing', which brings about both positive and negative impacts on socio-economic development in rural China.

Traditional Chinese agriculture is labor-intensive. With agricultural modernization and a contraction of the agricultural sector, due to further trade liberalization, a large percentage of the rural labor force will no longer be needed in this sector. These people will inevitably have to migrate into urban areas and find jobs in manufacturing industries, construction and household service sectors. In the long run, however, these rural migrants will not go back to the rural areas but will try to stay in the towns and cities, where incomes are higher and living conditions are better. Since most of these migrants are young and better educated than the remaining rural population, there is a significant rural to the urban brain-drain. The countryside is losing its most active population segment (Long et al. 2010). As such, the major negative effects of the population and settlement change involve the loss of rural human capital, a huge waste of land resources due to large areas of rural construction land being unused or idle, and the increased social costs arising from the loss of infrastructure in small or remote villages in rural areas. However, there are also some positive effects resulting from this change; for example, it boosts farmers' incomes since out-migration increases scale economies in agriculture and total factor productivity growth, and shows tremendous potential for development that would assist in achieving an optimal allocation of rural land resources, releasing and protecting agricultural land and helping to strike a balance between the demands of food security and urbanization (Long et al. 2012).

3.4.4 Discussion and Conclusions

The rural population and settlement change in China has been driven by economic, institutional-managerial, socio-cultural and environmental factors. The rapid urbanization and industrialization of the country have drawn migrants from rural to urban areas, leading to a trend of rural population decline; yet most migrants are young and better educated. However, most of the 'floating' rural population cannot really settle down in the city because they have difficulty in being granted permanent household registration in cities and are not eligible for many social welfare entitlements, and need to keep their houses in the countryside. There is a dilemma in terms of rural development in China. Some highly productive farmland and rural dwellings were abandoned because the owners left to work and resettle in the city, and those staying on in rural areas are aged individuals, women, children, and sick or disabled men, which makes rural development partly paralyzed in a sense.

China is a nation with a huge population of 1.4 billion, a large proportion of which is rural. More importantly, China's national economic development is based on the development of rural areas. Although China is rapidly transforming itself from an agricultural country into an industrial nation, sustainable rural development will be an important theme in China's development for a long time in the future due to the obvious regional discrepancies, rural poverty, and rural land use issues (Long et al. 2010). Aiming at the current difficult position for rural development in China due to the change of population and settlement, the solutions may focus on paying more attention to curbing excessive rural housing and consolidating rural housing land and providing more job opportunities for the surplus rural labor by developing TVEs. In addition, some encouraging measures and policies need to be set down to stimulate the development of scale management in agriculture.

3.5 A Perspective of Hollowed Villages Evolution

3.5.1 *Background*

China has experienced drastic changes in land use since the initiation of economic reforms in 1978, especially through the loss of farmland to construction, driven primarily by urbanization and industrialization (Lin and Ho 2003; Liu et al 2008; Xie et al. 2005; Zhong et al. 2011). Between 1996 and 2008, total farmland in China decreased by approximately 8.32 Mha, or about 6.4% (MLRC 2009), whilst during the same period, China's population increased from 1224 million to 1328 million, with the proportion living in urban areas rising from 30.5 to 45.7% (NBSC 2009). As such the per capita area of farmland fell from 0.106 ha in 1996 to 0.092 ha in 2008, raising concerns about food security (Christiansen 2009; Yang and Li 2000). With the production capacity for grain and other key commodities falling with the loss of farmland, in spite of the booming population, the question posed by Brown (1995) of 'Who will feed China?' has become increasingly critical.

In response, the Chinese central government has introduced strict farmland protection objectives, with the aim of maintaining the total farmland area at no less than 1.8 billion mu (120 million ha) in the period to 2020. However, as the agricultural land area had already fallen to only 1.826 billion mu (121.7 million ha) by 2007, the target leaves little room for further loss of farmland, particularly as the trends of urbanization and industrialization show no signs of abating. The pressure on farmland come both from urban expansion (Wang and Scott 2008; Yu and Ng 2007), and from the changing use of land in rural communities (Huang et al. 2011). According to the statistical data of the Ministry of Land and Resources of China (MLRC), 'Rural construction land' (that is, land in rural areas occupied by housing and other nonagricultural buildings) in China amounted to 16.56 million ha in 2006 or just over half of all built-up land in China. Significantly, the area of rural construction land has continued to expand marginally in spite of a decreasing rural population, such that the volume of rural housing land per capita increased by 18.28% between 1996 and 2007 according to the statistical data of MLRC, and at 228 m^2 per head in 2007, which is considerably above the national standard of 150 m^2 per head (MCC 1993). In part, this reflects a process of housing modernization and improved living standards in rural communities.

3.5.2 Identifying and Classifying Hollowed Villages

3.5.2.1 Methods and Principle of Classifying Hollowed Villages

Hollowed villages are the results of altered rural land use patterns and landscape induced by human activities (Liu et al. 2011). The landscape functions shaped by the man-land interactions are necessary to be underlined when classifying hollowed villages. Accordingly, the impacts of human behaviors on rural landscapes are viewed as the fundamental principle. Paying attention to the functional characteristics, and starting with their structure to fully reflect the spatial differentiation and organizational connection of hollowed villages is highly important for this work.

There are some basic guidelines for classifying hollowed villages. (1) Current conditions should be highlighted: The status quo of the hollowed village is the result of utilization activities of human being, and it can fully reflect the basic characteristics and essential differences of the diversified villages; (2) The identity of process and morphology: Each type of hollowed villages has its own process of occurrence and evolution. The identity of occurrence and evolution of similar hollowed villages brings them common characteristics. (3) Consistency of management measures and utilization directions: Different hollowed villages have different suitability and limiting factors for different utility patterns under certain conditions, which makes it have a corresponding unique function. (4) Dominance: The factors affecting the evolution of hollowed villages are diversified, but there is often a major one, which can reflect its own functions.

3.5.2.2 Regional Types of Hollowed Villages Evolution

The hollowed village is the product of regional economic and social development. Specific socio-economic development stages correspond to certain types of hollowed villages evolution. In addition, the evolution of hollowed villages is affected by terrain and location. Regional socio-economic level and natural conditions jointly contribute to the differentiated patterns of hollowed villages evolution. Generally speaking, villages at the urban fringe are relatively economically advanced, and that in the plain region, mountainous areas, farming areas and nomadic region all have

Table 3.6 Regional types of hollowed villages (Adapted from Long 2012)

Regional type	Major influential factors and characteristics of hollowed villages	Consolidation mode
Urban-rural fringe area	Urbanization, rapid agricultural transformation, non-agricultural employment of the rural population, management vacancy	Unified allocation and planning of urban and rural land; strengthening management of rural housing land; Entire-village consolidation; curbing the construction of uncompleted-right houses
Farming plain area	Disorder construction due to lack of planning; Rampant phenomenon of "one household with multiple housing lands"	Coordinated development of urban and rural areas; encouraging the construction of central villages to promote rural construction land consolidation
Mountainous area	Mostly affected by natural factors and infrastructures, the hollowed villages are not common as plain areas	Coordinating man-land interrelations, encouraging the construction of small-scale ecological towns to relocate part of peasants
Nomadic region	Natural factors play a dominant role; Climate change and environmental degradation cause settlements hollowing	Developing ecological economy and promoting concentrated residence

their own characteristics, which determine the evolution trend of different hollowed villages. As a result, four regional types of hollowed villages evolution are proposed (Table 3.6).

3.5.2.3 Assessment and Identification Hollowed Village Evolution

Considering the multiple factors affecting the evolution of hollowed villages, institution and management, technological advancement, and the performance of social mainstream behavior have universal consistency at the national level. While the natural factors present strong regional heterogeneity. Economic development and demographic change are closely related to regional conditions, and largely determine the behavior of the land-use behavior mainstream. Hence, these two factors can be used as the dominant indicators to classify the hollowed villages, which are summarized as three categories-"non-local employment induced, agriculture dominated, and non-agriculture industry dominated". The hollowing degree and the concentration of vacant and abandoned parcels will greatly affect the countermeasures and utilization directions of hollowed villages (Table 3.7).

The hollowing degree can be measured by population density, the proportion of the permanent resident population, and the ratio of vacant rural housing land, and it is

Table 3.7 Indicators of measuring the degree of hollowed villages (Adapted from Long 2012)

Index	Calculation method	Description
Registered population density	The number of registered population density/The area of rural settlements in a village	The smaller the value, the higher the degree of hollowing
The ratio of permanent resident population	The number of permanent resident population/Total population	The smaller the value, the higher the degree of hollowing
The ratio of abandoned houses	The area of abandoned houses/The area of rural settlements in a village	The larger the value, the higher the degree of hollowing
The ratio of vacant houses	The area of vacant houses/The area of rural settlements in a village	The larger the value, the higher the degree of hollowing
The agglomeration degree of vacant and abandoned parcels	$\left\{\frac{g_{ii}}{\max - g_{ii}}\right\} \times 100\%$ *	The higher the value, the more concentrated the vacant and abandoned parcels

Note (*) McGarigal K, Cushman S A, Neel M C, Ene E, 2002. FRAGSTATS: Spatial Pattern Analysis Program for Categorical Maps

generally divided into two classes-high and low levels. The concentration degree of vacant and abandoned parcels reflects the difference in the current hollowing pattern, which can be categorized as centralization and decentralization pattern. Accordingly, based on the above analysis, we construct the classification indicators in terms of industry development, hollowing degree, and the concentration degree of vacant and abandoned parcels (Table 3.8).

3.5.3 Characteristics of Different Evolution Types

China's vast territory and obvious regional differences in socio-economic development have led to the co-existence of different types of hollowed village evolution. Due to the different locations and the intervention of external forces, the development, and evolution of hollowed villages present different characteristics. However, the complete interpretation of the hollowed villages at different development stages on the original site is most likely to occur in the urban-rural fringe, because during the urban-rural integration process, where the villages serve the city in the process of population transfer and land use change. In China, the most typical hollowed villages are in the plain farming area, while that in mountainous and pastoral is not representative. Therefore, this study mainly discusses the characteristics of hollowed villages in the rural-urban fringe and plain areas.

Table 3.8 Diagnosis of evolutive types of hollowing villages (Adapted from Long 2012)

Types	Characteristics of industry			The degree of hollowing village		The agglomeration degree of vacant and abandoned parcels	
	Non-local employment induced	Agriculture dominated	Non-agriculture industry dominated	High	Low	Concentrated	Decentralized
Non-local employment induced high-level and concentrated hollowing village	Δ			Δ		Δ	
Non-local employment induced high-level and decentralized hollowing village	Δ			Δ			Δ
Non-local employment induced low-level and concentrated hollowing village	Δ				Δ	Δ	
Non-local employment induced low-level and decentralized hollowing village	Δ				Δ		Δ
Agriculture dominated high-level and concentrated hollowing village		Δ		Δ		Δ	

(continued)

Table 3.8 (continued)

Types	Characteristics of industry			The degree of hollowing village		The agglomeration degree of vacant and abandoned parcels	
	Non-local employment induced	Agriculture dominated	Non-agriculture industry dominated	High	Low	Concentrated	Decentralized
Agriculture dominated high-level and decentralized hollowing village		Δ		Δ			Δ
Agriculture dominated low-level and concentrated hollowing village		Δ			Δ	Δ	
Agriculture dominated low-level and decentralized hollowing village		Δ			Δ		Δ
Non-agriculture industry dominated high-level and concentrated hollowing village			Δ	Δ		Δ	

(continued)

Table 3.8 (continued)

Types	Characteristics of industry			The degree of hollowing village		The agglomeration degree of vacant and abandoned parcels	
	Non-local employment induced	Agriculture dominated	Non-agriculture industry dominated	High	Low	Concentrated	Decentralized
Non-agriculture industry dominated high-level and decentralized hollowing village			Δ	Δ			Δ
Non-agriculture industry dominated low-level and concentrated hollowing village			Δ		Δ	Δ	
Non-agriculture industry dominated low-level and decentralized hollowing village			Δ		Δ		Δ

3.5.3.1 Theoretical Analysis of the Evolution Phases of Hollowed Villages

Demographic changes, employment opportunities, and land use patterns have far-reaching meanings on rural sustainability. The formation and evolution of hollowed villages are closely related to the population changes and the spatial dynamics of their employment. The reason why villages are hollowed out is that people living in the center of the village migrate out or to the fringe of the village, which is somewhat similar to suburbanization. The dynamic pattern of land use is the result of various driving forces of human.

Evidence from Europe suggests two possible models for the longer-term evolution of hollowed villages, presented here as the 'cyclical evolution model' and the 'vanishing evolution model'. The 'cyclical evolution model' draws on European experiences of urbanization, suburbanization, and counter-urbanization (Antrop 2004; Champion 2001), to propose that village-hollowing will follow a cyclical pattern moving through four phases (Fig. 3.14). In the first stage, the centralizing phase, the initial economic buoyancy produced by liberalization stimulated dramatic rural population growth that was reflected in the expansion of village centers as people migrated in from outlying properties to support cooperative agricultural production and implement household responsibility system. The progression of economic restructuring in the second stage, however, has contracted opportunities in rural communities, stimulating urbanization and the slowing down of rural population growth. This translates to a 'sub-hollowing' phase in which increasing house-building on

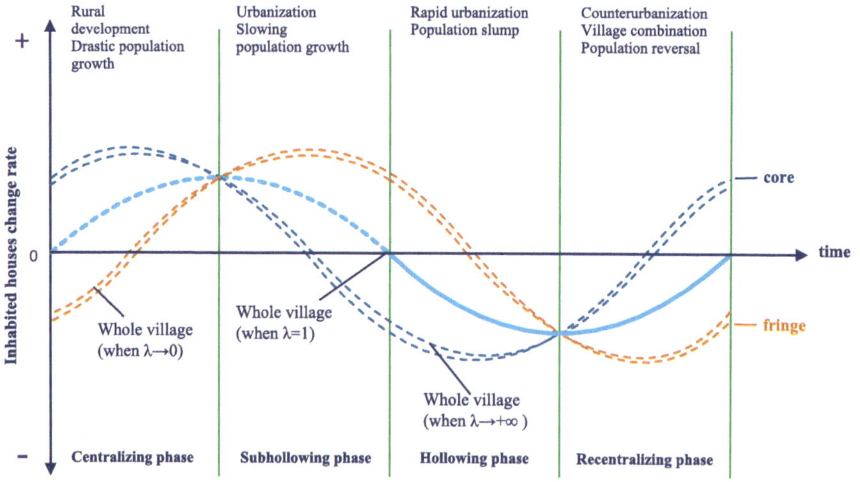

$$\frac{\text{Initial area of inhabited houses in the core}}{\text{Initial area of inhabited houses in the fringe}} = \lambda \quad \lambda \in (0, +\infty)$$

Fig. 3.14 Cyclical evolution model of the development stages of hollowed villages in China (Adapted from Long et al. 2012) village fringes is mirrored by the vacating of properties in the village center. Generally, competition between agricultural and non-agricultural land uses is at its most intense at the interface between the urban and the rural, frequently resulting in a succession of land uses at any one location over time (Robinson 2004). In the aspects of village-hollowing, the main factors driving such successions are the fluctuating strengths of urban and rural economies, which alter the bid rents (Robinson 2004), and cause the depreciation of rural housing relative to urban housing in the process of urbanization under the influence of the rent law (Li 2002). This will speed up the process of village-hollowing and translate to the third phase, the 'hollowing' phase, in which rapid urbanization leads to a sharp slump in the overall village population, with decreasing occupancy of houses in both the village center and the village fringe. This is arguably the situation faced by many rural communities in the last decade. However, the model also projects a fourth 'recentralizing' phase, in which a broader trend of counter-urbanization contributes to return migration and a re-growth in village populations, potentially supported by migration from other rural communities. Significantly, the model envisages that the recentralizing phase will see increases in the occupancy of houses in the village core exceeding the construction of new houses on the village fringe, reflecting both the reoccupation of properties by return migrants and the attraction and availability of cheaper, vacant property for redevelopment in village centers.

The fulfillment of the 'cyclical evolution model' for hollowed villages is contingent on both the development of appropriate social and economic conditions in China that will permit counter-urbanization and the adoption of appropriate policy instruments to direct redevelopment into the village cores and restrict new development on village fringes. However, even if these conditions are realized, not all hollowed villages may be regenerated. Smaller and more remote villages may find that they continue on a trajectory of population decline to the point of extinction, as represented by the second, 'vanishing evolution model' (Fig. 3.15). In this, they will be replicating the experience of villages in peripheral regions of European countries such as France and Ireland that has in effect ceased to exist through depopulation and the abandonment of housing. In some cases, this will result from a natural withering away of the population over time, yet in China, it is likely that the abandonment of some villages will be assisted by settlement rationalization and village combination programs, with remaining residents, relocated to neighboring communities in order to remove the costs of supporting separate villages and to free up land for agricultural production or nature rehabilitation.

3.5.3.2 Characteristics of the Evolution of Different Types of Hollowed Villages in Traditional Farming Areas and Empirical Studies

Considering that hollowed villages in the plain areas are the most representative, this study focused on the evolution of different types of hollowed villages in the Yucheng city, Shandong Province. According to the measurement index of the hollowing degree (Table 3.9), a comprehensive assessment was carried out by tak-

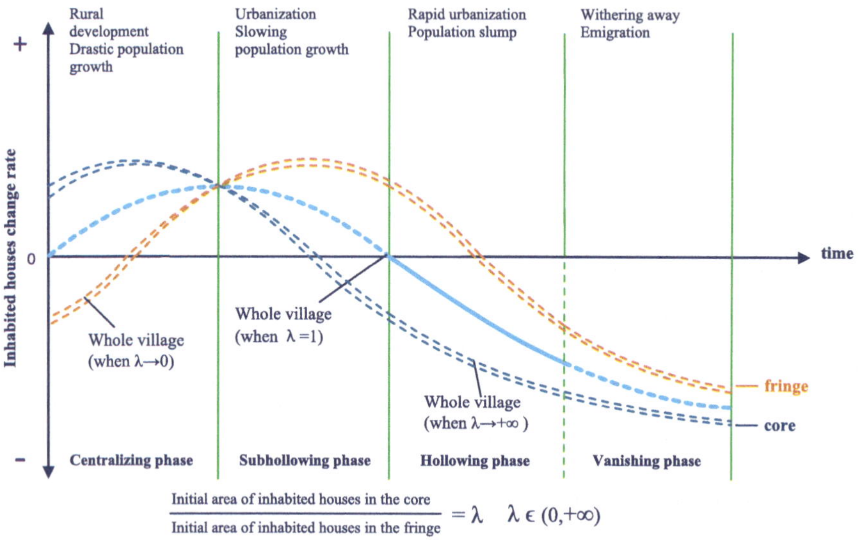

Fig. 3.15 Vanishing evolution model of the development stages of hollowed villages in China (Adapted from Long et al. 2012)

ing the actual conditions of typical areas into consideration. The results showed that Yucheng mainly presents four types-"non-local employment induced high-level and concentrated hollowing village", "non-local employment induced high-level and decentralized hollowing village", "agriculture dominated low-level and concentrated hollowing village" and "agriculture dominated high-level and concentrated hollowing village". Based on the high-resolution aerial remote sensing image (0.25 m) and household survey data, the current status of rural housing land in four representative villages (Paizi, Dongyuchang, Xingdian, and Lizhuang) was obtained (Fig. 3.16). By comparison, the phase characteristics and leading driving factors are sorted out (Table 3.9).

3.5.4 Village-Hollowing: Causal Factors and Driving Mechanism

There is consequently a major tension between urban development and farmland protection, in which the position of 'hollowed villages' is pivotal. 'Hollowed villages' are communities in which depopulation and housing modernization have led to the abandonment of a significant number of properties, spread throughout the settlements (Long et al. 2009). Although there is a physical resemblance to the abandoned farms and cottages found in many parts of rural western Europe (I. García and Ayuga 2007), or perhaps more closely, to the 'dying villages' of parts of rural eastern Europe, village-hollowing in China is a unique rural settlement morphology that has been

Table 3.9 Comparative analysis of evolutive characteristics and major driving factors of hollowing village types in farming plain area (Adapted from Long 2012)

Stage		Non-local employment induced high-level and concentrated hollowing village (**Pazi Village**)	Non-local employment induced high-level and decentralized hollowing village (**Dongyuchang Village**)	Agriculture dominated low-level and concentrated hollowing village (**Xingdian Village**)	Agriculture dominated high-level and concentrated hollowing village (**Lizhuang Village**)
Stage characteristics of socio-economic development and hollowing process	Before 1980	Mud-brick house; The overall village layout is more concentrated but irregular	Mud-brick house; The layout of the overall village is relatively concentrated and regular	Mud-brick house; The layout of the overall village is irregular	Mud-brick house; The layout of the overall village is relatively concentrated but irregular
	1980–1986	Rural economic reform promotes farmers' income growth; About 60 houses were built, most of which were outside the village; Some of the old houses were abandoned, and the village hollowing started	Rural economic reform promotes farmers' income growth; 15 new houses were built at the west and southwest part of the previous village center	Rural economic reform promotes farmers' income growth; 50 more new houses were built at both central and peripheral areas of the village	Rural economic reform promotes farmers' income growth; 10 more new houses were built in the south and west part of the previous village
	1987–1996	Substantial increase in agricultural production capacity; Farmers tend to engage in part-time work, which gave rise to the income increase; Approximately 110 more houses were built, which were mostly distributed in peripheral region; Most of the original old houses are vacant, showing anabatic hollowing degree	Substantial increase in agricultural production capacity; Farmers tend to engage in part-time work, which gave rise to the income increase; More than 50 houses were built at village fringe; The number of vacant homes increased and partially abandoned	Agricultural production capacity increased substantially; More than 120 houses were built, which were dispersedly distributed at village fringe; Farmers tend to build new houses but not demolish the old ones; The old buildings were gradually left idle	Agricultural production capacity increased substantially; About 40 houses were built, which were mainly distributed at the village fringe; Part of the old houses began to be left vacant

(continued)

Table 3.9 (continued)

Stage		Non-local employment induced high-level and concentrated hollowing village (**Pazi Village**)	Non-local employment induced high-level and decentralized hollowing village (**Dongyuchang Village**)	Agriculture dominated low-level and concentrated hollowing village (**Xingdian Village**)	Agriculture dominated high-level and concentrated hollowing village (**Lizhuang Village**)
	1997–2003	The proportion of people who got employed in off-farm work outside increased enormously; House building activities were still intensive; More than 30 new houses were built; Unused and un-repaired old buildings gradually abandoned and collapsed	The proportion of people who got employed in off-farm work outside increased enormously; House building activities were still intensive; 20 more houses were newly built at village fringe; The degree of village hollowing is intensified, and the partial houses were abandoned	High-efficiency crop farming and breeding industry promoted the growth of farmers' income; 90 more houses were newly built at village fringe; The degree of village hollowing is intensified, and more and more houses were abandoned	Farmers' income level increased slowly; 20 more houses were newly built, which were mainly distributed at the village fringe; The old houses were gradually left vacant or abandoned
	2004–present	Agricultural production support policies and general part-time employment brought about income growth; New house building activities occurred at both the central village and the peripheral region. Abandoned and vacant houses were increasingly serious	Policies supporting agricultural production and part-time employment brought about income growth. There were 9 scattered homesteads at the village fringe. Abandoned and vacant houses were increasingly serious	Farmers' income continued to increase; About 40 houses were newly built; Rows of old houses were left vacant	Farmers' income increased rapidly; About 15 houses were newly built; Rows of old houses were left vacant
Current conditions of village development	Farmland	Abundant (0.27 hm^2/person)	Moderate (0.14 hm^2/person)	Moderate (0.14 hm^2/person)	Moderate (0.16 hm^2/person)

(continued)

Table 3.9 (continued)

Stage		Non-local employment induced high-level and concentrated hollowing village (**Pazi Village**)	Non-local employment induced high-level and decentralized hollowing village (**Dongyuchang Village**)	Agriculture dominated low-level and concentrated hollowing village (**Xingdian Village**)	Agriculture dominated high-level and concentrated hollowing village (**Lizhuang Village**)
	Characteristics of production and employment	Traditional agricultural production; The ratio of farmers who worked outside the village increased	Traditional agricultural production; The ratio of farmers who worked outside the village increased	High-efficiency agriculture production; The ratio of farmers who worked outside the village is relatively low	Traditional agricultural production; The ratio of farmers who worked outside the village is relatively low
	Per capita income	Higher level	Average	High level	Average
	Per capita rural housing land area	167 m^2	132 m^2	148 m^2	171 m^2
	The degree of village hollowing	High-level of hollowing (21.8%) The ratio of abandoned houses is larger than that of vacant houses	High-level of hollowing (21.2%) The ratio of vacant houses is larger than that of abandoned houses	Low-level of hollowing (15.1%) The ratio of vacant houses is larger than that of abandoned houses	High-level of hollowing (24.3%) The ratio of abandoned houses is larger that of vacant houses
	The agglomeration degree of vacant and abandoned parcels	Concentrated (centralized abandonment, scattered vacancy)	Decentralized (scattered abandonment, scattered vacancy)	Concentrated (rows of vacancy, scattered abandonment)	Concentrated (scattered vacancy, rows of abandonment)

(continued)

Table 3.9 (continued)

Stage	Non-local employment induced high-level and concentrated hollowing village (**Pazi Village**)	Non-local employment induced high-level and decentralized hollowing village (**Dongyuchang Village**)	Agriculture dominated low-level and concentrated hollowing village (**Xingdian Village**)	Agriculture dominated high-level and concentrated hollowing village (**Lizhuang Village**)
Differences of socio-economic development and hollowing process	The old houses had many inherent defects; Part-time employment is extremely common; Per capita farmland is more, and people have weak sense of saving land; Farmers have higher demand and capacity of building houses; The control of house building is not strict; Building new houses without demolishing old ones cause severe waste of land resources	The conditions of old houses were relatively good; Part-time employment is common; Per capita farmland is less, and people have strong sense of saving land; Seasonal population migration led to more vacant houses	The conditions of old houses were average; Rapid development of planting and breeding industries; The demand for building houses is large; Due to the scarcity of farmland, new houses were mostly built on the original sites by demolishing the old ones; The degree of village hollowing is relatively low	The old houses had many inherent defects; Although the cultivated land resources were not abundant, the traditional agricultural production made the cost of building houses on occupied cultivated land relatively low; Rows of old houses were left abandoned; Part-time and seasonal employment resulted in scattered vacant houses
Dominant driving factors	The old houses had many inherent defects; Production development and employment transformation; Homestead management system was absent	Production development and employment transformation; Farmers have strong sense of saving land due to limited land resources; Homestead management system was absent	High-efficiency agriculture production; Economic leverage of land use promoted intensive use of homesteads	The old houses had many inherent defects; Production development; Homestead management system was absent

Fig. 3.16 Rural housing land-use survey of typical types of hollowing villages in Yucheng city, Shandong province (Adapted from Long 2012)

shaped by the 'dual-track' structure of rural-urban development (Wu and Ma 2006), including the contribution of economic, socio-cultural, institutional-managerial and environmental factors, as described below (Long et al. 2009).

Economic factors: Prior to 1978, the rationale of economic planning policy in China had been to eradicate regional disparities through an even distribution of productive materials and activities. The centrally planned economy compelled the rural population to primarily working on collective farms, with output shared more-or-less equally between members. Rural households were hence tied to low incomes, making improvements to housing conditions impossible. Economic liberalization after 1978, however, including the growth of TVEs, permitted greater wealth accumulation by individuals. As the affluence of farmers increased they invested in new more spacious and more comfortable housing, often at the expense of investment in agricultural production. Together with the effects of wider economic restructuring, low agricultural investment meant that returns from agricultural land decreased, such that it became attractive for farmland on the edge of villages to be converted to housing, replacing dwellings in village centers which in turn became vacant.

Socio-cultural factors: The 'dual-track' policy of rural-urban development in China dictates that the millions of rural migrants to cities are not granted permanent household registration in cities and are not eligible for many social welfare entitlements and employment opportunities that are reserved for permanent urban residents (Shen 2002). This temporary, unofficial status in cities and the bifurcated social security system, together with the lack of security of urban employment in a fluctuating global economy (Wang 2010), encourages many migrants to retain vacant rural housing and idle rural land to give the opportunity of returning home. Absenteeism is hence a key factor in the disparity between the rural resident population and the rural housing stock. Additionally, ambiguous property rights and limited understanding of the law act as deterrents to residents who have built new houses acting to sell or demolish their former homes, as does nostalgia and superstition.

At the same time, improved economic conditions in rural regions and diversification of employment away from agriculture have reduced the traditional dependency on the 'big family'. Traditional models of the 'coalition family' and the 'stem family', with diverse family members forming a single inter-generational household, are being replaced by smaller, 'nuclear family' households. The number of rural households in China increased from 173.5 million in 1978 to 256.6 million in 2008, even against the trend of a falling rural population since 1995 (Fig. 3.7). This proliferation of smaller households has been a key driver of rural house-building (Sargeson 2002).

Institutional and managerial factors: Village-hollowing is also influenced by a number of factors pertaining to legal and institutional frameworks and the management of rural land. Firstly, restrictions on the sale of the property and ambiguous property rights with respect to collective land limit the opportunity for farmers to sell poor-returning farmland (Ding 2003). Secondly, a historic legacy of poor land-use planning, including a lack of effective road planning, has left many communities with unwieldy and impractical morphologies. Village centers may be over-crowded, whilst peripheral farmland may fall into disuse because access roads are insufficient for modern agricultural vehicles. These factors encourage new house-building. Thirdly, the absence of rules and regulations aimed at dealing with vacant or abandoned housing means that local governments are limited in their capacity to act to address the problem, whilst the inspection and approval process for new house-building is unsound, e.g., heads of villages usually make money from selling lands to farmers who want to build a new house.

Environmental factors: Finally, in some cases, village-hollowing has responded to both natural and human-induced local environmental conditions. Liberalization has enabled rural residents to relocate housing to avoid risks from natural hazards such as flooding, debris flow and landslides. At the same time, relocation has in some instances been necessitated by human-induced environmental problems, often resulting from the unregulated activities of TVEs. For example, drinking water pollution and the lowering of water tables through over-extraction have led in some instances to the abandonment of established village centers and the development of village fringes where water supplies are clean and plentiful.

The above factors combine to drive the phenomenon of village hollowing in rural China. The engagement of farmers in excessive house-building on village fringes whilst failing to release vacated properties in village centers may be the core driver of the process, but this behavior is conditioned by the interaction of a complex web of background influences, and facilitated by the lack of regulatory control, as Fig. 3.17 shows. As such, the process of village-hollowing is embedded in the wider context of economic restructuring and rural-to-urban migration in China and hence is susceptible to temporal and spatial variations in these broader trajectories. Thus, village-hollowing will evolve differentially over time and may lead to different end results in different places.

3.5.5 Discussion and Conclusions

Economic factors are usually the primary driver of land use change, which indeed plays an important role in the formation and evolution of hollowed villages in China. For example, the increase in farmers' income and the low land cost of building houses are important factors contributing to shaping hollowed villages. However, due to the difficulty in realizing complete competition in the land market and the externalities of land use, it is impossible to clarify the mechanism of land use change perspective of economics. The market failure is more prominent in the process of regulating "hollowed villages", which requires the interpretation of the impact of social mainstream behavior on the formation and evolution of hollowed villages in terms of laws, regulations and policies.

Land use change is usually driven by three dimensions of socio-economic, biological nature, and land management. The analysis of the formation and evolution mechanism still needs to be combined with the evolutionary stage, and the three-dimensional forces from different scales should be considered in the historical and cultural background at different scales. To deepen the understanding, it is necessary to decompose the relevant driving factors into three types: socioeconomic, physical-biological, and management according to the differences in their action modes and functions. Then, driving forces of each type according to space and time scales should be analyzed, and how the driving factors are related and the socio-economic-biological-land management three-dimensional driving mechanism should be further explored.

The factors affecting the formation and evolution of hollowed villages mainly involve economic, natural, social and cultural, institutional and management aspects. The differences in regional economic, social, and natural conditions determine the differentiated types of the hollowed village -"urban-rural fringe area, farming plain area, mountainous area and pastoral area". Accordingly, the characteristics of the

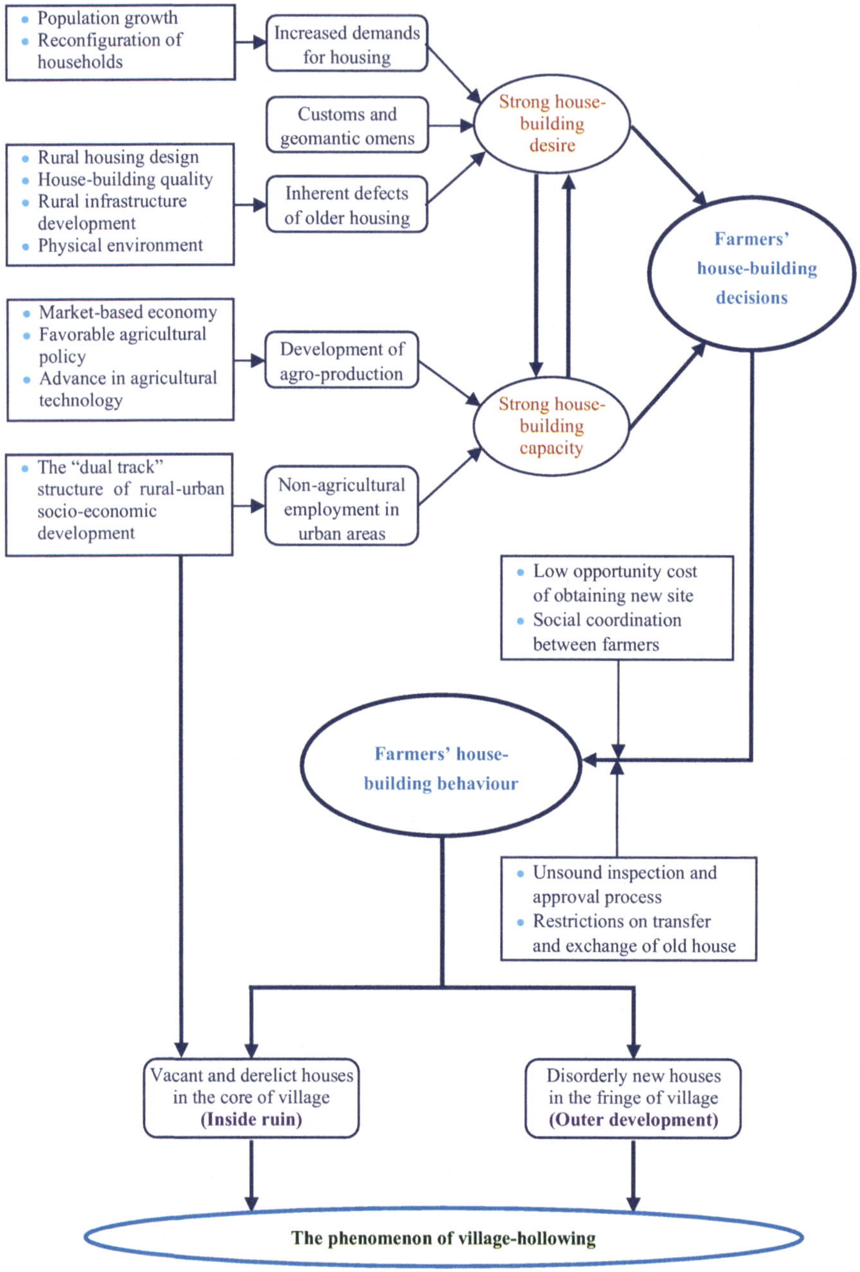

Fig. 3.17 Driving mechanisms of the evolution of hollowed villages in China (Long et al. 2012)

rural industry, the degree of hollowing, and the agglomeration degree of vacant and abandoned parcels are all should be considered when classifying hollowed villages.

The hollowed village is the product of regional economic and social development to a certain stage, and its development and evolution can be included in the research context of land use transition. The specific evolutionary stage of hollowed villages corresponds to a certain level of socio-economic development. China's vast territory and obvious regional differences in socio-economic development have led to the existence of different types of hollowed villages evolution at the same time at province scale or even a larger scale. Due to the different locations and the intervention of external forces, the development and evolution process has different characteristics. Hollowed villages in some places may directly enter the recession period from the growth period or even the emergence period, and some regions may not even have the obvious sign of hollowing.

The formation and evolution of hollow villages are closely related to the spatial dynamics of urbanization development, rural population change, and employment in different regions. The hollowed villages in the urban-rural fringe area will complete the four stages of centralizing, sub-hollowing, hollowing, and recentralizing. There are four main types of hollowed villages in plain areas: "non-local employment induced high-level and concentrated hollowing village", "non-local employment induced high-level and decentralized hollowing village", "agriculture dominated low-level and concentrated hollowing village", and "agriculture dominated high-level and concentrated hollowing village".

Taking Yucheng city in the Huang-Huai-Hai Plain as an example, combining with the high-resolution aerial photographs and household survey data, the evolution characteristics of different types of hollowed villages in the traditional farming areas were discussed. A comprehensive evaluation of the type of hollowed villages was carried out by combining the hollowing measurement indicators by considering the actual conditions of typical areas. Results indicate that the increase in farmers' desire and capacity to build a house driven by the increase demand, and the absence of corresponding regulatory control policies, coupled with the regional development stage, contribute to form unreasonable behaviors of farmers' building and eventually created a variety of characteristics and types of village hollowing evolution.

3.6 A Multi-scale Analysis

3.6.1 Background

With the rapid industrialization and urbanization, the rural population has been confronting with significant changes in demographic structures, employment opportunities, community organization, lifestyles and standard of living, accessibility and rural culture (Long 2014; Mukherjee and Zhang 2007). These changes, especially the family structure miniaturization and improved living standards, raised the demand for rural housing land (Li et al. 2014; Long et al. 2007). In the absence of long-term planning and land use control, these changes led to the sprawl of rural housing land (Deng et al. 2009). During 2000–2008, China's rural resident population decreased from 808.37 million in 2000 to 721.35 million in 2008, with a decrease of 87.02 million and by 10.8%; however, rural housing land expanded from 16.53 million ha in 2000 to 16.66 million ha in 2008, with an increase of 0.13 million ha and by 0.8% (MLRC 2009; NBSC 2009).

The research of land use transitions has been a hot topic since the turn of the new millennium (Ferreira et al. 2015). With the introduction of land use transitions to the academic circle of China, related researches combining with the characteristics of China's socio-economic development have been carried out extensively (Gong et al. 2014; Wang et al. 2015). Recently, issues related to rural housing land transition in China have attracted interest among a wide variety of researchers, ranging from those who favor theoretical hypothesis and verification (Long et al. 2007), morphology evolution of rural settlements (Fang and Liu 2014), hollowed villages, rural housing land consolidation (Qu and Long 2018), functional transition of the rural settlements (Zhang et al. 2019), urban village (Liu et al. 2010), to those who try to understand the Spatio-temporal patterns, driving forces and coupling mechanism of rural housing land transition Since there exists obvious human-land conflicts, the researches on rural housing land in China just focusing on land itself are incomplete, to some extent, and the "rural population-rural housing" joint research framework need to be paid more attention. Compared with the increasing rural housing land, the rural population decreased with the key trends including mass migration from rural to urban areas and the development of villages in-city' (chengzhongcun) of rural migrant workers (Chen et al. 2014; Song et al. 2008; Long et al. 2012). During 2000–2008, the per capita rural housing land (PRHL) of China expanded from 204.5 to 231.0 m^2 with an annual growth rate of 1.6%, which was far greater than 0.1% that of rural housing land (MLRC 2009; NBSC 2009). This "population decrease but rural housing land increase" phenomenon resulted in the formation of hollowed villages (Long et al. 2012).

Based on the understanding of land use transitions (Long et al. 2012), there are two categories of rural housing land transition (RHLT). One is the changes of dominant morphology of rural housing land, which means the quantity, structure and spatial pattern of rural housing land; another is the changes of recessive morphology, which includes the quality, property rights and management mode of rural housing land. The change of PRHL belongs to the change of the recessive morphology (i.e., property rights) of rural housing land, which is one of the important components of rural housing land transition under rapid urbanization. However, the research of this kind of rural housing land transition has received much less attention. Currently, the analysis of the relationship between rural housing land and rural population changes mostly concentrated on the provincial scale constrained by data availability and its processing technology. It is imaginable that the research outcomes based on the data at the provincial scale are not competent to guide the practice of rural housing land consolidation at township or village level. The aims of this section are: (1) to analyze the spatiotemporal pattern of RHLT especially in the aspect of PRHL based on 1 km grid data; (2) to detect the physical and socio-economic driving forces of RHLT with the aid of spatial econometric regression model at both grid-scale and county scale in Bohai Rim using high-resolution Landsat TM data in 2000 and 2010, socio-economic data and physical environment data; and (3) to discuss some of the major implications for carrying out rural housing land consolidation and pushing forward China's new-type urbanization.

3.6.2 *Materials and Methods*

3.6.2.1 Study Area

This study selects the Bohai Rim (including Beijing, Tianjin, Hebei, Liaoning and Shandong five provinces) as a study area (Fig. 3.18). It covers about 52.20 million ha, which accounts for 5.5% of the national land area. The north temperature monsoon climate dominates this region year-round, with an average annual temperature of 9.5 °C, average annual rainfall of 590.6 mm, all of which are beneficial for agricultural production. Therefore, the Bohai Rim is also an important agricultural base with an arable land area of 2.7 million ha, which accounts for a quarter of the total arable land in China and produces 23% of national grain outputs.

As one of the economic growth poles of China, the Bohai Rim had undergone rapid urbanization during 2000–2010. The urbanization rate of Bohai Rim has increased from 39.5% in 2000 to 54.9% in 2010, which were higher than 36.2 and 50.0% of national level which is in 2000 and 2010, respectively. Its per capita GDP increased from 10,332 yuan in 2000 to 54,506 yuan in 2010. Recently, the Bohai Rim was considered as China's "third engine" of economic growth, besides the Yangtze River

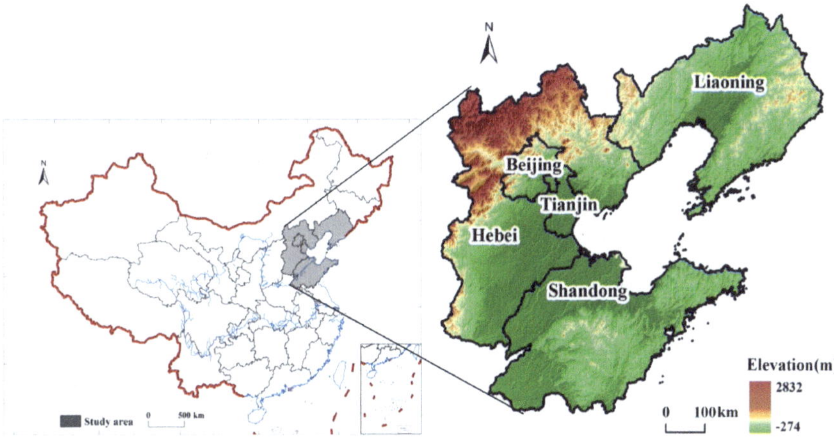

Fig. 3.18 Location of the study area, Bohai Rim (Li et al. 2015)

Delta (which has Shanghai as the central city) and the Pearl River Delta (which has Guangzhou as the central city) (Yang et al. 2015).

Accompanied by the rapid urbanization, the elements of agricultural production achieved a rapid non-agricultural transition with the phenomenon of non-agricultural employment of rural population and expansion of rural construction land. Under this circumstance, the PRHL changed dramatically and brought about rapid rural housing land transition. Meanwhile, the developed land amounted to 84% of the total study area, which indicated that the reserved land resources were very scarce since most of the undeveloped land belonged to ecological land. The contradiction between the limit of land resources and the increasing need for construction land due to rapid socio-economic development was increasingly intensified; however, which can be released by carrying out rural housing land consolidation. As such, the Bohai Rim is one of the typical areas to carry out the research on rural housing land transition under rapid urbanization.

3.6.2.2 Data Source

The rural housing land data was obtained through detection analysis of historical Landsat TM (Thematic Mapper) satellite images in 2000 and 2010 from China National Environmental Monitoring Center. The 1 × 1 km gridded demographic data was provided by the Data Sharing Infrastructure of Earth System Science (www.geodata.cn). The socio-economic data of 2000 and 2010 were mainly derived from the "China statistical yearbook for the regional economy", "China county statistical yearbook", "Tabulation on the 2000 population census of China" and "Tabulation on the 2010 population census of China". The missing data was supplemented from the Statistical Yearbook and the rural survey data of the five provinces in Bohai Rim. The 90 m resolution DEM was obtained from USGS. The digital maps of soil PH, total N, total K and soil organic matter were derived from the Ministry of Agriculture of China. The average annual temperature and rainfall were provided by the China Meteorological Administration. The vector data of faults was provided by the Data Sharing Center of Land Resource Science of the Chinese Academy of Sciences. In this study, these basic data such as administrative boundaries, government headquarters, roads and rivers were downloaded from the National Geomatics Center of China.

3.6.2.3 Data Process

The process of 1 km gridded PRHL data is shown as Fig. 3.19, specifically including four steps. The step one consists of developing 1 × 1 km raster data for rural housing land and verifying the 1 km gridded demographic data. The former refers to first extracting the rural housing land polygons from the interpreted remote sensing data, then intersecting and dissolving them with 1 km gridded fishnet, and finally converting them to raster data, i.e., the 1 × 1 km raster data of rural housing land. The latter

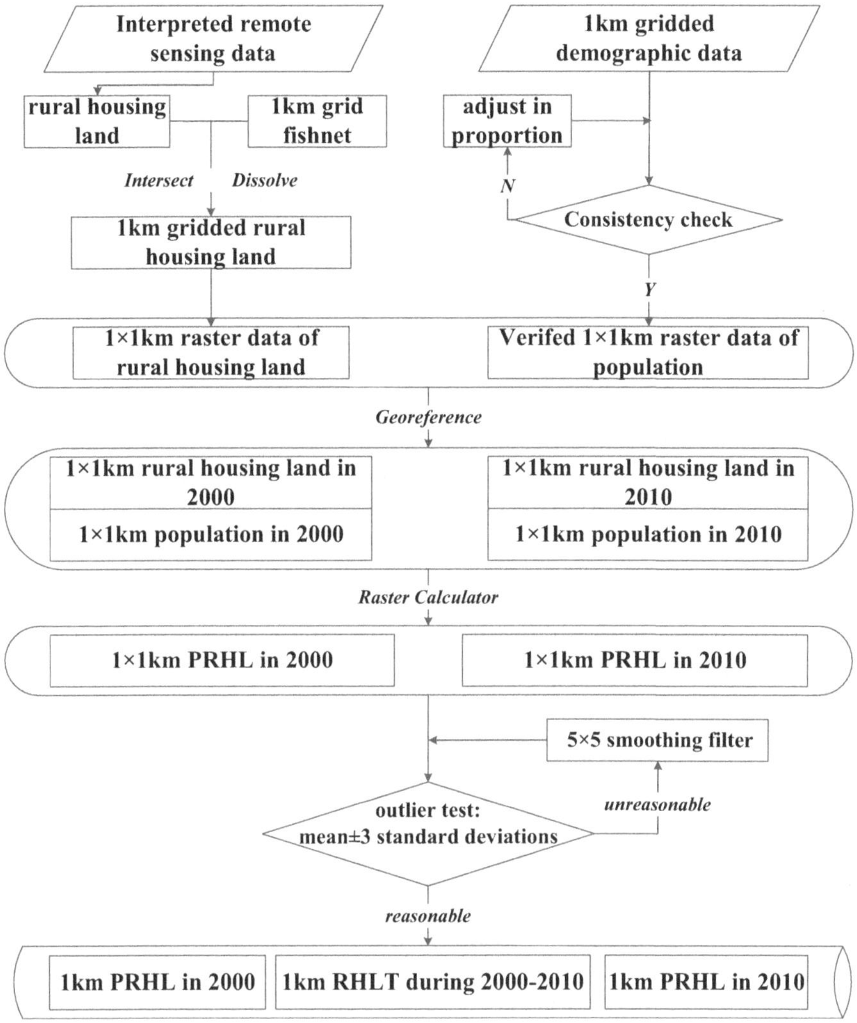

Fig. 3.19 The process of obtaining 1 km gridded PRHL data (Li et al. 2015)

aims to check the consistency of the population census data at county level with the 1 km gridded demographic data. If they are equal at county level, the population of each grid in the county will be kept; otherwise, the 1 km gridded demographic data in corresponding county will be adjusted in proportion. Step two is to use Georeferencing tools in ArcGIS software to adjust 1 × 1 km raster data of population and rural housing land obtained from step one so that they will match. After that, we got the 1 × 1 km raster data of the population in 2000 and 2010, respectively, which had been calibrated. Step three is to get the 1 km gridded PRHL data. With the help of the "Raster Calculator" tool in ArcGIS software, each grid's PRHL was obtained

Table 3.10 The number of grids of different scales in the Bohai Rim (Li et al. 2015)

Scale	1 km	2 km	3 km	4 km	5 km	6 km
Number of grids	521882	131630	59095	33577	21710	15200

through divided the rural housing land area by population, then we got the 1 × 1 km PRHL in 2000 and 2010, respectively. Step four is also known as the correction of an outlier. If the value of one grid is within 3 standard deviations of the mean, it will be considered reasonable. Otherwise, it will be smoothed using a 5 × 5 smoothing filter until the value is in the reasonable range. In the end, we got the tested 1 km gridded PRHL in 2000 and 2010, respectively, and the dynamic data of 1 km gridded PRHL during 2000–2010.

Multi-scale gridded spatial econometric regression analysis (SERA) aims at studying the causality between various independent variables (physical conditions, transportation, and location) and the RHLT at five different scales, i.e., 2, 3, 4, 5, and 6 km. The reasons for such design are as follows (Paquette and Domon 2003). First, the SERA is limited by the data process capacity of GeoDa software. As shown in Table 3.10, the number of grids at a 1 km scale is 521,882, over the upper limit of GeoDa software. However, the number of grids at other scales is within the range. Thus, the method of scale-up is taken to solve the problem of GeoDa's limited data processing capacity. Secondly, the multi-scale SERA can successfully avoid the contingency of the result caused by the single-scale SERA. So, the results obtained by the multi-scale SERA are more convincing.

The data processing of multi-scale SERA contains the processes of independent variables and dependent variables. The dependent variables refer to RHLT in the aspect of PRHL during 2000–2010 at 2–6 km grid scales. The independent variables refer to seven categories consist of 17 indicators (Fig. 3.20 and Table 3.11), which in turn are terrain, soil, climate, geology, hydrology, transportation, and location. The data processing of the multi-scale variables is as follows:

(1) The terrain can be characterized by elevation and slope. The slope and elevation data were generated from the 90 m DEM through ArcGIS10.1® Spatial Analysis toolbox and Resample toolbox with a 1 km resolution.
(2) In China, the rural housing land generally expands in the peripheral area of a village, and it is inevitable to occupy the high-quality arable land. Therefore, the RHLT may be related to soil quality. Considering the data availability, we selected the soil PH, total N, total K and soil organic matter as the indicators of soil condition, which were derived by the respective maps using vectorization tool and Polygon to Raster tool with 1 km resolution.
(3) Usually, the climate is characterized by average annual temperature and rainfall, which were integrated on the basis of average data from 1970 to 2000 through kriging interpolation with a 1 km resolution.
(4) The data processor of the geological conditions, hydrological conditions, transportation and location were similar. The hydrological conditions and geological conditions were respectively defined as the shortest distance to rivers

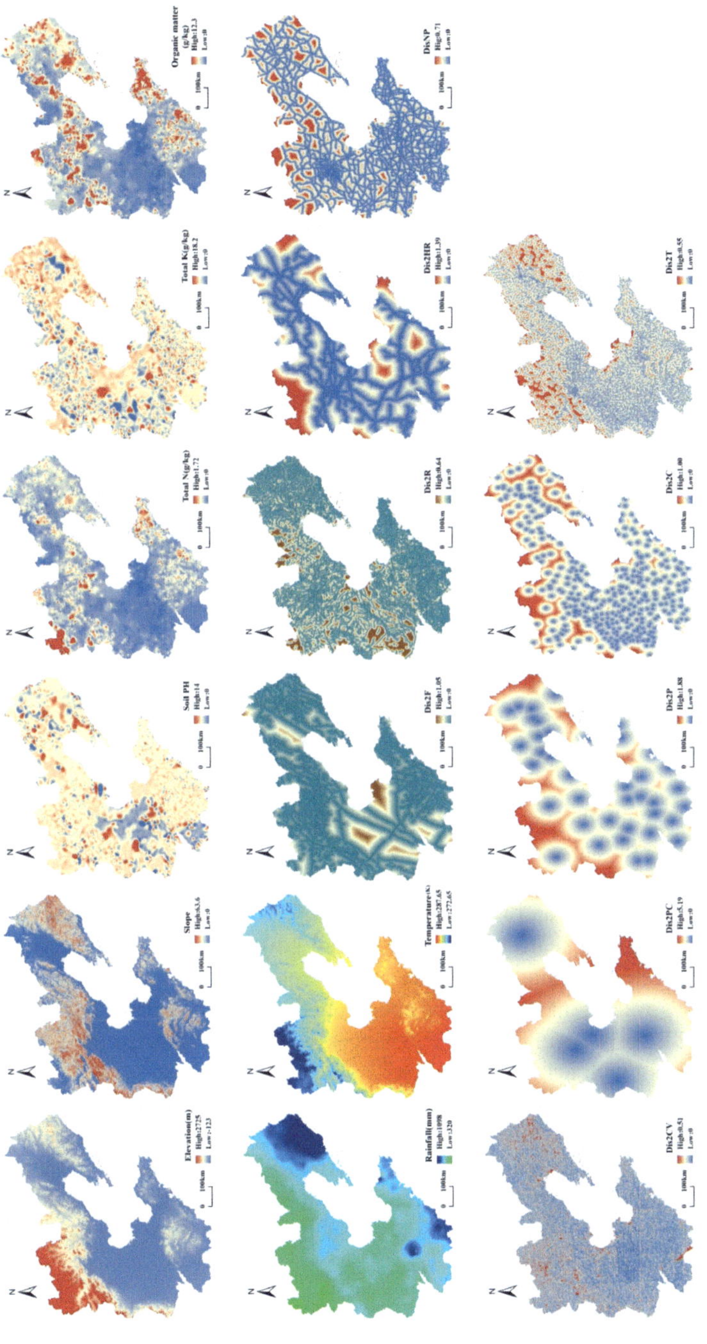

Fig. 3.20 1 × 1 km raster variables (Li et al. 2015)

Table 3.11 The indicator system used in SERA at the gridded scale (Li et al. 2015)

Variable category	Variable	Description	Unit
Dependent variable	RHLT	Rural housing land transition	m^2
Terrain	Elevation	Average elevation	m
	Slope	Average slope	–
Soil	Soil PH	Average Soil PH	–
	Total N	Average Total N	g/kg
	Total K	Average Total K	g/kg
	Organic matter	Average organic matter	g/kg
Climate	Rainfall	Average annual rainfall	mm
	Temperature	Average annual temperature	K
Geology	Dis2F	Shortest distance to faults	–
Hydrology	Dis2R	Shortest distance to rivers	–
Transportation	Dis2HR	Shortest distance to highway/railroads	–
	Dis2NP	Shortest distance to national/provincial roads	–
	Dis2CV	Shortest distance to county/village roads	–
Location	Dis2PC	Distance to nearest provincial capital government site	–
	Dis2P	Distance to the nearest prefecture government site	–
	Dis2C	Distance to the nearest county government site	–
	Dis2T	Distance to the nearest township government site	–

and faults with the corresponding abbreviated name of Dis2R and Dis2F. The transportation was defined as the shortest distance to highway/railroads, national/provincial roads and county/village roads with the abbreviated names of Dis2HR, Dis2NP and Dis2CV, respectively. The location was presented by the distance to the nearest provincial capital government site, the nearest prefecture government site, the nearest county government site and the nearest township government site with the abbreviated names of Dis2PC, Dis2P, Dis2C and Dis2T, respectively. All the distances were calculated by the Euclidean Distance tool with a 1 km resolution.

(5) The final step, also called data scaling up, was mainly to convert the above data and the RHLT from 1 km resolution to 2–6 km resolution using the Raster-to-Point tools and Analysis tools provided by ArcGIS. The latter was to respectively intersect the 1 × 1 km data with the 2–6 km gridded fishnet. Ultimately, five sets of data with a 2–6 km resolution with a gap of 1 km were generated. These data are used in the SERA analysis.

3.6.2.4 Methodology

The calculation of 1 km gridded PRHL is given as:

$$\text{PRHL} = \frac{RHArea}{PopSize} \times 100\% \quad (3.4)$$

where PRHL is the area of per capita rural housing land in each grid (m^2/person), is the area of rural housing land in each grid (m^2), is the population size in each grid (persons).

The hot spot analysis calculates the Getis-Ord Gi* statistic for each feature in a dataset to demonstrate where features with either high (hot spots) or low (cold spots) values cluster spatially. In this study, the method was used to detect the hot spots (or cold spots) of changes in rural housing land, population and PRHL during 2000–2010. The Getis-Ord Gi* is given as:

$$G_i^* = \sum_i W_{ij} Y_j / \sum_j Y_j \text{ and } Z(G_i^*) = (G_i^* - E(G_i^*))/\sqrt{VAR(G_i^*)} \quad (3.5)$$

where G_i^* is a local spatial autocorrelation index, $Z(G_i^*)$ is a test score, $E(G_i^*)$ and $VAR(G_i^*)$ is the expected value and the variance of G_i^*, W_{ij} is spatial weight matrix, Y_j is the observed value of the j grid. For statistically significant positive (negative) Z scores, the larger (smaller) the Z scores, the more intense the clustering high (low) values.

It is very important to take spatial dependence into account when dealing with spatial data. In this study, the spatial econometric regression models include Spatial Lag Model (SLM) and Spatial Error Model (SEM). SLM assumes that the value of RHLT depends on the values of it in neighboring regions. It is thus a formulation of the idea of a spatial spillover. An SLM is expressed as follows (Anselin 1988):

$$y = \rho W y + X\beta + \varepsilon \quad (3.6)$$

where y is RHLT in each grid, y contains a set of explanatory variables for the regression relationship with β as the corresponding regression coefficient, ρ is a spatial autoregressive coefficient indicating the extent to which variations in the observations y are explained by the average of neighboring observations values, ε is the model error terms and W is spatial weights matrix.

Compared with SLM, the spatial dependence in the SEM enters through the errors rather than through the neighboring values of the dependent variable in the SLM (Anselin 2002) SEM is specified as

$$y = X\beta + \varepsilon \text{ and } \varepsilon = \lambda W \varepsilon + \mu \quad (3.7)$$

where y, X, β, W are the same as them in Eq. (3.6), λ is a spatial autoregressive coefficient, β is regression residuals, μ is normally distributed random error.

Ordinary least squares (OLS) models the relationship between a dependent variable and a collection of independent variables. The value of the dependent variable is defined as a linear combination of the independent variables plus an error term:

$$y_i = \beta_0 + \sum_{j=1}^{k} \beta_j x_{ij} + \varepsilon_i \tag{3.8}$$

where y_i are the dependent variables, x_{ij} are the independent variables, β_0 is a constant term, β_j are the regression coefficients, ε_i is a normally distributed errors of prediction.

3.6.3 Results

3.6.3.1 Spatio-Temporal Pattern of RHLT

According to the analysis results, the area of rural housing land in Bohai Rim was 3.30 million ha in 2000 and 3.33 million ha in 2010. It expanded 0.03 million ha from 2000 to 2010. The spatiotemporal pattern of rural housing land change was showed as follows. To some extent, the distribution of the rural housing land was characterized by vertical zonality, i.e., the proportion of rural housing land decreased gradually with the increase of elevation forareas beyond the sea level (elevation > 0 m) (Table 3.12). The plain area (elevation within the range of 0–100 m) was the best location choice for rural housing land, and almost 79.0% rural housing land distributed there, such as Liaohe Plain, Liaotung Peninsula, Hebei Plain, Yellow River flood plain and southwestern Shandong. While there was little rural housing land distributed in areas with high elevations, such as western Liaoning highlands, eastern Liaoning hills, Taihang Mountain, Yan Mountain, and Taishan Mountain (Fig. 3.21).

There was a substantial increase in the proportion of rural housing land within the elevation range of 0–100 m, increased by 0.4%, from 78.6% in 2000 to 79.0% in 2010. However, there was only a slight change in the proportion of rural housing land within other ranges of elevation, especially for the mountainous-hilly area (elevation $\geq$ 500 m). Increased rural housing land and lost rural housing land during 2000–2010 were fancy of plain areas (within the elevation range of 0–100 m), the proportion of which amounted to 66.8% and 65.0%, respectively, since the intensity of human activities is high in plain areas (Table 3.12).

Rural housing land was more sensitive to slope than to elevation and there was a phenomenon of agglomeration. More than 90% of the rural housing land concentrated in slight slope areas (slope < 5°) and the proportion of rural housing land there increased by 1.1%, from 93.4% in 2000 to 94.5% in 2010 (Table 3.13). On

Table 3.12 Elevation and rural housing land (RHL) distribution in Bohai Rim (%) (Li et al. 2015)

Elevation (m)	<=0	0–100	100–200	200–300	300–400	400–500	500–600	600–700	700–800	≥800
Total land area	3.3	47.3	11.7	7.2	5.2	4.2	3.6	2.8	2.1	12.6
RHL in 2000	1.7	78.6	10.4	3.4	1.5	0.9	0.8	0.5	0.3	1.9
RHL in 2010	1.5	79.0	10.5	3.1	1.3	0.9	0.9	0.5	0.3	2.0
Increased RHL during 2000–2010	1.6	66.8	13.4	5.3	2.7	1.9	1.7	1.1	0.7	4.8
Lost RHL during 2000–2010	1.7	65.0	14	6.4	3.4	2	1.5	1.1	0.7	4.2

Note The row of increased RHL during 2000–2010 means the proportion of the area of the increased rural housing land in different elevation zone in the total area of all increased rural housing land. The same as the row of lost RHL from 2000 to 2010

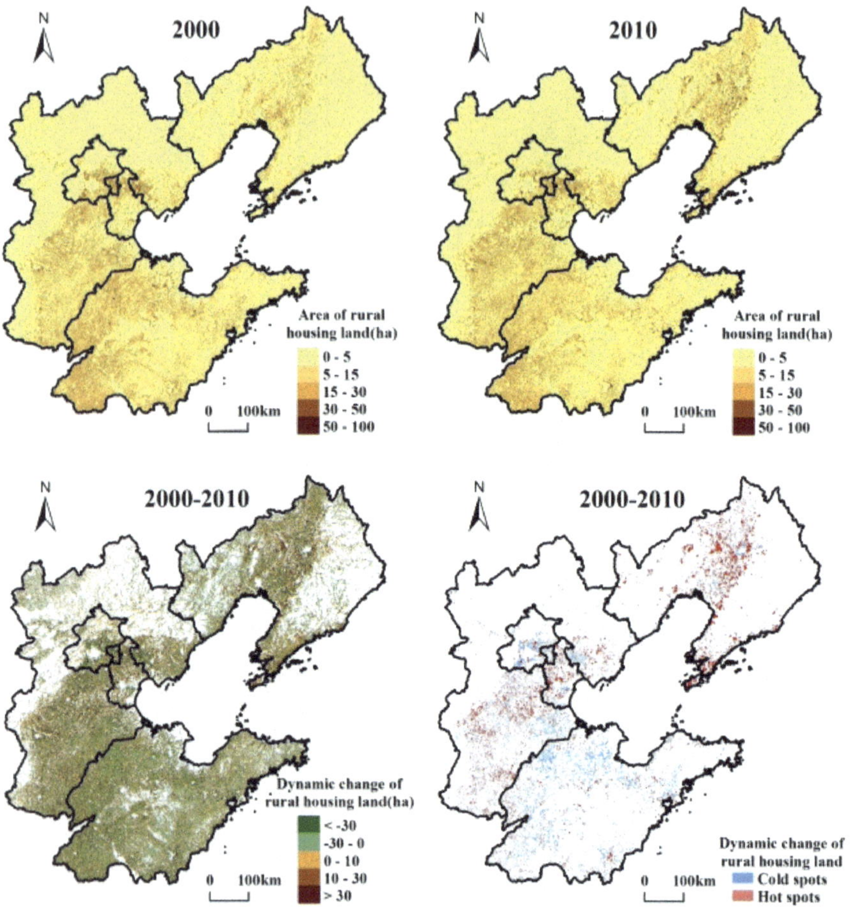

Fig. 3.21 The spatio-temporal pattern of RHLT in the aspect of rural housing land area during 2000–2010 (Li et al. 2015)

the contrary, the rural housing land in other ranges of slope accounted for a limited proportion, and with a trend of decrease or remain unchanged. Overall, the proportion of rural housing land decreased obviously with the increase in slope. This was reasonable because the steep site conditions could easily trigger geological disasters that might destroy rural housing land and do great harm to farmers' lives, property and safety. Increased rural housing land and lost rural housing land during 2000–2010 majorly focused on the slight slope area, the proportion of which amounted to 91.2 and 93.0%, respectively.

The PRHL was generally high in Bohai Rim and the regional differences were significant. The average PRHL increased from 226 to 258 m^2 during 2000–2010, which was far beyond 150 m^2 the national highest standard of rural housing land use. This showed great potential for rural housing land consolidation. Compared

Table 3.13 Slope and rural housing land (RHL) distribution in Bohai Rim (%) (Li et al. 2015)

Slope	Total land area	RHL in 2000	RHL in 2010	Increased RHL during 2000–2010	Lost RHL during 2000–2010
<5°	67.0	93.4	94.5	91.2	93.0
5°–8°	7.8	2.9	2.7	3.6	3.3
8°–15°	14.2	2.7	2.2	3.7	2.8
15°–25°	9.2	0.9	0.6	1.4	0.9
25°–35°	1.7	0.1	0.1	0.1	0.1
>35°	0.1	0.0	0.0	0.0	0.0

Note The row of increased RHL during 2000–2010 means the proportion of area of the increased rural housing land in different slope zone in the total area of all increased rural housing land. The same as the row of lost RHL during 2000–2010

with that in 2000, the standard deviations of PRHL in 2010 doubled to 229 m^2 within a decade, which indicated that the regional differences of PRHL distribution in Bohai Rim were more obvious. As shown in Fig. 3.22, the hot spots of RHLT mainly distributed in Liaohe Plain, Beijing suburbs and the border region of Hebe and Shandong. While the cold spots concentrated in eastern Liaoning hills, Taihang Mountain, Yan Mountain, and Taishan Mountain. During 2000–2010, the proportion of grids with decreasing and increasing PRHL accounted for 54% and 46%, respectively. There was the regional disparity in the pattern of hot spots: concentrated in a zone in Liaotung Peninsula, scattered across the Liaohe plain, distributed along a line in Taihang Mountain frontier in southern Hebei, and again scattered across Shandong. The cold spots concentrated in Liaohe plain, suburbs of Beijing and Tianjin, northwestern Hebei and the border region of Hebei and Shandong (Fig. 3.22).

Usually, the RHLT was influenced by both rural housing land change and rural population migration. The result of hot spots analysis showed the spatial mismatch of rural housing land change and rural population migration, which can be summarized as two models. For the first model, the rural population decrease was accompanied by the increase of rural housing land. With the rural-to-urban migration, decreased rural population and improved rural income levels were helpful for the rural people to improve their living conditions, which resulted in the expansion of rural housing land. This RHLT model is primarily concentrated in the three plains in Bohai Rim. Conversely, the second model represented some suburban areas of rapidly developed big cities with the increasing rural population and decreasing rural housing land, such as Beijing, Tianjin, Shenyang, and so on. On the one hand, the big cities attracted a lot of migrants gathering to the suburban areas; on the other hand, the expansion of urban areas swallowed up the rural housing land around the cities. Besides these two models, some areas presented the synchronous increase or decrease of rural housing land and the rural population. The former mainly located around Baoding where the PRHL decreased because the rural housing land expanded at a slower pace than that

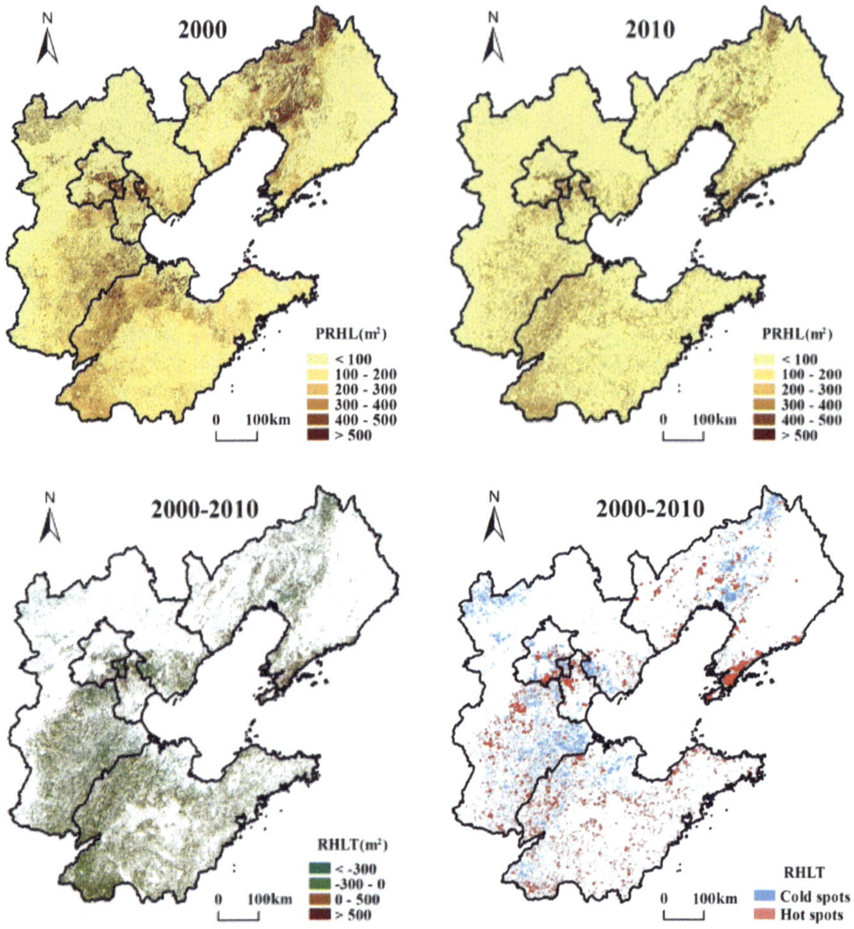

Fig. 3.22 The spatio-temporal pattern of RHLT in the aspect of PRHL during 2000–2010 (Li et al. 2015)

of the rural population. While the latter mainly located in Shandong except for its northern part where there was a widespread increase of PRHL.

3.6.3.2 Driving Forces and Mechanism of RHLT

The dynamic pattern of RHLT is the result of the interaction of physical conditions and socio-economic factors. The physical conditions have a long-term influence on the RHLT, while the socioeconomic factors determine the speed and scale of RHLT in a relatively short period. This study analyzes the physical and socioeconomic driving forces of the RHLT with the aid of spatial econometric regression model at grid-scale and county scale, respectively. Based on the above analysis, the driving mechanism of RHLT is further discussed.

1. **Driving forces analysis of RHLT at grid scale**

The raster-based explanatory model is one of the approaches to analyze the relationship between independent variables and dependent variables. Based on the SERA, the multi-scale gridded explanatory model was constructed to explore the quantitative relationship between physical conditions and RHLT and reveal the direction and strength of driving forces acting on the RHLT. This study selects terrain, soil, geology, climate, hydrology, transportation and location to analyze the driving forces.

First, the spatial correlation test based on multi-grid scales of the OLS, SLM, and SEM was carried out to determine which model will be more desirable. Table 3.14 showed that no matter at which scale, the Lagrange Multiplier (error) is more significant than the Lagrange Multiplier (lag) especially for the high-level grid scale. Furthermore, the Robust Lagrange Multiplier (error) is more significant than the Robust Lagrange Multiplier (lag). Based on the above analysis, we think that the SEM is more suitable for the analysis of the driving force of RHLT at grid-scale (Table 3.15). Compared with the multi-scale regression analysis results of a 2–6 km grid, it was found that except for the result of a 3 km grid, the spatial econometric regression analyses of other scales were effective. The increasing R^2 with scale-up illustrated the validation of interpretation enhanced when the scale went up from 1 to 6 km. The significant positive Lambda pointed out that the RHLT in one grid was not only influenced by the socio-economic conditions and physical conditions itself but also influenced by the RHLT, socio-economic conditions and the physical conditions of neighboring grids.

In addition to Dis2HR, Dis2NP, and rainfall, all other factors had significant effects on RHLT. According to the size of regression coefficients, the descending order of factors is Dis2CV, Dis2T, Dis2R, Dis2C, total N, Dis2F, Dis2P, organic matter, Dis2PC, slope, total K, temperature, PRHL in 2000, soil PH, and elevation. Generally speaking, transportation and location have a great effect on RHLT.

(1) Transportation. There were differences between the influences of various levels of transportation conditions on RHLT. As the high-level transportation, such as railways, highway, the national roads and provincial roads, are relatively

Table 3.14 Results of spatial correlation test based on multi-grid scales (Li et al. 2015)

Test	2 km	3 km	4 km	5 km	6 km
Lagrange Multiplier (lag)	733.07***	0.00	199.99***	119.94***	82.96***
Robust LM (lag)	314.06***	88.07***	169.11***	158.95***	161.536***
Lagrange Multiplier (error)	2805.40***	44.74***	1209.22***	983.00***	898.12***
Robust LM (error)	2386.40***	132.80***	1178.34***	1022.01***	976.69***

*Significance at 10%. **Significance at 5%. ***Significance at 1%

Table 3.15 Results of SEM at multi-grid scales (Li et al. 2015)

Variable	2 km	3 km	4 km	5 km	6 km
Constant	536.35***	340.49***	494.67***	478.20***	481.55***
PRHL 2000	−1.01***	−1.05***	−0.97***	−0.98***	−0.97***
Dis2HR	−1.67	52.39***	10.07	−0.89	−13.48
Dis2NP	−3.75	43.13	−30.13	7.69	0.57
Dis2CV	−2292.97***	−2666.28***	−1671.60***	−1532.54***	−1501.28***
Elevation	−0.19***	−0.21***	−0.19***	−0.17***	−0.17***
Slope	−4.92***	−1.44**	−3.71***	−5.00***	−4.19***
Soil PH	−0.87***	−0.88***	−0.73**	−0.55**	−0.83**
Total N	59.20***	−69.86	61.96***	45.63***	65.78***
Total K	3.46***	−0.43	2.62***	1.46	2.64**
Organic matter	−6.89***	−0.60	−6.78***	−6.28***	−7.38***
Rainfall	0.00	−0.01***	0.00	0.00	0.00
Temperature	−1.87***	−0.12	−1.63***	−1.52***	−1.41***
Dis2R	−117.83***	−128.61***	−38.02	−91.13	−101.23**
Dis2PC	−6.09***	−2.45*	−3.09*	−1.39	−3.27
Dis2P	15.87***	−9.23*	6.79	13.78**	4.86
Dis2C	78.03***	50.98***	68.89***	55.43***	49.15***
Dis2T	−60.31	148.42***	−222.11***	−136.57**	−186.68***
Dis2F	25.73***	−1.83	24.05*	36.47**	28.76*
Lambda	0.27***	0.05***	0.30***	0.33***	0.33***
R^2	0.589873	0.4894	0.640141	0.684296	0.706594

*Significance at 10%. **Significance at 5%. ***Significance at 1%

enclosed and they have little effects on rural housing land expansion and rural population migration. While the county/village roads led to RHLT by affecting the siting of rural housing. The new rural housing was built along the roads and eventually formed a phenomenon called "along-roads residence".

(2) Location. Dis2C had a significant positive relationship with RHLT which means that the higher the Dis2C, the faster increase in PRHL. The growth of PRHL was increasing along the transect "urban-suburbs-exurbs-rural"; Dis2T had a negative correlation with RHLT, so the grids near the township government site would have dramatic RHLT. The villages around towns generally have a higher level of socio-economic development, and the sources of farmer's income are various. So the ability to improve their living condition is strong. In this case, the rural housing land expanded rapidly which resulted in the increase of PRHL dramatically; Dis2PC and Dis2P mainly through absorbing surplus rural labor to affect the spatial distribution of rural population, which led to the RHLT.

(3) Hydrological conditions. Since water is indispensable for the survival of mankind, the siting of the settlements must ensure access to enough water. The

negative correlation between hydrological conditions and RHLT meant closer to the river, the faster growth of PRHL. The regions near rivers are easy to access water for farming and living, so there are a large number of settlements and high-quality arable land along the river. In the context of cultivated land abandoned, the farmland becomes the preferred site for the expansion of rural housing.

(4) PRHL in 2000. There was a significant negative correlation between PRHL in 2000 and RHLT during 2000–2010. The larger the PRHL 2000 was, the slower the RHLT, and vice versa. The rural housing supply in some areas with low PRHL in 2000 cannot meet the rising demand generated by the population growth and improving living conditions. So, the rural housing land will expand and the PRHL grows rapidly.

(5) Environmental conditions. Other conditions, such as geological conditions, terrain, climate, and soil, almost have little effects on the RHLT. On the one hand, in a short period, little changes occurred to these conditions and they will have little influences on the changes of rural housing land and the demographic change, which is closely related to socioeconomic development. On the other hand, with the enhancement of building technology, the physical conditions are already not considered as an important factor for the siting of rural housing. However, the lost rural housing occurred in the areas with a steep slope, extremely high or low elevation, or poor geological conditions.

2. **Driving forces analysis of RHLT at the county scale**

In China, the county is the smallest unit to obtain the overall socio-economic data. The results based on county scale analysis can provide a theoretical basis for policy-making. The same methods were selected to determine which SERA model will be suitable for the analysis of the driving force of RHLT at the county scale. Lagrange Multiplier (lag) (22.08) is more significant than Lagrange Multiplier (error) (19.42). The Robust LM (lag) is significant while the Robust LM (error) is not significant. So we chose SLM for the analysis of the driving force of RHLT at the county scale (Table 3.16). The county scale SERA showed that the per capita net income of farmer (PCNI) and rural non-agricultural employment rate (RNER) had significant impacts on RHLT, while the GDP per capita, urbanization and industrialization were not significant as expected. As we expected, the coefficient of PCNI in 2000 was positive (Table 3.16), which demonstrated the increase of PRHL was directly proportional to PCNI. The high PCNI enhanced the farmers' ability to improving the living conditions. Thus, the overcrowding degree was relieved and the PRHL inevitably increased significantly. Reflecting the degree of non-agricultural employment in rural areas, RNER's coefficient was significantly negative as shown in Table 3.16 In fact, RNER and RHLT might be related in a positive way or a negative way. On the one hand, the increase of RNER made the sources of farmer's income diversity and stimulated the desire of improving living conditions, which led to an increase of PRHL. On the other hand, with the increase of RNER, lots of farmers migrated to cities to seek jobs, which prevented the expansion of rural housing land and led to the decrease

Table 3.16 Results of SLM at the county scale (Li et al. 2015)

Variable	SLM
Constant	5.414**
GDP per capita 2000	0.001
PCNI 2000	0.002**
Industralization 2000	0.081
Urbanization 2000	0.050
RNER 2000	−0.176**
GDP per capita 2010	0.000
PCNI 2010	0.000
Industralization 2010	−0.193
Urbanization 2010	0.188
RNER 2010	−0.061
W-RHLT	0.336***
R^2	0.144

*Significance at 10%. **Significance at 5%. ***Significance at 1%

of PRHL. The ultimate relationship between RHLT and RNER is the result of the combined action of the above two cases. However, in the Bohai Rim, the relationship between them is negative. The factors of urbanization and industrialization are not significant, which was not consistent with the previous estimate. As to urbanization, the reason is most likely caused by the typical type of statistical unrealistically high urbanization. As to industrialization, maybe there is no direct but indirect causality between it and RHLT. The GDP per capita is not significant at all. GDP per capita is a comprehensive indicator measuring socio-economic development in both urban and rural systems but not the only rural system, so there is no direct causality between it and RHLT.

3. **Mechanism analysis of RHLT**

The analysis of driving forces of RHLT at grid-scale and county scales shows that the dynamic pattern of RHLT is the result of interaction among various factors, including physical conditions, socio-economic conditions, transportation, location and the PRHL in 2000. These factors finally led to the RHLT, affecting the rural population migration and rural housing land change. As shown in Fig. 3.23, solid arrow lines (dotted arrow lines) represent each factor has significant (no significant) effect on RHLT, and the "+" ("−") represents the two variables linked by the arrow line have a positive (negative) correlation. The rural housing land change and rural population migration have positive and negative effects on RHLT, respectively.

With the development of science and technology, physical conditions in the aspects of geology, terrain, climate, soil, and hydrological factors, have little impact on RHLT. The socio-economic factors, however, have significant and complex effects on RHLT. PCNI and RNER are the key factors driving the RHLT, while the industrialization

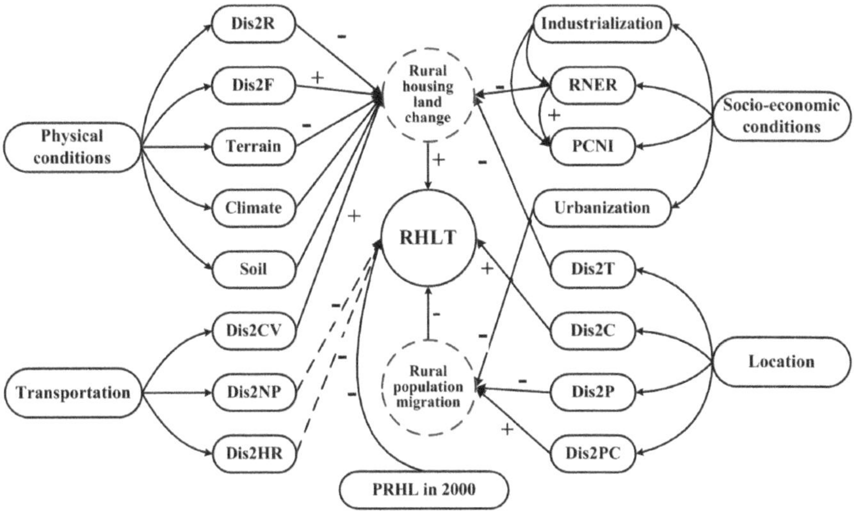

Fig. 3.23 Driving mechanism of RHLT (Li et al. 2015)

and urbanization have indirect effects on it through changing farmers' income and employment structure. More remarkably, the transportation and location conditions play an important role in driving RHLT and their interactions with RHLT vary in strength and direction depending on the level of transportation and location. The study found that Dis2CV, Dis2C and Dis2T have great influences on RHLT, which demonstrated the development and construction of towns (including chengguan town that county government site locates in) should be paid more attention in establishing a future urban-town system of China. In addition, the RHLT also was affected by PRHL in the base period to a certain extent. The less the PRHL of 2000 is, the more dramatically the RHLT during 2000–2010, and vice versa. Therefore, the implementation of rural housing land consolidation needs to take into account both the status quo and its changing trend of PRHL.

3.6.4 Discussion and Conclusions

This section uses grid technology with the advantage of the precise location to analyze the RHLT in Bohai Rim, the results of which may provide the parameters for potential evaluation of and practical pieces of advice for rural housing land consolidation. Gridded research of RHLT perfectly meets the needs of village-town construction and planning under the background of rapid urbanization. The results of this study indicated that the distribution of rural housing land was characterized by vertical zonality, i.e., the proportion of rural housing land decreased gradually with the increase of elevation, and was more sensitive to slope than to elevation. The plain areas with the elevation range of 0–200 m and the areas with a slight slope (<5°) were the best location choice for the rural housing. The PRHL is generally high in Bohai Rim which was far beyond the national highest standard of rural construction land use, which shows there is a great potential for carrying out rural housing land consolidation.

The results of hot spots analysis showed that the spatial mismatch of rural housing land change and rural population migration made the regional differences of RHLT more remarkable. The rural housing land consolidation should adopt the corresponding reconstruction models and chronological sequence according to the regional differences. Generally speaking, there are four regional types of rural housing land consolidation followed by "rural population increase vs. rural housing land decrease", "rural population decrease vs. rural housing land increase", "synchronous decrease of rural housing land and rural population" and "synchronous increase of rural housing land and rural population".

As to PRHL, with the introduction of series of laws and regulations about "one homestead one house", the sprawl of rural housing land has been controlled in the study area, so the RHLT will be mainly affected by rural population migration. The towns have the advantage of low barriers to entry, particularly for the relatively low living costs and social costs, so they will be the main carriers for workers moving from rural areas to cities, which was mentioned in National New-type Urbanization Plan (2014–2020). That will change the recent phenomenon of rural population flowing to big cities to some extent. At present, the population of many towns is decreasing, which demonstrates there is enough space and capacity to carry the migration from rural to urban.

Towns are effective hubs for linking the urban system and the rural system. In the process of pushing forward China's new-type urbanization, it is better to get the farmers to migrate to towns instead of big cities. For this, it is necessary to further improve the urban and rural household registration system, to integrate the urban and rural social safeguard system, and to reform the urban and rural land institution. Specifically, there are urgent needs to establish the withdrawing mechanism of rural housing land, the compensation mechanism of rural housing land use, and the multiple guarantee mechanism for farmers whose rural housing land is expropriated. Only by this way can the relationship between out-migrated farmers and rural housing land be orderly and completely removed, so as to smoothly pushing forward rural housing land consolidation.

3.7 Conclusions

This chapter aims to theorize and verify the rural housing land transition, as well as reveal the mechanism from a comprehensive perspective. In examining the linkages between economic growth and rural housing land change, theoretical and empirical studies have indicated that rural housing in every region will undergo specific stages—the proportion of rural housing land in the increase of total construction land will decline gradually with the development of the local economy, and the end of the transition corresponds to a new equilibrium between rural housing and other construction activities. In order to deal with the lack of long-term statistical data of land use in China, the concept of Transects was adopted to associate land-use change with socio-economic development relying on China's regional differentiation of socio-economic development, which is demonstrated to a novel and valuable research tool for land use research.

Demographic changes open a window for investigating rural housing land transition. The results of the systematic analysis of changing regional patterns of China's rural population and settlements showed that a positive pattern of co-evolution of population and residential land in rural areas has not appeared, and land utilization tends to be extensive, especially in northern and eastern coastal provinces. Rapid industrialization and urbanization in China have resulted in a severe phenomenon of 'village-hollowing' caused by the dual-track structure of socio-economic development separating rural and urban areas, a process also underpinned by institutional-managerial, cultural and environmental factors. Besides, based on the characteristics of rural housing land-use, an evolutive model of the development stages of hollowing villages was constructed. Usually, there are four evolutive stages in a normal development process of hollowing villages at the urban-rural fringe, i.e., centralizing, sub-hollowing, hollowing and recentralizing, which corresponds with the temporal characteristics of China's socio-economic development.

Based on the above analysis, spatio-temporal patterns and the underlying driving forces and mechanism were detected by using multiple-source data. Results signified that the distribution of rural housing land was characterized by vertical zonality, i.e., the proportion of rural housing land decreased gradually with the increase of elevation, and was more sensitive to slope than to elevation. The results of hot spots analysis showed that the spatial mismatch of rural housing land change and rural population migration made the regional differences of rural housing land transition more remarkable, based on which four rural housing land consolidation regional types were differentiated-"rural population increase vs. rural housing land decrease", "rural population decrease vs. rural housing land increase", "synchronous decrease of rural housing land and rural population" and "synchronous increase of rural housing land and rural population".

Accordingly, the Central Government needs to define uniform regulations for rural housing according to regional socio-economic developmental level, physical conditions, and rural housing land transition phase. curbing excessive rural housing, consolidating rural housing land, providing more job opportunities for surplus rural labor is necessary to cope with the challenges under the context of rapid urbanization.

References

Anselin L (1998) Spatial econometrics: methods and models. Kluwer Academic Publisher, Dordrecht

Anselin L (2002) Under the hood Issues in the specification and interpretation of spatial regression models. Agr Econ 27:247–267

Antrop M (2004) Landscape change and the urbanization process in Europe. Landscape Urban Plan 67:9–26

Brown LR (1995) Who will feed China? Wake-up call for a small planet. WW Norton & Company, New York

Champion T (2001) Urbanization, suburbanisation, counterurbanisation and reurbanisation. In: Paddison R (ed) Handbook of urban studies. Sage, London, pp 143–161

Chen R, Ye C, Cai Y et al (2014) The impact of rural out-migration on land use transition in China: past, present and trend. Land Use Pol 40:101–110

Christiansen F (2009) Food security, urbanization and social stability in China. J Agrar Change 9:548–575

Deng J, Wang K, Hong Y et al (2009) Spatio-temporal dynamics and evolution of land use change and landscape pattern in response to rapid urbanization. Landscape Urban Plan 92:187–198

Ding C (2003) Land use policy reform in China: assessment and prospects. Land Use Pol 20:109–120

Fang Y, Liu J (2014) The modification of North China quadrangles in response to rural social and economic changes in agricultural villages: 1970–2010s. Land Use Pol 39:266–280

Ferreira MP, Alves DS, Shimabukuro YE (2015) Forest dynamics and land-use transitions in the Brazilian Atlantic Forest: the case of sugarcane expansion. Reg Envir Chang 15:365–377

García AI, Ayuga F (2007) Reuse of abandoned buildings and the rural landscape: the situation in Spain. T ASABE 50:1383–1394

Gong J, Chen W, Liu Y et al (2014) The intensity change of urban development land: Implications for the city master plan of Guangzhou, China. Land Use Pol 40:91–100

Grainger A (1995a) National land use morphology: patterns and possibilities. Geography 80:235–245

Grainger A (1995b) The forest transition: an alternative approach. Area 27:242–251

Haberl H, Batterbury S, Moran E (2001) Using and shaping the land: a long-term perspective. Land Use Pol 18:1–8

Huang Q, Li M, Chen Z et al (2011) Land consolidation: an approach for sustainable development in rural China. Ambio 40:93–95

Irwin EG, Geoghegan J (2001) Theory, data, methods: developing spatially explicit economic models of land use change. Agr Ecosyst Envir 85:7–24

Kanbur R, Zhang X (1999) Which regional inequality? the evolution of rural-urban and inland-coastal inequality in China from 1983 to 1995. J Comp Econ 27:686–701

Lambin EF, Geist HJ, Lepers E (2003) Dynamics of land-use and land-cover change in tropical regions. Annu Rev Env Resour 28:205–241

Li S, Hou Y (2001) 我国经济发展阶段特征与“十五”时期产业发展的主要任务 (The distinctiveness of China's various economic growth stages and the main targets of industrial development during the 10th five-year-plan). Manage World 2:95–101

Li T, Long H, Liu Y et al (2015) Multi-scale analysis of rural housing land transition under China's rapid urbanization: the case of Bohai Rim. Habitat Int 48:227–238

Li X (2002) 土地利用变化的解释 (Explanation of land use changes). Prog Geogr 21(3):195–203

Li Y, Liu Y, Long H (2010) 中国农村人口与农村居民点用地的时空变化 (Spatio-temporal analysis of population and residential land change in rural China). J Nat Resour 25(10):1629–1638

Li Y, Liu Y, Long H et al (2014) Community-based rural residential land consolidation and allocation can help to revitalize hollowed villages in traditional agricultural areas of China: evidence from Dancheng county, Henan province. Land Use Pol 39:188–198

Lin GC, Ho SP (2003) China's land resources and land-use change: insights from the 1996 land survey. Land Use Pol 20:87–107

Liu J, Liu M, Zhuang D et al (2003) Study on spatial pattern of land-use change in China during 1995–2000. Sci China 4:373–378
Liu Y (2007) 中国东部沿海地区乡村转型发展与新农村建设 (Rural transformation development and new countryside construction in eastern coastal China). Acta Geogr Sinica 62:563–570
Liu Y, He S, Wu F et al (2010) Urban villages under China's rapid urbanization: unregulated assets and transitional neighbourhoods. Habitat Int 34:135–144
Liu Y, Long H, Chen Y et al (2011) 中国乡村发展研究报告: 农村空心化及其整治策略 (Research Report on Rural Development in China: Hollowed Village and Its Renovation Strategy). Science Press, Beijing
Liu Y, Wang L, Long H (2008) Spatio-temporal analysis of land-use conversion in the eastern coastal China during 1996–2005. J Geogr Sci 18:274–282
Long H (2006) 中国农村宅基地转型的理论与证实 (Rural housing land transition in China: theory and verification). Acta Geogr Sinica 61(10):1093–1100
Long H (2012) 中国乡村转型发展与土地利用(Land use and rural transformation development in China). Science Press, Beijing
Long H (2014) Land use policy in China: introduction. Land Use Pol 40:1–5
Long H, Heilig GK, Li X et al (2007) Socio-economic development and land-use change: analysis of rural housing land transition in the transect of the Yangtse River, China. Land Use Pol 24:141–153
Long H, Li Y, Liu Y et al (2012) Accelerated restructuring in rural China fueled by 'increasing vs. decreasing balance' land-use policy for dealing with hollowed villages. Land Use Pol 29:11–22
Long H, Li Y, Liu Y et al (2013) Population and settlement change in China's countryside: causes and consequences. In: Cawley M (ed) The sustainability of rural systems: global and local challenges and opportunities. The Commission on the Sustainability of Rural Systems (CSRS) of the International Geographical Union (IGU) and the Whitaker Institute, Galway, pp 123–133
Long H, Liu Y, Li X et al (2010) Building new countryside in China: a geographical perspective. Land Use Pol 27:457–470
Long H, Liu Y, Wu X et al (2009) Spatio-temporal dynamic patterns of farmland and rural settlements in Su-Xi-Chang region: implications for building a new countryside in coastal China. Land Use Pol 26:322–333
Long H, Woods M (2011) Rural restructuring under globalisation in eastern coastal China: what can be learned from Wales? J Rural Community Dev 6:70–94
Long H, Zou J, Pykett J et al (2011) Analysis of rural transformation development in China since the turn of the new millennium. Appl Geogr 31:1094–1105
Massey D (1999) Space-time, 'science' and the relationship between physical geography and human geography. T I Brit Geogr 24:261–276
Mather AS (1990) Global forest resources. Belhaven Press, London
Mather AS (2005) Assessing the world's forests. Glob Environ Change 15:267–280
Ministry of Construction of China (MCC) (1993) 村镇规划标准 (Village/small-town planning standards-GB 50188-93). Ministry of Construction of China, Beijing
Ministry of Land and Resources of China (MLRC) (2009) 全国土地利用变更调查报告-2008 (Reports on China's land-use survey and update in 2008). China Land Press, Beijing
Mukherjee A, Zhang X (2007) Rural Industrialization in China and India: role of policies and institutions. World Dev 35:1621–1634
Myers N (1980) The present status and future prospects of tropical moist forests. Environ Conserv 7:101–114
National Bureau of Statistics of China (NBSC) (2009) 中国统计年鉴 (China statistical yearbook). China Statistics Press, Beijing
Paquette S, Domon G (2003) Changing ruralities, changing landscapes: exploring social recomposition using a multi-scale approach. J Rural Stud 19:425–444
Putterman L (1997) On the past and future of china's township and village-owned enterprises. World Dev 25:1639–1655
Qu Y, Long H (2018) The economic and environmental effects of land use transitions under rapid urbanization and the implications for land use management. Habitat Int 82:113–121

Robinson GM (2004) Geographies of agriculture: globalisation, restructuring and sustainability. Pearson, Harlow
Sargeson S (2002) Subduing "the rural house-building craze": attitudes towards housing construction and land use controls in four zhejiang villages. China Quart 172:927–955
Shen J (2002) A study of the temporary population in Chinese cities. Habitat Int 26:363–377
Song Y, Zenou Y, Ding C (2008) Let's not throw the baby out with the bath water: the role of urban villages in housing rural migrants in China. Urban Stud 45:313–330
Unger J, Chan A (1999) Inheritors of the boom: private enterprise and the role of local government in a rural South China township. China J 42:45–74
Walker R (2001) Urban sprawl and natural areas encroachment: linking land cover change and economic development in the Florida Everglades. Ecol Econ 37:357–369
Wang M (2010) Impact of the global economic crisis on China's migrant workers: a survey of 2,700 in 2009. Eurasian Geogr Econ 51:218–235
Wang X, Hui E, Choguill C et al (2015) The new urbanization policy in China: which way forward? Habitat Int 47:279–284
Wang Y, Scott S (2008) Illegal farmland conversion in China's urban periphery: local regime and national transitions. Urban Geogr 29:327–347
Woods M (2007) Engaging the global countryside: globalization, hybridity and the reconstitution of rural place. Prog Hum Geog 31:485–507
Wu F, Ma L (2006) Transforming China's globalizing cities. Habitat Int 30:191–198
Xie Y, Mei Y, Guangjin T et al (2005) Socio-economic driving forces of arable land conversion: A case study of Wuxian City, China. Glob Environ Change 15:238–252
Yang H, Li X (2000) Cultivated land and food supply in China. Land Use Pol 17:73–88
Yang R, Liu Y, Long H et al (2015) 基于格网的农村居民点用地时空特征及空间指向性的地理要素识别—以环渤海地区为例 (Spatial-temporal characteristics of rural residential land use change and spatial directivity identification based on grid in the Bohai Rim in China). J Geogr Sci 25(5):559–572
Yu X, Ng CN (2007) Spatial and temporal dynamics of urban sprawl along two urban-rural transects: a case study of Guangzhou, China. Landscape Urban Plan 79:96–109
Zhang Y, Long H, Tu S et al (2019) Spatial identification of land use functions and their tradeoffs/synergies in China: implications for sustainable land management. Ecol Indic 107:105550
Zhong T, Huang X, Zhang X et al (2011) Temporal and spatial variability of agricultural land loss in relation to policy and accessibility in a low hilly region of southeast China. Land Use Pol 28:762–769

Chapter 4
Coupling Analysis of Farmland and Rural Housing Land Transitions in China

Abstract Farmland and rural housing land are the two major land-use types closely related to human livelihood and production activities. It is well known that the urban is originally developed from the rural, and its formation and further development is closely linked with the transitions of farmland and rural housing land. Dramatic quantity variation in rural housing land and farmland has taken place in China induced by the economic growth, lifestyle transformation, and living standard improvement in rapidly urbanized rural areas. In essence, these two types of land use compete with each other, and demonstrate constant interactions and tradeoffs in different socio-economic context. Attempting to identify the mutual coupling relations between rural housing land and farmland, this chapter not only reveals the characteristics of their spatio-temporal coupling, but also qualitatively and quantitatively discussed and analyzed the mechanism, which contributes to the knowledge about man-land interrelations. Thereafter, the coupling relations are empirically examined in economically burgeoning regions-the Su-Xi-Chang Region and the rapidly urbanizing traditional areas-the Huang-Huai-Hai Plain. The author argues that it is fundamental to lay out a scientific urban-rural integrated development planning for restructuring and revitalizing the countryside, which needs to pay more attention to making the rural have certain functions serving for the urban. In addition, measures regarding promoting the reclamation and redevelopment of inefficient and unused urban-rural construction land, reducing inefficient occupation of farmland resources, and developing appropriate scale management of agriculture are needed to ensure food security in farming areas.

4.1 Introduction

Much importance is attributed to the search for knowledge about the impact of human societies on our environment as rapid population growth and economic development intensify the stresses that human beings place on the biosphere and ecosystems. Farmland and rural housing land are the two major land-use types closely related to human livelihood and production activities. It is well known that the urban is originally developed from the rural, and its formation and further development is

H. Long, *Land Use Transitions and Rural Restructuring in China*,
https://doi.org/10.1007/978-981-15-4924-3_4

closely linked with the transitions of farmland and rural housing land. In general, socio-economic development affects land use morphology, which also makes a counteraction to socio-economic development. Dramatic quantity variation in rural housing land and farmland has taken place induced by the economic growth, lifestyle transformation, and living standard improvement in rapidly urbanized rural areas. In essence, these two types of land use compete with each other, and demonstrate constant interactions and tradeoffs in different socio-economic context. Yet, how farmland will respond to rural housing land transition hasn't been clearly clarified, and vice verse.

What's more, the interaction between land use morphology and socio-economic development facilitates the land use transition. Therefore, the changes of the regional land use morphology may be basically reflected by the changes of farmland and rural housing land, two major land-use types affecting human production activities. Farmland and rural housing land, as the two major sources of land use transitions, play an important role in the process of facilitating land use transition. That is also the main reason why this chapter focuses on the analysis of farmland and rural housing land transitions. Undoubtedly, analyzing the spatio-temporal coupling characteristics and mechanism of farmland and rural housing land transitions, to a great extent, will be helpful in promoting the understanding the coupled "land system".

This chapter have 4 parts: the first section aims to depict the trend of the spatio-temporal coupling of farmland and rural housing land transitions; then, the coupling mechanism will be analyzed and discussed. In the following two parts, the coupling relations will be empirically examined in economically burgeoning regions-the Su-Xi-Chang Region and the rapidly urbanizing traditional areas-the Huang-Huai-Hai Plain. Finally, the conclusions are presented.

4.2 Spatio-Temporal Coupling

4.2.1 Background

Land-use and land-cover change, as the core of coupled human-environment systems, has become a potential field of land change science (LCS) in the study of global environmental change (Liu et al. 2010). The goal of Global Land Project is to measure, model and understand the coupled socio-environmental terrestrial system, which is also referred to as "land system". In this kind of land system, farmland and rural housing land are tightly inter-related with human production activities, and are treated as the two most important land-use types depicting urban-rural development in China (Long et al. 2009). Currently, farmland and rural housing land change in China has attracted the interest in academic circles, ranging from farmland loss due to urbanization and farmland preservation (Tan et al. 2009), rural settlement and housing land expansion (Su et al. 2011), spatio-temporal dynamic pattern of land-use change (Long et al. 2009; Tian et al. 2007), to driving forces analysis and simulation

(Xie et al. 2007). However, the research on the coupling interaction of farmland and rural housing land transition has found much less attention.

Since the turn of the new millennium, the implementation of a series of socio-economic and regional development policies has resulted in accelerated economic growth and hence caused considerable land-use change in China, which was characterized by the reduction of farmland and the rapid expansion of urban-rural construction land, two-thirds of which are rural housing land. Accordingly, it may be hypothesized that, to some extent, there is a coupling relationship between farmland and rural housing land transitions in the early 21st century in China. The aims of this section is to examine the coupling relationship between farmland and rural housing land transitions in China.

4.2.2 Materials and Methods

4.2.2.1 Data Source

This section analyzes the spatial and temporal coupling characteristics of farmland and rural housing land transition using high-resolution Landsat TM (Thematic Mapper) data in 2000 and 2008 (lack of the data of Shaanxi and Qinghai provinces) from the Ministry of Environmental Protection of China, as well as the data from Reports on China's Land-use Survey and Update at the provincial level from 2000 to 2008 released by the Ministry of Land and Resources of China. The two sets of data collected from different sources have their respective advantages in revealing the spatio-temporal laws of land use transition. The integrated application of the analysis results from the two sets of data is beneficial to explore the spatio-temporal coupling characteristics of farmland and rural housing land transition.

4.2.2.2 Methods

(1) **Spearman rank correlation analysis**

The correlation coefficient of farmland versus rural housing land change reflects the coordinative relation of farmland and rural housing land change over a certain period of time. The higher the absolute value of the coefficient, the stronger the coordinative relation of farmland and rural housing land change, and vice versa. When it is negative or positive, the change of the farmland and rural housing land shows a negative correlation or positive correlation, respectively. Spearman rank correlation coefficient is used to measure the correlationship between the ordinal variables and its range is $[-1, 1]$. The positive value of the spearman rank correlation coefficient illustrates the positive rank correlation, and vice versa. The higher the value is, the higher the rank. When the value equals 1 or -1, the rank of two variables is equal or totally different, respectively. In this section, Spearman rank correlation analysis is used to analyze the correlationship between the change rate of rural housing land and the correlation coefficient of farmland versus rural housing land change.

(2) **Coupling coefficient of the farmland and rural housing land change**

In order to quantitatively analyze the coupling relationship of the farmland and rural housing land transition, this section calculates the proportion of mutual transformation area between farmland and rural housing land to the total area of farmland change and rural housing land change during the study period. Therefore, the section uses the coupling coefficient of farmland and rural housing land change (FHCC) to reflect the spatial coupling degree of the farmland and rural housing land transition, according to Eq. (4.1):

$$FHCC = \frac{C_{FH} + C_{HF}}{C_F + C_H - C_{FH} - C_{HF}} \times 100\% \tag{4.1}$$

where $FHCC$ is the coupling coefficient of farmland and rural housing land change; C_{FH} is the area of farmland transformed to rural housing land; C_{HF} is the area of rural housing land transformed to farmland; C_F or C_H is the total area of the overlayed changed patches whose attribute information in the early period and later period of the research contains farmland or rural housing land, respectively. The higher the value of coupling coefficient, the higher the spatial coupling degree of farmland and rural housing land change.

Geographic information system (GIS) analysis was adopted for the research. In particular, the ESRI's ArcGIS 9.2 spatial analysis module was used to aggregate, synthesize and analyze the databases, and to identify spatial relationships.

4.2.3 Results

4.2.3.1 Temporal Coupling of Farmland and Rural Housing Land Transition

The result of the correlation analysis of China's farmland and rural housing land change during 2000–2008 indicates that: during 2000–2005, the area of farmland decreased dramatically, but the area of rural housing land had an obvious increase; during 2005–2008, the area change of farmland and rural housing land gradually tended to be stable and the area of rural housing land even had a small decrease; during 2000–2008, the correlation coefficient of farmland versus rural housing land change is −0.921, which shows that the change pattern of farmland and rural housing land is uncoordinated (Fig. 4.1).

To carry out the Spearman rank correlation analysis of the change rate of rural housing land and the correlation coefficient of farmland versus rural housing land change, it is needed to ensure the rank order of the two variables. Because there is a negative correlation between farmland and rural housing land change in China, it

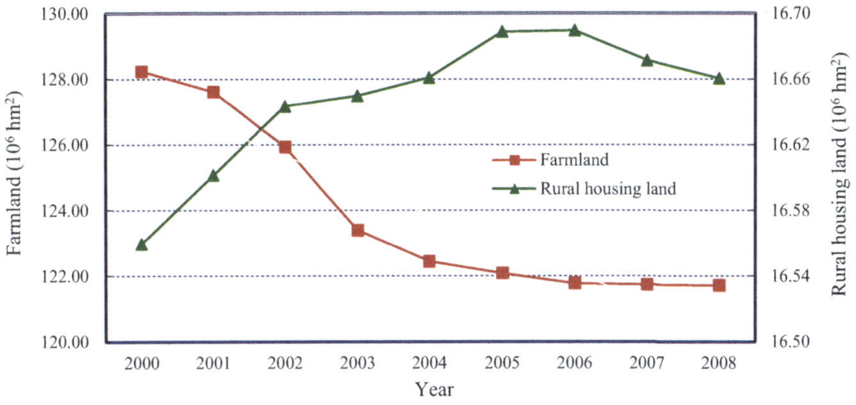

Fig. 4.1 The change of China's farmland and rural housing land during 2000–2008 (Long and Li 2012)

assumed that the lower the value of correlation coefficient of farmland versus rural housing land change, the stronger the correlation and the higher the rank. The rank of the change rate of rural housing land is determined by its value in descending order, and the higher the value, the higher the rank. Table 4.1 shows the ranks of the two variables.

The result of Spearman rank correlation analysis using the two sets of rank series data shows that the Spearman rank correlation coefficient of the two variables is 0.808, which is close to 1 and is significant at the 1% level (Table 4.2). So, it can be concluded that there is a significant positive rank correlation between the two variables. The higher the change rate of rural housing land, the lower the correlation coefficient of farmland versus rural housing land change (the higher its rank is) and the stronger the correlation between farmland and rural housing land change. The result also shows that rural housing land change has played a major role in the mutual transformation of farmland and rural housing land.

4.2.3.2 Spatial Coupling Characteristics of Farmland and Rural Housing Land Transition

Basing on the calculation of the coupling coefficient of farmland and rural housing land change at provincial level, it shows a high-degree spatial coupling between farmland and rural housing land change in the southeast of "HU Huanyong Line" during 2000–2008 (Fig. 4.2). The inclined line in Fig. 4.2 is called "HU Huanyong Line", a "geo-demographic demarcation line" discovered by Chinese population geographer HU Huanyong in 1935. The imaginary Heihe (in Heilongjiang)-Tengchong (in Yunnan) Line divides the territory of China into two parts: northwest of the line covers 64% of the total area but only 4% of the population; however, southeast of the line covers 36% of the total area but 96% of the population (Hu 1935). The southeast

Table 4.1 The change rate of rural housing land and the correlation coefficient of farmland versus rural housing land change (Long and Li 2012)

Province	Correlation coefficient of farmland versus rural housing land change	Rank of correlation coefficient	The change rate of rural housing land (%)	Rank of the change rate
Beijing	−0.017	17	0.01	19
Tianjin	−0.988	1	1.14	1
Hebei	−0.889	9	0.44	7
Shanxi	−0.936	7	0.30	8
Inner Mongolia	−0.791	10	0.10	16
Liaoning	−0.712	13	0.14	15
Jilin	0.696	27	−0.07	23
Heilongjiang	0.275	21	0.06	17
Shanghai	0.462	23	−0.90	31
Jiangsu	0.201	19	0.02	18
Zhejiang	−0.919	8	0.47	6
Anhui	0.752	29	−0.20	27
Fujian	−0.736	12	0.24	11
Jiangxi	0.618	25	−0.20	28
Shandong	0.257	20	−0.06	22
Henan	0.712	28	−0.08	25
Hubei	-0.760	11	0.17	12
Hunan	0.548	24	−0.03	21
Guangdong	−0.968	3	0.17	13
Guangxi	−0.939	6	0.28	9
Hainan	−0.651	14	0.00	20
Chongqing	0.911	31	−0.21	30
Sichuan	0.410	22	−0.07	24
Guizhou	−0.946	5	0.16	14
Yunnan	−0.974	2	0.28	10
Tibet	−0.513	15	0.48	5
Shaanxi	0.650	26	−0.14	26
Gansu	0.837	30	−0.20	29
Qinghai	0.141	18	0.94	2
Ningxia	−0.959	4	0.94	3
Xinjiang	−0.407	16	0.79	4

Table 4.2 Spearman rank correlation coefficient and related test (Long and Li 2012)

		Correlation coefficient of farmland versus rural housing land change	The change rate of rural housing land
Correlation coefficient of farmland versus rural housing land change	Correlation coefficient	1.000	0.808***
	Two-tailed test of significance		0.000
	Number of samples	31	31
The change rate of rural housing land	Correlation coefficient	0.808***	1.000
	Two-tailed test of significance	0.000	
	Number of samples	31	31

*** Significance at 1%

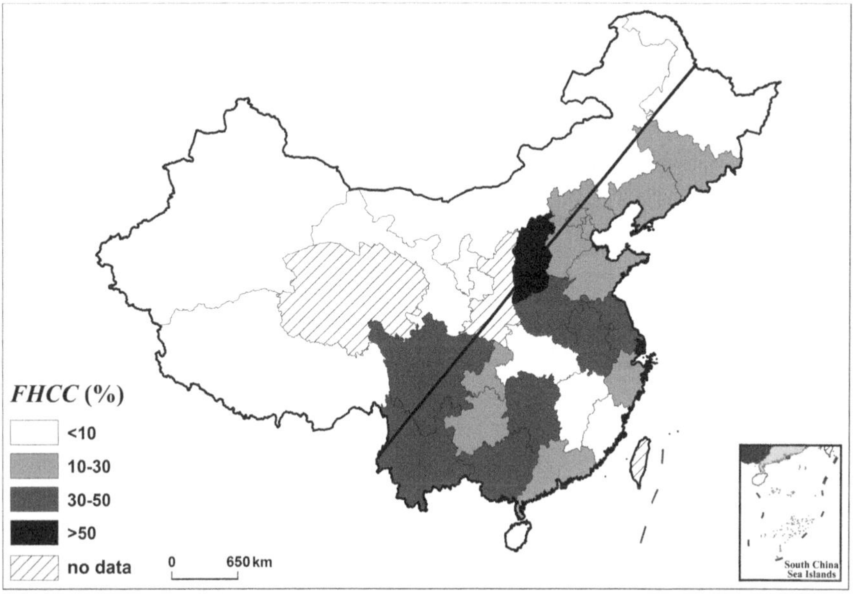

Fig. 4.2 The coupling coefficient of China's farmland and rural housing land change at provincial level (Long and Li 2012)

of "HU Huanyong Line" has been affected by the most frequent human activities, which caused obvious spatio-temporal coupling relationship between agricultural

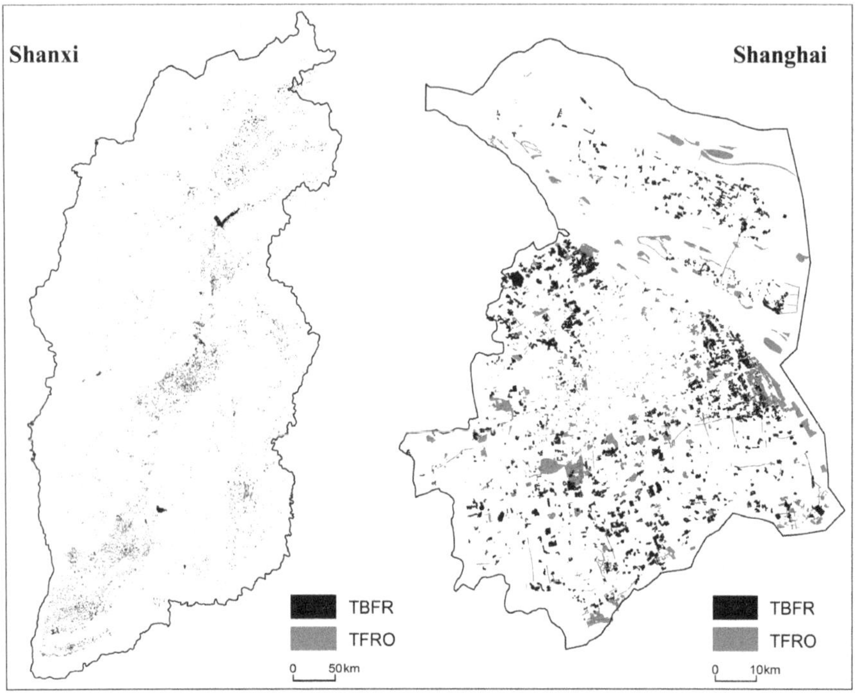

Fig. 4.3 The coupling of farmland and rural housing land change during 2000–2008 in Shanxi and Shanghai based on interpreted remote sensing data (Long and Li 2012) "TBFR" refers to transformation between farmland and rural housing land; "TFRO" refers to transformation of farmland or rural housing land from/to other land-use types

laborer transfer and farmland transition. Shanxi and Shanghai have the highest coupling coefficients of farmland and rural housing land change, which are more than 50% (Fig. 4.3).

4.3 Coupling Mechanism

4.3.1 Driving Forces

Usually, land-use change is driven by socio-economic, bio-physical and managerial three-dimensional driving factors. It is better to divide the driving factors into three driving forces according to the three dimensions mentioned above to analyze the coupling mechanism of farmland and rural housing land transition, by examining how the driving factors interact with each other and finally drive land use transition through the socio-economic, bio-physical and managerial three-dimensional dynamical mechanism.

(1) **Socio-economic driving forces**

(1) Economic growth and population change. Industrialization and urbanization are two different processes of modern socio-economic development. The process of industrialization and urbanization in China has been gradually accelerated with the reform of rural economic system and the adjustment of urbanization strategy. It was reported that: during 1996–2001, China's urbanization level increased by about 1.43% per year; during 2000–2005, the urban population increased from 456 million to 562 million, an increment of 106 million and an average annual increment of 21 million; during 1997–2000, the average annual area of farmland occupied by construction land was 1800 km^2, and the annual area amounted to 2187 km^2 during 2001–2005. The reasons for this phenomenon can be summarized into two aspects. On the one hand, the existing urban planning is unreasonable because it did not fully take the coordinated development of rural-urban area and industry-agriculture production into account, which caused low-efficient urban expansion and unnecessary non-agriculturalization of farmland. On the other hand, the potential of stock land has not been well tapped, which leads to a serious waste of urban land and aggravates the non-agriculturalization of farmland. Recently, the flourish of the "development zone fever" and "new district fever"[1] is a major aspect of industrialization and urbanization. It is worth mentioning that the lost farmland due to industrialization and urbanization was usually located at densely populated area or along the trunk roads, with a high quality. In addition, with rapid industrialization and urbanization, farmer's income has been greatly raised, which promoted the expansion of rural housing land. Low opportunity cost of occupying farmland further leads to the loss of farmland. The demographic structure change and migration will bring a series of influences on land-use change, especially on farmland and rural housing land change. Since the adoption of economic reform and open-door policy, the change of farmland and rural housing land has been greatly affected by the miniaturization of rural household scale, the flow of the rural population between the rural and the urban, and the rural migrant's settling down in the city.

(2) Social behavior and mainstream behavior of land user. In the early 1990s, the phenomena of "development zone fever" and "real estate fever" caused a lot of farmland was transformed to construction land. Then the "macro urban planning fever" was flourished although the farmland loss issues mentioned above have not yet been solved. Recently, the "new district fever" and "urban agglomeration

[1]The national level "new district" approved by the State Council refers to the regions with national special preferential policy in urban planning area. They have the property of governmental functional departments and are deputy provincial units. According to China's related industrial policy, the tax relief and other preferential policies for encouraging projects of the direct investment enterprises, are issued to attract various capital and entity investment to set up factories or companies. Usually, the planning areas of these new districts are relatively large, and the supporting facilities are completed. In these areas, attracting investment can improve the local economy development. There are some famous new districts, i.e., Shanghai Pudong New District, Tianjin Binhai New District, Chongqing Liangjiang New Area, Zhejiang Zhoushan Islands New District, and Xi'an-Xianyang New District.

fever"[2] have sprung up. To some extent, these "fevers" as a kind of social behavior have exerted great influences on national or regional land-use change.

The mainstream behavior of land user, such as the understanding and application of related policies, the choice of industry, allocation of funds and how to coordinate the relationship with the community and the government, also has a great impact on land-use change. For example, some counties did not follow the land-use planning when they changed the land-use mode and failed to form an effective management mechanism to the planning and distribution of non-agricultural construction land, which resulted in the orderless expansion of urban and rural construction land. Some local governments take the penalties for illegal land occupancy and over-standard house-building as a source of revenue, which caused some wealthy farmers take the opportunity to take up more land.

(3) Technology development. The land-use structure is determined by the socio-economic and technological conditions, and by the rationality, feasibility and assurance degree of technology application. Based on the development of science and technology, it is meaningful to protect and manage the existing land, improve its productivity and appropriately develop reserved land resources, so as to mitigate the human-land contradiction, and make the land-use change coordinated with local socio-economic development.

(2) **Bio-physical driving forces**

In the short term, the bio-physical driving factors influencing land-use change mainly refer to the natural disasters, topography and slope.

(1) Natural disasters. Usually, desertification, flooding and soil erosion are the major natural disasters influencing farmland use. In China, desertification leads to the lower farmland productivity and large-scale farmland degradation. Reclamation of lake results in environmental degradation of the lake and poor buffer capacity for flood prevention, and increases the hazards of flooding. Unreasonable reclamation of the upper reaches leads to soil erosion and siltation of reservoir and river way, which increase the burden of flood control. According to the analysis and estimate, farmland destroyed by natural disasters reached 1.662 million hm^2, and the direct grain production loss reached 6.055 million t during 1988–2008 in China.

(2) Topography and slope. The reclamation of marginal land mainly occurred in some regions with special physical factors, e.g., topography and slope. In the process of implementing the "Grain for Green" program in China, topography and slope the two physical factors are mainly taken into account for the selection of farmland to be returned to forest or grassland. In addition, the expansion of rural housing land in China, to some extent, is affected by topography and slope.

[2]For example, some urban agglomerations have been approved by the State Council recently, such as Wuhan urban agglomeration, Changsha-Zhuzhou-Xiangtan urban agglomeration, Wanjiang urban Belt, and so on.

(3) **Land management driving force**

Land management, the direct decision-making factor of land use, drives land-use change through property rights system, price system and land managerial policy.

(1) Property rights system. The land property rights system plays a basic role in driving land-use change. Since the 1980s, with the deepening of reform, the existing unstandardized and unstable farmland property rights system has led to some problems exposed, such as the incomplete land ownership, ambiguous land-use right, unrealized land transfer right and so on, which affect land-use change in China to a great extent. Currently, the farmer only has the right of use instead of dealing with his house, which resulted in the difficulty in transferring and exchanging his house, also the difficulty in implementing new village construction planning. That is an important factor causing the phenomenon of hollowed villages (Long et al. 2012).
(2) Price system. One of the characteristics of market economy is profit maximization, which makes the allocation of resources affected by the price system. The change of farmland is also affected by the price system. Recently, industrialization and urbanization in China have apparently triggered massive farmland loss for the benefits of market-oriented farming and non-agricultural development (Long et al. 2007a). The market mechanism only concerns the economic benefit, instead of environmental benefit and social benefit. As such, it is inevitable that the farmland with low comparative benefit will be occupied if the market mechanism is allowed to work blindly. The farmland transition is promoted by the disparity of comparative benefit in two aspects: firstly, it induces the transfer of agricultural laborer force to second or tertiary industry, which creates conditions for farmland transfer; secondly, according to market rules, competition among various economic departments drives land-use change.
(3) Land managerial policy. The current land managerial mechanism in China is distempered even somewhat unreasonable. The unreasonable land income distribution forms a mechanism that different groups vie with each other to occupy more farmland. In 1999, the Ministry of Finance of China and the Ministry of Land and Resources of China jointly enacted "Measures for the Administration of the Collection and Use of Land-Use Fees for Newly Added Construction Land", in which 70% of the land-use fees for newly added construction land belongs to local governments. Hereafter, 60% or 70% revenue of China's local governments is from land deals.

    Therefore, the city, county and township governments try their best to benefit more from land deals through requisitioning and selling farmland. Current distempered land managerial mechanism in China made it difficult for the central government to restrict the land-use behavior of local government through implementing farmland protection policy. Driven by the partial and local interests, even individual interests, local governments and officials tend to convert farmland to construction land, causing massive farmland loss. Currently, lack of related rules and regulations aiming at dealing with idle or vacant rural houses and the distempered examination and approval institution for new rural housing site made it difficult for the governments to supervise and adjust rural housing land.

4.3.2 Mechanism

In general, farmland and rural housing land transition in China is driven by socio-economic, bio-physical and managerial three-dimensional driving factors through the interactions among rural population, farmland and rural housing land. At the beginning of the economic reform in China, the great demand for rural housing increased with the development of rural economy and the increase of rural population, which caused the expansion of rural housing land. The farmers usually prefer the flat terrain with convenient facilities to build rural houses, where are often the distribution area of high-quality farmland. Together with the relaxed land managerial policies and low opportunity cost of farmland occupation, a large number of farmland has been occupied by rural housing construction. Meanwhile, driven by the social behavior of blindly developing "new town", "new district" and "urban agglomeration", as well as the low price of farmland requisition, much farmland was occupied by urban construction. Under the nibbling of rural-urban construction, the farmland tends to be fragmental, which is not in favor of scale management and causes the decrease of comparative benefit of farming. With the rapid urbanization in China, massive rural laborers migrate to urban area. That leads to the farmland management transferred to an extensive model even abandoned model, and the morphology of farmland is transformed from A to B (Fig. 4.4). With the rural population migrating to the city, due to the distempered withdrawing mechanism of rural housing land and the defect of property rights system, the morphology of rural housing land is transformed from A to B (Fig. 4.4), accompanied with the phenomenon of village-hollowing. When the issue of food security threatened by the continued decrease of farmland is severe enough to draw the public's concern, the institutional system will adjust the rural-urban land-use pattern through the resources and environmental managerial instruments, e.g., laws, regulations and policies.

Aiming at dealing with idle rural housing land and the contradiction between the demand and supply of urban construction land, several powerful policies related to optimally allocating idle rural housing land have been enacted, such as "homestead exchange for house", "reconstruction of villages-in-the-city", and "increasing versus decreasing balance" land-use policy to balance increases in urban construction land with a reduction in rural construction land. Part of the suburban idle rural housing land has been transformed into urban construction land, but most of them have been converted to farmland through carrying out land consolidation and reclamation. As such, by building new rural houses with unified planning and making full use of the

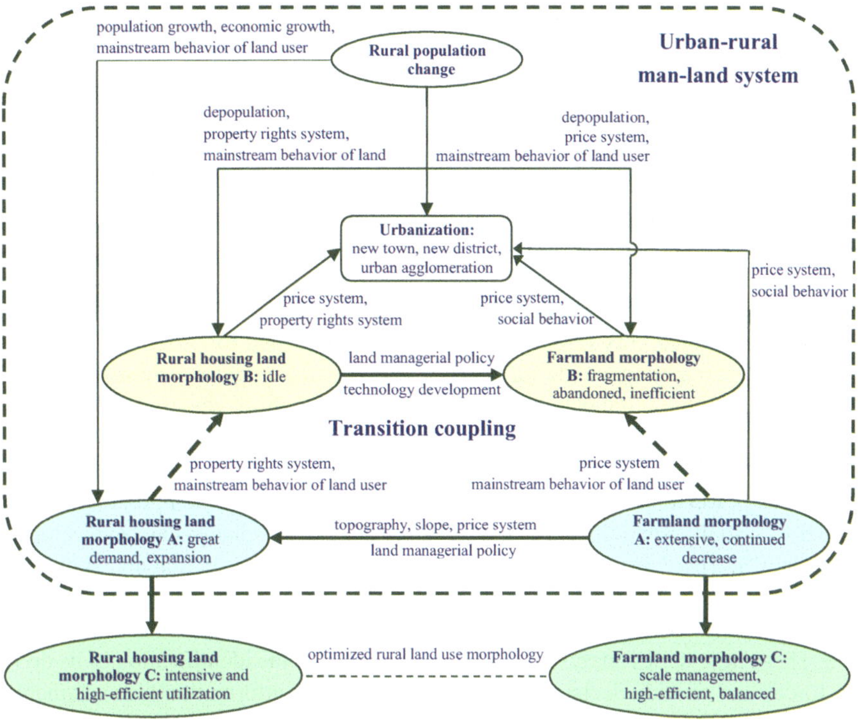

Fig. 4.4 The coupling mechanism of farmland and rural housing land transitions in the process of rural transformation development in China (Long and Li 2012)

original site, the morphology of rural housing land is transformed to an intensive and high efficient one (C) (Fig. 4.4). With the implementation of rural housing land reclamation project and high-standard farmland construction project, the fragmented farmland and abandoned rural housing land are converted to high-standard farmland, which cause the morphology of farmland transformed to a scale management, high-efficient, and balanced one (C) (Fig. 4.4). The coupling of farmland and rural housing land transition in urban-rural man-land system made rural land use morphology further optimized, as evidenced by that the area change of farmland and rural housing land gradually tended to be stable and the area of rural housing land even had a little decrease during 2005–2008 (Fig. 4.1), which will smoothly promote the rural transformation development (Fig. 4.4).

4.4 Case of Socio-Economic Developed Area: The Su-Xi-Chang Region

4.4.1 Background

Land change science, an interdisciplinary field seeks to understand the dynamics of land use and land cover (LULC) as a coupled human–environment system to address theory, concepts, models, and applications relevant to environmental and societal problems, has emerged as a fundamental component of global environmental change and sustainability research. Recently, issues related to LULC change have attracted interest among a wide variety of researchers, ranging from those who favor modeling spatio-temporal patterns of land conversion to those who try to understand the coupled system—causes, impacts, and consequences (Verburg et al. 1999).

Remote sensing (RS) and geographic information systems (GIS) have been recognized as powerful and effective tools for detecting the spatio-temporal dynamics of LULC. Usually, spatially explicit time series of land-use change can be developed based on RS. There are widely used approaches of change detection and statistical analysis, which enable us to discover the structural variation among different land-cover patterns and to diagnose land-use change based on time-series socio-economic data. These time-series analyses of land-use change and the identification of its driving forces can provide decision-makers with important information for sustainable land management and regional development.

Since China initiated economic reforms and an open-door policy in 1978, tremendous changes in China's development philosophy and regional economics have occurred (Wei and Fan 2000). As a result, many coastal regions of China such as the Yangtze River Delta region and Pearl River Delta region experienced dramatic economic and spatial restructuring, which resulted in tremendous land-use change (Seto and Kaufmann 2003). Land-use changes, while restricted by physical conditions, are mainly driven by socio-economic factors, and they can be mainly characterized by the changes of farmland and construction land, which are tightly inter-related with human production activities. In China's rural areas, farmland and rural settlements[3] are the two most important land-use types depicting rural development.

[3]Rural settlements, also called rural housing land, refers to the land utilized by rural residents for dwelling and living, i.e. land for building house and other structures or affiliated facilities. Since the early 1980s, the household responsibility system has provided strong incentives for rural towns and villages to diversify and grow their economies by developing nonagricultural enterprises, which often involves small-scale, individual, privately owned, nonagricultural land use, termed "rural construction" in official Chinese statistical yearbooks, with most construction registered as rural housing. But the functional uses of such rural construction are diverse; many individual houses are in mixed use, based on small factories, craft and other retail shops, restaurants, and related privately owned and operated businesses (Xie et al. 2007). In coastal China it is a universal phenomenon.

The conversion of farmland to non-agricultural use during the last two decades in China has been arguably the most widespread in the country's history, and in coastal China the process has been more intense than any other regions. The accelerated industrialization and urbanization following economic reforms and population increases have greatly affected land-use change through the increase of rural settlements and urban sprawl. The fragmentation of farmland can be also observed due to construction on land of the country-side (Fei 1996). In addition, the adoption of market principles has resulted in the internal restructuring of agricultural land use from traditional paddy production to more diversified agricultural activities such as growing cash crops, fruits and aquaculture. A study of Land-sat photographs of one south Jiangsu region revealed that because rural settlements are scattered and use a large amount of land, the shift in land use has not been restricted to a few major cities but has been widely dispersed (Ho and Lin 2004). Rural housing development may have contributed greatly to the loss of farmland, and there is considerable debate on the extent and magnitude of Chinese rural land-use change since the onset of reforms in 1978 (Fei 1996), largely because of the complexity of land-use change regarding its process, dynamic and driving forces. With the continuous growth of China's economy, massive farmland loss for the benefit of market farming and non-agricultural development may occur without appropriate planning and management of existing land resources in coastal China.

Competition between agricultural and non-agricultural land uses is at its most intense at the interface between the urban and the rural, frequently resulting in a succession of land uses at any one location over time, and the changing directions of agricultural policy and countryside planning policies are the main factors driving such successions as well as the fluctuating strengths of urban and rural economies. Changes in land and ecosystems and their implications for global environ- mental change and sustainability are a major research challenge for the human–environmental sciences. While there are numerous studies analyzing the loss of farmland in China due to urban sprawl and rural settlements construction (Long et al. 2007b), China's rural land-use change and its policy dimensional analysis has found much less attention. The aims of this section are: (1) to briefly introduce China's "building a new countryside," an epoch-making countryside planning policy; (2) to examine the dynamic patterns of rural land (focused on farmland and rural settlements) in The Su-Xi-Chang Region of coastal China experienced dramatic economic and spatial restructuring, using high-resolution Landsat TM (Thematic Mapper) data in 1990, 1995, 2000 and 2006, and socio-economic data from both research institutes and government departments; and (3) to discuss some of the major implications for building a new countryside in coastal China.

4.4.2 China's "Building a New Countryside" Policy

Usually, the development of countryside will be paid more attention to when the industrialization and urbanization of a country attain a certain phase, in the form of laying out a series of favorable policies. However, the segmentation of rural policy has become more and more unsustainable with the progress of social and economic restructuring, and a more integrated rural policy needs to be developed because of the diverse character of the contemporary countryside. For example, several general terms to all government bodies having a duty of care in relation to the countryside were referred in the Countryside Act (1968) of UK (Mather et al. 2006). The 'Rural White Papers' for England, Scotland and Wales (1995) were integrated statements of the government's policies on a wide range of issues relating to rural areas, from agriculture to telecommunications, housing to village halls, and forestry to sport (Woods 2004). The policies and funds associated with the basic agricultural laws of 1960 and 1962 have transformed many aspects of the French countryside over the past half-century. In the 1970s, South Korea carried out "New Countryside Campaign" to speed up the development of rural areas.[4] Japan implemented similar campaign in the 1960s, and established corresponding laws and regulations, which were in favor of rural development.[5]

At present, both rural development and urban development in China are in a transition period—the transformation of a traditional agricultural society into a modern industrial and urban society, and the economy is changing from a traditional planned economy to a modern market system. In the process of industrialization and urbanization since the economic reforms of 1978, agriculture and the countryside have made a big contribution to, to some extent, the huge sacrifices for the development of industries and the cities in China. As a result, a series of problems that hamper the social and economic development of China occurred (Xu and Tan 2001). First, the amount of cultivated land shrunk from 138.38 Mha in 1978 to 122.07 Mha in 2005. Lacking irrigation facilities, many farmers are forced to depend on nature. Both scientific and financial supports on the agricultural sector are weak. All of these make the country's agricultural sector lack a solid foundation for speeding up the social and economic development of China and upgrading people's living standards in the countryside. Second, the income gap between the rural and urban residents has been widened from 2.57 to 1 in 1978 to 3.22 to 1 in 2005. Third, building a harmonious society demands that the achievements and responsibilities of economic and social development be shared by all people, but the widening gap between urban and rural quality of life will make a harmonious society harder to achieve. Finally, rural residents, who make up nearly 56% of China's population, have contributed less than one-third of the total retail output. The slow increase in farmers' income is hampering the overall economy from developing any faster.

In 2003, China's per capita gross domestic product (GDP) reached 1090 US dollars, which indicates China's economic and social development enters an important, brand-new period. China's fast-growing economy and its stronger position in the

[4]The new countryside campaign of South Korea (http://news.xinhuanet.com/world/2006-02/16/content4189182.htm).

[5]"New countryside campaign" of South Korea and "55 years system" of Japan (http://dycj.ynet.com/article.jsp?oid=7520089&pageno=3).

global community make itself be ready for providing an atmosphere in which industries support agriculture and cities support the countryside. Furthermore, under the pressures of both the recently widening income gap between the rural and urban population and the problems related to farmers, agriculture and rural areas, the central government of China recently mapped out a long-term development strategy on "building a new countryside," which was established as one of the primary objectives of the 11th Five-Year (2006–2010) Plan, and was expected to solve above problems through coordinating urban and rural development. The Guidelines of PRC's 11th Five-Year Plan for National Economic and Social Development suggests that a new countryside means advanced production, improved livelihood, clean and tidy villages, a civilized social atmosphere and efficient management. To some extent, the policy of building new countryside in China could be labeled 'post-productivism' (Mather et al. 2006). Not only agricultural production, social and environmental objectives were also now explicitly included within it.

Developing modern agriculture and strengthening the productive forces of the countryside are the major elements of advanced production. First of all, national food security was paid a high attention, and the total grain productive capacity is expected to amount 500 million tons by the end of 2010, through strengthening the productive capacity building of main grain productive areas, and enhancing the yield per unit area, quality and efficiency of grain production. The policies concerning farmland protection will be severely carried out, so as to ensure that the total number of basic farmland not to be cut down, and the quality of that not to be lowered. Improving traditional cultivation techniques and popularizing agricultural standardization are also the important contents of developing modern agriculture and strengthening the productive forces of the countryside. Second, the strategic adjustment of agriculture structure will be promoted by optimizing the industrial structure, the product structure and regional distribution of agriculture.

There is an obvious gap between China's urban and rural areas. In 2005, the disposable income of urban dwellers was 10493 RMB¥ (US$ to RMB¥: 1–8.0), which is 3.22 times over the per capita net income of Chinese farmers (3255 RMB¥). In order to increase the income of farmers, three measures were put forward in the 11th Five-Year Plan as follows: (1) developing more profitable agricultural products with improved breeds and individual characteristics, and prolonging the industrial chains, such as storage and process, fresh-keeping, transportation, and so on; (2) increasing farmers' non-agricultural income by promoting the structural adjustment and institutional innovation of the Township and Village Enterprises (TVEs); and (3) improving subsidiary policy and lessening farmers' burdens.

In most China's current villages, "dirty, disorderly and bad" is the general impression to the visitors. The "disorderly" was due to two reasons. On the one hand, with the implementing of market based economy after 1978, many farmers were becoming affluent and they began to prefer multi-functional, more comfortable or spacious houses, which resulted in that rural housing has been expanding sharply. Farmers prefer to build their houses in the valleys (instead of the hillside), and they prefer good location and settle close to roads and other available infrastructure. However, there was no scientific village plan. On the other hand, with the progress of urbanization, some village houses are vacant in China's developed rural areas (especially in eastern/coastal China), either because their owners have two or more houses or because they are rural-urban migrators and live permanently in urban area, which resulted in a serious waste of land resources and contributed to the disorder. The blame of "dirty and bad" falls entirely on the lack of process system for the living sewage, living garbage, and non-point sources pollution of excrement and urine from livestock and poultry breeding fields. The objectives of building clean and tidy villages involve: (1) efforts will be made to expand the use of clean fuels such as marsh gas and solar energy in rural areas; (2) the status quo of more than 60% of rural households do not have access to flush toilets will be gradually changed; and (3) to gradually improve the rural houses according to scientific village plan.

4.4.3 Materials and Methods

4.4.3.1 Study Area

Suzhou, Wuxi and Changzhou three municipalities (usually called the Su-Xi-Chang Region), the study area, is situated at the lower reaches of the Yangtze River in the southern Jiangsu province, one of the fastest growing regions in China (Fig. 4.5). It covers about 1.76 million ha with average elevation below 50 m. The northern subtropical monsoon climate dominates this region year-round, with average annual temperature of 17.5 °C, mean annual rainfall of 1055 mm, all of which are beneficial for agricultural production. The Su-Xi-Chang Region has a population of 14.29 million, and it is unique because of its high population density. In 2006, the population density in this region was up to 809 persons/km^2, which was much higher than the average population density (137 persons/km^2) of China at the same period.

The growth of Shanghai, the most important industrial, commercial and financial center of China, has pushed the Su-X-Chang Region into the industrial age since the early 1980s. Unprecedented changes in the local economy have taken place in recent decades, and by the mid 1980s the Su-Xi-Chang Region provided a model, also called "Sunan model," for the development of rural industries based on diversified collective enterprises run by the local municipality model (Xie et al. 2007). The term "Sunan model" was originally put forward by the eminent sociologist Fei (1996) to refer to the rapid development of the regional economy in southern Jiangsu (Sunan) based on the massive surge of collectively owned TVEs that prospered from the 1980s to the early 1990s. More specifically, the model refers to the speedy growth of the rural economy in the Su-Xi-Chang Region of southern Jiangsu where TVEs flourished widely under the direct control of the local county and township governments that ran their enterprises like diversified corporations.

In 2006, the GDP per capita of the Su-Xi-Chang Region amounted to 67833 RMB¥, which was much higher than the GDP per capita of China (16042 RMB¥) at the same period. In 2006, the primary industry only provides 2.1% of the total GDP

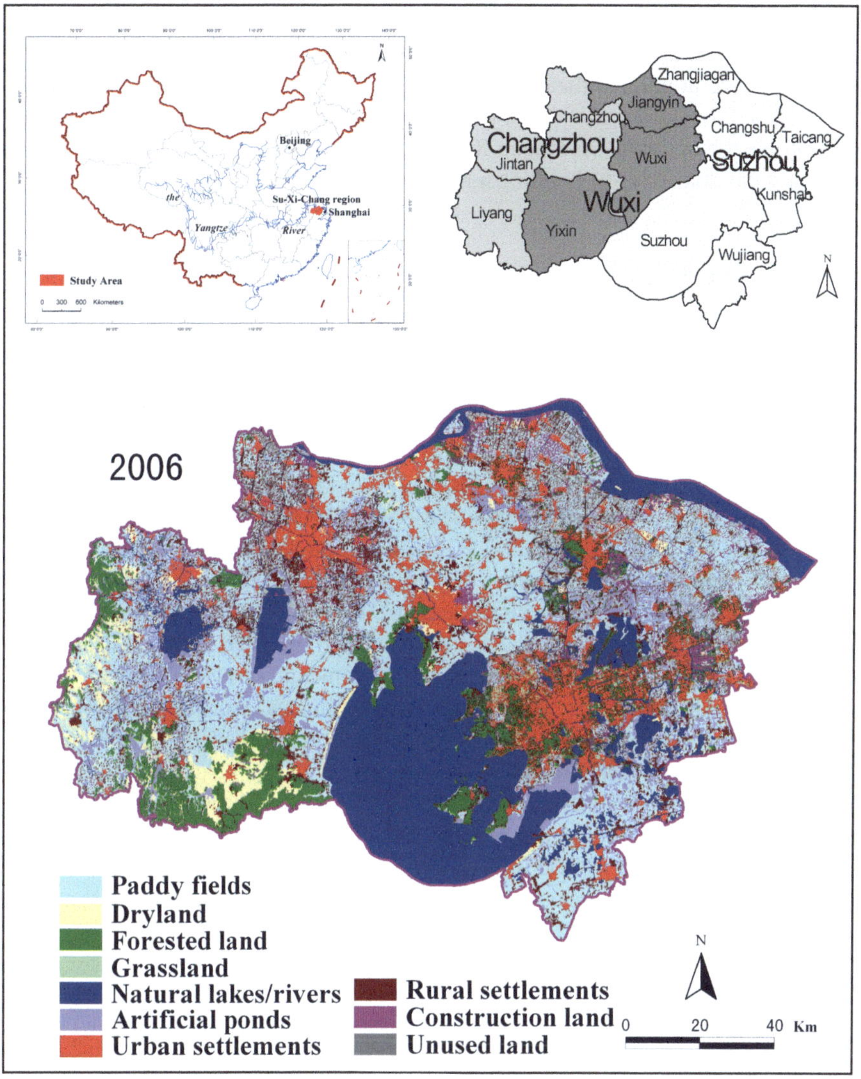

Fig. 4.5 Location and land use of study area, the Su-Xi-Chang Region (Long et al. 2009)

(969 billion RMB¥) in the Su-Xi-Chang Region; however, 62.6 and 35.3% are provided by the secondary industry and tertiary industry[6] respectively. However, with the growth of population and economic activities, land use in the Su-Xi-Chang Region

[6]In China, economic activities are categorized into the following three strata of industry: primary industry refers to agriculture, forestry, animal husbandry and fishery and services in support of these industries; secondary industry refers to mining and quarrying, manufacturing, production and supply of electricity, water and gas, and construction; tertiary industry refers to all other economic activities not included in the primary or secondary industries.

has changed dramatically as evidenced by the continuous decline of farmland. Moreover, due to loose and neglected protection of the environment, the consequences of uncontrolled rural industrialization in this region are particularly serious (Xie et al. 2007). Rapid loss of farmland and ongoing deterioration of local environment are major concerns in the Su-Xi-Chang Region, which should take the lead in China in achieving the objectives of building a new countryside.

4.4.3.2 Methods

(1) **Data source and processing**

The vector data of land-use and land-cover change was obtained through detection analysis of historical Landsat TM (Thematic Mapper) satellite images in 1990, 1995, 2000 and 2006, from the Institute of Geographic Sciences and Natural Resources Research of the Chinese Academy of Sciences and China National Environmental Monitoring Center. An efficient classification system was drafted and an effective research team was organized to work on remote sensed data through human-machine interactive interpretation to guarantee classification consistency and accuracy. Based on these Landsat TM data, four land-use maps in 1990, 1995, 2000 and 2006 were classified into 10 LULC types: paddy fields, dryland, forested land, grassland, natural lakes and rivers, artificial ponds, urban settlements, rural settlements, unused land, and construction land mainly for industry, mining and transportation. Then, we measured the variations among the ten different LULC types by converting the four LULC vector format maps into raster format with a spatial resolution of 100 m × 100 m using ESRI's ArcGIS spatial analyst module. In addition, some time-series socio-economic data on land use and industry from 1990 to 2006, were collected from the local governments and used to analyze potential driving forces resulting in land-use change in study area.

(2) **Method**

The four gridded LULC maps were mainly used to detect the internal variations of LULC in The Su-Xi-Chang Region between three different periods: from 1990 to 1995, from 1995 to 2000, and from 2000 to 2006. For each pair of gridded datasets, a change matrix was constructed. Then, for each LULC category *i* in a change matrix A, the change between the three periods was calculated according to the following Eq. (4.2), and for each LULC type we calculated the percentage of "conversion loss to" or "conversion gain from," in relation to the total "loss or gain" conversion of a LULC type according to Eq. (4.3).

$$CH_i = (p_{i.} - p_{.i})/p_{.i} \times 100 \tag{4.2}$$

where

CH_i Is the change of LULC in row *i* relative to the previous compared year;
$p_{i.}$ Is the row total of grid cells for category *i*;

$p_{.i}$ Is the column total of grid cells for category i

$$\begin{cases} P_{loss(i),j} = (p_{j,i} - p_{i,j})/(p_{i.} - p_{.i}) \times 100 \ i \neq j \\ P_{gain(i),j} = (p_{i,j} - p_{j,i})/(p_{i.} - p_{.i}) \times 100 \ i \neq j \end{cases} \tag{4.3}$$

where

$P_{loss(i),j}$ Is the percentage taken by type j in the total "conversion loss" of category row i;

$P_{gain(i),j}$ Is the percentage taken by type j in the total "conversion gain" of category row i;

$p_{i,j}$ and $p_{j,i}$ Is the individual entry in a change matrix A

To examine the spatial patterns of farmland (including paddy fields and dryland) and rural settlements and their change over time, we identify a set of pattern metrics that capture different dimensions of land fragmentation (Table 4.3): (i) maximum patch area, (ii) minimum patch area, (iii) mean patch size, (ivc) patch density, (v) landscape shape index, (vi) patch cohesion index, and (vii) aggregation index, which are calculated using FRAGSTATS, a spatial pattern analysis program for categorical maps. Considering the limitation of only 4 years of LULC data (derived from remotely sensed images), we used the annual series of farmland data to explore the possible driving forces contributing to land-use change in study area. *R* (R Development Core Team), a language and environment for statistical computing, is used to explore the correlations between different socio-economic variables.

4.4.4 Results

4.4.4.1 Land-Use Change in the Su-Xi-Chang Region

Land use has changed significantly over the whole period from 1990 to 2006 in the Su-Xi-Chang Region (Tables 4.4, 4.5, and 4.6). During the three periods of from 1990 to 1995, from 1995 to 2000, and from 2000 to 2006, paddy fields and dryland decreased by 6.14% and 16.52%, 2.79% and 3.52%, and 7.96% and 3.99%, respectively. In contrast, urban settlements, rural settlements and construction land increased by 76.65%, 41.23% and 52.65%, respectively, from 1990 to 1995. These trends continued in the periods of from 1995 to 2000 and from 2000 to 2006. From 1990 to 1995 and from 1995 to 2000, artificial ponds expanded by 6.29% and 4.22%, respectively; however, it shrunk a little (by 1%) over the period between 2000 and 2006. The change trends of forested land, grassland, natural lakes and rivers, and unused land are not obvious over the whole period between 1990 and 2006, except for an increase of about 23.5% for grassland and unused land from 2000 to 2006 (Table 4.6). To a large extent, land-use change from 1990 to 2006 in the Su-Xi-Chang

Table 4.3 Explanation of pattern measures (Long et al. 2009)

Pattern measure	Definition	Explanation
Maximum patch area		Units = hectares
Minimum patch area		Units = hectares
Mean patch size	$\frac{CA_i}{n_i}$	CA_i = total class area of land use i; n_i = total number of patches in land use i; units = hectares
Patch density	$\frac{n_i}{A}$	A = total landscape area; units = 1/km^2
Landscape shape index	$\frac{e_i}{\min e_i}$	e_i = total length of edge (or perimeter) of class i in terms of number of cell surfaces; min e_i = minimum total length of edge of class i in terms of number of cell surfaces; ≥ 1, without limit
Patch cohesion index	$\left[1-\frac{\sum_{j=1}^{n} p_{ij}}{\sum_{j=1}^{n} p_{ij}\sqrt{a_{ij}}}\right]\left[1-\frac{1}{\sqrt{A}}\right]^{-1}100$	p_{ij} = perimeter of patch ij in terms of number of cell surfaces; a_{ij} = area of patch ij in terms of number of cells; A = total number of cells in the landscape; varies between 0 and 100
Aggregation index	$\left[\frac{g_{ii}}{\max - g_{ii}}\right]100$	g_{ii} = number of like adjacencies (joins) between pixels of patch type (class) i based on the single-count method; max-g_{ii} = maximum number of like adjacencies (joins) between pixels of patch type (class) i based on the single-count method; varies between 0 and 100

Region was characterized by a serious replacement of farmland with urban and rural settlements, construction land, and artificial ponds.

4.4.4.2 Dynamic Patterns and Driving Forces of Farmland Change

Fragmentation is the obvious characteristics of paddy fields change in the study area (Fig. 4.6), and we find substantial and significant increases in mean fragmentation

Table 4.4 Change matrix of each compared LULC type in 1990 and 1995, and its changes in 1995 (ha) (Long et al. 2009)

LULC type in 1995	LULC type in 1990											Changes in 1995 (%)
	PF	DL	FL	GL	LR	AP	US	RS	CL	UL	Total	
PF	908085	5			16	622	7	16			908751	−6.14
DL	3	95152	1				1	1			95158	−16.52
FL		1	96611				1	1			96614	−0.19
GL				4229							4229	−0.09
LR	680				359542						360222	0.18
AP	6026	146	2			87198					93372	6.29
US	23078	15433	13		0	5	50375	56	42		89002	76.65
RS	28763	3130	124	4	0	21	0	77460			109502	41.23
CL	1555	123	43						3147		4868	52.65
UL	8									682	690	1.17
Total	968198	113990	96794	4233	359558	87846	50384	77534	3189	682	1762408	

Note PF = paddy fields, DL = dryland, FL = forested land, GL = grassland, LR = natural lakes and rivers, AP = artificial ponds, US = urban settlements, RS = rural settlements, CL = construction land, UL = unused land

Table 4.5 Change matrix of each compared LULC type in 1995 and 2000, and its changes in 2000 (ha) (Long et al. 2009)

LULC type in 2000	LULC type in 1995											Changes in 2000 (%)
	PF	DL	FL	GL	LR	AP	US	RS	CL	UL	Total	
PF	881290	357	30		4	29	4	10			881724	−2.97
DL	80	91569	123			37	1				91810	−3.52
FL	64	117	94210					23			94414	−2.28
GL	10		11	4103	177						4301	1.70
LR	1				359257						359258	−0.27
AP	3372	193	161	15	624	92941		10			97316	4.22
US	9028	875	188	21	16	144	88992				99264	11.53
RS	13436	1423	997	49	49	211	5	109459			125629	14.73
CL	1470	624	852	41	95	10			4868		7960	63.52
UL			42							690	732	6.09
Total	908751	95158	96614	4229	360222	93372	89002	109502	4868	690	1762408	

Note PF = paddy fields, DL = dryland, FL = forested land, GL = grassland, LR = natural lakes and rivers, AP = artificial ponds, US = urban settlements, RS = rural settlements, CL = construction land, UL = unused land

Table 4.6 Change matrix of each compared LULC type in 2000 and 2006, and its changes in 2006 (ha) (Long et al. 2009)

LULC type in 2006	LULC type in 2000											Changes in 2006 (%)
	PF	DL	FL	GL	LR	AP	US	RS	CL	UL	Total	
PF	801132	32	472		581	3638	2086	3158	48	413	811560	−7.96
DL	106	85665	517		30	302	396	1122		5	88143	−3.99
FL	1909		91304	83	560	1137	17	108		23	95141	0.77
GL	370			3993	740	49	105	52		6	5315	23.58
LR	282	76	65		351487	250	3				352163	−1.97
AP	4478	31	123	15	3615	87334	40	669	6	35	96346	−1.00
US	34632	1821	270	29	232	617	96615	3965		200	138381	39.41
RS	35186	1408	777	17	1508	3299		116361		19	158575	26.22
CL	3419	2777	240	164	492	689	2	161	7906	31	15881	99.51
UL	210		646		13	1		33			903	23.36
Total	881724	91810	94414	4301	359258	97316	99264	125629	7960	732	1762408	

Note PF = paddy fields, DL = dryland, FL = forested land, GL = grassland, LR = natural lakes and rivers, AP = artificial ponds, US = urban settlements, RS = rural settlements, CL = construction land, UL = unused land

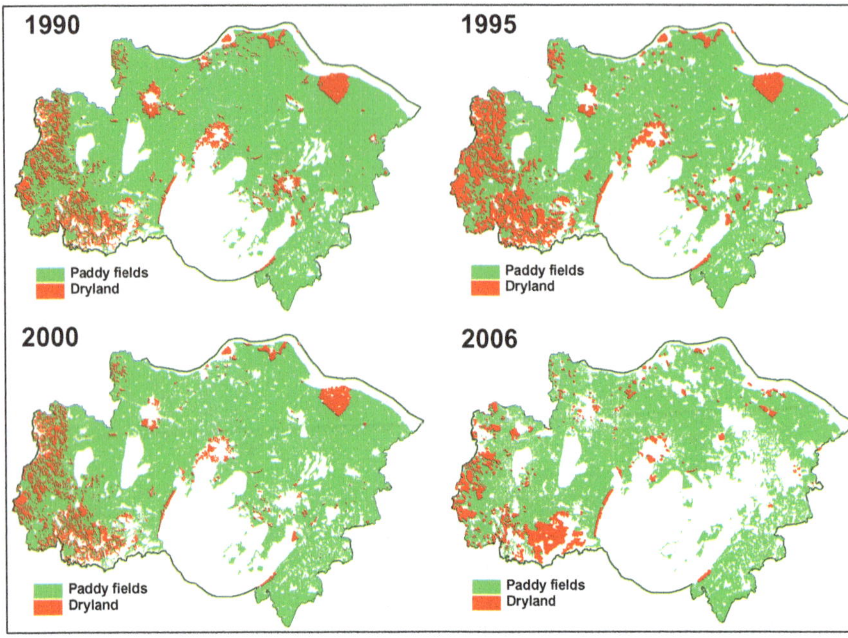

Fig. 4.6 Change pattern of farmland in the Su-Xi-Chang Region (Long et al. 2009)

values of paddy fields over the whole period from 1990 to 2006 in the Su-Xi-Chang Region (Table 4.7). In 1990, the mean patch size of paddy fields was 1204.4 ha, and by 2006 the size was only 138.8 ha. The patch density of paddy fields increased by 331.5%, and it changed from 0.033 in 1990 to 0.144 in 2006. From 1990 to 2006, the fragmentation of paddy fields can also be proved by the continuous decline of aggregation index and rapid increase of landscape shape index (Table 4.7). Nevertheless, the most interesting result we find is the decreases in mean fragmentation values of dryland from 1990 to 2006 in the study area, as evidenced by the values change of mean patch size, patch density, aggregation index and landscape shape index (Table 4.7).

The accelerating industrialization and urbanization process may increase the degree of fragmentation and structural complexity of LULC. Our statistical analysis also suggested that population growth, rapid industrialization and urbanization are the major driving forces of farmland change in the Su-Xi-Chang Region from 1990 to 2006 (Fig. 4.7).

Rapid industrialization and urbanization in the Su-Xi-Chang Region is a very significant factor of land-use change, which can be illustrated by the strong negative relationship between farmland and total industrial output value (Fig. 4.7a), and between farmland and population (Fig. 4.7b). In addition, industrialization and urbanization have apparently triggered massive farmland loss for the benefits of

Table 4.7 Aggregate land-use pattern measures (Long et al. 2009)

Pattern measure	Paddy fields				Dryland				Rural settlements			
	90	95	00	06	90	95	00	06	90	95	00	06
Maximum patch area	87847	82960	77509	86193	15475	13875	13430	15252	142	369	504	1261
Minimum patch area	0.020	0.017	0.014	0.005	0.009	0.021	0.021	0.005	0.010	0.011	0.011	0.007
Mean patch size	1204.4	1409.3	1017.2	138.8	75.2	85.7	76.0	112.4	8.4	12.0	11.4	12.1
Patch density	0.033	0.029	0.042	0.144	0.075	0.057	0.062	0.029	0.511	0.499	0.585	0.646
Landscape shape index	76.7	85.8	92.6	137.6	68.1	64.9	66.1	39.0	123.5	125.3	134.7	153.4
Patch cohesion index	99.96	99.95	99.95	99.78	97.19	97.38	97.36	96.93	71.56	76.98	78.42	87.61
Aggregation index	92.30	91.09	90.23	82.44	80.05	79.22	78.41	83.49	55.85	62.31	62.17	60.22

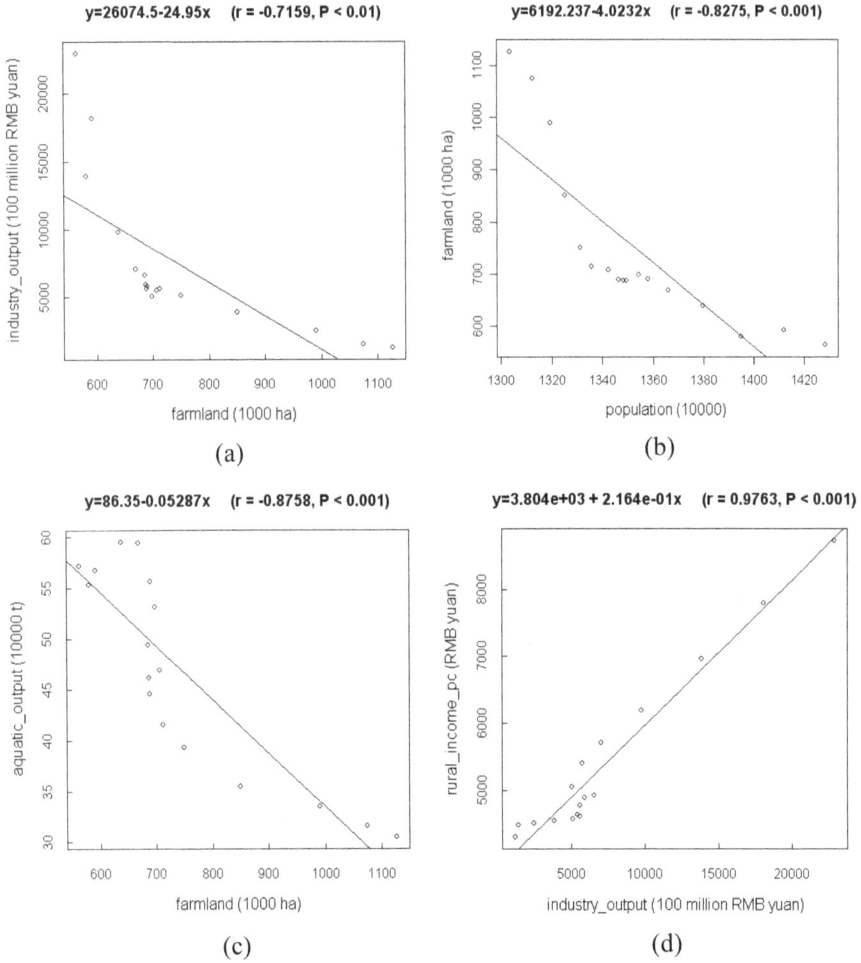

Fig. 4.7 Farmland versus industrial output value (**a**), population (**b**) and aquatic products (**c**), and industrial output value versus net income per capita of rural households (**d**) of the Su-Xi-Chang Region between 1990 and 2006 (Long et al. 2009)

market-oriented farming and non-agricultural development. According to Fig. 4.7c, there was a strong negative relationship between farmland and aquatic products.

Table 4.6 presents the land-use change resulted from industrialization and urbanization. Urban settlements expansion accounted for 38.81%, 33.39% and 46.39% of the decrease of paddy fields, and for 81.95%, 26.11% and 38.86% of the decrease of dryland, during the periods of from 1990 to 1995, from 1995 to 2000 and from 2000 to 2006, respectively. Similarly, the expansion of rural settlements occurred mainly at the expense of paddy fields and dryland, which accounted for 48.36%, 49.68% and 45.65% of the changes in paddy fields, and for 16.62%, 42.5% and 7.8% of the decrease of dryland, during the periods of from 1990 to 1995, from 1995 to 2000 and from 2000 to 2006, respectively. During the period between 2000 and 2006, the expansion of construction land accounted for 75.73% of the decrease of dryland. The expansion of artificial ponds occurred mainly at the expense of paddy fields, which accounted for 9.09% and 12.37% of the changes in paddy fields during the periods of from 1990 to 1995 and from 1995 to 2000, respectively. In addition, the decreases in mean fragmentation values of dryland may be due to the demands for developing modern agriculture and the scale management of agriculture.

4.4.4.3 Dynamic Patterns and Driving Forces of Rural Settlements Change

From 1990 to 2006, rural settlements change in the Su-Xi-Chang Region has a dual characteristic of coexisting fragmentation and aggregation (Fig. 4.8), as evidenced by the changes of mean fragmentation values of rural settlements over the whole period from 1990 to 2006 (Table 4.6). On the one hand, the fragmentation of rural settlements can be proved by the continuous increase of landscape shape index and patch density from 1990 to 2006 (Table 4.6). On the other hand, during the same periods, the maximum patch area is continuously increased double or more, and the value of patch cohesion index has been keeping on increasing (Table 4.6). Patch cohesion index measures the physical connectedness of the corresponding patch type.

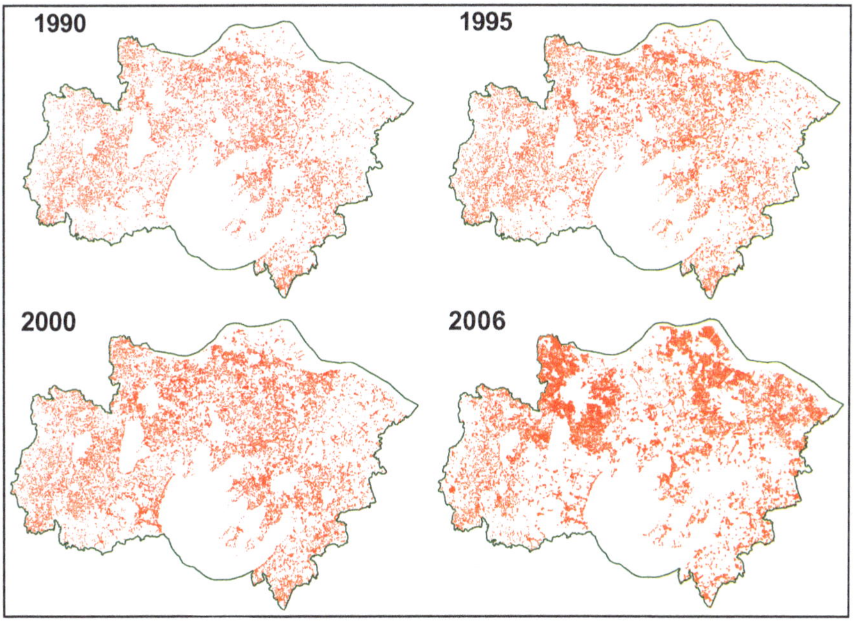

Fig. 4.8 Change pattern of rural settlements in the Su-Xi-Chang Region (Long et al. 2009). *Note* All the patches in the maps are rural settlements

Below the percolation threshold, patch cohesion is sensitive to the aggregation of the focal class. Patch cohesion increases as the patch type becomes more clumped or aggregated in its distribution; hence, more physically connected. The expansion of rural settlements occurred mainly at the expense of farmland, as evidenced by more than 92% of the increased rural settlements are from farmland (Table 4.6).

The China's economic reforms played an important role in the transformation of rural settlements. The implementation of the responsibility system has aroused Chinese peasant's enthusiasm, liberated Chinese countryside from the self-sufficient status to open up a new prospect with the emerging of TVEs. China's TVEs are widely regarded as the primary form of rural industrialization. To some extent, the expansion of construction land is largely a result of rural industrialization in China. Numerous factory buildings can be easily seen in the residential backyards of the villagers in coastal China. In addition, rural industrialization contributes significantly to the increase in rural income levels and employment by making full use of the local resources, utilizing the capital scattered in the peasant's hands, developing the wisdom of the skilled craftsmen, and rising up the peasant's income, which can be illustrated by the very strong positive relationship between industrial output value and net income per capita of rural households (Fig. 4.7d).

Usually, peasants in China want to build another new multi- functional, more comfortable or spacious house if they become affluent, although present housing can fundamentally meet their needs. Farmers prefer to build their houses close to roads and other available infrastructure. The houses usually have small-scale household (in which only three persons or so live) and larger courtyards. There is a short cycle of replacing old house with a new one. As a result, there is more idle rural housing land, vacant rural housing, with unoccupied land in villages. That can explain the fact of the expansion of rural settlements (Tables 4.4, 4.5, and 4.6), although there was a decline trend in rural population and the number of rural households, which decreased 214 thousand, from 2945 thousand in 2000 to 2731 thousand in 2006. Over the period from 2000 to 2006, we can find an obvious transformation from rural settlements to paddy fields and dryland, which amounted to 4280 ha (Table 4.6). However, from 1990 to 2000, only 27 ha rural settlements was changed to farmland (Tables 4.4 and 4.5). That is mainly due to the new policies of building a new countryside and carrying out rural construction land consolidation.

4.4.5 Discussion and Conclusions

The rapid industrialization and urbanization in coastal China increased the demand for non-agricultural land, which diminished farmland and withered the traditional agriculture. The industrialization pushed forward the development of TVEs, which led the expansion of rural settlements and the pollution of local environment. Hence, the focuses of building a new countryside in coastal China need to be concentrated on protecting the farmland, developing modern agriculture, and building "clean and tidy villages."

In coastal China with rapid urbanization and industrialization, local further development needs land, which is very scarce. However, local government grasped the opportunity of an innovative land management policy called "linking up increased urban construction land with decreased rural construction land".[7] Land displacement and centralized settle-down is a principal mode, also a successful mode that local governments thought, of building new countryside in coastal China.

Usually, the model is operated in this way: at first, village collectives jointly with some enterprises with strong economic base invest to construct a brand-new community (Fig. 4.9); then, try to persuade the peasants to give up their farmland and rural housing land and move into the new community by promising to provide new apartments (or subsidies) and employment in the enterprises for those who moved in, thus the peasants can live the 'city life'. In most cases, peasants submitted to the village collectives (or local governments) driven by immediate interests, and gave up their land. The village collective put together the let farmland and subleased them to some plant companies or individuals who want to manage a lot of land; in return, the peasants who found a job in enterprise and gave up their land could get

Fig. 4.9 A new rural community in Kunshan city of the Su-Xi-Chang Region (Long et al. 2009)

[7]The policy "linking up increased urban construction land with decreased rural construction land" means that, in a county level, the increased construction land for future urban development should be no more than the decreased rural construction land, which will be turned into cultivated land through carrying out rural construction land consolidation, so as to keep the total construction land in this county not to increase, to keep the quantity and quality of local farmland not to decrease, and to form a rational land-use patterns. The actual potential of existing rural construction land in a county needs to be investigated and corresponding plan for carrying out rural construction land should be laid out before implementing this policy (Long et al. 2012).

limited money for their land shares. Much 'new' land is displaced out by carrying out rural settlements consolidation. However, most are allocated as construction land instead of returning to arable land according to the innovative land management policy. The enterprises who invested for building new community get enough land for expanding their production scale. The village collective builds some standard rental factory workshops on the 'new land', and promised that the rent will be used for the expenditure of village public affairs and capital bonus assigned to villagers. To great extent, this kind of model contributed to the aggregation of rural settlements from 2000 to 2006 (Fig. 4.9 and Table 4.6).

This model of building new countryside in coastal China brought about obvious land-use change, and the impacts of which on the livelihood of rural residents were far-reaching. Although the farm- land per capita was declined from 0.052 ha in 2000 to 0.04 ha in 2006, the livelihood of rural residents is improved greatly. Guohao Liu, the village head of Taixing village in Shaxi town of Taicang city in the northeast of the Su-Xi-Chang Region, told us that the increased villagers' net income per capita in 2006 from collective-owned land was 1880 RMB¥, which was composed of 560 RMB¥ capital bonus of collective assets, 900 RMB¥ rent of farmland subleased to Ltd. Company of Suzhou Urban Greenbelt, and 420 RMB¥ laborage for working at the Company.

However, in general, there are some latent problems in the back of the so-called successful model. To some extent, the village and township governments are selling off land to developers in a disguised form just to feed their staff, because township and village governments are paralyzed from lack of funding since the central government's decision two years ago to abolish the agricultural tax, although it can bring some help for village public affairs. But they did not care of the peasants' future livelihoods. First of all, those peasants who lack of special skills and are employed in the enterprises will confront the situation of unemployment to a great extent, because most of the enterprises what they are working at are labor-intensive industries, which are experiencing the phase of industrial upgrade. At present, lots of polluted, labor-intensive industries in eastern/coastal China are being gradually transferred to central region and western region. But the pace of peasants' skill improvement cannot chase up that of industrial upgrade, which may cause their unemployment in the near future. In addition, the enterprises, in order to save more land for their industrial development, try to build high-rise apartment buildings (some even with 35 floors) without considering the carrying capacity of local resources and environment, which brings a lot of inconvenience for the new dwellers. Cultural landscape should not be overlooked in land consolidation. Lacking of original rural landscape and rural culture, some interviewed older peasants, who have moved into the new apartments, usually miss the life living in their old houses, which were already not existed.

Rural construction land consolidation may play an important role in protecting the farmland and building "clean and tidy villages." First, farmland may be increased or 'new' construction land may be available for meeting the needs of future economic development, which can also reduce the conversion of farmland in the future. Second, consolidating rural construction land accompanied with the campaign of building new countryside is an effective prescription of changing the appearance of "dirty, disorderly and bad" in rural China. Finally, it can improve the villagers' living conditions.

Accordingly, consolidating rural construction land needs to be carried out by the following ways: (1) local governments should first register idle rural housing land, vacant rural housing, idle workshop of insolvent TVEs, unoccupied land in villages, and cases where one household owns two or more houses according to local conditions, so as to estimate the actual potential of existing construction land, then set down and implement rational plans of consolidating rural construction land according to the general land- use planning and local developing strategy, in order to enhance the intensity of land use in town and village and save land for further needs; (2) putting together the dispersed rural housing (needs to be improved) and TVEs to establish new residential areas and new industrial areas with unified planning and design and perfect infrastructure (roads, water, electricity, sewage, gas, heating and telecommunications), for the purpose of improving villagers' living conditions and building a clean and tidy village. From 2000 to 2006 rural settlements consolidation contributed to the increase of 4280 ha farmland (Table 4.6). However, it should be clear that rural construction land consolidation will confront many obstacles. For example, most peasants have strong reminiscent mood and are not willing to move to live in another area; China's land legal and managerial system is somewhat problematic involving rural housing land consolidation in the aspect of land-use rights; and the benefit allocation system of stakeholders concerning rural construction land consolidation needs to be established, which cannot come to an agreement in a short time.

The study of Li and Wang (2003) showed that the low incentive for raising agricultural land-use intensity may more seriously threaten food security than the shrinking cultivated land area or the low technological potential in China. At present, China's farmers have seldom interest in pursuing grain production, largely because it generates very low net income[8] which resulted in millions of farmers being out of countryside and the continuous decline of the percentage of primary industry in the GDP of the The Su-Xi-Chang Region, from 15.35% in 1990 to 2.11% in 2006 (Fig. 4.10). On the contrary, it creates some opportunities for developing modern agriculture and the scale management of agriculture. During the period between 2000 and 2006, many modern agricultural gardens were established. Through carrying out cultivated land consolidation, the maximum patch area of paddy fields and dryland in study area increased from 77 509 ha and 13 430 ha in 2000 to 86 193 ha and 15 252 ha in 2006, respectively (Table 4.7).

Frankly, cultivated land consolidation may contribute to developing modern agriculture and strengthening the productive forces of the countryside in the following aspects: (1) to enhance land productivity and labor productivity through improving

[8]In 2008, the author's household interview in a village of coastal China showed that a farmer can only obtain profit of 580 RMB¥, which still includes his man-days pay, from planting 1 mu (1 mu = 0.067 ha) rice: the annual yield is 600 kg, the price is 1.48 RMB¥ per kg, and the total income from planting rice is 888 RMB¥; however, the total expenditure is 308 RMB¥, which includes 100 RMB¥ for fertilizer, 160 RMB¥ for pesticide, and 48 RMB¥ for plowing land. But he can earn 40 RMB¥ per day if he pursues other labor activities on building site.

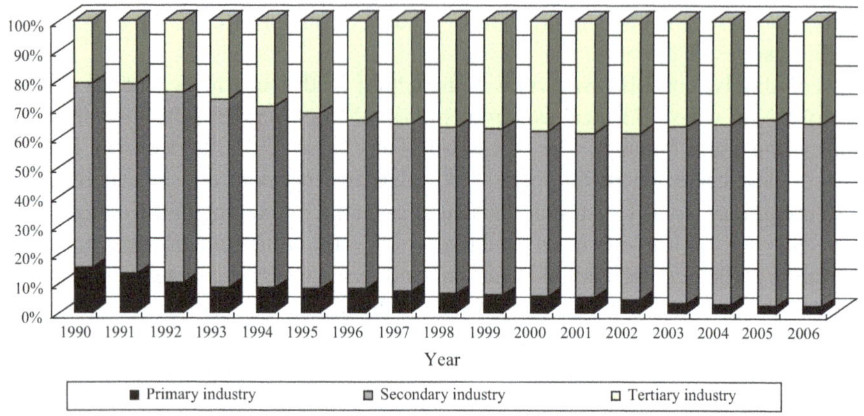

Fig. 4.10 Change in industrial structure of the Su-Xi-Chang Region between 1990 and 2006 (Long et al. 2009)

land quality, irrigation conditions and machinery accessibility; (2) to reduce land fragmentation and improve the efficiency of agricultural production; (3) to form rational, efficient and intensive land-use structure by reallocating land resources; (4) to facilitate adjustments to the crop structure through general improvements in land conditions; and (5) to promote the progress of agricultural industrialization and scale management of agriculture.

To achieve coordinated and balanced development between urban and rural areas in The Su-Xi-Chang Region, it can be seen from the dynamic patterns of farmland and rural settlements in Figs. 4.5, 4.6 and 4.8 that an urban-rural integrated development planning is necessary. Urban sprawl is increasingly considered a significant and growing problem that entails a wide range of social and environmental costs. China's rural area used to pay the bill for these costs due to lacking early planning. When some loads resulting from urban expansion were shifted upon rural area, the rural had no choice but to take over the heavy burden, although without any preparation. It is fundamental to lay out a scientific plan for building a new countryside, and the plan should coincide with local land-use planning, urban planning and land consolidation planning. Hence, for the long run, the plan for building new countryside needs to pay more attention to making the rural have certain functions serving for the urban, e.g. accepting some transferred industries, providing ecological and other kinds of services for urban population. In addition, we need to reserve and respect the cultural elements of idyll and the rural landscape, in the process of building a new countryside in coastal China, instead of building a new countryside, which looks more like a city.

4.5 Case of Traditional Farming Area: The Huang-Huai-Hai Plain

4.5.1 Background

China's total agricultural available resource is large, but the per capita possession is less. The per capita farmland area (PCFA) of China is 0.1 ha, only 12% that of the United States (Xiao et al. 2017) The farmland area per China's household is 0.38 ha, which is below the world average, even lower than that of India, Japan and South Korea (Song et al. 2014). The proportion of households with a total farmland land area of less than 0.6 ha is 83%. Since the new millennium, the urban-rural transformation development in China has entered a critical period (Long and Woods 2011). The extensive and accelerated urbanization has brought more challenges to scarce land resources.

In the process of urbanization, urban construction land and rural housing land are increasing rapidly, but the efficiency of land development and utilization of construction land both in the urban and the rural is limited (Long and Li 2012). Under the stimulation of "land revenue" and the pursuit of GDP, local governments are accustomed to planning large development areas far beyond the actual needs, and the inefficient or idle land use phenomena is very common. In 2003, there were 6866 economic development zones in China, covering an area of 3.86 Mha, of which 1.3 Mha land resources was idle (Cartier 2001).

During 2003–2013, urban built-up area increased by 78.5%, while the urban population increased by only 46% (Bai et al. 2014). This phenomenon led to an enclave sprawl of urban, which brought about the serious waste of farmland resources and the disorderly sprawl of the urban.

During 2000–2015, China's urbanization rate increased from 36.22 to 56.1%, the number of rural resident population decreased from 808 million to 603 million, with a decrease of 205 million. Simultaneously, China's rural housing land increased from 18.12 Mha in 2000 to 19.11 Mha in 2015, with an increase of 0.99 Mha. The per capita rural housing land of China increased from 204.5 to 231 m^2 from 2000 to 2008, and the phenomenon of "decreasing population versus increasing rural housing land" occurred in rural areas (Long et al. 2012). Along with the population outflow, the issue of rural hollowing became a hot research topic (Long et al. 2012), the efficiency of rural housing land use is low, combined with farmland fragmentation and abandoned.

Land use transitions research provides theoretical support and technical guidance for rational allocation and optimization of urban-rural land resources (Long and Qu 2018). At present, many scholars has launched an in-depth study on urban-rural land use transitions, such as forest transition, rural housing land transition, farmland transition, urban construction land transition, etc. The impacts of land use transitions on the agricultural production system and the safety of food production under urbanization process have also contributed to more research. With the rural-to-urban migration, the plight of land use transitions in the process of urbanization of China is obvious. The urbanization progress affects the food production and consumption system, which gives the farm-land protection with more severe challenges. As the world's most populous country, revealing its coupling mechanism between per capita construction land area (PCCA) and PCFA has an important practical significance for ensuring the food security of China who bears the more people and less farmland. The change and evolution of per capita construction land transition (PCCT) and per capita farmland transition (PCFT) are the core elements of regional man-land interrelations, which can reveal the process of demographic transition (urban-rural population distribution and employment structure) and urban-rural transformation development under urbanization.

Farmland is the core resource ensuring food security. Enlarging agricultural production, we can intensify and expand the farmland utilization. No matter what methods, adequate farmland resource is a prerequisite. The Huang-Huai-Hai Plain (HHH) accounts for 4.3% of China's land area, while the share of farmland, population and grain yield of total China is 20.6%, 22.5% and 30.8%, respectively, which is a main grain-producing area of China. Bren d'Amour et al. (2017) predicted that the HHH is one of the fastest growing areas of urbanization in the world, which includes the Beijing-Tianjin-Hebei urban agglomeration, Shandong Peninsula urban agglomeration and the Central Plains urban agglomerations. Therefore, taking the HHH as an example, the coupling relationship research between PCCT and PCFT is of great significance for revealing the law of farmland transition and protecting regional food security.

Accelerated urbanization and corresponding human activities are triggering tremendous land use transitions in China. Farmland and construction land are tightly inter-related with human livelihood and production activities, and are treated as the two most important land-use types depicting urban-rural development in China. Land use transitions in China under rapid urbanization were mainly characterized by the shrinkage of farmland and the rapid expansion of urban-rural construction land. So, farmland and construction land play an important role in facilitating land use transitions. To some extent, probing the coupling relationship between PCCT and PCFT in the HHH may be helpful to find a solution for ensuring food security of China. Therefore, based on the major grain-producing areas with rapid urban-rural transformation development and most population density areas of the HHH, this section deeply studies the coupling relationships between PCCT[9] and PCFT and the related influential factors, then, reveals the implications of land use transitions for regional food security.

[9]In China, almost two-thirds of urban-rural construction land are rural housing land. Accordingly, analyzing the coupling relationships between PCCT and PCFT can represent the coupling analysis of farmland and rural housing land transitions, to a great extent.

4.5.2 Materials and Methods

4.5.2.1 Study Area

The HHH, one of traditional major grain-producing areas of China, is undergoing rapid urbanization, and the contradiction between urban-rural transformation development and farmland protection and grain production is being intensified (Fig. 4.11). During 2000–2015, the total population of the HHH rose to 328 million from 276 million in 2000, of which the urban population increased from 63 million to 157 million, and the urbanization rate increased by 24.9 percentage points to 47.47% in 2015. Under the background of accelerated urbanization, farmland protection pressure and ecological environment bearing pressure continuously emerged. How to evaluate and forecast the transition of the man-land interrelations, it is not only related to the general pattern of urban-rural transformation development, but also to guarantee food security. Analysis of the coupling relationships between construction

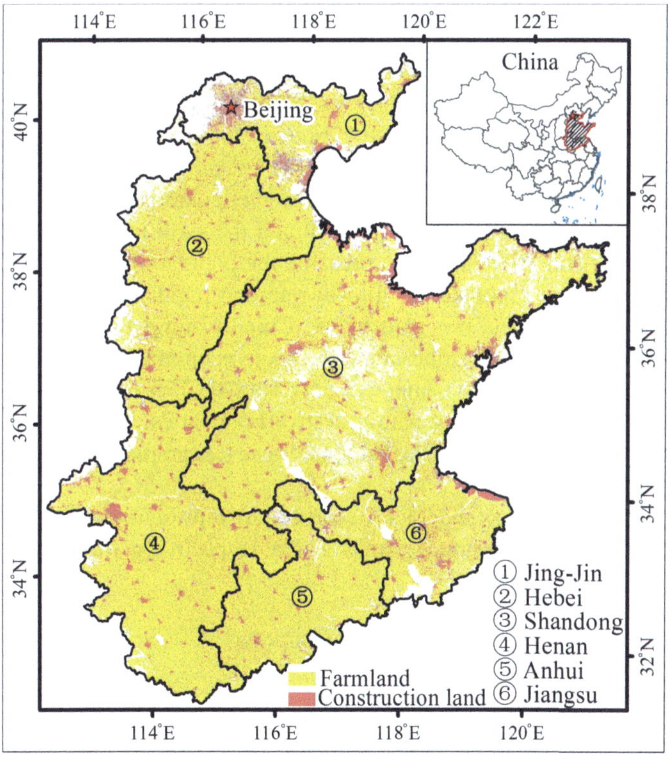

Fig. 4.11 Map of the HHH in China. The distribution pattern of construction land and farmland is in 2015 (Long et al. 2018) In accordance with the administrative boundary, the HHH is divided into six parts: Jing-jin, Hebei, Shandong, Henan, Anhui and Jiangsu

land transition and farmland transition of the HHH may play an important role in coordinating the regional urban and rural development and ensuring food security.

With the rural-to-urban migration, the rural development and agricultural production system of the HHH have undergone dramatic changes. The rural population in the HHH decreased from 0.21 billion in 2000 to 0.17 billion in 2015, with an average annual decrease of 2.26 million, and the number of rural labor force decreased from 77 million in 2000 to 56 million in 2015, among which rural nonagricultural employment rate increased from 33.33 to 59.85%. The spatio-temporal coupling process between demographic transition and land use transitions drive PCCT and PCFT, which further changes the man-land interrelations schema in the HHH.

4.5.2.2 Materials and Methods

(1) **Data processing of PCCT and PCFT**

In this section, the land-use data is obtained from Landsat TM and ETM true color composite image after geometric rectification and supervised classification in 2000, 2005, 2010 and 2015. Construction land, including urban construction land, rural housing land and other construction land, the data spatial resolution of which is 30 m × 30 m. Town-level demographics for 2000, 2005, 2010 and 2015 are derived from GPWv4 which developed by CIESIN of Columbia University (http://www.sedac.ciesin.columbia.edu/data/collection/gpw-v4/sets/browse/). This dataset contains 6580 town-level resident population data and population density sampling data in the HHH. The data are based on the fifth and the sixth town-level census data of China in 2000 and 2010, and then presume the population density in 2005 and 2015 relying on natural population growth rate and population migration trends. 30 m resolution DEM data provided by the China Natural Resources Data Center (http://www.data.ac.cn). Furthermore, the socio-economic statistical data of the county- level are derived from China's County Statistical Yearbook. In this section, basic geographic data including administrative boundaries, administrative center, roads and rivers are derived from the National Basic Geographic Information Center (http://ngcc.sbsm.gov.cn/).

The PCCA (m^2/person) and PCFA (ha/person) represent the per capital construction land area and per capital farmland area, respectively, which are calculated according to Eqs. (4.4) and (4.5):

$$PCCA = \frac{Conarea}{Popdensity} \tag{4.4}$$

$$PCFA = \frac{Farmarea}{Popdensity} \tag{4.5}$$

where *Conarea and Farmarea* are the construction land area and farmland area in each 1 km × 1 km grid, respectively, *Popdensity is the* population size in each grid (person). PCCT = $PCCA_{t2} - PCCA_{t1}$ and PCFT = $PCFA_{t2} - PCFA_{t1}$ ($PCCA_{t2}$ and

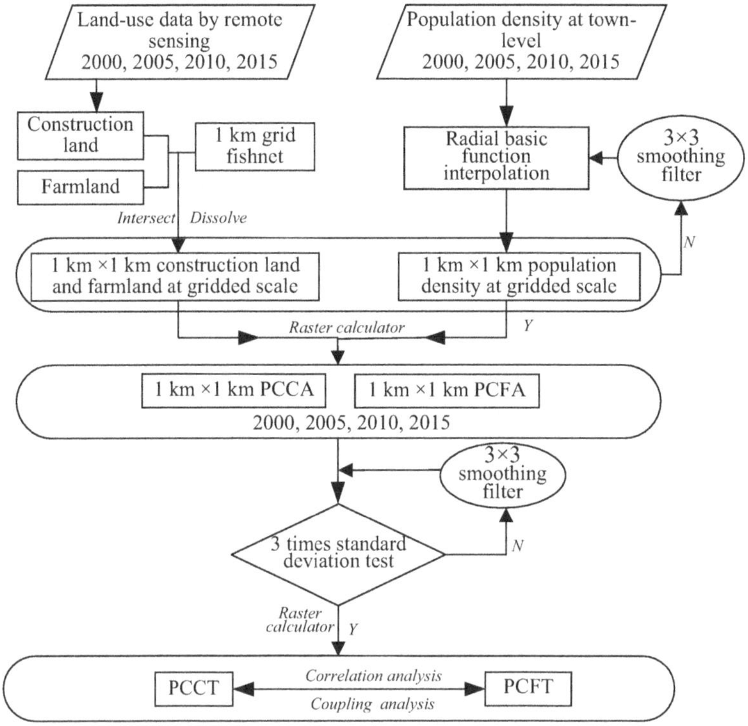

Fig. 4.12 Data processing flow (Long et al. 2018)

$PCFA_{t2}$ represent the data at the end period, $PCCA_{t1}$ and $PCFA_{t1}$ represent the data at the base period, respectively). The calculation process of PCCT and PCFT at 1 km grid-scale is as follows (Fig. 4.12).

Step one is to get construction land and farmland data in 1 km gridded. Extracting all construction land and farmland layer based on 2000, 2005, 2010 and 2015 vector land-use data, respectively, and then intersecting and dissolving them with 1 km fishing net data; Step two is to calculate 1 km gridded population density data. Based on the population density point data of each period, 1 km gridded population density data are calculated by using the "Radial basic function interpolation" method of ArcGIS software. After that, if the value of each pixel is within 3 standard deviations of the mean value, it is ok. Otherwise, it will be smoothed by 3 × 3 low-pass filters until the value is within the reasonable range. Step three is to obtain 1 km gridded data of the PCCA and PCFA. The PCCA and PCFA of 2000, 2005, 2010 and 2015 are calculated by "Raster calculator" tool through construction land and farmland divided by population density of each period, respectively. Step four is to test the PCCA and PCFA data, and further calculate the PCCT and PCFT data. The second step of the data test method is also used to process the PCCA and PCFA for each period, until all the data is reasonable. Afterwards, PCCT and PCFT data at 5-year intervals are calculated by using PCCA and PCFA data for each period.

(2) **Multi-layer buffer analysis**

In order to further analyze the spatio-temporal coupling process of PCCT and PCFT, the multi-layer buffers are generated by transportation elements and location elements. During urbanization, the distance to the transportation line and city has deeply affected the transformation process of population migration and regional economic development. Using zoning statistics of grid-scale by different distance buffers, we can get a better understanding of the coupling relationships between PCCT and PCFT at grid-scale. According to the previous analysis (Li et al. 2015), the distance to the city center and the main traffic line are important factors that affect the land use change. Therefore, according to the mean value of each factor, four different types of buffers are built: the distance buffer of Dis2P (0–42 km) is separated from the prefecture-level city; distance buffer of Dis2C (0–16 km) is separated from the county-level city; distance buffer of Dis2HR (0–10 km) is separated from the high-speed roads and railway; and distance buffer of Dis2NP (0–5 km) is separated from the national road and provincial roads (Fig. 4.13).

(3) **Grid scale correlation analysis of PCCA and PCFA**

Based on the time series data of PCCA and PCFA at each grid in 2000, 2005, 2010 and 2015, grid-scale coupling relationships between PCCT and PCFT are detected by gridded correlation analysis. In this section, the MATLAB tool was used to realize the correlation analysis of 1 km grid-scale with p-value test.

The correlation coefficient could be estimated by Eq. (4.6):

$$R_i = \frac{\sum_{t=1}^{n} (x_{it} - \overline{x_t})(y_{it} - \overline{y_t})}{\sqrt{\sum_{t=1}^{n} (x_{it} - \overline{x_t})^2 \sum_{t=1}^{n} (y_{it} - \overline{y_t})^2}} \tag{4.6}$$

where R_i is the correlation coefficient between PCCT and PCFT at grid i, x_{it} is the value of PCCA at grid i in t-th year, $\overline{x_t}$ is the mean value of PCCA in t-th year; y_{it} is the value of PCFA at grid i in t-th year, $\overline{y_t}$ is the mean value of PCFA in t-th year; R_i is in the range of [-1 to 1], and the closer to 1 (-1), the higher positive (negative) correlation between PCCA and PCFA.

(4) **Spatial econometric regression analysis**

This section uses the spatial econometric regression models to analyze the influential factors of the coupling relationships between PCCT and PCFT at grid-scale, and further reveals the coupling law. The models used in this section mainly include spatial effects (spatial correlation and spatial differences), including Spatial Lag Model (SLM) and Spatial Error Model (SEM). The SLM model is used to evaluate the impact of the PCCT and PCFT by neighboring grids. SLM is expressed as follows:

$$Y = \rho W_y + X\beta + \varepsilon \tag{4.7}$$

where Y is PCCT or PCFT of the HHH in each grid; X is the independent variable matrix; β is the corresponding regression coefficient; W is the spatial weight matrix;

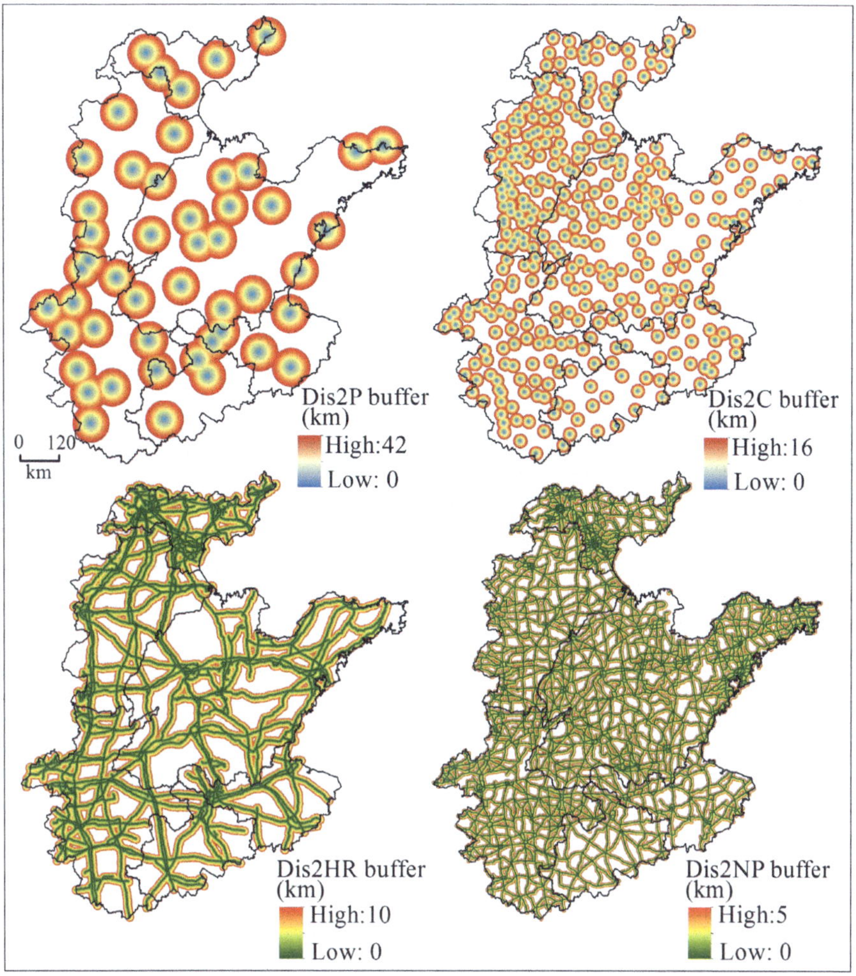

Fig. 4.13 Four types multilayer buffers (Long et al. 2018) The distance buffer of Dis2P, Dis2C, Dis2HR and Dis2NP

W_y. is the weighted mean of the surrounding dependent variable, which is regarded as the spatial lag dependent variable; ρ. is spatial autoregressive correlation coefficient; ε. is the model error term.

The SEM is used to reflect the interaction of the independent variables by the spatial dependence of the disturbance term. SEM is expressed as follows:

$$Y = X\beta + \varepsilon \tag{4.8}$$

$$\varepsilon = \lambda W_{\varepsilon} + \mu \tag{4.9}$$

where Y, X, β and W are the same meaning as those in SLM. ε is the regression residual; λ is a spatial autoregressive coefficient; μ is normally distributed random error W_ε is a spatial lag perturbation term. In order to select the appropriate algorithm for spatial econometric regression, we used the Lagrange Multiplier diagnostics to test the regression equation

Based on the previous works (Li et al. 2014), we select the physical factors, transportation factors, location factors and socio-economic factors such as 4 class including 15 elements as independent variables system that these factors closely related to construction land and farmland transitions (Fig. 4.14). Physical factors mainly take into account the Elevation (Elevation) and Slope (Slope); Location factors include nearest distance to the provincial capital government site (Dis2PC), nearest distance to the prefecture government site (Dis2P), nearest distance to the county government site (Dis2C), nearest distance to the township government site (Dis2T) and nearest distance to river (Dis2R); Transportation factors include nearest distance to highways and railways (Dis2HR), nearest distance to national and provincial roads (Dis2NP), nearest distance to county-level and township-level roads (Dis2CV). Socio-economic factors include urbanization rate at county-level (Urbanization), per capita net income of farmers at county-level (PCNI), per capita Gross Domestic Product at county- level (PGDP), per capita added value of the primary industry at county-level (PPAV) and per capita added value of the second industry at county-level (PSAV); physical factors are calculated by 30 m resolution DEM; the socio-economic data were converted from county statistics to 1 km gridded data; the 1 km resolution location and transportation data are calculated by the Euclidean Distance tool by ArcGIS.

4.5.3 Results

4.5.3.1 Construction Land Transition and Farmland Transition

During 2000–2015, the area of farmland in the HHH decreased from 29.88 to 27.31 Mha, with a shrink of 2.57 Mha. During the same period, the construction land increased from 5.52 to 8.15 Mha, with an average annual increase of 0.16 Mha (Fig. 4.15a and e). Among them, the proportion of urban construction land in the total construction land increased from 15.85 to 21.5% (Lambin and Meyfroidt 2011). In addition, the rural housing land increased by 14.87%. However, the proportion of rural housing land in the total construction land decreased by 14 percentage points. During this period, the construction land occupied 3.63 Mha of farmland, in which the urban construction land occupied 0.73 Mha and the rural residential area occupied 1.93 Mha (Fig. 4.15b). By analyzing the direction of land use transitions, it is found that the total contribution rate of farmland reduction is 72.25% by construction land occupation. The hot spots (Li et al. 2015) of urban construction land growth are mainly concentrated in the periphery of the urban (Fig. 4.15f), especially in Beijing and Tianjin, and urban agglomeration in Shandong peninsula. The hot spots of rural

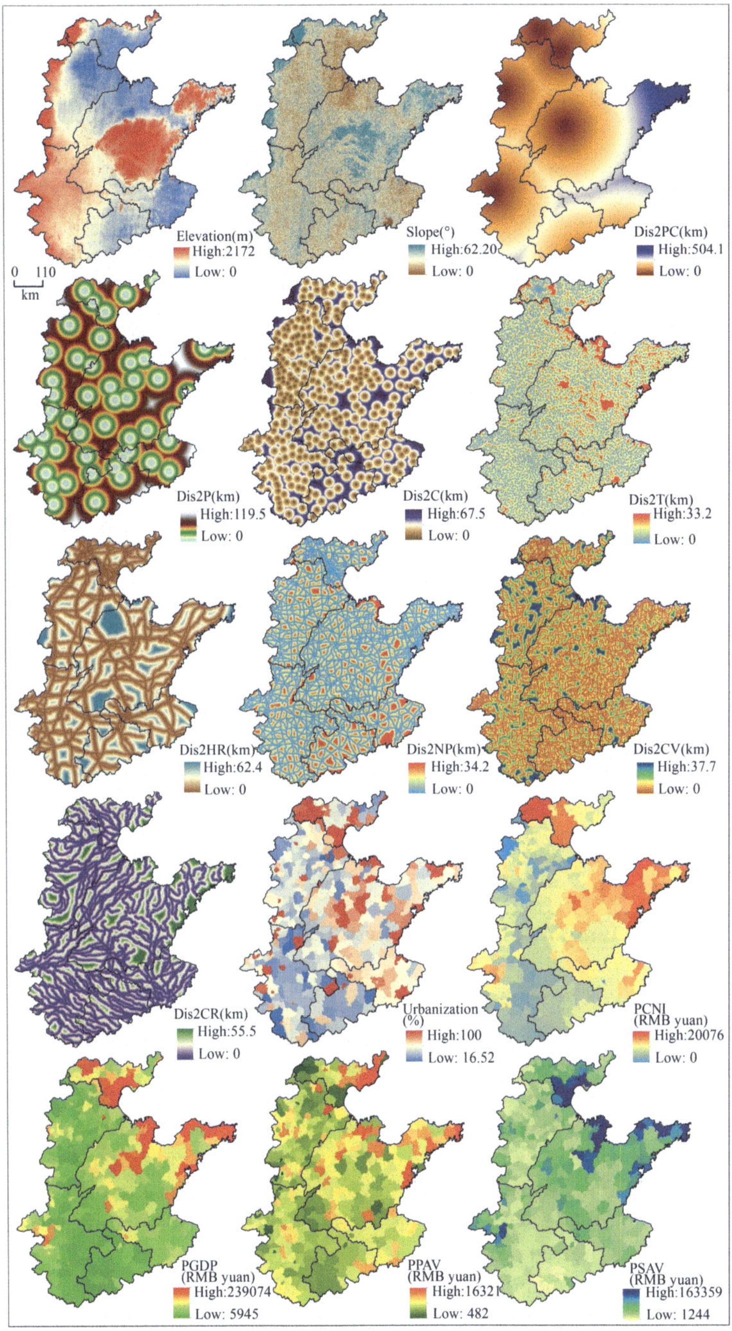

Fig. 4.14 Independent variables at 1 km grid-scale (Long et al. 2018)

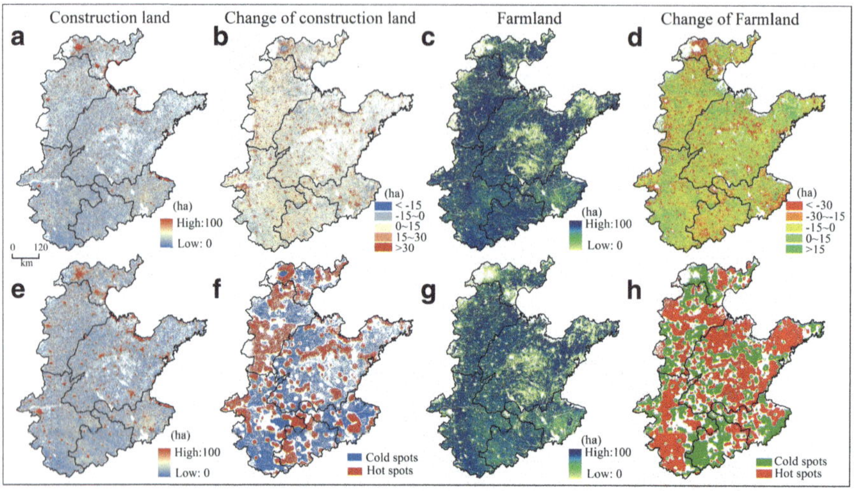

Fig. 4.15 Spatio-temporal pattern of construction land transition and farmland transition at 1 km grid (Long et al. 2018) Distribution of construction land in 2000 (**a**) and 2015 (**e**). Change of construction land during 2000–2015 (**b**). Hot spots analysis of change of construction land during 2000–2015 (**f**). Distribution of farmland in 2000 (**c**) and 2015 (**g**). Change of farmland during 2000–2015 (**d**). Hot spots analysis of change of farmland during 2000–2015 (**h**)

housing land growth mainly concentrated in northern Jiangsu and Anhui, and northeastern Henan and other plains areas of southern HHH. The above construction land transition and farmland transition indicate that the farmland has dropped sharply in the HHH by a doubly squeezed demand of construction land from the urban and the rural (Fig. 4.15d and h).

4.5.3.2 Spatio-Temporal Pattern of PCCT and PCFT

We find that the PCCA in the core area of urban development is correspondingly lower, because of the high population density. On the contrary, the PCCA in the plain farming area is higher, due to the organizational patterns of rural settlements and population distribution status (Li et al. 2015). Among them, the PCCA in the northwest of Shandong, northern Jiangsu is higher than that of other regions (Fig. 4.16a and e). In addition, the PCCA is limited in central Shandong mountainous areas and the eastern part of the Taihang Mountains, due to the less total area of construction land. Compared with the PCCA, the PCFA is mainly related to the area of farmland and population density in rural areas (Fig. 4.16c and g).

There are a large number of land use types in the surrounding areas of the urban area, but the number of farmland resources per unit area is small. The hot spots of PCCT are mainly located in Laizhou bay and Jiaodong peninsula (Fig. 4.16b and f). While the cold spots of the decrease region of PCFT are distributed in the Jing-Jin rapid urbanization areas, the adjacent region of Hebei and Shandong. Southeastern

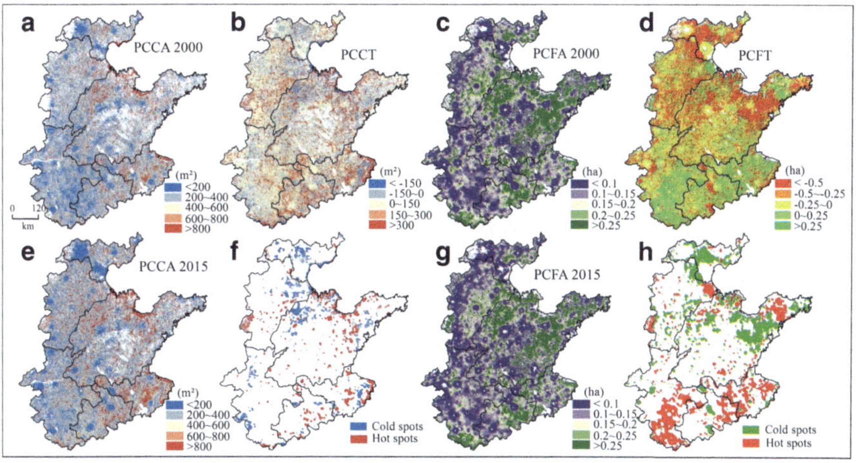

Fig. 4.16 Spatio-temporal pattern of PCCT and PCFT at 1 km grid-scale during 2000–2015 (Long et al. 2018) Distribution of PCCA in 2000 (**a**) and 2015 (**e**). PCCT during 2000–2015 (**b**). Hot spots analysis of PCCT during 2000–2015 (**f**). Distribution of PCFA in 2000 (**c**) and 2015 (**g**). PCFT during 2000–2015 (**d**). Hot spots analysis of PCFT during 2000–2015 (**h**)

Henan and northern Jiangsu are the hot spots region of growth of PCFT (Fig. 4.16d and h). From the perspective of spatial coupling, we can see that "the richer PCFA is, the more PCCA is" in traditional plains of farming areas, such as northern Jiangsu plain and northwest of Shandong pain; In contrast, "the poorer PCFA is, the less PCCA is" (Tan and Li 2013), such as the eastern foothills of Taihang Mountains. In addition, in the surrounding areas of urban, both the PCCA and PCFA are less.

4.5.3.3 Spatial Coupling Relationships of PCCT and PCFT by Multi-layer Buffer

We find that PCCA and PCFA are sensitive to the change of the distance to city and road (Li et al. 2015). We can see that the PCCA experienced a rapid decline, and then gradually increased from the city center to the periphery (Fig. 4.13). While the PCFA shows a rapid increase, and then gradually stabilized. The PCCA of each buffer in 2015 was higher than that in 2000, which indicates that the PCCA increased significantly in each buffer (Long et al. 2012). Interestingly, the trend of the PCFA was the opposite in the same period, which means that PCFA decreased of each buffer during 2000–2015. We also detect the spatial coupling between PCCT and PCFT that they have a high degree of symmetrical coupling relationship in space, that is to say, the PCCA and PCFA change in the opposite direction (Fig. 4.17). It can explain that if the PCCA increases, the PCFA will decrease in the neighboring space (vice versa).

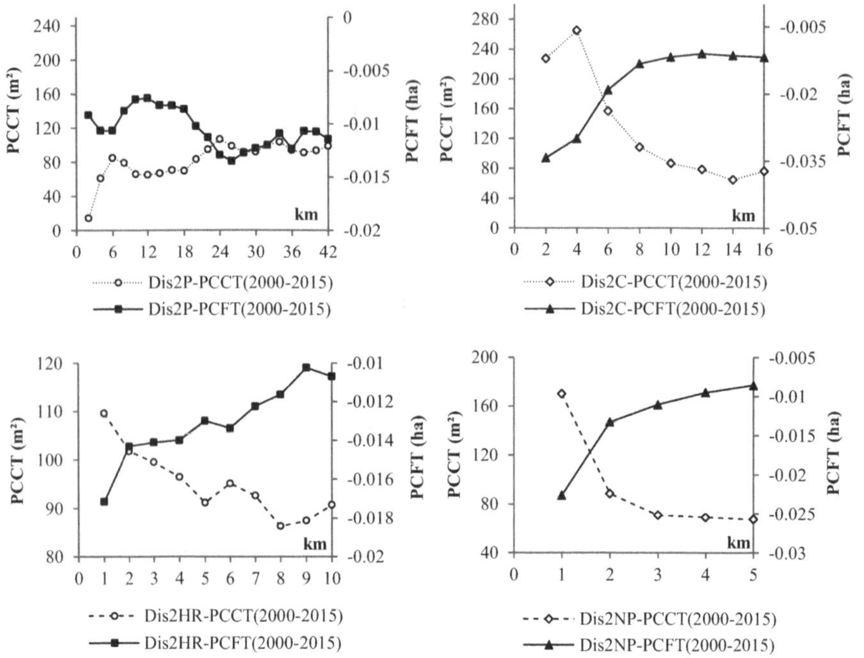

Fig. 4.17 Spatial coupling pattern of PCCT and PCFT within multi-layer buffer at 1 km grid-scale during 2000-2015 (Long et al. 2018)

4.5.3.4 Correlation Relationships Between PCCT and PCFT

We find that 61565 pixels pass the $p < 0.1$ significant test, which account for 17.96% of all the grid involved in the correlation analysis between PCCT and PCFT (Fig. 4.18b). The mean correlation coefficient of the HHH is 0.07, but the correlation is significant in the aspect of regional difference. The zoning statistics shows that the mean value of correlation coefficient is positive in Shandong and Jiangsu, which is 0.11 and 0.06, respectively. While others regional correlation is negative, the mean value of correlation coefficient is 0.14, 0.17, 0.21 and 0.26 for Anhui, Henan, Jing-Jin and Hebei, respectively (Fig. 4.18a and b). The positive correlation areas are mainly concentrated in the mountainous area of central Shandong and hilly areas of Jiaodong Peninsula, which mainly express the coupling characteristics of double increasing of PCCA and PCFA. The negative correlation areas are mainly located in the periphery of the city and Hebei, Henan and Anhui, which show a significant coupling characteristic of the increasing of PCCA and the decreasing of PCFA. Using the four multi-layer buffers for statistical analysis, the correlation coefficients of them are increased from the negative correlation to weak correlation during the elongating distance from the center of human economic activities to that far from it (Fig. 4.18).

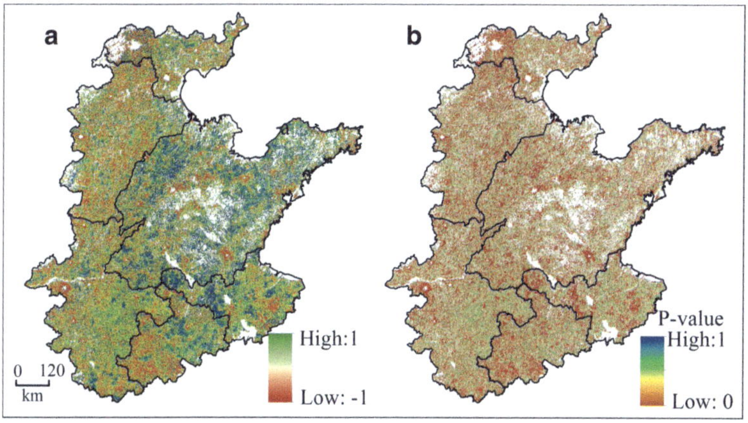

Fig. 4.18 Spatial pattern of correlation relationship between PCCT and PCFT at 1 km grid-scale (Long et al. 2018) Correlation coefficient between PCCT and PCFT from 2000 to 2015 (**a**). P significant test value of the correlation coefficient (**b**)

4.5.3.5 Mechanism of PCCT and PCFT

The distance to multi-level cities and rivers has a converse effect on PCCT and PCFT, which shows symmetrical coupling characteristics. That is to say, if the independent variable has a positive effect on PCCT, correspondingly has a negative effect on PCFT. The closer the distance to the river, the slower PCCT, and the faster PCFT. The PCCA in 2000 has a significant negative effect on PCCT and PCFT, the lower PCCA, the higher PCCT and PCFT. The site selection of high- level roads generally avoids highly developed areas, and most high-level roads are closed. So, the closer Dis2HR, the lower PCCT. The national and provincial roads generally cross the densely populated area, and they are open roads. Most rural settlements are along the road. As a result, the closer to these roads, the faster PCCT (Li et al. 2015). However, there is a clear positive effect on PCFT on different grade roads, which means that the closer distance to traffic line, the slower PCFT (vice versa).

PCCT and PCFT are affected by regional socio-economic development, and there are significantly different (Table 4.8). The level of urbanization has positive effect on both PCCT and PCFT, i.e., the higher the urbanization level, the faster PCCT and PCFT. Per capita net income of farmers (PCNI) and Per capita primary industry added value (PPAV) have negative effects on PCFT, which is in line with the law of rural migration that farmer leaves the rural area to city in order to improve the quality of life. In addition, there is a significant negative cor- relation between Per capita GDP (PGDP) and PCCT. The faster growth of PCCA in the region with lower PGDP confirms the development phenomenon of extensively utilizing valuable land resources in this region. The effects of Per capita secondary industry added value (PSAV) on the PCCT and PCFT are opposite, indicating that rapid urbanization and industrialization caused a large number of land resources occupied with low

Table 4.8 The result of spatial econometric regression analysis at grid-scale (Long et al. 2018)

Variable	PCCT (SEM)	PCFT (SLM)
CONSTANT	141***	−0.3***
PCCA 2000	−3254.2***	−2.8***
Elevation	−514.3***	0.9***
Slope	41.2	−0.2**
Dis2PC	−11.9	0.1***
Dis2P	250.2***	−0.08**
Dis2C	−110.8**	0.04**
Dis2T	−291.4***	0.4***
Dis2R	390.1***	−0.2***
Dis2HR	360.6***	0.02**
Dis2NP	−440.7***	0.2***
Dis2CV	−338.7***	0.2***
Urbanization	132.3***	0.04**
PCNI	4.3	−0.2***
PGDP	−396.3***	0.3***
PPAV	32.5	0.1***
PSAV	212.9**	−0.4***
ρ		0.8***
λ	0.9***	
R^2	0.7	0.6
AIC	1354010	310903
SC	1354170	311074
Log likelihood	−676989	−155434
Lagrange multiplier (lag)	135160.7***	115984.8***
Robust LM (lag)	62.8***	1595.6***
Lagrange multiplier (error)	136325.8***	114411.9***
Robust LM (error)	1228***	22.8***

*Significance at 10%. **Significance at 5%. ***Significance at 1%. In order to further reveal the relationship between PCCT and PCFT, the PCCA in 2000 is added as the independent variable

development intensity (Table 4.8). The negative effect of PSAV on PCFT mainly reflects the characteristics of regional industrial structure with limited employment absorption capacity. The lower PSAV it is, the more population outflow and the more growth of PCFA. In addition, the positive effect of PPAV on PCFT shows that benign rural development is beneficial to the protection of farmland.

4.5.4 Discussion and Conclusions

Based on the gridded data of construction land and farmland, the spatial distribution of PCCA and PCFA, the gridded correlation analysis and the driving mechanism analysis between PCCT and PCFT are clearly revealed. There is a high degree of the coupling relationship between PCCT and PCFT. The spatio-temporal evolution of the coupling relationship between them is an important mean to study the interaction between construction land, farmland and population change in the process of urbanization (Bai et al. 2014). The changing man-land interrelations in the HHH are the important features of urban-rural transformation development under urbanization. Combined with the special status of the HHH in China's grain production, it is important to reveal the profound implications of construction land transition and farmland transition for food security.

The contribution rate of construction land occupancy for farmland decrease is over 70% from 2000 to 2015 in the HHH. The total area of farmland occupied by urban construction land and rural housing land exceeded the net reduction of farmland. During 2000–2015, the increasing area of PCCA is 66.09% in the HHH, only 33.91% area is decreasing. At the same time, the increasing area of PCFA is only 31.48% in the HHH, while 68.52% area is decreasing. The overall coupling analysis can be seen that the PCCT and PCFT are obviously negative correlated. The fluctuation increasing of PCCA and the decreasing of PCFA are important factors that threatened the safety of grain production in the HHH. In the context of rapid urbanization, the total amount of farmland resources and PCFA are all declining, which becomes a serious contradiction on the issue of farmland protection under the strict protection policy for farmland. Therefore, more attention should be paid to how to protect the farmland in the HHH with high-speed urbanization area, high rural population density area, and grain production core area (Long and Qu 2018).

In the process of rapid urbanization, there is a large number of inefficient and unused occupied farmland. By analyzing the mechanism of PCCT and PCFT, it is found that the urbanization level, PGDP and PSAV confirmed from different sides that urbanization characterized by a high-speed industrialization obviously occupied amounts of farmland. Large-scale construction of the development zone and inefficient use of occupied farmland have caused serious damage to the scant farmland resources in the HHH (Chen et al. 2014). At the same time, there is an obvious phenomenon of de-agriculturalization of the rural population in the HHH, including their living space structure and employment structure. The problem of inefficient utilization of rural housing land and farmland in hollowed villages is also serious. The construction land both in the urban and the rural has a serious impact on farmland resources. Therefore, the man-land interrelations in the HHH depict the urgent crisis of farmland protection and rural transformation. The in-depth analysis of coupling relationship between PCCT and PCFT creates conditions for optimizing regional man-land interrelations, regulating regional grain production models and ensuring regional food security.

There are four kinds of theoretically coupling relationship between PCCT and PCFT, i.e., double increasing, double decreasing, PCCA increasing and PCFA decreasing, and PCCA decreasing and PCFA increasing. It is found that there are two coupling relationships in the HHH (Fig. 4.19). One is double increasing of PCCA and PCFA (Fig. 4.19 coupling model 1) and another is PCCA increasing and PCFA decreasing (Fig. 4.19 coupling model 2). The coupling type with decreasing of PCCA and increasing of PCFA has not yet emerged (Fig. 4.19 coupling model 3). In addition, it lacks ideal conditions for the theoretically coupling type of PCCA and PCFA both decreasing in the HHH (Long et al. 2012). The above coupling relationships are important for improving food production safety. The decline of per capita farmland resources is a serious challenge to ensure the safety of grain supply in China (Tilman et al. 2011). The above coupling relationship between PCCT and PCFT gives important implications for food security. Under the condition of China's large population base, the PCFA is especially limited. So, the decline of PCFA is a serious challenge to ensure the safety of grain supply in China.

In the process of urbanization, how to optimize the utilization of construction land and explore the utilization value of stock land resources between urban and rural? The effective coordination of the elements contained in the mode of man-land interrelations in rural areas, such as land use, population, capital and technology, provides an optimized path for ensuring regional food security. For example, promoting the reclamation and redevelopment of the inefficient and unused land in the low-intensity development zone and rural residential area are effective means to protect the farmland in the future (Long et al. 2016). In the process of rapid urbanization and the rural-to-urban migration, new requirements for improving the land resources management system (Bren d'Amour et al. 2017), social security system and other

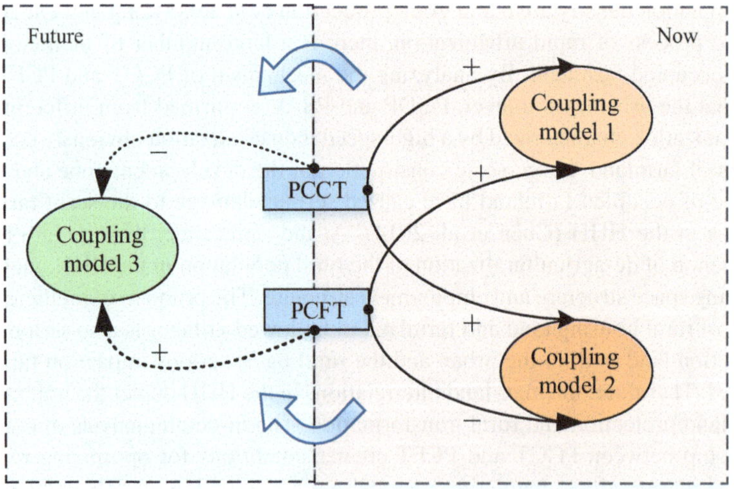

Fig. 4.19 Coupling relationship transition between PCCT and PCFT (Long et al. 2018). "+" and "−" represent the increase and decrease in quantity, respectively

national governance system are put forward. The government needs to take feasible measures to improve the urbanization rate of the floating population, and gradually reduce the PCCA. Meanwhile, it is helpful to reduce the number of rural population and increase PCFA, which will contribute to the formation of coupling mode 3 (Fig. 4.19). In addition, changing the traditional small-scale agricultural production mode (Lambin and Meyfroidt 2011) and developing modern agricultural production system are also the important tasks of ensuring food security and promoting urban and rural transformation development in China's farming areas.

Based on the coupling relationships between PCCT and PCFT, the evolution process between construction land transition and farmland transition in the process of urbanization in the HHH was analyzed. In addition, the coupling types and their correlation relationships between PCCT and PCFT at grid-scale were also discussed. Using spatial econometric regression model, the different coupling mechanisms between PCCT and PCFT were detected, and the challenges and implications of the coupling relationships for food security were further analyzed. The results show that the contribution rate of construction land to farmland loss is 72.25%, and PCCA fluctuation increasing and PCFA fluctuation decreasing coexisted during 2000–2015. There is obviously a negative correlation between PCCT and PCFT at grid-scale, and the spatial pattern is different. The multi-layer buffer analysis finds that they have symmetrical coupling characteristics in space. Spatial econometric regression models analysis shows that PCCT and PCFT are affected by physical factors, location factors, transportation factors, and socio-economic development status with obvious spatial spillover and spatial error characteristics.

The above results bring some implications for ensuring food security in the HHH. The government needs to enhance the exploitation level of urban-rural stock land resources, reduce PCCA, and control the amount of farmland occupied by the construction (Foley 2017). Furthermore, aiming at the national strategy of new-type urbanization, it needs to promote the rural floating population stably settled down in the city, and improve the urban and rural social security system. After that, it is helpful to increase PCFA, promote the appropriate scale management of grain production, and develop modern agricultural production system in favor of ensuring food production security.

4.6 Conclusions

Attempting to identify the mutual relations between rural housing land and farmland, this chapter not only revealed the characteristics of their spatio-temporal coupling, but also qualitatively and quantitatively discussed and analyzed the mechanism, which contributes to the knowledge about man-land interrelations. Outcomes of the analysis of spatio-temporal interaction indicate that the correlation coefficient of farmland versus rural housing land change is −0.921, and it shows that the change pattern of farmland and rural housing land is uncoordinated. Rural housing land change has played a major role in the mutual transformation of farmland and rural housing land. Besides, there exists a high-degree spatial coupling between farmland and rural housing land changes in southeast China during 2000–2008. But, what factors shape their coupling relations? In general, farmland and rural housing land transitions in China are driven by socio-economic, bio-physical and managerial three-dimensional driving factors through the interactions among rural population, farmland and rural housing land. However, the spatio-temporal coupling phenomenon and mechanism of farmland and rural housing land transition in China are largely due to the "dual-track" structure of rural-urban development.

In examining spatio-temporal dynamic patterns of farmland and rural settlements from 1990 to 2006 in The Su-Xi-Chang Region of coastal China experienced dramatic economic and spatial restructuring, the results showed that to a large extent, land-use change from 1990 to 2006 in The Su-Xi-Chang Region was characterized by a serious replacement of farmland with urban and rural settlements, construction land, and artificial ponds. Population growth, rapid industrialization and urbanization are the major driving forces of farmland change, and China's economic reforms played an important role in the transformation of rural settlements. China's "building a new countryside" is an epoch-making countryside planning policy.

With regard to fast-growing traditional farming areas, during 2000–2015, the farmland decreased by 8.59, 72.25% of which were occupied by construction land. There are two major coupling types between per capita construction land transition (PCCT) and per capita farmland transition (PCFT), one is the double increasing of per capita construction land area (PCCA) and per capita farmland area (PCFA); another is the increasing of PCCA and the decreasing of PCFA. The fluctuant increasing of PCCT and decreasing of PCFT coexisted and presented symmetrical coupling characteristics in space. Physical, location, transportation and socio-economic factors play significantly different roles in driving PCCT and PCFT.

Based on above results, proper policy implications are proposed. It is fundamental to lay out a scientific urban-rural integrated development planning for building a new countryside, which needs to pay more attention to making the rural have certain functions serving for the urban. In addition, the cultural elements of idyll and the rural landscape need to be reserved and respected in the process of building a new countryside in coastal China, instead of building a new countryside, which looks more like a city. Measures regarding promoting the reclamation and redevelopment of inefficient and unused urban-rural construction land, reducing inefficient occupation of farmland resources, and developing appropriate scale management of agriculture are needed to ensure food security in farming areas.

References

Bai X, Shi P, Liu Y (2014) Society: realizing China's urban dream. Nature 509:158–160

Bren d'Amour C, Reitsma F, Baiocchi G (2017) Future urban land expansion and implications for global croplands. P Natl Acad Sci 114(34):8939–8944

Cartier C (2001) 'Zone Fever', the arable land debate, and real estate speculation: China's evolving land use regime and its geographical contradictions. J Contemp China 10:445–469
Chen X, Cui Z, Fan M et al (2014) Producing more grain with lower environmental costs. Nature 514:486–489
Fei X (1996) 爱我家乡 (Love my hometown). Qunyan Press, Beijing
Foley J (2017) Living by the lessons of the planet. Science 356(6335):251–252
Ho SP, Lin GC (2004) Converting land to nonagricultural use in China's coastal provinces: evidence from Jiangsu. Mod China 30:81–112
Hu H (1935) 中国人口之分布——附统计表与密度图 (The distribution of population in China with statistics and maps). Acta Geogr Sinica 2(2):33–74
Lambin EF, Meyfroidt P (2011) Global land use change, economic globalization, and the looming land scarcity. P Natl Acad Sci 108(9):3465–3472
Li T, Long H, Liu Y et al (2015) Multi-scale analysis of rural housing land transition under China's rapid urbanization: the case of Bohai Rim. Habitat Int 48:227–238
Li X, Wang X (2003) Changes in agricultural land use in china: 1981–2000. Asian Geogr 22:27–42
Li Y, Liu Y, Long H et al (2014) Community-based rural residential land consolidation and allocation can help to revitalize hollowed villages in traditional agricultural areas of China: evidence from Dancheng county, Henan province. Land Use Pol 39:188–198
Liu J, Zhang Z, Xu X et al (2010) Spatial patterns and driving forces of land use change in China during the early 21st century. J Geogr Sci 20:483–494
Long H, Ge D, Zhang Y et al (2018) Changing man-land interrelations in China's farming area under urbanization and its implications for food security. J Environ Manage 209:440–451
Long H, Tang G, Li X et al (2007a) Socio-economic driving forces of land-use change in Kunshan, the Yangtze River Delta Economic Area of China. J Environ Manage 83(3):351–364
Long H, Heilig GK, Li X et al (2007b) Socio-economic development and land-use change: analysis of rural housing land transition in the Transect of the Yangtse River, China. Land Use Pol 24:141–153
Long H, Li T (2012) The coupling characteristics and mechanism of farmland and rural housing land transition in China. J Geogr Sci 22:548–562
Long H, Li Y, Liu Y et al (2012) Accelerated restructuring in rural China fueled by 'increasing vs. decreasing balance' land-use policy for dealing with hollowed villages. Land Use Pol 29:11–22
Long H, Liu Y, Wu X et al (2009) Spatio-temporal dynamic patterns of farmland and rural settlements in Su-Xi-Chang region: Implications for building a new countryside in coastal China. Land Use Pol 26:322–333
Long H, Qu Y (2018) Land use transitions and land management: a mutual feedback perspective. Land Use Pol 74:111–120
Long H, Tu S, Ge D et al (2016) The allocation and management of critical resources in rural China under restructuring: problems and prospects. J Rural Stud 47:392–412
Long H, Woods M (2011) Rural restructuring under globalisation in eastern coastal China: what can be learned from Wales? J Rural Community Dev 6:70–94
Mather AS, Hill G, Nijnik M (2006) Post-productivism and rural land use: cul de sac or challenge for theorization? J Rural Stud 22:441–455
Seto KC, Kaufmann RK (2003) Modeling the drivers of urban land use change in the Pearl River Delta, China: Integrating remote sensing with socioeconomic data. Land Econ 79:106–121
Song Y, Qi G, Zhang Y et al (2014) Farmer cooperatives in China: diverse pathways to sustainable rural development. Int J Agr Sustain 12:95–108
Su S, Zhang Q, Zhang Z et al (2011) Rural settlement expansion and paddy soil loss across an ex-urbanizing watershed in eastern coastal China during market transition. Reg Envir Chang 11:651–662
Tan M, Li X (2013) The changing settlements in rural areas under urban pressure in China: patterns, driving forces and policy implications. Landsc Urban Plan 120:170–177
Tan R, Beckmann V, DenBerg L et al (2009) Governing farmland conversion: comparing China with the Netherlands and Germany. Land Use Pol 26:961–974

Tian G, Yang Z, Zhang Y (2007) The spatio-temporal dynamic pattern of rural residential land in China in the 1990s using landsat TM images and GIS. Environ Manage 40:803–813

Tilman D, Balzer C, Hill J et al (2011) Global food demand and the sustainable intensification of agriculture. Pro Natl Acad Sci 108(50):20260–20264

Verburg PH, Veldkamp A, Fresco LO (1999) Simulation of changes in the spatial pattern of land use in China. Appl Geogr 19:211–233

Wei Y, Fan C (2000) Regional inequality in China: a case study of Jiangsu province. Prof Geogr 52:455–469

Woods M (2004) Rural geography: processes, responses and experiences in rural restructuring. Sage, London

Xiao Y, Wu X, Wang L et al (2017) Optimal farmland conversion in China under double restraints of economic growth and resource protection. J Clean Prod 142:524–537

Xie Y, Batty M, Zhao K (2007) Simulating emergent urban form using agent-based modeling: Desakota in the Suzhou-Wuxian Region in China. Ann Assoc Am Geogr 97:477–495

Xu W, Tan K (2001) Reform and the process of economic restructuring in rural China: a case study of Yuhang, Zhejiang. J Rural Stud 17:165–181

Part II
Rural Restructuring

Chapter 5
Understanding Rural Restructuring: A Theoretical Approach

Abstract Based on the researches by Woods (2005), this chapter develops the concept and connotations of rural restructuring, and establishes a theoretical framework of that by melting the elements of China's background condition. Rural restructuring is a process of reshaping socio-economic morphology and spatial pattern in rural territory in respond to the changes of elements both in kernel system and external system of rural development, by optimally allocating and efficiently managing the material and non-material elements in the two systems. It aims at ultimately optimizing the structure and promoting the function within rural territorial system as well as realizing the coordination of structure and complementation of function between urban and rural territorial systems. The evolution of "elements-structure-function" of rural territory has broken utterly the traditional spatial structure and socio-economic morphology in rural China, which triggers rural restructuring and brings about a series of challenges to sustainable rural development. It is urgent for local participants to take actions timely to restructure rural space, economy and society in rural territory system. Accordingly, the approaches of rural restructuring include the three aspects of spatial restructuring, economic restructuring and social restructuring. The interactions between multiple elements and subsystems inside and outside the rural territory promote the process of rural restructuring, through synthesizing and dispersing, pulling force and pushing force, transformation and feedback under the combined action of inducing mechanism, supporting mechanism, constraining/promoting mechanism, guiding mechanism and engine mechanism. Finally, the progress of rural restructuring research in China is summarized.

5.1 Introduction

Rural territory is an open system with certain structure and function, which is composed of diversified elements including natural resources endowments, geographical conditions, economic base, human resource, cultural customs, etc. Structurally, rural territorial system is composed of kernel system and external system (Zhang and Liu 2008; Long et al. 2016). With the advancement of globalization, urbanization, industrialization and informatization, the rural territory continuously exchanges the

H. Long, *Land Use Transitions and Rural Restructuring in China*,
https://doi.org/10.1007/978-981-15-4924-3_5

material, energy and information with external urban system (Wu 2001). Inevitably, under the integrated effects of the changes of rural development elements, e.g., vast out-flow of rural population, de-agriculturalization of industry, low-efficient rural land use, and the external regulation factors, etc., there are a variety of interactions between material and non-material elements in the aspects of the rural production and living behaviors. These interactions will subsequently lead to the changes in employment structure, consumption structure, land use structure and social organization structure, and the constantly evolving of the function of rural territorial system in the aspects of living, production, ecology and culture.

The evolution of the "element-structure-function" of the rural territory requires local participants (including local government, enterprisers, rural elites and villagers) to make respond and readjustment corresponding to the processes and outcomes of changes in socio-economic morphology and regional spatial patterns, and with the intervention of multiple actors of "top-down" and "bottom-up", the rural areas have experienced restructuring with different degrees. The process of rural restructuring contains a wide range of transformations, such as the industrial structure, employment mode, the rural settlements morphology and social cultural landscape, the relationships between the urban and the rural as well as agriculture and industry, etc., which determine that rural restructuring is a systematic project in practice. Especially, influenced by special physical and socio-economic conditions, e.g., notable territorial differences, the dual-track structure of urban-rural development and the prominent government intervention, it may exacerbate the volatility and complexity of rural restructuring practice in rural China. So, there is an urgent need for probing related theories to guide the practice of rural restructuring in China.

This chapter has five parts. Firstly, the development background of rural restructuring is analyzed. Then, the concept and connotations of rural restructuring are developed by melting the elements of China's background conditions based on the concept put forward by Woods (2005). Further, a theoretical framework of rural restructuring is established. Thereafter, the approaches of rural restructuring are probed from the aspects of space, economy and society, and the mechanism of rural restructuring is elaborated from the perspective of the "elements-structure-function" in rural territorial system. Finally, the progress of rural restructuring research in China is summarized.

5.2 Background

Rural restructuring is more or less interlinked with rural decline, an inevitable process associated with globalization and urbanization (Woods 2005). Most developed countries including United States, Canada, Japan, South Korea, and some European countries have experienced this kind of decline as well as rapid rural restructuring (Woods 2013). The experiences of rural restructuring in developed countries show that challenges and subsequent improvements coexisted during the process.

On the one hand, this restructuring process has brought about a series of challenges. For instance, the uneven impact of regional migration and restructuring of employment is transforming rural regions in Europe in different ways, and is bringing about challenges facing rural regions with regard to social services provision (Manthorpe and Livsey 2009). The territorial restructuring occurs at multiple scales to facilitate the investment in and sale and export of natural resource commodities, and the construction of rural communities usually blends the restructuring of places and spaces and the resistance from peasants' struggle for land reform to their indigenous alliance in defense of territory (Brent 2015). In some European countries, the agricultural and rural capital markets are also facing specific credit constraints related to agricultural assets and rural fixed asset specificities (Bojnec 2012). The European experiences showed that land consolidation as an isolated tool for rural restructuring in transitional environments may trigger some unfavorable side-effects (Djanibekov et al. 2012).

On the other hand, some endearing improvements have been achieved. For example, in developed countries, consumption turns to be the driving force behind growth and change in many 'post-productivist' rural areas, and landscapes are actively produced to capitalize upon local amenities. Currently, rural restructuring in the European countryside is shifting from an agricultural and manufacturing-based economy towards a service-centred economy in which the countryside is considered a place of consumption instead of a place of production (Rogge et al. 2013). Rural economic restructuring and the expansion of low-density exurban areas and growth of amenity-based rural areas have led to a new form of urban-rural space, which is characterized by the merging of a rural landscape form with urban economic function (Irwin et al. 2009). The process of agricultural restructuring in Europe was characterized by agricultural multifunctionality and entrepreneurial farmers (Morgan et al. 2010). The achievements of South Korea's new countryside movement and Japan's "one village one trademark" revolution have provided useful references for China's rural restructuring.

In China, a typical developing country, rapid and far-reaching development transition has triggered corresponding rural restructuring especially since the turn of the new millennium (Long et al. 2011; Long and Liu 2016). With the influences of globalization and the implementation of the national strategies on industrialization, urbanization, informatization and agricultural modernization (Li et al. 2014), vast rural China has been profoundly changed due to the recombination of regional development factors and the reshaping of regional industries. It has taken on the status quo of non-agriculturalization and concurrent occupation of rural population, non-agriculturalization and non-grain preference of rural land use, and non-agriculturalization and diversification of rural industry (Long 2014). Furthermore, it has brought about some problems in part of the rural areas, e.g., the flow away of young laborers, the weakening of rural development mainstream, the abandoning of rural land, the hollowing of villages, the shortage of public infrastructure, the desalinating of consanguineous relation as well as the disappearing of cultural memory symbols. Meanwhile, driven by the changes of rural internal factors and the demand of external system, the re-allocation of rural production elements has brought about unprecedented opportunities for rural development. Aiming at the challenges and opportunities during the period of social transformation, the central government has clearly put forward a series of strategic guiding policies concerning realizing rural restructuring and promoting rural development since 2004, e.g. "Building New Countryside" (Long et al. 2010) and "Implementing New-Type Urbanization" (Fang and Yu 2016).

Since the 1990s, rural restructuring has been a hot topic of rural studies (Wilson 1995; Hoggart and Paniagua 2001; Woods 2005; Long and Woods 2011). Rural restructuring was defined by Woods (2005) as "the reshaping of social and economic structures in rural areas during the late twentieth century and early twenty-first century produced by various, interconnected processes of change including the declining economic significance of agriculture, the rise of the service sector, urban to rural migration, and so on". Zhang et al. (2006) argued that rural restructuring is one of the rural development strategies as a whole set of economy, society and space for the purpose of realizing urban-rural integrated development in terms of the practice of rural China. Based on the above research, this chapter attempts to establish a theoretical analysis framework for rural restructuring concerning its connotations, approaches and mechanism.

5.3 Theoretical Framework

5.3.1 "Elements-Structure-Function" of Rural Territorial System

As mentioned above, rural territory is an open system with certain elements, structure and function. With regard to the elements, it contains natural resources endowments, geographical conditions, economic base, human resource, cultural customs, etc. (Long et al. 2016). Structurally, the rural territorial system may be divided into kernel system and external system as well as object system and subject system. On the one hand, the kernel system consists of some subsystems, e.g., natural resources, ecological environment, economic development and social development (Wu 2001). However, the external system mainly includes the regional development policy, industrialization and urbanization (Zhang and Liu 2008). On the other hand, the object system consists of natural endowments, geographical conditions, economic base and other objective factors which influence the development of rural area; and the subject system is composed of some behavioral mainstreams such as local governments, enterprisers, elites and farmers. Usually, local behavioral mainstreams integrates relevant factors of the object system to promote the coordinated development between the subsystems within the kernel system, and the continuous exchange of material, energy and information between the kernel system and the external system. This kind of exchange of material, energy and information keeps the structure and function of rural territorial system gradually optimized, which forms the driving forces of rural development (Wu 2001).

Rural territory has multi-functions in the aspects of living, production, ecology and culture, which are constantly evolved and varied with the integration of system elements, the optimization of system structure, the change of external environment and the driving of social demand. In the traditional sense, the rural territory carries a certain number of people, and meets the needs of rural areas itself through the production and processing of primary agricultural products. Besides, it also sustains the running and developing of the city by providing it with resources. Living and agricultural production functions are the primary functions of rural territory. With the increasing frequency of the interaction between urban and rural areas, and the accelerating flow of economic development factors, the rural area begins to share some of the functions of industrial production and service supply, and the connotation of the production function of rural area is gradually enriched. However, large consumption of water and land resources and subsequent environmental pollution problems grow fast together with the advance of industrialization and urbanization. The ecological function of rural territory has become increasingly prominent because of its high vegetation coverage and low population density, which are pivotal to the mitigation and regulation of natural disasters and the maintenance of biological diversity. And with the growing regional cultural exchanges and increasing cultural homogeneity, the characteristic style of rural settlements, simple folk customs and the unique cultural characteristics have aroused more and more people's attention, and the rural territory begins to bear the function of inheriting culture.

5.3.2 Rural Restructuring in China Based on "Elements-Structure-Function"

During 1990–2014, China's urbanization rate rose from 26.41 to 54.77%, with an annual increase rate of 1.18%, and the continuous out-flow of rural population is one of the major driving forces of rapid urbanization. According to the Peasant Laborer Monitoring Report in 2014, the total number of migrant workers in China was 274 million, and the average age of migrant workers was 38.3 years old, of which 20–49 years old migrant workers accounted for 79.4%. With the continual boost of rapid urbanization, the evolution of "elements-structure-function" of rural territory has broken utterly the traditional spatial structure and socio-economic forms in rural China, which triggers rural restructuring (Tu and Long 2017).

Affected by the rising opportunity cost of farming owing to non-agricultural employment of rural population, coupled with the influence of increasingly high cost of grain-growing, the economic value of farmland has been reduced relatively (Long et al. 2016), which has caused regional de-agriculturalization, non-grain preference and extensive utilization of farmland, and many of the plentiful commodity grain producing base has irreversibly evolved into urbanized areas as a consequence (Long

2014). Along with the outflow of massive rural population, the rural housing land has not shrunk as expected due to the urban-rural dual institutional barriers in household registration, land use policy and social security, together with the undefined collective property rights. Instead, the issue of derelict, abandoned and inefficient use of rural housing land has become increasingly prominent in rural China, which has further generated obstacles for the layout of the industry, the optimization of public infrastructure and the allocation of social services. With the change of production elements like labor force, technology and so on, there is an urgent need to reallocate the land use right among different management mainstream. However, hampered by underdeveloped non-agricultural industries, the fragmentation of cultivated land and unclear property rights of rural land, it is difficult to realize the transfer and large-scale management for farmland, as well as reuse and circulation for vacant and derelict rural housing land (Long et al. 2016). In addition, water and soil are contaminated and damaged rapidly in some rural areas owing to the disordered layout of township and village enterprises, less strict control of pollutants emission and the improper handling of agricultural production and living pollution in the past years, which has directly resulted in the severe deterioration of the rural ecological space (Long 2014). In a word, the spatial morphology of rural China takes on a trend of decentralization of living space, disordering of production space and deterioration of ecological space.

Under the background of the reducing of farmland resource and the decreasing of grain production, as well as the negative effects of large-scale outflow and concurrent occupation of rural labor force on manpower input and science-technology popularizing of agricultural development, the traditional function of agricultural production in some areas has been gradually declined (Liu 2011). At the same time, the rural industry has presented non-agricultural and diversified trends. With the acceleration of the flow of economic elements between urban and rural areas as well as different regions, labor-intensive industries have emerged in certain rural areas with good location and better economic base. Especially in recent years, some villages, relying on the original ecological rural and natural landscape together with the varied traditional customs, take advantage of location being adjacent to the famous scenic spots or urban edge to develop rural tourism, the ecological function and cultural function have become increasingly prominent in those rural areas (Tu et al. 2015). Accompanied with the evolution of economic forms, production functions of rural territory have transformed from the traditional single agricultural production gradually to the industrial, leisure services and other multifunction accordingly.

The continuous unilateral outflow of rural labor has changed the population structure (mainly including the age, gender, education level) in the rural areas, which further resulted in the problems such as aging population, low-educated labor force, lack of talents and the decline of grassroots organizations in rural areas (Long et al. 2016). Meanwhile, with the changes of marginal productivity brought about by the introduction of new production tools, various forms of agricultural moderately scale management, such as helping to plough and sow, land trusteeship and stock cooperation, are gradually developing, which greatly promote the emergence of family farms, professional investors, farmer professional cooperatives and other new business entities. The phenomenon shows that the differentiation of social class is increasingly apparent in rural China. According to the statistical data from the Ministry of Agriculture, there were 877 thousand family farms totally in China by the end of 2012, with operating cultivated land 11.7 million ha, accounting for 13.4% of China's total farmland. The average management scale of family farm reached 13.3 ha, and was nearly 27 times that of contracted farmland per household in China (Long et al. 2016).

To summarize, under the background of globalization, industrialization, urbanization and informatization, elements-structure-function of rural territory have undergone dramatic evolution triggered by the flow of urban-rural development elements, the innovation of production tools, the change of dominated economic value and the national existing policy system, which profoundly changed the spatial pattern, economic forms and social relations in rural areas, and inevitably brought about a series of challenges to the sustainable rural development. To cope with the changes and outcomes, it is urgent for the local participants to take actions timely to restructure rural space, economy and society through integrating elements, recombining structure and optimizing function in rural territory system (Yang et al. 2016), and ultimately build a new platform to promote the implementation of new-type urbanization, agricultural modernization and other national strategies concerning integrated urban-rural development (Fig. 5.1).

5.3.3 Connotations of Rural Restructuring

There are two different ways of understanding the connotations of rural restructuring in academic circle: one is regarding it as an evolutionary process of rural territory, focusing on analyzing the various objective features shown in the process; another is taking it as a strategic means of rural development, emphasizing the human intervention to effectively promote the positive evolution of rural areas in a predetermined way (Feng 2012). Due to the blurring definition, together with the fact that rural restructuring involves various contents, the term of "rural restructuring" has been abused in the available literature, which resulted in related research lacking of hierarchy and rationality, and hampered the further study on rural restructuring. In essence, rural restructuring is a process of reshaping socio-economic morphology and spatial pattern in rural territory in respond to the changes of elements both in kernel system and external system of rural development, by optimally allocating and efficiently managing the material and non-material elements in the two systems. It aims at ultimately optimizing the structure and promoting the function within rural territorial system as well as realizing the coordination of structure and complementation of function between urban and rural territorial system (Long et al. 2016).

Generally, the development and evolution of rural territory occurs in an interaction framework of the kernel system and external system at any time. The intertwined effects of internal and external development elements and the interaction of the

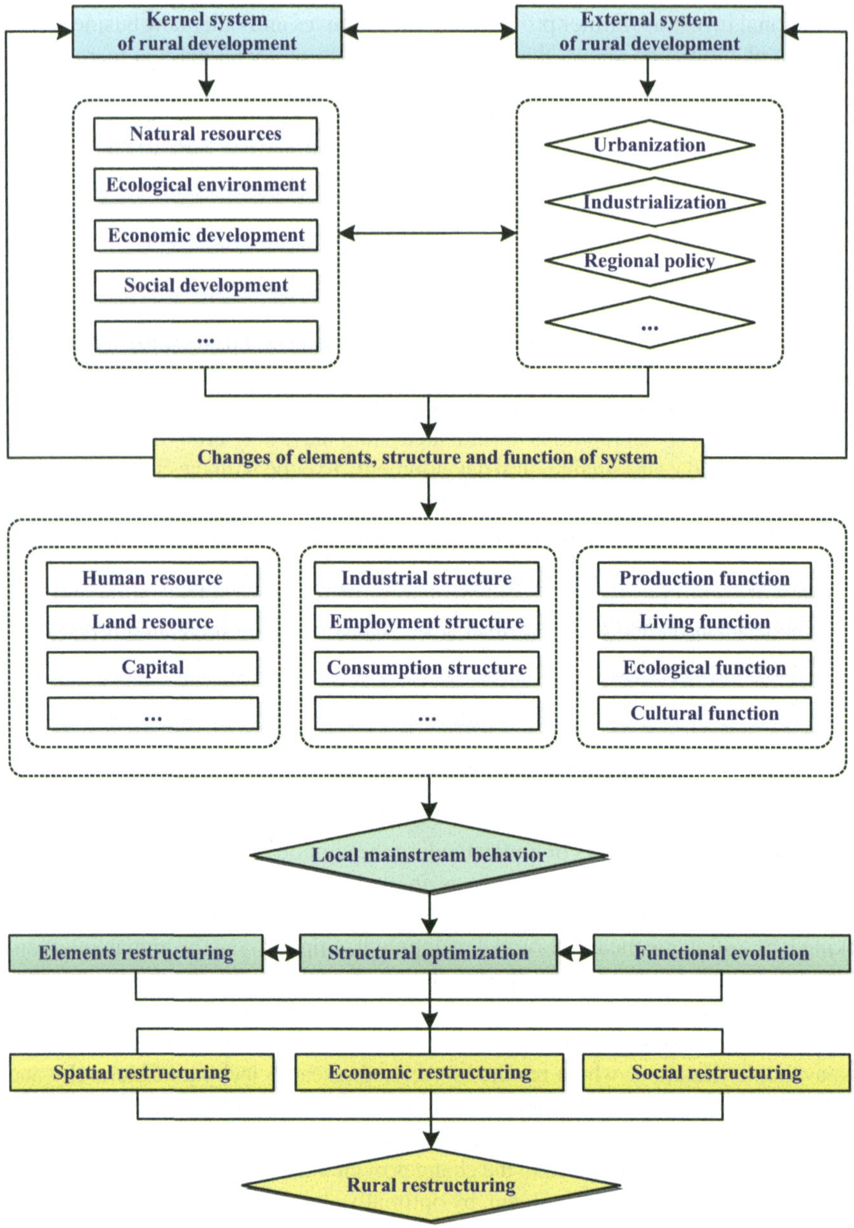

Fig. 5.1 A theoretical framework of rural restructuring based on "elements-structure-function" (Long et al. 2016)

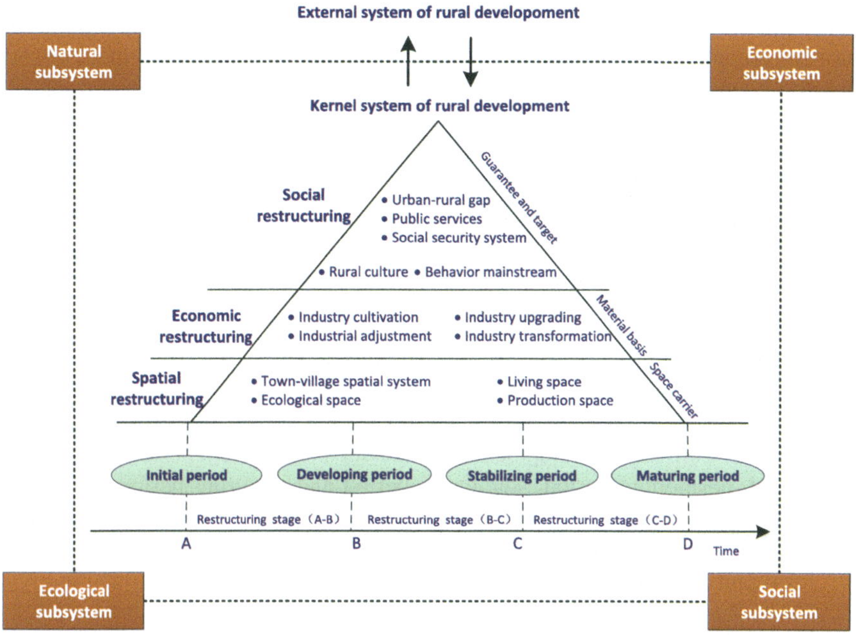

Fig. 5.2 The concept and connotations of rural restructuring (Tu and Long 2017)

subsystems on the rural territory make rural restructuring with the characteristics of comprehensive multidimensionality and temporal evolution (Fig. 5.2).

(1) Comprehensive multidimensionality. Usually, the socio-economic changes are the main storyline of the development of rural territory, and the land use is a mirror of the human socio-economic development. The socio-economic evolution in the rural areas will inevitably change the spatial carrier-land, including its utilization and allocation pattern, and cause rural spatial restructuring (Long 2014). Therefore, the contents of rural restructuring include spatial restructuring, economic restructuring and social restructuring, three mutually reinforcing and restrictive dimensions (Tu and Long 2017).

Initially, as the core of the economic restructuring, industry cultivation exerts significant impacts on broadening the channels of farmers' employment, achieving the non-agricultural transfer of surplus labor force, enhancing the value of land use and releasing the land consolidation potential, which can also lay down a material basis for the rural spatial and social restructuring with the enhancement of rural self-development ability (Tu et al. 2015). Additionally, cultivating the social behavioral mainstream of rural development, improving the public service and social security system, inheriting local culture, and narrowing the gap between urban and rural areas not only constitute the important parts of social restructuring, but also become the ultimate goal of rural restructuring. Among the numerous factors affecting rural restructuring, the perfect public service and social security system ensure the effective organization of spatial and economic restructuring. As the most dynamic factor in rural restructuring, social behavioral mainstream can provide intellectual support for rural economic restructuring and push forward rural spatial restructuring via playing a role of organization, coordination and demonstration in the process of restructuring (Tu et al. 2015). Last but not the least, rural space, as the carrier of socio-economic activities, the restructuring of which is of particular significance to provide space for developing scale farm management and non-agricultural industries as well as improving the living conditions and public service facilities via optimizing the town-village hierarchy system and integrating production, living and ecological spaces.

(2) Temporal evolution. The evolution of rural territorial system is a complex non-linear process, and rural restructuring in certain region may appear fluctuations in a certain period of time. Usually, a complete rural restructuring process usually consists of different development stages divided by different periods, i.e., initial period, developing period, stabilizing period and maturing period (Fig. 5.2). However, with the changes of internal elements and external demands in rural areas, it is possible for rural territory to start a new round of evolution and restructuring process.

As a result of the comprehensive influences of natural resources endowments, economic basis, technical level and other factors, the process of rural restructuring takes on regional differences in terms of experiencing stage and its stage characteristics. In some rural areas, the initial period may last a long period failing to enter the next stage of rural restructuring due to lack of driving force of restructuring, or the restructuring process may appear the end just after the initial period. On the contrary, it is possible for parts of the rural areas not to go through the initial period, and directly jump into the developing period, which presents the characteristics of cross-stage in the process of restructuring. Rural restructuring in different stages has its typical characteristics and mechanism. It is necessary to take human intervention to boost the whole process of restructuring through grasping the characteristics of different stages and the temporal evolution rules of rural restructuring.

There is a close relationship among rural evolution, rural restructuring and rural development transformation. Rural evolution refers to the statuses both positive promotion and negative degradation in a certain rural territory, such as slow development, leap progress, transient recession, renaissance and other situation (Tu and Long 2017). While rural restructuring is a positive evolution process, which mainly involves two types of cases from the angle of the statuses before and after restructuring: one is the qualitative transformation from recession to revival through effective human intervention for under-developed rural areas; another is a process of upgrading development by taking advantage of the external environment and integrating the internal elements for developed rural areas. Besides, the connotations of rural restructuring are similar to rural development transformation, but there are some differences. Compared with rural development transformation, rural restructuring emphasizes the human intervention and regulation such as integrating critical development elements and optimizing spatial structure with the purpose of changing the rural territorial system from a non-benign state to a benign state, or achieving function upgrading and enhancing the quality of development. In other words, rural restructuring is a process of realizing rural development transformation, and the latter is the result of the former.

5.4 Approaches

5.4.1 Spatial Restructuring

The scattered spatial distribution of Chinese traditional rural settlements and the phenomenon of idle rural housing land at present have affected the flowing process of production elements, and increased the cost of economic development considerably. Moreover, owing to lacking of scale benefit and output efficiency for public resources investment, the allocation of educational, medical and public infrastructures have been insufficient in some parts of rural China, which has exerted the far-reaching influences to the production space, living space and ecological space in rural China. Rural spatial restructuring is a process of achieving optimal adjustment and even fundamental change of rural territorial space by optimizing town-village spatial system and restructuring the rural production, living and ecological spaces.

(1) Establishing town-village spatial system at county level. Rural spatial restructuring initially relies on constructing the spatial agglomeration axis and the structure network of town-village planning taking into account the overall rural settlements scale, traffic network, regional cooperation, industrial development, employment characteristics, farming radius, the services scope of public facilities and other factors, which contribute to forming a three-level rural settlements system composed of central town, major town and central village for the purpose of strengthening the effective convergence and mutual support of the urban-rural territory in the aspects of space and function. Moreover, considering the node role of central town and major town in the energy transmission chain of urban-rural territory system, there should be a greater concern to accelerate the restructuring of central town and major town, by intensifying the cultivation of characteristic industry, improving public service facilities as well as enhancing comprehensive service function, in order to strengthen economic agglomeration function and exert the radiation and driving roles in the development of rural hinterland.

(2) Restructuring the production, living and ecological spaces at village level. Taking the spatial distribution characteristics of scattered, idle and disordered rural settlements into account, it is necessary to probe the diversified regional models of pushing forward rural settlements spatial restructuring, such as agglomeration model through village relocation and combination, intensive model through village inner integration and preservation model for characteristic village, according to natural environment, socio-economic development, cultural customs and

other features of territorial differences. After that, more attentions are needed to be paid to strengthen rural infrastructure construction through the scientific assessment on the allocation suitability of public facilities and infrastructures in order to optimize a livable rural space. In view of the status quo of the fragmentation of cultivated land operation and the disorderly distribution of township enterprises, promoting the construction of large-scale high standard basic farmland as well as strengthening the centralized layout of non-agricultural industry is an effective way to restructure intensive and efficient rural production space, which have been favored to provide spaces for the development of modern agriculture and the cultivation of new rural economic forms.

Besides, it is significant to restructure the environmental friendly and harmonious rural ecological space by gradually implementing pollution-free agricultural production, establishing ecological interception system, strengthening the comprehensive treatment of pollutants, improving ecosystem corridor and other means, so as to provide healthy and beautiful rural ecological space for the rural industrial development and rural residents.

5.4.2 Economic Restructuring

Owing to its close correlation with absorbing rural employment, increasing farmers' income, improving land use efficiency and building rural infrastructures etc., the prosperity of rural economy usually plays a leading role in realizing rural spatial and social restructuring. Therefore, remodeling the new impetus of rural economic development, with industrial cultivation as the core, is the fundamental issue to be solved in rural restructuring. While rural economic reshaping rests on opening up the flow channels of urban-rural development elements, so as to make internal system of rural territory full of vitality and generate output value (unique rural culture, ecological environment amenity) to external system (urban areas). Under the background of the changes of intrinsic development elements, as well as the emergence of new technologies and new economic modes, it is an indispensable way to reshape the rural economy by making full use of the new elements of economic development, innovating rural traditional industries and cultivating new rural economic forms.

(1) Innovating rural traditional industries. Chinese traditional agriculture has the features of decentralized management, inefficient production and low organization level, which led to lack of competitiveness with the advance of international allocation of the agricultural products in the context of globalization. Consequently, it is calling for urgently innovating rural traditional industries through optimizing agricultural industry structure (the proportion of the planting, animal husbandry, forestry, fishery and other production departments), product structure (e.g., the ratio of grain, oil, vegetables, fruits and other products) and variety and quality structure (the proportion of different variety type, e.g., durum wheat and ordinary wheat), as well as accelerating the construction of

agricultural product market information system, quality standard system and quarantine system, so as to facilitate the establishment of modern agricultural industry system. Moreover, it is essential to strengthen the organic combination of production, post-processing, market circulation and brand cultivation, so as to realize the specialization, standardization, intensive and large-scale production by promoting the ownership registration and certification issuing of rural contracted farmland, actively developing various forms of moderate scale management, supporting leading enterprises and characteristic industry bases, and establishing the benefit sharing mechanism that connects leading enterprises with farmer cooperation and peasants. At the same time, high attentions should be paid to protect the water and soil resources, which are the foundation of agricultural production, and promote the key technology such as new varieties of breeding, cultivation, soil conservation, water-saving irrigation, disaster reduction and prevention etc., so as to form a modern science and technology support system guaranteeing the safety of national grain production and the quality of agricultural products.

(2) Actively fostering new forms of rural industry. Owing to its high vegetation coverage, low-density population distribution as well as the main preservation of traditional and regional culture, cultural and ecological functions should be emphasized in the process of industrial reshaping in rural areas (Tu et al. 2015). Currently, to cope with the demands of personalized and diversified consumption market, exporting rural cultural and production elements with scarce value to the urban, is of great importance to activating the rural economic system. Taking the differences between urban and rural as well as the multi-functional value of rural areas into account, it is suggested that new forms of rural industry like providing services for retired and old people, preserving people's health, as well as eco-tourism industry should be encouraged strongly in order to pour vitality and power into rural economic revitalization.

5.4.3 Social Restructuring

Accompanied with the advance of rapid urbanization as well as the implementation of village relocation and combination, China's traditional village community characterized by agricultural civilization, acquaintance society and clan relations have been disassembled, and rural social development is facing the decline of autonomous organization, the weakening of rural management mainstream and the disappearance of cultural memory symbols. Especially along with the cancellation of the agricultural tax nationwide in 2006, the public affairs service function of rural grassroots administrative organization taking the village committee and the villagers' congress system as the mainstream, has been further weakened. Therefore, it is crucial to restructure rural society by means of establishing the rural organization system, cultivating new-type rural management mainstream, and protecting the characteristic rural cultural landscape.

(1) Improving the rural organization system and cultivating new-type rural development mainstream. To some extent, the rural rejuvenation depends on giving full play to the autonomy function of grassroots governance mainstream through establishing diversified rural organization system, such as village committee, economic industry organizations, social intermediary organizations, public service organizations etc. In particular, it should pay more attention to innovating the village cadres' selection system by encouraging local elite to govern the village, which would not only contribute to improving the management ability of the rural autonomous organization, but also help to promoting rural economic development. At the same time, the perfecting of the rural governance organization also relies on the establishment of the balance mechanism of rural democracy at the grassroots through building the villagers' congress system, so as to effectively play the participation and supervision role of farmers in public affairs and administrative management. Facing to the new problems of weakening traditional agricultural management mainstream as well as the new situation of farmers differentiation, it is urgent to speed up the system construction to develop a new-type professional farmer, so as to actively cultivate the new management mainstream engaged in modern agricultural production. Additionally, it is also important to strengthen the policy support for fostering various types of rural elite, e.g. developing normalized training mechanism by giving full play the role of entrepreneurial guidance of civil economic organizations to rural elites, as well as guaranteeing the legitimate rights of rural entrepreneurs by forming the equity incentive and other related benefit sharing mechanism, which are beneficial to play all kinds of rural talents' demonstration and leading effects on entrepreneurship development.
(2) Protecting the unique rural cultural landscape, and enhancing the rural cultural function. The formation of cultural landscape is influenced deeply by various factors like natural environment, production mode and social culture. As the carrier of regional culture as well as the fossil of social and cultural evolution in a certain period, characteristic rural settlements and folk customs exhibit the function of cultural heritage in rural areas. Therefore, protecting the rural material and non-material cultural landscape, particularly the traditional rural settlements, is an effective way to conserve the historical and cultural context, inherit rural memory and retain nostalgia (Liu 2015). As for the protection of rural cultural landscape, on the one hand, it is essential to strengthen the guidance of related law, policies, planning and management mechanism at national level, e.g. speeding up the formulation of the policies and documents regulating the implementation range of village relocation and combination as soon as possible, compiling scientific planning based on the principle of obeying different types and grades as well as constructing the financing mechanism and the normalized regulatory mechanism to protecting the characteristic villages.

On the other hand, it should highlight the significance of developing cultural and creative industries by taking advantage of characteristic rural culture on the basis of improving support ability of infrastructure construction and public service, which is an inevitable approach to seeking the economic driving force to protect characteristic culture, and also meet the emotional attachment and spiritual needs for traditional rural culture, local memories and rural nostalgia.

5.5 Mechanism

Based on the elements and structure of rural territorial system, the factors affecting rural restructuring can be divided into exogenous factors and endogenous factors. Exogenous factors can induce, catalyze, promote or hinder the process of rural restructuring via external environment, including policy, market and technical means. During the process of socio-economic development, industrialization, urbanization, informatization and agricultural modernization drive the restructuring of the elements of urban and rural territory, which constitutes the macro-background of rural restructuring; rural development policies and related institutions hinder or promote the process of rural restructuring by affecting the efficiency of resource allocation; market demand promotes the continuous evolution and improvement of rural territorial functions and economic morphology. Natural resources, location conditions, economic basis, social actors and cultural characteristics are endogenous factors affecting rural restructuring, which directly determine the choice of rural industrial development path, the level of development and the speed of restructuring. The social mainstream behavior runs through the whole process of rural restructuring, and plays a key role in the process of rural restructuring by utilizing exogenous factors and integrating internal factors. The interactions between multiple elements and subsystems inside and outside the system promote the process of rural restructuring through synthesizing and dispersing, pulling force and pushing force, transformation and feedback under the combined action of inducing mechanism, supporting mechanism, constraining/promoting mechanism, guiding mechanism, and engine mechanism (Long and Tu 2018).

(1) Inducing mechanism. The advancement of globalization, industrialization, urbanization and informatization, as well as the advance of technologies in communication, transportation and agricultural production during the process of socio-economic development, drive the flow and reallocation of production elements between urban and rural areas and different regions, resulting the change in many aspects such as industrial structure, employment structure, lifestyle and value orientation in rural areas, and exerting an important influence on rural socio-economic morphology and spatial pattern. All these actions constitute the inducing mechanism of rural restructuring. However, due to the diverse abilities of attracting, accepting, digesting and absorbing various production elements

in different regions, the impact direction and intensity of urbanization in different regions and different stages of development on rural territory in terms of agricultural development transformation, land resources utilization, social organization system construction, and inheritance of local culture are obviously different, thus exerting different influences on the process of rural restructuring.

(2) Supporting mechanism. The kernel system of rural territory composed of natural resources, location conditions, economic basis and cultural traits, forms the supporting mechanism that influences rural restructuring. Among them, mineral resources, land, climate, water resources and biological resources are the natural endowments and spatial carrier for the development of rural territorial system. Location conditions determine the frequency and efficiency of interactions of material flow, energy flow and information flow between rural territory system and urban as well as among different rural regions. Industrial structure, development level and other economic foundations, to some extent, determine the current level of economic development and the potential of future economic growth in rural territory through path dependence (Tu et al. 2015); The cultural traits shaped over a long history, which are reflecting the values, ways of thinking, open consciousness, innovation consciousness and competition consciousness of different cultural groups, have exerted a subtle influence on rural restructuring. Among them, the cultural traits of unity, enterprising, pragmatic, inclusive and innovative play a positive role in the process of regional economic development and rural restructuring, while that of conservatism, moderation, complacency and lack of contractual spirit have a negative effect on rural restructuring.

(3) Constraining/promoting mechanism. Rural development policies and related institutional arrangements influence rural restructuring by affecting the efficiency of factor flow and resource allocation. Policies and institutions adapting to socio-economic development have promoted the practice of rural restructuring, such as the new-type urbanization, building new countryside, beautiful village construction and other related strategic measures implemented at the national level in recent years, which have improved the living environment of rural residents to a certain extent and contributed to the promotion of rural socio-economic and spatial restructuring. While the dual household registration system, land and financial system in the urban and the rural seriously constrain the optimal allocation of urban and rural development elements, and hinder the process of rural restructuring. Meanwhile, the relevant policies have been revised and improved based on the practice of rural development, and the practice of rural restructuring has also reacted on the formulation of the national macro-economic development strategy and the development of urban-rural interrelations.

(4) Guiding mechanism. Market is the invisible hand to adjust the allocation of resources and economic development. Market demand has changed the traditional resources value via the interaction of supply and demand, price and competition, which promotes the elements flow of rural primary agricultural products, industrial raw materials, ecological environment and local culture,

and guides the evolution of rural industry, the choice of economic development path and the evolution of rural territorial functions.

(5) Engine mechanism. The local behavior mainstream consisting of local governments, rural elites, rural autonomous organizations and ordinary farmers runs through the whole process of rural restructuring and constitutes the engine mechanism of rural restructuring. Among them, the local governments, as the mainstream of management and service, play a fundamental role in the planning, organizing and allocating resources during rural socio-economic development. The rural elites who have vision and courage for scientific development positioning and comprehensive leading ability are the key to the rural restructuring practice. While the credibility, combat effectiveness and cohesion of rural autonomous organizations play an important role in stimulating the enthusiasm of ordinary farmers to participate in the practice of rural restructuring.

The essence of rural restructuring is the process of exerting influences on socio-economic morphology and spatial pattern through integrating and allocating the elements of population, land and industry in rural territory by the local behavior mainstream (engine mechanism), based on the evaluation of local resources endowments, location conditions, economic basis, cultural traits and other development conditions (supporting mechanism), so as to adapt to the changes of rural development elements (inducing mechanism) under the background of market demand of rural-urban development elements (guiding mechanism) and relevant rural development policies (constraining/promoting mechanism) (Fig. 5.3).

Rural restructuring in practice is affected by a variety of factors inside and outside the system, and the influence of each factor on rural restructuring is dynamic in different restructuring stages and modes. Generally, economic basis and market demand play a regular role in the process of rural restructuring, relevant policies and institutional arrangements always play a macro-basic role, natural environment and location conditions have a decisive impact on the choice of rural restructuring paths and modes, especially in the initial period of rural restructuring, and social factors such as local behavior mainstream and cultural traits are becoming increasingly active. With the evolution of the restructuring stages, the development and transformation of rural territory are more dependent on the scientifically judging and dynamically steering the macro-development environment by behavior mainstream, which are helpful for quickly adjusting and optimizing development mode in time. The comprehensive interactions of development elements inside and outside the rural territory system promote rural restructuring. However, the results of rural restructuring will also be part of the rural "physical-ecological-economic-social" system, which react upon and drive the changes of rural development elements, while the changed elements will promote the rural further restructuring in a new way. Meanwhile, the characteristics of socio-economic morphology and spatial pattern of rural restructuring determine the advantages and constraints of industrialization, modern-

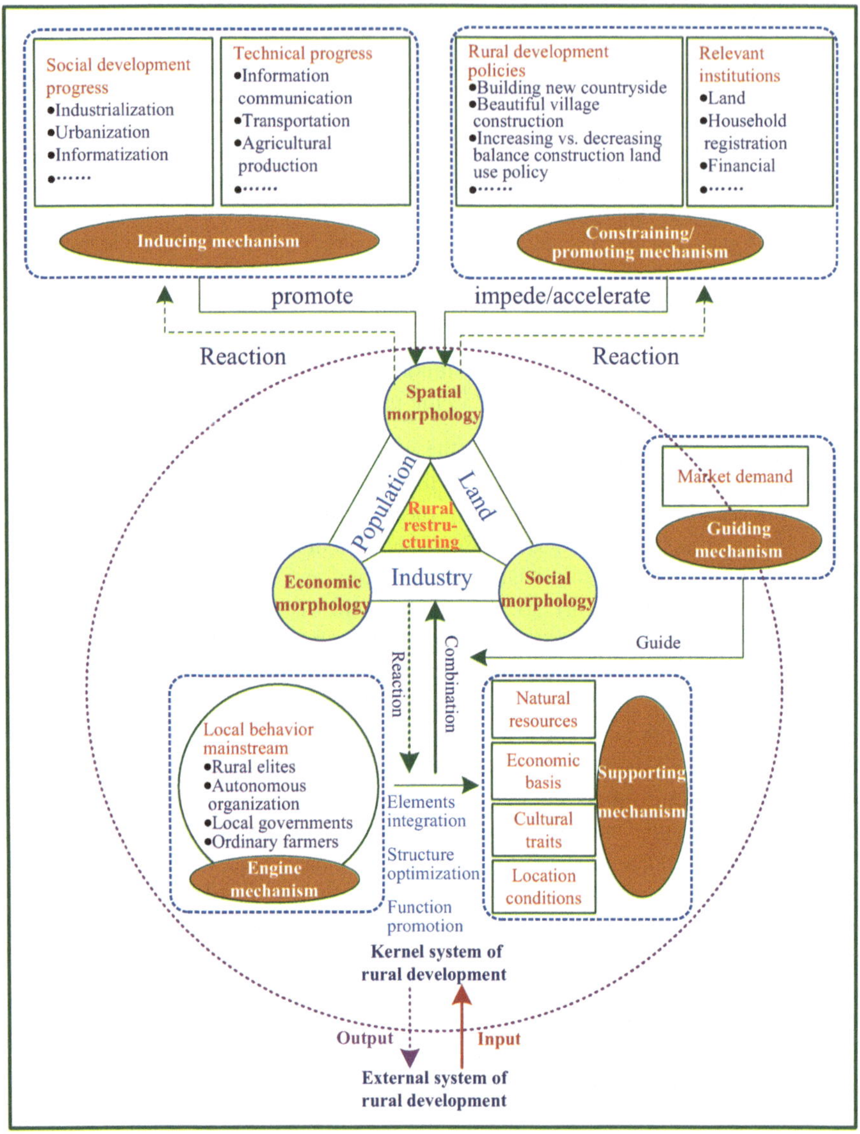

Fig. 5.3 The action mechanism of pushing forward rural restructuring (Adapted from Long and Tu 2018)

ization and informatization, and then affect the process of regional social-economic development. Therefore, rural restructuring is a process of progressively evolution in the interactive feedback of "action-reaction" between the kernel system and the external of rural development.

5.6 Research of Rural Restructuring in China

Currently, rural restructuring in China has increasingly attracted the interest of numerous scholars in various disciplines such as geography, sociology, architecture, management, who have carried out systematic research from different perspectives and time-space scales, and already obtained a large number of enlightening research outcomes by comprehensively using various methods like case analysis and theoretical research. The main topics range from the spatial evolution of rural settlements (Zhang 1999; Long et al. 2009; Qiao 2011; Lei 2009; Fang et al. 2016), functional differentiation and evolution of rural territory, the pattern and process of rural development and transformation (Long 2012; Liu 2007; Yang et al. 2016), the layout and optimization of rural settlements (Long et al. 2012), the mechanism and mode of rural spatial restructuring in typical regions (Li et al. 2019; Li 2016; Feng 2012; Liu 2011), the social order change and the remodeling of rural governance system (Yu 2014), to the paths of rural restructuring (Wang et al. 2016) etc. *Journal of Rural Studies* published a special issue on 'Rural Restructuring in China' (Long and Liu 2016), which highlighted the latest research achievements focusing on the rural restructuring practice in contemporary China, and contains 25 academic papers related to the allocation and management of key resources in rural China, the spatial distribution and optimization of rural settlements, the spatial and temporal changes of urban-rural land use, the transferring of rural labor forces, the allocation of rural educational resources, the declining of traditional rural settlements, dynamic change and spatial transformation in suburban village and other topics (Long and Liu 2016).

In this special issue, firstly, aiming at the increasing trend emphasizing the role of regional resources in formulating rural development policy and restructuring rural areas, Long et al. analyze the rural restructuring in China affected by the allocation and management of critical resources including human resource, land resource and capital, by establishing a theoretical framework of "elements-structure-function" of rural territorial system.

Rural spatial restructuring is an optimization and adjustment even utterly change process of rural production, living and ecological space accompanied by the reshaping of socio-economic structure in rural areas owing to an integrated result of endogenous development need and exogenous driving forces of rural development, under the pressure of rapid industrialization and urbanization, and it is also an integrated approach aiming to optimizing urban–rural space organization and promoting coordinated urban–rural development. Liu et al. investigate the spatial patterns, temporal processes and driving mechanisms of urban–rural land use transformation based on the case study of Xuanhua district of Hebei province, and argue that industrialization and urbanization are the basic driving factors of urban–rural land use transformation, and local, regional and national socio-economic transformation deeply and widely affect its developments. Using the methods of nearest neighbor distance analysis and geographical detector analysis, Yang et al. conclude the spatial distribution of rural settlements in China with significant regional differences, i.e., denser in the southeastern regions compared to the northwestern regions, high density and mainly random and disperse in plains, and low density and mainly clustered in regions such as cold alpine areas and desert fringes. Yao and Xie examine rural spatial restructuring in an ecologically fragile mountainous area of southern China at a village-township level based on a set of methods concerning landscape security pattern construction as well as a model of rural residential land expansion using an ant colony optimization algorithm and scenario simulation, which were proved to be a useful technical toolkit supporting rural spatial restructuring in the ecologically fragile mountainous areas of southern China.

Land consolidation is a spatial problem-solving land management instrument and is also regarded as an indispensable way of spatial restructuring in rural China (Long 2014). Liu et al. investigate the drivers of land coverage under the program of community-based agricultural land consolidation aiming at restructuring the agricultural sector by pooling fragmented land and leasing it as consolidated plots or employing labors to farm the land, with a particular interest in the role of local elites and village characteristics. Fang et al. analyzes the means and ends of the practices and villagers' attitudes and behaviors in response to rural construction land consolidation based on a comparison of a failed case and a successful case in Changchun City, northeast China, and suggest that rural construction land consolidation should follow a trial and error approach and villagers must have real opportunities to take part in the decision-making that will influence their futures. Aiming at the "land bill system" concerning rural construction land consolidation innovated by Chongqing governments, Yep and Forrest provides an account of the genesis, implementation and evaluation of this new scheme, and points out that this institutional innovation seems to provide a win-win situation for all stakeholders, but peasants may suffer in the long run and the impact on rural governance may also be damaging. Wang et al. analyze the elements, mechanisms and operation of a peasant household symbiotic system in the process of rural settlement restructuring based on the case study of Fengsi village in Chongqing, and develop the spatial pattern by establishing a symbiotic relationship among which resources, benefits and responsibility for environmental construction are shared. Lo et al. examines poverty alleviation resettlement, one of China's key poverty reduction initiatives, from the perspective of spatial restructuring through a household survey conducted in Shanxi and Shaanxi provinces, and describe the significant challenges remain to be addressed for short-distance resettlement and long-distance resettlement, including establishing two-way communication between villagers and the government about resettlement plans and providing better financial support for the resettlers, safeguards for the livelihoods of non-movers, and post-resettlement support programmes to help resettlers adjust to their new environments.

Since industrial development imposes great effects on widening farmers' employment channel, transforming land use mode, enhancing rural economic strength and self-developing ability, economic restructuring based on industrial reshaping is becoming increasingly important in rural restructuring. The improvement of rural economic strength may ensure the implementation of rural spatial restructuring and social restructuring. Hence, restructuring the rural economic engine with the core of industrial cultivation is one of the fundamental issues of rural restructuring. Based on Hirschman's 'exit-voice' theory, Li et al. investigate the way in which local stakeholders respond to processes of rural decline based on the case studies of Xiaoguan village in China and Sweden, and reveal the effectiveness of bottom-up revitalization initiatives in combating rural decline, by showing how local stakeholders' strong "voices" in these places held people and groups together, encouraging them to work together with shared values and attitude. Aiming at the booming of Taobao villages which specialize in e-commerce and are changing the pattern of the rural economy as well as the social foundation of rural daily life throughout China, Lin et al. explore how everyday life in rural China has changed in response to the development of the network economy. The case study of Junpu village in Guangdong province shows that the development of e-commerce and network technology has changed the common values held towards rural life, the affection felt for the village, and the pace of life of local people.

Under the background of rural transformation development, vast amounts of land have been abandoned due to labor loss, especially in mountainous areas. Zhang et al. examine how two main farmland parcel attributes, farmland-to-housing distance and land quality, by affecting farming costs and benefits, and thereby profits, determine farmland abandonment, by establishing a semi-empirical crop profit model based on output benefits and input costs derived from household survey data from mountainous Wulong county, Chongqing. Using a qualitative comparative analysis approach, Qin and Liao conduct a systematic review and meta-analysis of recent case studies of labor out-migration and agricultural change in rural China, and reveal a general contextual pattern of migration impacts on agriculture. Qian et al. explore the key impacts of migration on agricultural restructuring in Jiangxi province by considering household farm activities choice and crop acreage adjustment behavior. The results show that, amid China's unique institutional background, the farmers left behind will be more likely to turn to less labor-intensive subsistence grain production instead of more capital-intensive livestock cultivation, or to even abandon their farmland. Zhang et al. quantitatively analyzes the impact of labor and capital productivity on wheat production in Shandong and Henan provinces, two leading provinces for grain production also with the highest population in China, and the consequent effects on exported and imported wheat volumes in other provinces, using the Enormous Regional Model.

With the advancing of globalization, urbanization and industrialization, the foundation maintaining Chinese traditional rural society consists of agricultural civilization, village culture and family clanship has been broken, rural human resource tends to be low level, and the mainstream of rural development has been weakened continuously, and the unfairness and structural imbalance of public goods and capital allocation owing to rural-urban dual system cause rural development resources further occupied by the urban, and rural self-development ability further weakened. There is an urgent need to reconstruct rural social structure emphasizing particularly on the organization system and rural management mainstream. Kan treats the Chinese village as a historical entity emerging from socialist collectivization and examines how the socialist institutions of collective property and redistributive mechanisms have continued to persist in the reform-era village. Rao and Ye explores the historical transformation of the rural education system and the mapping of rural schools in ancient, modern and contemporary China to explain the problems of the Rural School Mapping Adjustment policy. He et al. adopts a method that combines economics and geography, and conducts dynamic monitoring and evaluations of the efficiency of rural public goods investments in specific mountainous areas based on long-term data from tracking surveys. The resilience and adaptability of rural communities to environmental uncertainties events is one of the major areas that present key challenges for rural areas in the early 21st century (Woods 2012). Jiang et al. analyze the adaptive capacity incorporating a variety of geological disaster susceptibilities in prefecture-level cities in the mountainous rural Yunnan province by using socio-economic data and geological disaster data. Aiming at the situation that traditional dwellings in rural areas are gradually disappearing as part of the residents' long-term daily practices, Wang et al. discusses the changes in the traditional cave dwelling called "sunken courtyard" (known as "di keng yuan" in Mandarin) in Shanxian county, Henan province, and explores the reasons behind these changes, based on the literature and preliminary investigation as well as a formal interview. Zacharias and Lei examine the community relations among villagers, artists and students in Xiaozhou, a peri-urban village in south China is regenerating by developing its own productive economy, in the context of threats to the village's long-term survival, based on structured and semi-structured questionnaires and in-depth interviews with artists. Wu et al. demonstrates the multiple dualities in the social-spatial dimension of villages in semi-urbanized suburbs of China's large cities have emerged during China's rapid urbanization. Chen et al. use multi-parameter income distribution function and decoupling theory to investigate rural poverty in China, focusing on how demographic structures and economic development have been restructuring the incidence of rural poverty.

5.7 Conclusions

In this chapter, some basic questions concerning rural restructuring have been elaborated: what is rural restructuring? why and how to carry out rural restructuring? Rural restructuring is a process of reshaping socio-economic morphology and spatial pattern in rural territory in respond to the changes of elements both in kernel system and external system of rural development, by optimally allocating and efficiently managing the material and non-material elements in the two systems. It aims at ultimately optimizing the structure and promoting the function within rural territorial system as well as realizing the coordination of structure and complementation of function between urban and rural territorial systems. With the continual boost of rapid urbanization, the evolution of "elements-structure-function" of rural territory has broken utterly the traditional spatial structure and socio-economic morphology in rural China, which triggers rural restructuring and brings about a series of challenges to sustainable rural development. It is urgent for the local participants to take actions timely to restructure rural space, economy and society through integrating elements, recombining structure and optimizing function in rural territory system. Accordingly, the approaches of rural restructuring include the three aspects of spatial restructuring, economic restructuring and social restructuring. The interactions between multiple elements and subsystems inside and outside the rural territory promote the process of rural restructuring through synthesizing and dispersing, pulling force and pushing force, transformation and feedback under the combined action of inducing mechanism, supporting mechanism, constraining/promoting mechanism, guiding mechanism and engine mechanism.

References

Bojnec S (2012) Agricultural and rural capital markets in Turkey, Croatia and the FYR of Macedonia. Agr Econ-Czech 58(11):533–541

Brent ZW (2015) Territorial restructuring and resistance in Argentina. J Peasant Stud 42(3–4):671–694

Djanibekov N, Assche VK, Bobojonov I et al (2012) Farm restructuring and land consolidation in Uzbekistan: new farms with old barriers. Europe-Asia Stud 64(6):1101–1126

Fang C, Yu D (2016) China's new urbanization. Science Press & Spriner Press, Beijing

Fang Y, Shi K, Niu C (2016) A comparison of the means and ends of rural construction land consolidation: case studies of villagers' attitudes and behaviours in Changchun City, Jilin province, China. J Rural Stud 47:459–473

Feng J (2012) 乡村重构：模式与创新 (Rural restructuring: models and innovations). The Commercial Press, Beijing

Hoggart K, Paniagua A (2001) What rural restructuring? J Rural Stud 17(1):41–62

Irwin EG, Bell KP, Bockstael NE et al (2009) The economics of urban-rural space. Annu Rev Resour Econ 1:435–459

Lei Z (2009) 整合与重构——关中乡村聚落转型研究 (Combination and restruction: study on transformation of rural settlements in the Guanzhong Area). Dongnan University Press, Nanjing

Li Y (2016) 村域发展机理与资源环境效应——不同类型地区的案例研究 (Towards the mechanism of village development and its effects on resources and environment: case studies in different types of regions). Science Press, Beijing

Li Y, Wang J, Liu Y et al (2014) Problem regions and regional problems of socioeconomic development in China: a perspective from the coordinated development of industrialization, informatization, urbanization and agricultural modernization. J Geogr Sci 24(6):1115–1130

Li Y, Fan P, Liu Y (2019) What makes better village development in traditional agricultural areas of China? Evidence from long-term observation of typical villages. Habitat Int 83:111–124

Liu Y (2007) 中国东部沿海地区乡村转型发展与新农村建设 (Rural transformation development and new countryside construction in eastern coastal area of China). Acta Geogr Sinica 62(6):563–570

Liu Y (2011) 中国新农村建设地理论 (Geography of new countryside construction in China). Science Press, Beijing

Liu P (2015) 新型城镇化建设中"留住乡愁"的理论与实践探索 (The theory and practice exploration of "remembering nostalgia" in the construction of new urbanization). Geogr Res 34(7):1205–1212

Long H (2012) 中国乡村转型发展与土地利用 (Land use and rural transformation development in China). Science Press, Beijing

Long H (2014) Land consolidation: an indispensable way of spatial restructuring in rural China. J Geogr Sci 24(2):211–225

Long H, Liu Y (eds) (2016) Rural restructuring in China. J Rural Stud 47(Part B):387–676
Long H, Tu S (2018) 乡村重构的理论认知 (Theoretical thinking of rural restructuring). Prog Geogr 37(5):581–590
Long H, Woods M (2011) Rural restructuring under globalization in eastern coastal China: what can be learned from Wales? J Rural Community Dev 6(1):70–94
Long H, Zou J, Liu Y (2009) Differentiation of rural development driven by industrialization and urbanization in eastern coastal China. Habitat Int 33(4):454–462
Long H, Li Y, Liu Y et al (2012) Accelerated restructuring in rural China fueled by "increasing vs. decreasing balance" land-use policy for dealingwith hollowed villages. Land Use Pol 29(1):11–22
Long H, Liu Y, Li X et al (2010) Building new countryside in China: a geographical perspective. Land Use Pol 27(2):457–470
Long H, Zou J, Pykett J et al (2011) Analysis of rural transformation development in China since the turn of the new millennium. Appl Geogr 31:1094–1105
Long H, Tu S, Ge D et al (2016) The allocation and management of critical resources in rural China under restructuring: Problems and prospects. J Rural Stud 47:392–412
Manthorpe J, Livsey L (2009) European challenges in delivering social services in rural regions: a scoping review. Eur J Soc Work 12(1):5–24
Morgan SL, Marsden T, Miele M et al (2010) Agricultural multifunctionality and farmers' entrepreneurial skills: a study of Tuscan and Welsh farmers. J Rural Stud 26(2):116–129
Qiao J (2011) 中国乡村社区空间论 (Rural community space of China). Science Press, Beijing
Rogge E, Dessein J, Verhoeve A (2013) The organisation of complexity: a set of five components to organise the social interface of rural policy making. Land Use Pol 35:329–340
Tu S, Long H (2017) Rural restructuring in China: theory, approaches and research prospect. J Geogr Sci 27(10):1169–1184
Tu S, Long H, Li T et al (2015) 中国村镇建设和农村发展的机理与模式研究(Study on the mechanism and models of villages and towns construction and rural development in China). Econ Geogr 35(12):149–156
Wang C, Huang B, Deng C et al (2016) Rural settlement restructuring based on analysis of the peasant household symbiotic system at village level: a Case study of Fengsi Village in Chongqing, China. J Rural Stud 47:485–495
Wilson OJ (1995) Rural restructuring and agriculture-rural economy linkages: a New Zealand study. J Rural Stud 11(4):417–431
Woods M (2005) Rural geography: Processes, responses and experiences in rural restructuring. Sage, London
Woods M (2012) New directions in rural studies? J Rural Stud 28(1):1–4
Woods M (2013) Regions engaging globalization: a typology of regional responses in rural Europe. J Rural Community Dev 8(3):113–126
Wu C (2001) 中国农业与农村经济可持续发展问题: 不同类型地区实证研究 (Problems of the sustainable development of agriculture and rural economy in China: case studies of agricultural area of different types). China Environmental Science Press, Beijing
Yang R, Xu Q, Long H (2016) Spatial distribution characteristics and optimized reconstruction analysis of China's rural settlements during the process of rapid urbanization. J Rural Stud 47:413–424
Yu F (2014) 社会转型与空间重构 (Social transformation and spatial restructuring). China Architecture & Building Press, Beijing
Zhang X (1999) 乡村空间系统及其演变研究: 以苏南为例 (Study on rural spatial system and its revolution: a case study of South Jiangsu region). Nanjing Normal University Press, Nanjing
Zhang F, Liu Y (2008) 中国区域农村发展动力机制及其发展模式 (Dynamic mechanism and models of regional rural development in China). Acta Geogr Sinica 63(2):115–122
Zhang Q, Wang H, Chen H et al (2006) 城乡统筹下的乡村重构 (Rural restructuring under urban-rural coordination development). China Architecture & Building Press, Beijing

Chapter 6
Analysis of Rural Transformation Development and Restructuring in China

Abstract This chapter analyzes the spatio-temporal characteristics of China's rural transformation development based on three dimensions of rural development level, rural transformation level, and urban-rural coordination level. Then, the degree of rurality in recent China at county level was evaluated using classical methods and successive national census data, so as to present a comparable picture of rurality and to improve our knowledge about current characteristics of China's rural and regional differences. Finally, the rural restructuring in China affected by the allocation and management of critical resources, including human resource, land resource and capital, was analyzed. As for the relationship between rural transformation development and rural restructuring, the author argues that rural restructuring is a process of realizing rural transformation development, which is the result of rural restructuring. Under rapid urban-rural transformation development, China's rural areas are facing a series of challenges, which are protruding in the following aspects: rural human resource allocation takes on the trend of low level and the mainstream of rural development is weakening; nonagriculturalization, non-grain preference and abandonment of farmland use together with the derelict and idle rural housing land result in low efficiency of rural land use; unfair urban-rural allocation of capital and its structural imbalance gradually weaken the self-development ability of rural China. Aiming at how to adapt to the challenges due to the changes of allocation of critical resources including human, land and capital, it is pivotal to restructure the rural development space, rural industry, and rural social organization and management mainstream.

6.1 Introduction

The relationship between rural transformation development and rural restructuring can be depicted as, rural restructuring is a process of realizing rural transformation development, which is the result of rural restructuring. Under rapid urban-rural transformation development, China's rural areas are undergoing considerable spatio-temporal changes, which increase the difficulty in our understanding regional development and thus brings about challenges for the making of feasible regional rural development policies and planning. Since the turn of the new millennium, the Chinese

H. Long, *Land Use Transitions and Rural Restructuring in China*,
https://doi.org/10.1007/978-981-15-4924-3_6

central government has focused significant attention on substantially improving rural residents' well-being and achieving the coordinated development of urban and rural areas. The core objective of rural restructuring and transformation development is to systemically establish a coupling pattern of various development elements including population, land and industry. And there has been an increasing trend emphasizing regional resources in formulating rural development policy and restructuring rural areas. Therefore, it is necessary to form a scientific cognition of rural transformation development and restructuring to promote the development of China's rural territory.

This chapter analyzes the rural transformation development based on three dimensions of the rural development level, the rural transformation level, and the urban-rural coordination level, and the spatio-temporal characteristics of China's rural transformation development from 2000 to 2008 were analyzed in relation to these three indicators. Then, the degree of rurality in recent China at county level was evaluated using classical methods and successive national census data, so as to present a comparable picture of rurality in China and to improve our knowledge about current characteristics of China's rural and regional differences, and thus provide valuable information for local planning and the innovation of rural development policy. Finally, this chapter analyzes the rural restructuring in China affected by the allocation and management of critical resources including human resource, land resource and capital.

6.2 Measuring Rural Transformation Development

6.2.1 Overview of Rural Development in China

In the late 20th and early 21st centuries, rural restructuring has been identified in Western Europe, North America, and Israel in the Middle East (Nelson 2001). At the same time, such transformational development has also taken place in the rural areas of developing countries, such as China, India, the Philippines, Zimbabwe, and Ecuador. This rapid and radical rural restructuring is often referred to as rural transformation development (RTD) (Liu 2007; Long 2012). In most developing countries, RTD is usually characterized by changes in agricultural intensity, crop selection patterns, farmland, land productivity and farm income, labor and technological productivity, and major improvements in rural housing and economic and social conditions resulting from industrialization and urbanization (Ali 2007; Cai and Smit 1994; Long et al. 2007a, b, 2009).

China is a nation with strong rural roots. Despite rampant urbanization, 40.42% of its population still lives in rural areas, and the national economy has been built on agricultural foundations since ancient times. However, China has transformed rapidly since Deng Xiaoping launched the economic reforms of 1978. The traditional centralized economy changed to a market-based economy, and the economy, which was once primarily agricultural, has become increasingly urban and industrial.

During this period, the RTD process and its results have been influenced by national macroeconomic development strategies. Since 1978, RTD in China has been linked to three macroeconomic development strategies: implementing a household responsibility system, developing township and village enterprises (TVEs), and building a new countryside. These strategies have resulted in tremendous changes in the rural population, lifestyles, employment structure, industrial structure, community organization, culture, and public transport accessibility (Goodman 2008; Long et al. 2010; Putterman 1997).

Since the economic reforms of 1978, agriculture and the countryside have contributed greatly to, and have made substantial sacrifices for, the development of industries and cities in China. This urban development has caused a series of problems that affect China's social and economic development (Cai and Smit 1994; Xu and Tan 2002; Long et al. 2010). The most notable issue is a shift in the economic balance and the widening of the prosperity gap between urban and rural areas. The income gap between rural and urban residents has increased from 2.57 to 1 in 1978 to 3.31:1 in 2008, and this trend has strengthened since the turn of the new millennium. On one hand, the income gap is reinforced by long-standing urban-rural inequalities. Farmers are also disadvantaged by the disparity between artificially low prices for agricultural products and the high cost of industrial goods. On the other hand, since the turn of the new millennium, China's economic and social development has entered an important new period. China's fast-growing economy and stronger position in the global community have enabled industries to support agriculture and cities to support the countryside. Under these circumstances, Chinese central government has implemented "urban-rural coordination development" (Chengxiang Xietiao Fazhan) since the turn of the new millennium, which has resulted in significant RTD.

Earlier, uncoordinated urban-rural development created problems with agricultural production, low farm incomes and the need for broader rural development, all of which have become major political issues. Since 2004, the Chinese central government has recognized the importance of rural development, and it devoted its Number One Policy Document to rural issues for each of the next seven years. These documents successively addressed the following: increasing farmers' incomes (2004); improving agricultural production capacity (2005); advancing the "building a new countryside" scheme (2006); developing modern agriculture (2007); resolving rural problems (2008); stabilizing agricultural prices and increasing farmers' incomes (2009); and strengthening agricultural and rural development (2010). Together, these policy actions contributed to the overarching goal of "building a new countryside", which targeted five major objectives (Long et al. 2010): advanced production; improved livelihoods; cultural and ethical progress; clean and tidy villages; and efficient management.

Rural development has traditionally been seen as a vague term that is difficult to specify, measure and evaluate. Today's rural economy and its social system are much more diverse, complex, sophisticated and global than those of the last century. Rural development is now seen as a multilevel, multiactor and multifaceted process that requires an understanding of the agricultural developmental model, the relationship between agriculture and society, the regional socio-economic structure and rural economic status, individual farm households and their behaviors, and local policies and institutions.

Issues related to measuring rural development have attracted the interest of a wide variety of researchers. During the 1970s and 1980s, an index of rurality for local government districts in England and Wales was developed to identify some of the differences between degrees of rurality. This index included such indicators as population, household amenities, occupational structure, commuting patterns, and the distance to urban centers, and was constructed using a range of statistics from the 1971 and 1981 censuses (Cloke 1977; Cloke and Edwards 1986). A similar rurality degree index (RDI) was recently established to distinguish degrees of rurality in eastern coastal China (Long et al. 2009). A corresponding index has also been developed to measure and explain both urban and rural development (Mann 2009).

In this chapter, related development indexes were established to examine rural transformation development in China and explore the future possibilities of achieving more coordinated and balanced urban-rural development.

6.2.2 Three Dimensions of Rural Transformation Development

A useful theoretical approach to rural development considers how local factors and external driving forces interact to determine the development of the rural economic system (Terluin 2003). This relationship between the local, rural system and the external system—in particular, the links between the rural and urban economies and cultures—are important strands in current rural development research (Murdoch 2000). For instance, rural agricultural development has been seen to play an important role in regional industrialization and urbanization by guaranteeing the supply of food and raw materials needed to sustain industrialization. This kind of role, in turn, leads to the transition and transformation of rural areas (McGee 2008).

RTD is, in essence, a term that captures changes in traditional rural industries, the employment consumption structure, and the social structure. These changes signify a transformation from previously isolated urban and rural economic structures toward more coordinated urban-rural development. Such transformation radically changes the urban-rural relationship and the relationship between agriculture and industry (Liu 2007). RTD assessment involves measuring three major components: the development of a distinctively rural economic system; the transformation of rural social, economic and consumption structures; and the improvement of the urban-rural relationship. Accordingly, this section establishes three evaluation dimensions to systematically diagnose the RTD in a given region: the rural development level (RDL), the rural transformation level (RTL), and the urban-rural coordination level (URCL).

The RDL reflects to both the rural socio-economic base level before RTD and the integrated effects of RTD on the rural system's social, economic, and cultural values.

The RTL reflects the degree and rate of change in rural socio-economic and consumption structures. The URCL reflects the relative allocation of regional resources to urban and rural systems and the integrated effects of RTD on socio-economic development at the regional level. The URCL can be used to judge the effectiveness of RTD; namely, well-ordered RTD can improve the URCL by promoting the optimal allocation of regional resources between urban and rural systems, while disordered RTD can intensify the relative deprivation of the rural system.

RTD is the result of rural socio-economic development at a certain stage, and the RTD of a given region needs to adapt to that regions RDL baseline. Generally, a low regional RDL means that the regional rural system has not accumulated enough developmental capital, agricultural technology, human capital, and rural production infrastructures and service facilities to effectively support developmental transformation. In this situation, if the RTD is pushed forward recklessly, the original endogenous rural development would be disturbed, and changes to the rural subsystem will not improve rural development capacity. As a result, urban and rural development will be uncoordinated, as the development of the rural system has been restricted. However, when the regional RTL is appropriate and the rural economic transformation mainly focuses on the agricultural infrastructure or production technology improvements, the regional RDL will improve significantly, and the regional URCL will also be positively impacted.

While some regional rural systems have experienced relatively rapid development and an improved RDL, the development of the rural system does not always keep pace with the existing development pattern. Under such conditions, RTD should be accelerated or emphasized to improve the rural system structure and ensure a more satisfactory speed for rural development. At the same time, some incentive policies (i.e., those concerning rural industrial development, employment support, and technological improvement) need to be introduced to optimize the rural system's socio-economic structure and promote coordinated urban-rural development.

With continuous socio-economic development, regional RDL is enhanced, thus promoting the transformation of the rural socio-economic structure, which will ultimately affect the progress of regional urban-rural coordination development. Accordingly, the initial RDL conditions can influence the consequent RTL and may ultimately change the urban-rural relationship and the regional development pattern (Fig. 6.1).

6.2.3 *Materials and Methods*

6.2.3.1 Data Source and Processing

Because regional RTD is composed of three dimensions (RDL, RTL, and URCL), indicator systems corresponding to each dimension were established to comprehensively measure them. The indicators for RDL measurement (see Table 6.1) reflect changes within the rural society, economy, culture, resources, and the environment

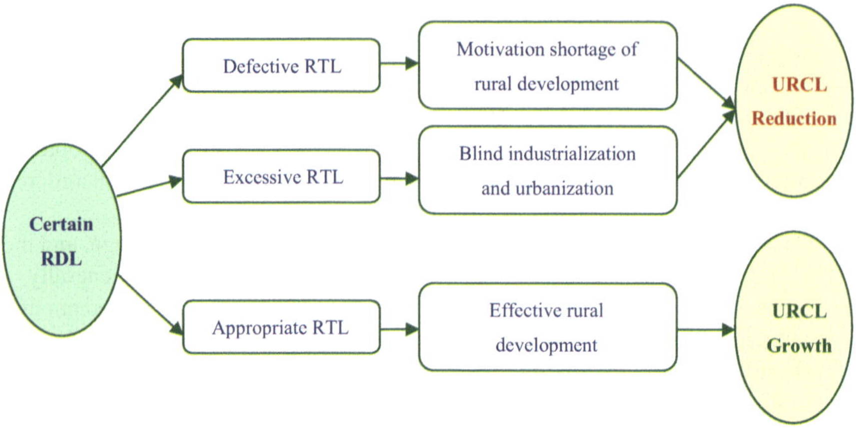

Fig. 6.1 Three dimensions measuring RTD: RDL, RTL, and URCL (Long et al. 2011)

Table 6.1 Indicator system for rural development level (RDL) assessment (Long et al. 2011)

Rule layer factors (weight)	Indicator layer factors (weight)	Definition
Rural economic development (0.383)	Agricultural output value level (0.542)	Gross output value of farming, forestry, animal husbandry and fishery per capita
	Productivity of rural labor (0.458)	Gross agricultural output value divided by the laborers employed in farming, forestry, animal husbandry and fishery
Agricultural production investment (0.263)	Power investment (0.344)	Gross power of farming mechanism per ha
	Fertilizer investment (0.310)	Gross fertilizer use per ha
	Irrigation index (0.346)	Irrigated farmland area divided by the total area of farmland
Rural livelihood (0.354)	Rural electricity consumption (0.283)	Electric power consumption per capita in the rural area
	Rural income level (0.363)	Per capita net income of rural households
	Rural consumption level (0.354)	Per capita consumption expenditure of rural households

(Long et al. 2009). Considering the availability of relative socio-economic data, we selected eight representative indicators belonging to three rule-layer factors: rural economic development, agricultural production investment, and rural livelihood. All of these factors can be analyzed using analytic hierarchy process (AHP) methods (Table 6.1). All of the indicators in Table 6.1 have positive effects on regional RDL assessment.

Essentially, rural transformation is marked by changes in the rural area's internal industrial, employment and consumption structures and corresponding changes in the rural population and land-use structure and intensity. Therefore, six indicators were chosen to assess RTL (Table 6.2). Three general and intuitive indicators were

Table 6.2 Indicator system for rural transformation level (RTL) assessment (Long et al. 2011)

Indicator (weight)	Definition	Explanation
Urbanization level change rate (0.209)	$\frac{UL_l - UL_e}{UL_e}$	UL_l = the proportion of the nonagricultural population in the total population for the later period; UL_e = UL for the early period. A positive indicator: the higher the value, the higher the RTL
Industrial structure change rate (0.204)	$\frac{IS_l - IS_e}{IS_e}$	IS_l = the proportion of the output value of primary industry in the total gross domestic product (GDP) for the later period; IS_e = IS for the early period. A negative indicator; the lower the value, the higher the RTL
Employment structure change rate (0.200)	$\frac{ES_l - ES_e}{ES_e}$	ES_l = the proportion of laborers employed in farming, forestry, animal husbandry and fishery among the total laborers for the later period; ES_e = ES for the early period. A negative indicator; the lower the value, the higher the RTL
Consumption structure change rate (0.174)	$\frac{CS_l - CS_e}{CS_e}$	CS_l = the Engel coefficient for rural residents for the later period; CS_e = CS for the early period. A negative indicator; the lower the value, the higher the RTL
Grain-farmland index change rate (0.123)	$\frac{GI_l - GI_e}{GI_e}$	GI_l = the proportion of grain-crop area in the total crop area for the later period; GI_e = GI for the early period. A negative indicator; the lower the value, the higher the RTL
Multi-cropping index change rate (0.090)	$\frac{MI_l - MI_e}{MI_e}$	MI_l = the proportion of the crop area in the area of farmland for the later period; MI_e = MI for the early period. A positive indicator; the higher the value, the higher the RTL

Table 6.3 Indicator system for urban-rural coordination level (URCL) assessment (Long et al. 2011)

Indicator (weight)	Definition
Rural-urban income gap (0.433)	Dividing the per capita net income of rural households by the per capita disposable income of urban households
Urban-rural consumption structure comparison (0.304)	Dividing the Engel coefficient of urban residents by that of rural residents
Industrial labor productivity comparison (0.263)	Dividing the productivity of laborers employed in farming, forestry, animal husbandry and fishery by that of laborers employed in other industries

chosen to comprehensively measure the degree of urban-rural coordination in society, economics, resources, and the environment, and these were used to assess the URCL (Table 6.3). All of the indicators in Table 6.3 have positive effects on regional URCL assessment.

Data for the indicators in Tables 6.1 through 6.3 are taken from China's regional economic statistical yearbook, in which prefecture-level city is the data collection unit, and naturally the prefecture-level city is the assessment unit of this study. Because we aimed to describe and compare all of these data from 2000 to 2008 for the whole country, we made sure that any data used were from unified statistical sources and that the economic data were based on comparable pricing, that is, all of the original economic data from 2008 were calculated using the constant price from 2000. Semi-structured interviews with geography and sociology experts and government officials for rural development were conducted to determine the weights for the rule-layer factors in Table 6.1 and all of the indicators in Tables 6.1 through 6.3.

6.2.3.2 Methods

Because the socio-economic data for the various indicators in Table 6.1 are in different units, they needed to be transformed into comparable common units by normalizing all measures, using Eq. (6.1):

$$X'_{ij} = \frac{X_{ij} - X_{i.\min}}{X_{i.\max} - X_{i.\min}} \tag{6.1}$$

where X'_{ij} is the standardized value of the indicator; ij means the indicator i in the rule layer j; X_{ij} is the value of the indicator ij; $X_{i.\max}$ is the maximum value of indicator ij for all prefectures; and $X_{i.\min}$ is the minimum value of indicator ij of all prefectures.

The indicators in Tables 6.2 and 6.3 are relative indices without dimensions. To render them comparable, we needed to range their values from −1 to 1 using the general normalization method, according to Eq. (6.2):

$$X_i' = \frac{X_i}{X_{i.\max}} \tag{6.2}$$

where X_i' is the standardized value of the indicator i; X_i is the value of the indicator i; and $X_{i.\max}$ is the maximum value of the absolute value of the indicator i for all prefectures.

To compare the RDL, RTL and URCL for all of the prefectures, the same data type from different prefectures between 2000 and 2008 were normalized using the same extremum. Using the same data type guarantees that the final RDL, RTL, and URCL for different prefectures are spatially and temporally comparable.

After multiplying each negative indicator by −1, we used the weight and normalized value of each indicator to calculate the RDL, RTL, and URCL scores for each prefecture, using Eqs. (6.3), (6.4), and (6.5):

$$RDL = \sum_{j=1}^{n} \left(\sum_{i=1}^{m} X_{ij}' \times W_{ij} \right) \times W_j \tag{6.3}$$

$$RTL = \sum_{k=1}^{t} X_k' \times W_k \tag{6.4}$$

$$URCL = \sum_{k=1}^{t} X_k' \times W_k \tag{6.5}$$

where X_{ij}' is the standardized value of the RDL indicator; W_{ij} is the weight for indicator layer factor ij; W_j is the weight of rule layer factor j; n is number of the rule layer factors; m is the number of indicators in each rule layer; X_k' is the standardized value of RTL or URCL indicator k; W_k is the corresponding weight of the RTL or URCL indicator; and t is the number of RTL or URCL indicators.

There are obvious regional discrepancies in physical conditions and socio-economic development throughout China (Long et al. 2010). According to Chinese statistical definitions, China has four regions: the eastern region, the central region, the western region, and the northeastern region (Fig. 6.2). The eastern region includes Beijing, Tianjin, Hebei, Shanghai, Jiangsu, Zhejiang, Fujian, Shandong, Guangdong, and Hainan; the central region includes Shanxi, Anhui, Jiangxi, Henan, Hubei, and Hunan; the western region includes Guangxi, Inner Mongolia, Chongqing, Sichuan, Guizhou, Yunnan, Tibet, Shaanxi, Gansu, Ningxia, Qinghai, and Xinjiang; and the northeastern region includes Liaoning, Jilin, and Heilongjiang. The four statistical regions described in Fig. 6.2 were adopted for the spatial analysis of China's RTD.

Fuzzy spatial clustering is commonly used for spatial data analysis and territorial type divisions, and both the spatial coordinates and attribute features of each unit form the basis of the clustering analysis. In this study, we used the statistical clustering of K-mean algorithms in SPSS 13 and the spatial coordinates extraction function

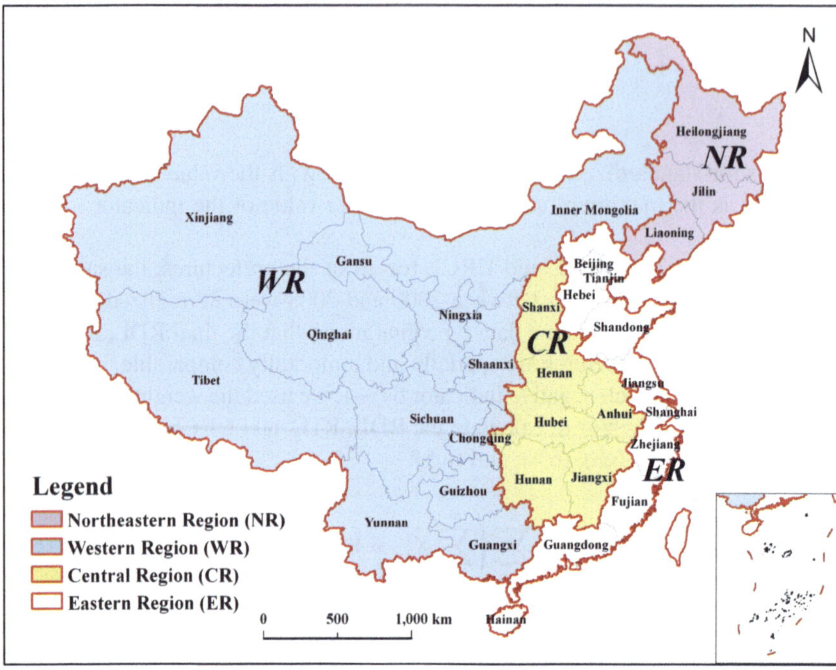

Fig. 6.2 The four statistical regions of China (Long et al. 2011)

in ArcGIS 9.2 to produce clustering based on both spatial and attribute data, and a high frequency of data iterative calculation was adopted to improve clustering accuracy. Finally, we used the clustering results and mapping function of ArcGIS 9.2 to determine the spatial-territorial types of RTD in China in the new century.

6.3 China's Rural Transformation Development

6.3.1 Spatio-Temporal Characteristics

Constrained by physical conditions and general socio-economic development, China's RDL shows a gradual, declining spatial pattern from east to west in 2000, except for the northwestern part, which had a relatively high RDL (Fig. 6.3). The regions with high RDL are mainly concentrated in eastern coastal China, due to its special advantages (optimal physical conditions, location, and socio-economic factors for the development of rural industry and agricultural production). Adequate financial and technological support for agricultural production is complemented by fertile farmland, distributed in plain areas, that is suitable for agricultural scale

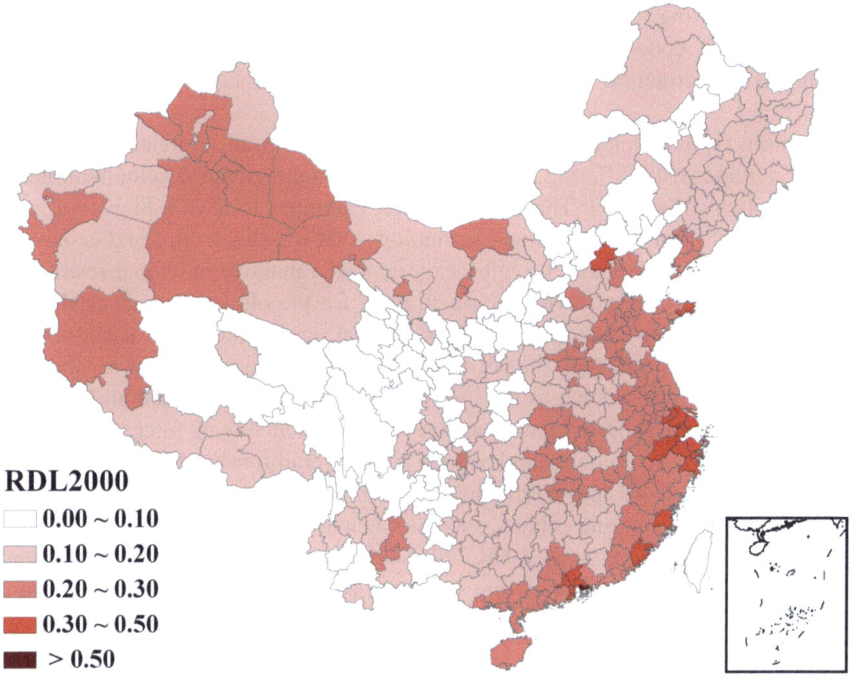

Fig. 6.3 The spatial pattern of China's rural development level (RDL) in 2000 (Long et al. 2011)

management. This support, together with the developed TVEs, has led to the high agricultural outputs and rural income levels that contributed to the high RDL.

The central region has long been one of China's main grain production bases. It has excellent agricultural production conditions, including good agricultural machinery, irrigating facilities, and capital accumulation, which have contributed to a high agricultural production level and relatively high RDL. However, due to relatively slow regional socio-economic development, its regional agricultural production lacked adequate and sustainable capital support, which gave rise to a low RDL compared with the eastern region. In the northeastern region, large amounts of fertile farmland and superior irrigation in the great plain have led to high agricultural output. However, the northeastern region is China's traditional heavy industrial base, and it faced economic depression and industrial restructuring during the 1990s. Regional socio-economic development was therefore unable to effectively support agricultural infrastructure improvements, resulting in nearly two decades of relative agricultural stagnation in this region and a lower regional RDL than in eastern region.

The agricultural production conditions in the western region are generally poor, due to the rugged mountainous terrain in southwestern China and extreme drought conditions in the northwest. In addition, socio-economic development in the western region has lagged far behind that of other regions, which has led to an extremely poor agricultural infrastructure and inferior conditions for developing rural industry, ultimately resulting in a low RDL. Of course, some areas have a high RDL, but those are mainly concentrated in socio-economic centers near big cities, or in areas with improved irrigation conditions or adequate financial support for agricultural production.

With the rapid development of China's economy, overall regional RDL shows a clear increase, with differing rates due to contrasting regional development models and policies (Fig. 6.4). The eastern region continues to have a high RDL rate due to support from the capital and technology that resulted from rapid economic development. The northeastern region experienced a massive industrial restructuring, guided by a national revitalization strategy for the old heavy industrial base, which brought about clear improvements in rural development and led to a high RDL growth. Compared with the northeastern region, national policies for the central and western region, such as the Central China Grow-up Program and Western China Development Program, have not generated sufficient driving forces for regional socio-economic development and have led to an overall lower RDL growth. However, in central and western China, higher RDL growth was found in some areas near the regional economic centers, such as Chengdu, Chongqing, Wuhan, Changsha, Ordos, and Urumchi. Since 2000, China has experienced rapid development and profound changes in socio-economic structure, which exerted a huge influence on China's RTD (Fig. 6.5).

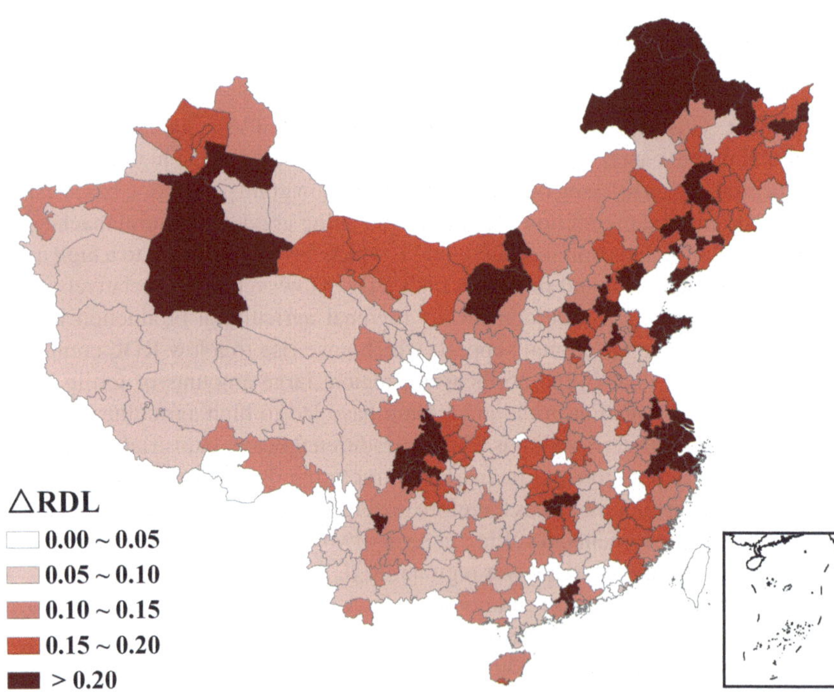

Fig. 6.4 The dynamic spatial pattern of China's RDL between 2000 and 2008 (Long et al. 2011)

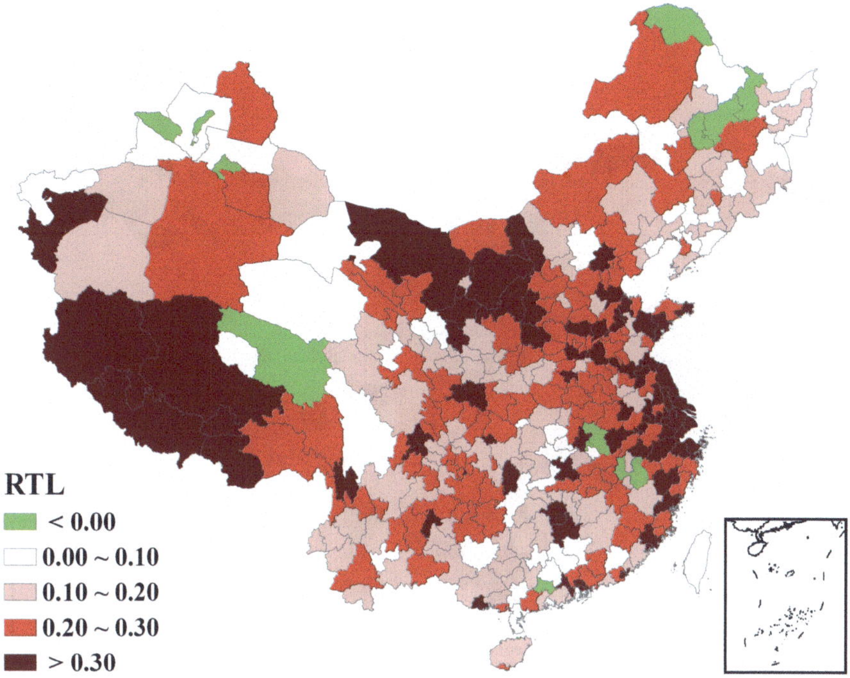

Fig. 6.5 The spatial pattern of China's rural transformation level (RTL) between 2000 and 2008 (Long et al. 2011)

In the Yangtze River delta economic area, significant reforms in TVEs development are intended to adapt to economic globalization, and an export-oriented development strategy has effectively promoted not only regional rural economic development, but also considerable progress in improving the rural system structure, which has resulted in an extremely high RTL. The governments of Beijing, Tianjin, Shandong, Hebei, and Henan actively pursued regional industrialization and urbanization, resulting in a relatively high RTL. The regions with rich energy and mineral resources achieved great progress in the mineral mining industry and pursued rapid economic growth, which led to a high RTL in the provinces of Shanxi, Shaanxi, Ningxia, Inner Mongolia, and Guizhou. In general, due to these completely different regional development models, China's RTL throughout 2000–2008 took on a complex spatial pattern (Fig. 6.5).

Driven by different economic development models and rural transformation intensities, the dynamic regional URCL in China from 2000 to 2008 does not show an obvious spatial pattern (Fig. 6.6). In general, China's urban-rural coordination development declined from 2000 to 2008; 57.9% of the regions experienced negative changes in the UCRL, and the URCL at national level decreased by 1.3%. The regions with UCRL reductions were mainly concentrated in areas with rapid economic development, mostly located in the eastern and central regions or in energy

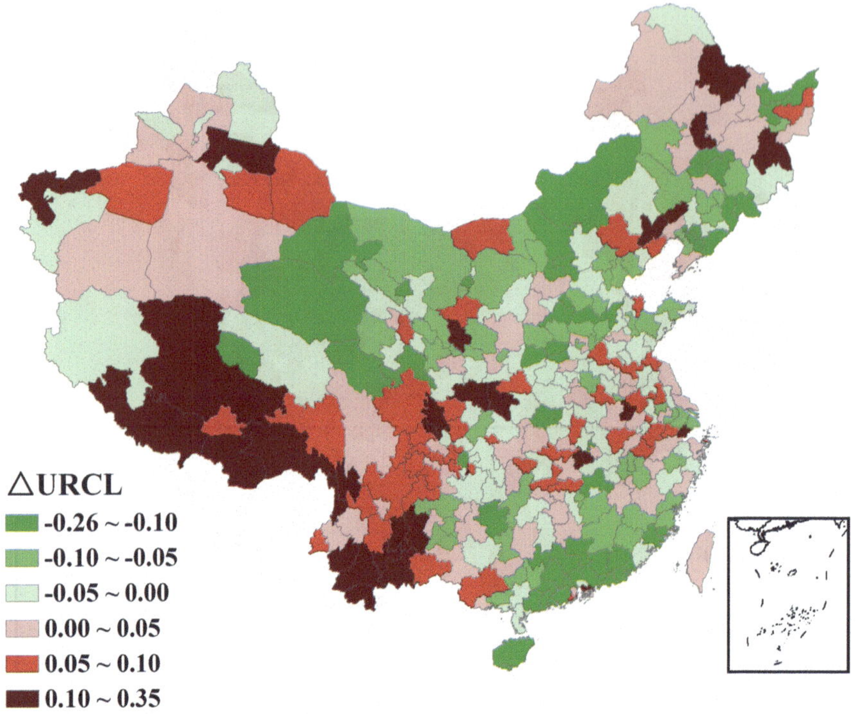

Fig. 6.6 The dynamic spatial pattern of China's urban-rural coordination level (URCL) between 2000 and 2008 (Long et al. 2011)

and mineral mining areas where socio-economic development has not achieved a relative balance between urban and rural areas.

6.3.2 Territorial Types

Due to the complex spatial patterns of China's RTL and URCL, it is necessary to examine transformation characteristics and regional differences in physical conditions and socio-economic development to analyze the internal mechanism of RTD in different regions. Therefore, it is useful to identify different territory types according to their characteristics, such as the regional RDL, RTL, and URCL, to analyze China's RTD patterns.

Using SPSS 13 and ArcGIS 9.2 software programs, the spatial coordinates, initial RDL, the change of RDL, the RTL, and the change of URCL were used as the clustering factors to examine the internal mechanism of China's RTD. On the basis of the clustering results, 10 RTD territorial types were identified (Fig. 6.7), and their respective regional transformation characteristics (listed in Table 6.4) followed the sequence of initial RDL value.

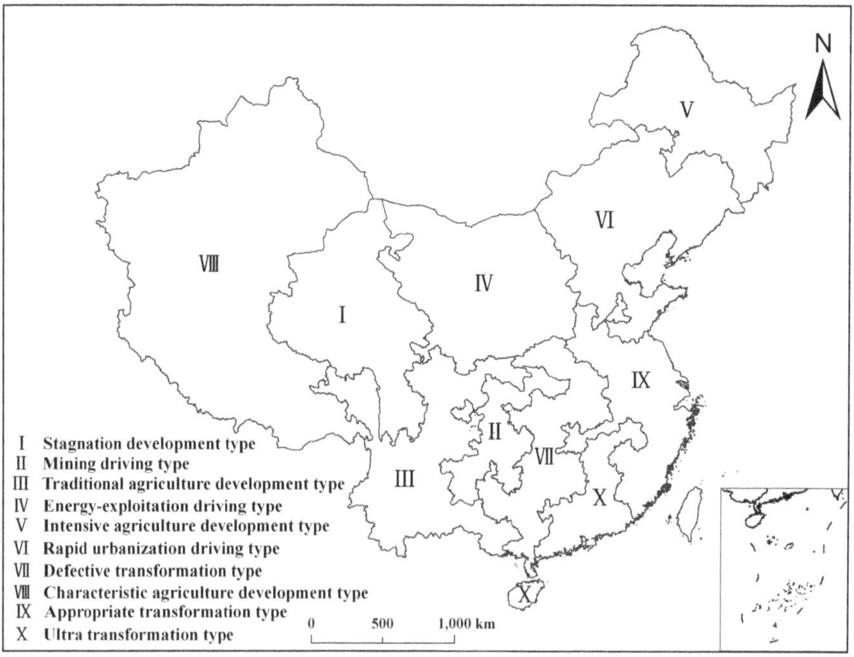

Fig. 6.7 The territorial types division of China's RTD (Long et al. 2011)

Usually, different regions adopt development approaches suited to their physical and socio-economic features to promote rapid socio-economic development. The initial regional RDL was generally low in China, meaning that capital accumulation for agricultural production or rural development was inadequate. Along with defective or excessive RTL, this inadequacy inevitably leads to uncoordinated urban-rural development. Accordingly, the regions with low initial RDL achieve different results with different developing approaches, thereby shaping the diverse territorial types of RTD.

The mining driving type and the energy-exploitation driving type had similar developments situations: a poor agricultural base with low-intensity production investment, scarce farmland resources facing serious soil erosion risks, slow rural socio-economic and TVE development, despite rich stores of energy and mineral resources. A similar development approach focusing on the heavy industries depended on the exploitation of resources, which promoted the rapid transfer of the employed population, social resources and capital from a rural, agricultural system to a heavy industrial urban system. As a result, rural and agricultural development stagnated because of a lack of external support, leading to low regional average RDL growth and an obvious reduction in URCL, despite a high RTL (Table 6.4).

Some RTD types have a relatively high RTL, but their transformation has mainly focused on changing the multicropping index and rural consumption structure, and their regional development strategies have focused on intensifying the agricultural

Table 6.4 The transformation characteristics of each RTD territorial type (Long et al. 2011)

RTD territorial type	Average initial RDL	Average RTL	Average △RDL	Average △URCL	Proportion of URCL growth units in the total units of the same type (%)	Total units of the same type
Stagnation development type	0.108	0.063	0.104	−0.073	14.3	7
Mining driving type	0.117	0.233	0.106	−0.029	25.0	24
Traditional agriculture development type	0.119	0.219	0.131	0.084	100.0	40
Energy-exploitation driving type	0.122	0.253	0.107	−0.043	23.3	43
Intensive agriculture development type	0.124	0.072	0.178	0.018	68.8	16
Rapid urbanization driving type	0.176	0.194	0.166	−0.053	14.5	55
Slow industrialization driving type	0.177	0.156	0.120	−0.003	42.9	42
Characteristic agriculture development type	0.177	0.180	0.113	0.062	70.0	20
TVEs dominated development type	0.235	0.312	0.159	0.002	50.7	69
Traditional inward-processing industry development type	0.271	0.199	0.124	−0.114	9.7	31
National level	0.175	0.216	0.135	−0.013	42.1	347

production base instead of blindly accelerating industrialization and urbanization. These approaches promoted urban-rural coordination development and caused an obvious growth in URCL. For example, the intensive agriculture development type focused on developing intensive agriculture with scale management. This approach was similar to the approach of traditional agriculture development type to increasing URCL (Table 6.4), which focused on improving traditional agriculture. The stagnation development type is unique; it has the lowest RDL and RTL due to extreme physical conditions and low socio-economic status. Under these conditions, blind industrialization and urbanization could not produce effective rural development and led to a low URCL.

The initial RDLs of the rapid urbanization driving type and the slow industrialization driving type were close to the average national level, but their RTLs were below the average national level, meaning they could not effectively promote regional socio-economic development. This low level ultimately led to a reduction in URCL, especially in the rapid urbanization driving type, which was characterized by rapid socio-economic development (ΔRDL = 0.166) but uncoordinated urban-rural development (ΔURCL = −0.053) (Table 6.4). However, there was also the special case of characteristic agriculture development type, which had a moderate RDL and a relatively low RTL; however, the development of locally appropriate agriculture promoted considerable progress in agricultural and rural development. This progress, together with appropriate industrialization and urbanization, led to a high URCL (Table 6.4).

Both the TVEs dominated development type and the traditional inward-processing industry development type belong to the regions with rapid economic development and universally high RDL that nonetheless faced a universal reduction in URCL, especially in the provinces of Guangdong and Hainan. In the TVEs dominated development type, the high RTL promoted rural development to a great extent, but accelerated industrialization and urbanization meant that URCL increased minimally (ΔURCL = 0.002) (Table 6.4). Although the traditional inward-processing industry development type had the highest initial RDL (0.271), the low RTL plus accelerated industrialization and urbanization led to a drastic reduction in URCL (ΔURCL = −0.114) (Table 6.4). Considering this result, systemic RTD should be pushed forward in the regions with developed rural and agricultural systems to avoid uncoordinated urban-rural development.

6.3.3 Internal Mechanism

To distinguish the internal mechanism of regional RTD from miscellaneous assessment results, we classified the initial RDL into four grades, and RTL into five grades, based on their mathematical statistical features (Table 6.5). Because the number of

Table 6.5 Specific classifications for statistical descriptions of China's RTD (Long et al. 2011)

Grade	Statistical standards	Initial RDL range	RTL range
Low	($-\infty$, Mean – 0.5 Std[a]]	($-\infty$, 0.135]	($-\infty$, 0.154]
Intermediate-low	(Mean – 0.5 Std, Mean]	(0.135, 0.175]	(0.154, 0.216]
Intermediate-high	(Mean, Mean + 0.5 Std]	(0.175, 0.216]	(0.216, 0.277]
High	(Mean + 0.5 Std, $+\infty$[a])	(0.216, $+\infty$)	(0.277, 0.339]
Extremely high	(Mean + Std, $+\infty$)	–	(0.339, $+\infty$)

[a]Std means the standard deviation. The statistical standard for a High RTL grade is "(Mean + 0.5 Std, Mean + Std)"

units with high initial RDL and RTL was comparatively large and the statistical features of these units were complex, a special grade of "extremely high" was established to describe RTL. Finally, we identified the statistical features of the corresponding grades (Table 6.6).

Table 6.6 shows △URCL's obvious tendency to increase with increases in RTL from the low grade to high or even extremely high grades; however, this tendency did not apply in cases with low initial RDL grades. This result shows that once the initial RDL has reached the intermediate-low grade, the intensification of rural transformation or RTL increases result in gradual improvements in rural development and URCL growth; that is, they promoting urban-rural coordination development. However, this law is not obvious in units with low initial RDL grades because different regions with a low initial RDL usually adopt development approaches uniquely suited to their respective physical and socio-economic conditions, which lead to different transformation development results, as shown in the above analysis of RTD territorial types.

The regions with a high initial RDL showed different urban-rural development patterns due to their universally low RTL and high RDL. From 2000 to 2008, a clear

Table 6.6 The statistical features of China's RTD (Long et al. 2011)

Initial RDL grade	RTL grade	Average △RDL	Average △URCL	Proportion of URCL growth units in the total units of the same grade (%)	Total units of the same grade
Low	Low	0.118	0.009	44.1	34
	Intermediate-low	0.119	0.005	56.1	41
	Intermediate-high	0.116	0.001	46.2	26
	High	0.120	0.028	59.1	22
Intermediate-low	Low	0.154	−0.052	28.6	14
	Intermediate-low	0.133	−0.030	37.0	27
	Intermediate-high	0.138	0.027	72.7	11
	High	0.154	0.061	70.0	10
Intermediate-high	Low	0.117	−0.019	44.4	18
	Intermediate-low	0.118	−0.018	31.3	16
	Intermediate-high	0.129	−0.011	55.6	9
	High	0.131	0.027	68.8	16
High	Low	0.125	−0.086	11.8	17
	Intermediate-low	0.132	−0.086	14.3	14
	Intermediate-high	0.144	−0.063	20.8	24
	High	0.167	−0.018	27.8	18
	Extremely high	0.207	−0.010	36.7	30

RDL increase occurred in the high and extremely high grade initial RDLs and RTLs, due to their adequate resources for agricultural and rural development; however, the URCL continued to grow negatively because regional RTD could not keep up with rapid industrialization and urbanization.

6.4 China's Rural Development: A Rurality Index Perspective

6.4.1 Background

6.4.1.1 Rurality and Rurality Index

An operational definition for 'rural area' is pivotal if proposals, policies and decisions aimed at optimizing the distribution of resources, closing the gap on inequity between areas and raising standards of living for the least advantaged populations are to be put in place (Ocañariola and Sánchezcantalejo 2005). Questions as to what is meant by a rural area, the identification of diagnostic features and attempts at understanding the nature and scope of rurality are continuing themes in the literatures (Cloke 1977; Blunden et al. 1998; Prietolara and Ocañariola 2010). It is widely acknowledged that rural is a fuzzy concept which is contested in terms of identifying the critical parameters of rural space (Halfacree 1993; Woods 2011). Low population density, abundance of farmland, and remoteness from urban agglomerations are characteristics that people typically associate with rural places. In fact, people frequently use the term "rural" to collectively express their perception of place characteristics that—in one way or another—typify rurality. As such, the meaning of rurality depends on the perception of each individual who integrates visions of rurality into everyday life (Halfacree 1995). Moreover, the developmental processes of social, economic and political restructuring in many countries are reshaping rural areas (Woods 2007, 2011, 2013), and rural has also been used in different contexts from developed countries to the underdeveloped ones. Thus, the same as rural, rurality remains an elusive concept (Woods 2010), and special attention on rural areas is necessary while focusing on the measurements of the differences in the degree of rurality (Cloke 1977).

In general, rurality has proved very difficult to define in an all-embracing manner for three important reasons involving functions, dynamics and variation (Cloke 2006). Moreover, rural areas are undergoing considerable spatio-temporal change due to social, economic and technological developments, and especially the interaction of various non-quantitative elements affecting rural development (Long et al. 2009). This kind of change, to some extent, increases the difficulty in our understanding of the rural and rurality. In broad terms, Cloke (2006) theoretically discussed functional concepts of rurality, political-economic concepts of rurality and social constructions of rurality, which have been influential in constructing conceptualizations of rurality. In their efforts to impose some order on the diversity of claims and conceptions related to the countryside, Frouws (1998) identifies three major strands of discourses, namely: agri-ruralism, utilitarianism and hedonism, while Lópezigelats et al. (2009) identify four discourses of rurality, namely: the conservationist, entrepreneurial, agriculturalist, and endogenous development.

In empirical studies, researchers and international organizations such as the Organization for Economic Co-operation and Development (OECD) and European Union (EU) have developed several typologies and different rural indicators in order to better understand the dynamics of rural areas and to develop relevant policies for rural areas (Cloke 1977; Woods 2013). Especially since the influential earlier work of Cloke (1977), several rurality indices have been developed specifically for different countries. The rurality index of England and Wales (Cloke 1977; Cloke and Edwards 1986), the rural indicator of OECD (OECD 1996), the MSU rurality index, USA, the accesibility/remoteness index of Australia, the scale measure of Urbanicity, USA, the general practice rurality index, Canada, and the rurality index for small areas in Spain (Prietolara and Ocañariola 2010) are some of the complex indices published in the international scientific literatures. As in China, the only precedent of a rurality compound index is the rurality degree index for the eastern coastal China (Long et al. 2009).

Overall, various ways of classification and definition in the literatures are derived to measure differences in the degree of rurality, including the level of population density, the rate of population loss or gain, settlement size, local economic structure, accessibility and landscape, etc. In line with the various perspectives of rurality, evidence abounds in literatures that rural areas could be characterized by: specific open landscape; a relatively low population density; the greater part of the population being associated with agriculture and forestry; traditional life styles and habits; agricultural and forest-related use of land; a scarcity of built-up areas and settlement that is dispersed; and a preponderance of inhabitants considering themselves country-dwellers (Madu 2010).

6.4.1.2 Spatial Perspective of China's Rural Areas

Since 1978, China has pursued sweeping economic changes in an officially sponsored transition from a centrally planned economy to a market economy. In practice, the economic reforms have set in train a process of potentially fundamental social and institutional change in rural China which is creating new socio-economic forces, shifting power in their direction, and raising the possibility of rural transformation (Xu and Tan 2002; Long et al. 2011, 2012). More and more rural young labors outmigrated to east coast China for off-farm work and to a large extent, promoted the development of the counties of destination (Fan 2005). Rural resident population experienced a process of increase and decrease, and now come into the period of fast decrease (Long et al. 2012). Specifically, the original counties are suffered more and more from the brain drain and consequent rural hollowing, these counties shows lower rate of economic growth (Long et al. 2012; Li et al. 2014).

The rural development in China has gradually attracted the interest in academic circles both at home and abroad (Long et al. 2009), ranging from rural–urban migration, rural industry and employment, rural associations and state corporatism, rural taxation and government regulation, rural poverty alleviation (Heilig et al. 2006; Unger 2002), rural transformation development (Unger 2002; Liu 2007; Long et al. 2011), rural gentrification, urban–rural equalized development, rural land-use change and building new countryside (Liu 2011; Long et al. 2010). Some studies has also depicted the development stage of China's rural development and explained the driving forces and breaking forces during different period (Xu and Tan 2002; Long et al. 2010). Overall, these studies have given more or less a history perspective or structure perspective on China's rural development. Currently, we still have little knowledge about the nature of pattern of rural development at a more fine scale except provincial. As such, we will give a brief depict about China's rural and its recent development from a spatial perspective, before analyzing China's rurality index.

The natural environment factor is one of the main indices for evaluating human habitats, sustained economic growth and ecological health status. The vast land expanses of China include plateaus, plains, basins, foothills, and mountains. Defining rugged plateaus, foothills and mountains as mountainous, they occupy nearly two-thirds of the land, higher in the west and lower in the east like a three-step ladder. The highest step of the typical 'ladder topography' including Qinghai province and Tibet autonomous region is formed by the Qinghai-Tibet Plateau at the average height of over 4000 m. On the second step are large basins and plateaus, most of which are 1000–2000 m high. The Daxing'an, Taihang, Wu and Xuefeng Mountains divide this step and the next lower one. Plateaus including Inner Mongolian, Loess, Yungui Plateaus, and basins such as Tarim, Junggar, and Sichuan Basins are situated here. The third step, abundant in broad plains, is dotted with the foothills and lower mountains, with average altitude of over 500 m. Here are located with famous plains: the Northeast Plain, the Huang-Huai-Hai Plain, and the Middle-Lower Yangtze Plain, neighboring with each other from north to south. These well-cultivated and fertile lands produce abundant crops. Using natural environment data including climate, hydrology, surface configuration and ecological conditions, Yang and Ma (2009) establish natural environment suitability index (NESI) of China. They find that the value of NESI is higher in the east and lower in the west of China, and the best natural environment area is the Yangtze River Delta region, while the worst are the northwest of Tibet and southwest of Xinjiang. These physical features largely shaped the patterns of local cultural and socio-economic development. Areas with better natural environment are more prone to be regions with higher population density and richer economy.

According to the state of economic development, Gross Domestic Product (GDP) per capita is one of the most frequently used indicators for measuring economic performance and comparing the state of development of different areas. Li and Qiao (2001) analyzed the economic disparities of China during 1990 and 1998 using nationwide county level data of GDP per capita. They found that the counties with faster growth rates than the national average were spread from several growth centers to outside. Consequently, three growth corridors gradually appeared, namely, the Coastal Corridor (along the nation's coastal line), Beijing—Guangzhou Corridor (along the railway from Beijing to Guangzhou), and the Yangtze River Corridor (along the Yangtze River from Chengdu to Shanghai). Overall, the less developed counties were mainly located in the western part of China. The distribution pattern of less developed counties is consistent with the disadvantageous development conditions in mountainous, cold and dry areas, as well as with the isolation in the provincial border regions. We collected the data of GDP per capita in 2010. As we can see clearly in Fig. 6.8, which has been divided into five grades by quintile, GDP per capita at county level shows huge regional disparities. Counties in eastern coastal China which have experienced rapid economic growth since the initiation of reform and opening up policies, and counties in north China with rich mineral resources and less population have higher GDP per capita. In contrast, inland counties especially in southwest China with mountain terrain and central China major in agricultural production have much lower GDP per capita. Furthermore, there are also some counties in eastern coastal China have lower GDP per capita and in west and central China, there are also some counties with higher GDP per capita. This economic development pattern can inevitably affect the local state of rural development.

With the rapid development of industrialization and urbanization as well as enhancement of geographic differentiation and diversity of man-earth areal system,

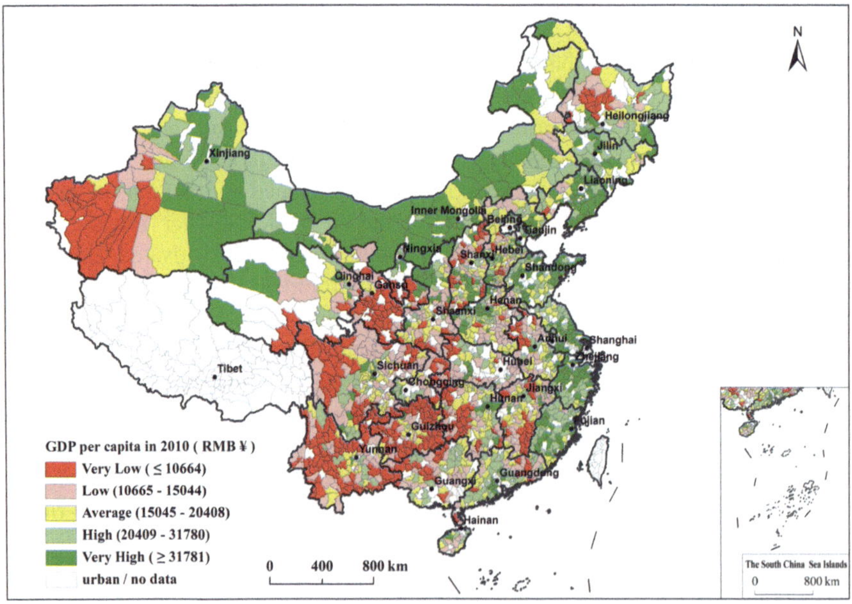

Fig. 6.8 Spatial pattern of GDP per capita in China at county level in 2010 (Li et al. 2015). *Note* The GDP per capita is divided into five grades by quintile; in 2010, 676.95 RMB Yuan = 100 USD

territorial functions and regional development orientations have shown an increasingly strong trend towards diversification. Liu et al. (2011a) evaluate and grade the functional areas of economic development, food security, social stability, environmental protection and comprehensive function of China at county level. The results show that economy-oriented functional areas are mainly distributed in eastern coastal developed areas and peripheral areas of the metropolitan regions, such as Pearl River Delta, Yangtze River Delta and Beijing-Tianjin-Hebei region. Grain-oriented functional areas are mainly distributed in the Northeast Plain, the Huang-Huai-Hai Plain, Sichuan Basin, central Hubei, eastern Hunan and other regions covered by a large area of plain. The social security function indexes are gradually weakened from coastal to inland areas and from north to south; Eco-conservation areas are concentrated in the Northeast China and southern Qinling Mountain-Huaihe River Line.

Obviously, the physical conditions and socio-economic development of rural China show great regional disparities, but our knowledge about this pattern is still very inadequate especially when making integrated and regionalized rural policies. So, this section will calculate China's rurality index at county level using national census data of 2000 and 2010, and reveal the spatial patterns and dynamics of rurality index. Moreover, the Pearson correlation coefficients between rurality index and major economic and geographical indicators will be calculated to examine to what extent the rurality index can characterize the pattern of China's rural areas and to provide scientific support for the making and improvement of rural policies.

6.4.2 Methods

6.4.2.1 Index System and Data Source

When establishing the index system of rurality evaluation, basically, the variables must be representative of the concept of rurality in China in line with the theoretical and empirical background we discussed above. At the same time, they can be both measured and quantified, especially be readily available and accessible for users at a reasonable cost-benefit ratio. Moreover, they must be easy to update at regular intervals (Ocañariola and Sánchezcantalejo 2005). Rural system is complex and hybrid. Rural development can be seen as the outcome of interactions between various rural system components (Marsden 2010) e.g., geographical and bio-geo-physical conditions, industrial development, rural population, and social and cultural characteristics. However, demographic characteristics are the most important and active factors in reflecting the change of rural development, as evidenced by some studies concerning land use transitions affected by rural migration, the spatio-temporal variations of driving factors of population change in rural, suburban, and urban areas, population pressure and agricultural intensification in rural systems, and implications of rural population change for policymaking in developing countries. As such, the national census data published every decade was selected as the main source of information for rurality evaluation, since many of the variables it includes meet these criteria (Cloke 1977; Prietolara and Ocañariola 2010).

Combining with previous studies and taking into account the basic characteristics of rural China and the availability of data, fifteen variables that could largely reflect the population density, age-structure, education level, employment, migration and housing condition of rural areas, have been chosen for evaluating rurality, and so as to explore the spatio-temporal pattern of China's rural development from the perspective of demographic characteristics and their changes. The fifteen variables are population density, size of household, young children index, senior citizen index, birth rate, mortality rate, in-migration, education level, illiteracy rate, professional and technical personnel, economic dependency index, agricultural employment, share of rural population, housing habitability index and share of self-built housing (Table 6.7). Our initial judgment is, to a large extent, that counties with higher rurality index may have lower population density, bigger size of household, higher young children index, lower senior citizen index, higher birth rate, higher mortality rate, lower in-migration, lower education level, higher illiteracy rate, lower professional and technical personnel, lower economic dependency index, higher agricultural employment, higher share of rural population, higher housing habitability index and higher share of self-built housing.

The data of the fifteen variables at county level in Table 6.7 were obtained from the national 2000 census and 2010 census. County-level socio-economic data were obtained from Chinese Counties (Cities) Socio-economic Statistical Yearbook (2000 and 2011) and Chinese Regional Economic Statistical Yearbook (2011). In order to ensure the accuracy of data, statistic yearbook of each province was used for revising and checking when preparing the dataset. DEM data for calculating relief degree of land surface and transportation data for calculating distance from nearest provincial capital, highway and railway were derived from National Resources and Environmental Database presented by Resources and Environmental Scientific Data Center, Chinese Academy of Sciences.

In terms of the number of counties for evaluation, according to China Statistical Yearbook of 2011, the number of districts under the jurisdiction of cities, cities at county level, counties and autonomous counties are 853, 370, 1461 and 117, respectively. For districts under the jurisdiction of cities could be seen as urban area, so the county level regions used for rurality evaluation would be cities at county level, counties and autonomous counties (in brief, we call the three kinds of administrative districts *counties*), with a total number of 1948. However, due to the adjustment of administrative divisions and a small number of counties lack socio-economic data, 81 counties are removed from the research. Specifically, for data availability, counties of Tibet are all removed. As a result, this research includes 1867 counties. The section does not include Taiwan, Hong Kong and Macau. Nevertheless, the 1867 counties consist of 96% of China's cities at county level, counties and autonomous counties, and thus has strong representativeness.

Table 6.7 Variables for measuring rurality index of China's counties (Li et al. 2015)

Variable name	Explanation	Unit	Direction
Population density	Number of inhabitants per square kilometer	Persons/km^2	–
Size of household	Average number of inhabitants living in one household	Persons	+
Young children index	Number of people aged 0–14 per 100 inhabitants	%	+
Senior citizen index	Number of people over the age 65 per 100 inhabitants	%	–
Birth rate	Total number of births per 1000 of a population during the year	‰	+
Mortality rate	Total number of deaths per 1000 of a population during the year	‰	+
In-migration	Number of immigrants from outside the county per 100 inhabitants	%	–
Education level	Average years of schooling of adults	year	–
Illiteracy rate	Percent of adults who can't read	%	+
Professional and technical personnel	Proportion of professional and technical personnel in total employment	%	–
Economic dependency index	Number of economically inactive per 100 persons of working age	%	–
Agricultural employment	Share of agricultural employment in total employment	%	+
Share of rural population	Percentage of population residing in rural areas	%	+
Housing habitability index	Percentage of households with exclusive use of running water, bathing facilities and inside w.c.	%	+
Share of self-built housing	Percentage of self-built house in rural area	%	+

Note '+' means the higher the value of variable, the higher the rurality index; in contrast, '–' means the higher the value of variable, the lower the rurality index

6.4.2.2 Statistical Analysis

Recent China's rurality at county level could be considered as a combination of variables listed in Table 6.7. Principal component analysis (PCA) is one of the most common methods used by data analysts to provide a condensed description and describe patterns of variation in multivariate data sets. Meanwhile, the PCA is a classic and widely used method in rurality studies (Cloke 1977; Madu 2010; Prietolara and Ocañariola 2010). As such, PCA was used in this study to summarize the information provided by the characteristics of the counties into a single rurality factor. In detail, the first principle component, which has the most comprehensive capacity, are chosen for calculating the component score, i.e., rurality index.

Exploratory Spatial Data Analysis (ESDA) has been employed to better reveal the spatial pattern of rurality (Anselin et al. 2007). More specifically, a local spatial autocorrelation analysis using a Local Indicator of Spatial Association (LISA) will help to describe and visualize spatial distribution and discover patterns of spatial association of rurality index (Griffith 2003). In addition, across the numerous studies of rurality index, few have quantitatively discussed the correlationship between the rurality index and typical socio-economic variables. In this section, Pearson correlation coefficient was employed to explore and examine the correlationship between rurality index and main socio-economic and geographical indicators.

As for the realization of above method, SPSS was the software used both for the PCA and correlation analysis. Scores for the rurality index were plotted on the county map of China and ArcGIS (ArcMap) was the Geographic Information System used for this purpose. The LISA analysis was also run in ArcGIS (ArcToolbox).

6.4.3 Performing Rurality Index

In the data of 2000, the first component accounted for 38.84% of the total variance. Table 6.8 demonstrates the loading scores for this combination. It is clear that the variables with large positive loading scores are those which correspond with characteristics which are evidently rural, such as a high level of agricultural employment. Similarly, variables with large negative loading scores are identified with more apparent urban characteristics, such as high levels of professional and technical personnel. Variables with small loading scores (between −0.5 and +0.5), which means contribute little to the differentiation between areas, were discarded (Cloke 1977). As a result, the less important of the population density, senior citizen index, mortality rate, and housing habitability index variables were discarded. And thus eleven variables remained, for which a new principal component was determined, accounting for 49.3% of the current total variance. As for the data of 2010, the same method has been employed to calculate loading score and revised loading score (Table 6.8). The first component accounted for 39.68% of the total variance. Further, four variables namely population density, senior citizen index, mortality rate and illiteracy rate were discarded because their loading scores were relatively small (between − 0.5 and +0.5). And thus also eleven variables remained, for which a new principal component was determined, accounting for 51.17% of the current total variance. Overall, the direction of each variable's affection on rurality index is in line with our initial expectations.

Table 6.8 Variable loading scores for the index of 2000 and 2010 (Li et al. 2015)

Variable	2000		2010	
	Loading score	Revised loading score	Loading score	Revised loading score
Population density	−0.177	–	−0.258	–
Share of rural population	0.820	0.834	0.870	0.871
Size of household	0.614	0.632	0.641	0.656
Young children index	0.622	0.623	0.713	0.702
Senior citizen index	−0.129	–	−0.135	–
Birth rate	0.619	0.590	0.623	0.626
Mortality rate	0.467	–	0.251	–
In-migration	−0.705	−0.728	−0.740	−0.768
Education level	−0.752	−0.715	−0.721	−0.660
Illiteracy rate	0.559	0.520	0.487	–
Professional and technical personnel	−0.645	−0.680	−0.703	−0.738
Economic dependency index	−0.718	−0.725	−0.555	−0.569
Agricultural employment	0.841	0.844	0.805	0.810
Housing habitability index	0.467	–	0.574	0.548
Share of self-built housing	0.714	0.762	0.809	0.839
Percentage of current variance explained	38.84%	49.30%	39.68%	51.17%

Concerning the loading changes exhibited by each individual variable, there has been a marked decline in the contribution of the education level variable between 2000 and 2010. In fact, the average years of schooling of adults in the most rural of districts is gradually being brought up to the standard of the more urban of districts, and so this variable has decreased in value as an indication of rurality distribution. Such change can also be seen on illiteracy rate, economic dependency index and agricultural employment. Other variables have shown increases in differentiation over the intercensal period. Especially, the contribution of variables closely linked with economic development such as in-migration and professional and technical personnel increased significantly. In the meantime, young children index and share of self-built housing also show increasing contribution on the variety of rurality.

These opposing trends have tended to counterbalance each other and have had little overall effect on the index. So, it seems that the index could effectively indicate the changing nature of rurality itself, as well as the spatial manifestation of this change. Such an analysis would provide an interesting insight into the reasons behind the varying contributions of different variables to the principal component of rurality.

Overall, share of rural population (0.834), in-migration (−0.728), economic dependency index (−0.725), agricultural employment (0.844) and share of self-built housing (0.762) are the top 5 variables most affecting rurality in 2000. As for 2010, the five variables were share of rural population (0.871), in-migration (−0.768), professional and technical personnel (−0.738), agricultural employment (0.810) and share of self-built housing (0.839). Obviously, from the perspective of social and economic development, education, professional skills, employment and urbanization are main factors contribute to the variety of rurality.

6.4.4 Spatial Pattern of Rurality

The index of rurality for 2000 and 2010 was calculated for each county by evaluating the reduced principal component value for each county; that is by substituting the values of the remaining eleven variables in the principal component equation. A highly positive index value denotes an extremely rural area. According to the value of rurality index, the rurality in China have been divided into five grades by quintile ranging from very low, through low, average and high, to very high. The fifth-grade counties has the highest rurality, while the first-grade has the lowest.

In 2000, the values of rurality index are between −4.83 and +3.17. It can be seen from Fig. 6.9 that constrained by physical conditions and general socio-economic development, China's rurality index shows a gradually increasing spatial pattern from east to west in 2000, except for the northern part, which has a very low rurality index. In term of rurality index in 2010, it ranges between −4.10 and +3.06. Driven by different but consolidated economic development models and rural transformation intensities, the dynamic regional rurality index in China from 2000 to 2010 does not show an obvious spatial pattern. In addition, the patterns of rurality index shown in Figs. 6.9 and 6.10 are very similar to the pattern of GDP per capita shown in Fig. 6.8. These patterns denote persistent spatial structures, and reveal, for instance, that the transition of China's rural policies since 2004 has not visibly altered the landscape of socio-economic conditions in rural areas.

Three major belts with "very low" rurality index areas are clearly visible from the index distribution (Figs. 6.9 and 6.10).

(1) The first belt is eastern coastal China. Due to its special advantages (optimal physical conditions, location, and socio-economic factors for the development of non-agricultural industry and urbanization), this region is the forefront and

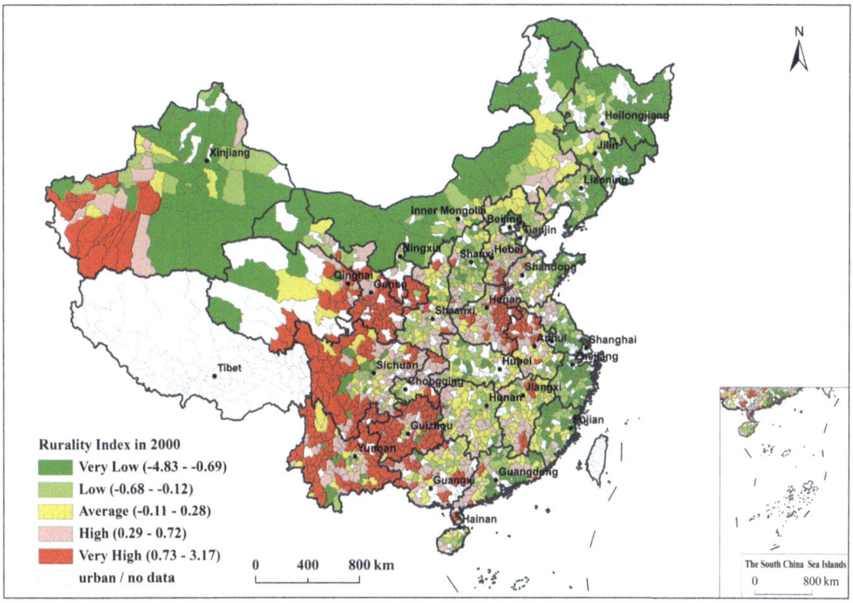

Fig. 6.9 Spatial pattern of rurality index in 2000 (Li et al. 2015)

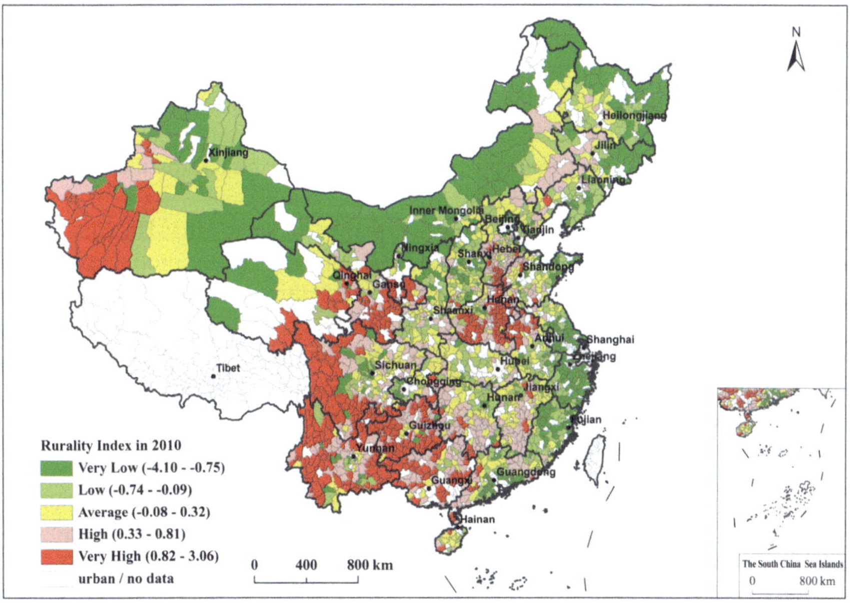

Fig. 6.10 Spatial pattern of rurality index in 2010 (Li et al. 2015)

engine of China's economic development, thus industrialization and urbanization dominate the process of socio-economic development but agriculture and rural landscape and culture are experiencing rapid decline.

(2) The second belt is northeast China. This region is both China's traditional heavy industrial base and national commodity grain base. This region faced economic depression and industrial restructuring during the 1990s, but has been experiencing a process of revitalization since the launch of Economic Rejuvenation Plan for Northeast China (Dunford and Li 2010). In addition, the level of modernization of agricultural production in this region is rather high and has led to high agricultural output. As such this region shows more urbanity and the rurality index in this region remains low.

(3) The third belt is northern China. Counties in this region usually have rich mineral resources, and vast resources exploration has dramatically changed local demographic characteristics, socio-economic structure and traditional culture, thus rurality index in this region declined significantly.

Moreover, three belts of "very high" rurality index areas can also be identified from Figs. 6.9 and 6.10.

(1) The first belt is southwest China. The development of non-agricultural industry and urbanization in this region is lagging behind compared to eastern coastal China, for relative poor physical conditions, location, and weak socio-economic foundation. Rural areas in this region are mountainous and current agricultural production is still extremely backward (Li et al. 2013).

(2) The second belt is northwest China, including southwest Xinjiang, northern Gansu and northwest of Sichuan. To a large extent, these regions are ethnic minority areas, with remote geographical, poor accessibility, backward socio-economic development, and are still struggling to transform the traditional subsistence agriculture.

(3) The third belt is central China or more exactly the Huang-Huai-Hai Plain. This region has long been one of China's most important grain production areas, which has excellent agricultural production conditions, and contribute a lot to national grain security. However, due to growing contradiction between population change and land use, and the strictly control of prices of agro-products under urban-rural dual structure, the socio-economic transition and urbanization of these region lagged behind (Li et al. 2014), thus have contributed to a high rurality index.

Over all, counties with "very low" value of rurality index mainly distribute in the eastern coastal, northeast and northern part of China, while counties with "very high" value of rurality index mainly distribute in the southwest of China and Huang-Huai-Hai plain. Counties with "low", "average" and "high" rurality index show a staggered distribution in the rest regions, more specifically, many counties in these three category are to be found on the periphery of inland urban areas.

6.4.5 Dynamic Change of Rurality

Areal variation is necessary to gain a widespread pattern of the distribution of rurality, and variation over time is likely to offer an interesting insight into the changing nature of rurality (Cloke 1977). The dynamics of rurality over time and space are further detected in three ways. Firstly is the correlation analysis. Figure 6.11 shows a scatter dot graph representing the rurality index in 2000 on the horizontal axis and that in 2010 on the vertical axis. Both the linear ascending trend in the cloud of dots and the high correlation between the two indices (regression coefficient (0.9155) almost equals 1; $R^2 = 0.838$, and $P < 0.01$) suggest little change in the rurality pattern of China's counties between 2000 and 2010.

Secondly is the changing matrix analysis. As shown in Table 6.9, in the year of 2000, 374 counties were classified as "very low" rurality. When compared to the year of 2010, among the 374 counties, there are 64, 8 and 3 counties became counties with "low", "average" and "high" rurality, respectively. That is, the rurality category of about 20 percent of the "very low" rurality counties in 2000 changed. In addition, the counties with "very high" rurality experienced nearly the same dynamics. However, the counties with rurality of "average", "high" and "low" experienced much more changes. In general, the rurality state of "very low" and "very high" counties experienced much less change in the category of rurality, while the other counties are more prone to get category of rurality changed.

Thirdly is the change of spatial pattern of rurality index. As shown in Fig. 6.12, counties with rurality index increased at least one level mainly distribute on Xinjiang, northeast China and the border areas of Henan, Guangxi and Hunan. To a large extent, this well reflects the large scale farmland reclamation and consequent agricultural development in northern China, and the dilemma of nearly stagnant socio-economic development in agricultural counties and mountainous counties of central China. While, counties with rurality index decreased at least one level mainly distribute in

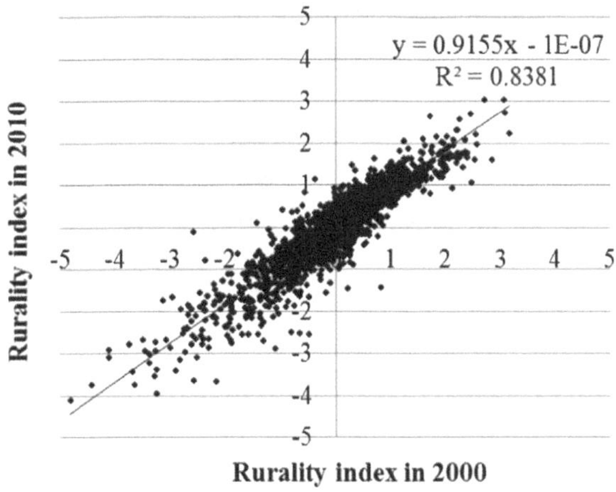

Fig. 6.11 Temporal persistence of rurality in China's counties (Li et al. 2015)

Table 6.9 The changing matrix of rurality index between 2000 and 2010 (Li et al. 2015)

Year 2000	Year 2010					Total (2000)	Loss	Changes (%)
	Very low	Low	Average	High	Very high			
Very low	299	64	8	3	0	374	75	20.05
Low	63	206	91	12	2	374	168	44.92
Average	10	85	168	101	9	373	205	54.96
High	1	18	96	188	70	373	185	49.60
Very high	1	1	10	69	292	373	81	21.72
Total (2010)	374	374	373	373	373	1867	714	38.24
Gain	75	168	205	185	81	714	–	–

Note According to the value of rurality index, the rurality in China have been divided into five grades by quintile ranging from very low, through low, average and high, to very high. For the total number of counties in this study is 1867, which is not divisible by5, the numbers of counties with very low, through low, average, high and very high rurality are 374, 374, 373, 373 and 373, respectively. This table demonstrates the change matrix of the number of counties with different level of rurality index. For example, in the row of "very low", compared to 2000, 299 counties remained "very low" in 2010, 64 counties became "low", 8 counties became "average", and 3 counties became "high". As a results, 75 counties with "very low" ruality index in 2000 became other types of rurality, which accounted for 20.05% the initial 374 counties. While, as shown in the column of "very low", 63 counties with "low" rurality index in 2000 became "very low" in 2010, and 10 counties with "average" rurality index, 1 counties with "high" rurality index and 1 counties with "very high" rurality index in 2000 became "very low" in 2010 too

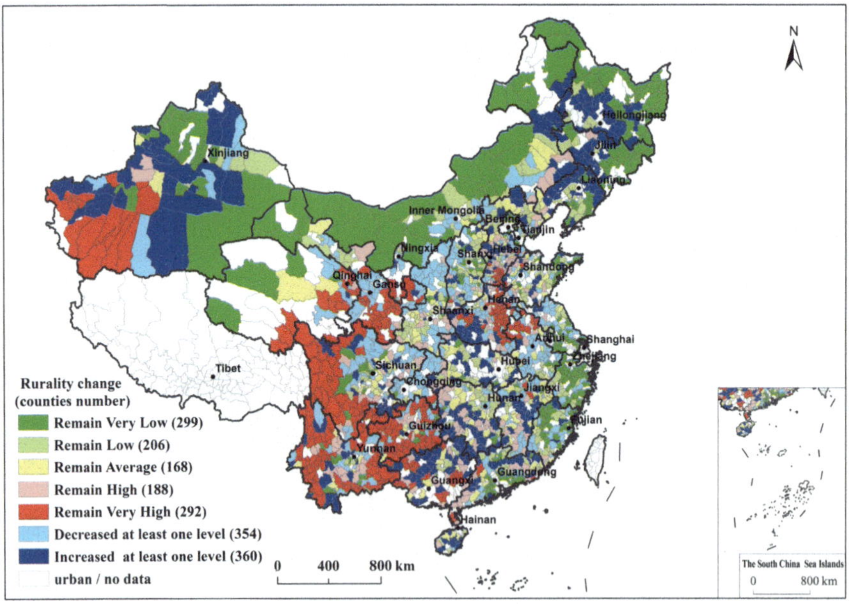

Fig. 6.12 The changing pattern of rurality index between 2000 and 2010 (Li et al. 2015)

agro-pastoral interlaced region around Yulin, periphery of Sichuan Basin, and the most part of Chongqing. This change is due to the vast mineral resources exploration in Yulin of Northern Shaanxi and the ambitious implementation of the Western Development Strategy in Sichuan Basin, which results in the development of non-agricultural economy.

6.4.6 Rurality and Socio-Economic Development

Rurality index has significant negative correlation with indicators reflecting the local ability to attract investment, output and value-added capabilities, local government financial strength and residents' income and savings levels (see Table 6.4 and Fig. 6.8). In the meantime, counties with higher rurality degree are prone to have higher relief degree of land surface, and have longer distances from nearest provincial capital, highway and railway. Therefore, high rurality counties have been marginalized both geographically and economically. Regional policies devote to promoting rural development and eliminating regional disparities should pay more attention to these marginal counties.

It is worth mentioning that, there was no significant correlation between the rurality index and per capita grain production and major agro-product per capita (Table 6.10). As shown in the last row of Fig. 6.13, per capita grain production and major agro-product per capita are higher in counties with moderate rurality index. In fact, China's major grain producing counties mainly located in plains of east part of China such as the Northeast Plain, the Huang-Huai-Hai Plain, the Sichuan Basin and the Yangtze River Basin. Compared to the counties with high rurality around them (Fig. 6.10), many of these counties especially in Sichuan Basin, and the Yangtze River Basin are experiencing gradually socio-economic transition driven by urbanization and industrialization and thus show a lower rurality. As such, rurality index can help to understand the overall pattern of China's rural development state effectively. However, due to the vast regional difference of county level physical geography condition and economic location, and China's unique agricultural and rural development policies under urban-rural dual structure, this index is less successful in revealing the agricultural production status quo alone.

6.5 Rural Restructuring in China: A Resources Allocation Perspective

Rural restructuring in China might mainly be described in three aspects, i.e., spatial restructuring, industrial reshaping and administrative reorganization (Long et al. 2012). China has undergone rapid economic growth and dramatic industrial restructuring, with the proportion of the primary, secondary and tertiary industry changed

Table 6.10 The correlation between rurality index and typical economic and geographical indicators (Li et al. 2015)

Types of indicators	Typical socio-economic indicators	2000	2010
The ability to attract investment	Per capita urban fixed asset investment	−0.331***	−0.436***
Output and value-added capabilities	Above-scale industrial output per capita	−0.433***	−0.543***
	Share of non-agro industrial value added	−0.538***	−0.500***
	GDP per capita	−0.612***	−0.597***
Local government financial strength	Local budget revenue per capita	−0.615***	−0.613***
	Local budget expenditure per capita	−0.352***	−0.294***
Residents' income and savings levels	Per capita net income of farmers	−0.567***	−0.653***
	Per capita savings deposits	−0.685***	−0.755***
Agricultural production	Per capita grain production	−0.012	0.014
	Major agro-product per capita	−0.031	0.017
Relief feature	Relief degree of land surface	0.301***	0.244***
Distance/marginalization	Distance from nearest provincial capital	−0.015	0.090***
	Distance from nearest highway	0.005	0.089***
	Distance from nearest railway	0.156***	0.191***

Note ***Significance at 1%

from 15%, 46% and 39% of GDP in 2000 to 10%, 44% and 46% in 2013, respectively. Research shows that the influential factors of China's industrial structure include domestic consumption propensity, urban-rural disparity, scale of the labor force and capital stock, property right protection, and administrative effectiveness. However, western scholars argued that human relations affecting different forms of embeddedness, a local or regional phenomenon that generally promotes the restructuring of rural areas, play an important role in industrial restructuring in rural areas (Fløysand and Sjøholt 2007).

As shown by the practices of rural restructuring of both developed and developing countries, there has been an increasing trend emphasizing regional resources in formulating rural development policy and restructuring rural areas (Zasada et al. 2015).

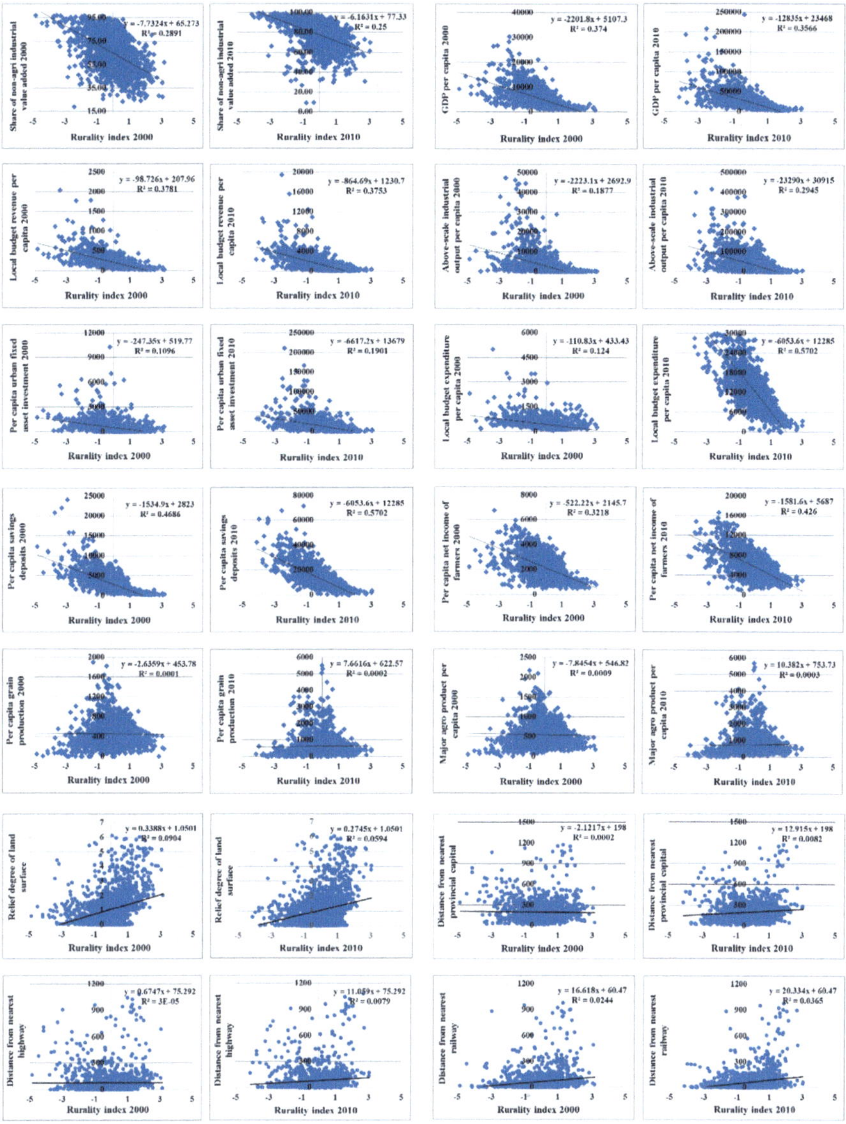

Fig. 6.13 The scatter map of rurality index and typical economic and geographical indicators (Li et al. 2015)

Among the resources affecting rural restructuring, the author argues that human resource, land resource and capital are critical. Migration plays an increasing role in China's economy by affecting economic restructuring, and results in a series of socio-economic changes including rural restructuring, balanced regional development, and laborer market changes (Li et al. 2014). The allocation policy on local resource use such as land and forest directly affected rural household livelihoods in Lao People's Democratic Republic (Fujita and Phanvilay 2008). Territorial capital investment has been considered as one of main cornerstones for rural development (OECD 2006). Tolbert et al. (2014) examined the changes in local ownership of traditional financial services and found that the rate of decline had been greatest in most rural counties.

It is crucial for rural restructuring to achieve the optimal allocation and effective management of critical resources through coordinating the development elements. This study is intended to analyze the rural restructuring in China affected by the allocation and management of critical resources based on a theoretical framework of "elements-structure-function" expounded in Chap. 5.

6.5.1 *Status Quo and Problems*

6.5.1.1 Human Resource

From 2000 to 2013, China's urbanization rate has increased from 36.22 to 53.73%, with an annual growth of 1.25%. The inflow of rural population is one of the main impetuses of this rapid advancement of urbanization. During the same period, rural resident population has decreased from 808.37 million to 629.61 million and rural registered population from 942.44 million to 876.01 million. Under China's special household registration system, the difference between rural registered population and rural resident population keeps on increasing (Fig. 6.14) with rural population flows into the urban area but without their household registration changed (Liu et al. 2011b). This number amounted to 246.4 million in 2013, accounting for 33.70% of urban resident population and 39.14% of rural resident population that year. The difference can also be used to represent the quantity of population who flow into the urban area but without their household registration changed. Enormous outflow of rural population has brought about changes on rural population structure and human resource deployment accordingly.

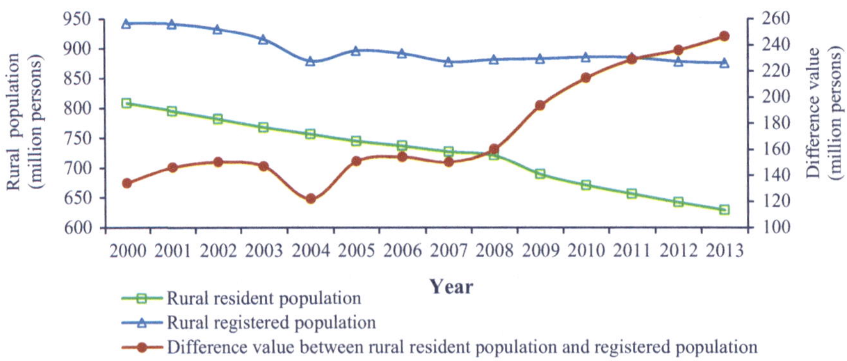

Fig. 6.14 Changes of rural registered population and rural resident population in China from 2000 to 2013 (Long et al. 2016)

(1) Changes of the structure of rural population

In 2014, there were 273.95 million peasant laborers[1] in China, including 105.74 million local peasant laborers[2] and 168.21 million migrant peasant laborers.[3] The latter took up 61.4% of the total number. The average age of peasant laborers was 38.3 with people aged 20–49 accounting for 79.4%; male took up 69% and female 31%.[4]

Most migrant laborers are young. Due to their migration, there is a clear tendency towards infantilization and aging of the population in rural area. For population aged 0–14, the proportion of rural area was 5.09% higher than that of urban area in 2010; for aged 20–49, 8.52% lower than that of urban area; aged 50–59, almost the same; aged above 60, 3.30% higher (Fig. 6.15). The outflow of rural population has exacerbated the aging process in rural area. In 2000, the population aged above 60 took up 10.91% in rural area, 1.23% higher compared with the proportion of 9.68% in urban area. In 2010, the proportion of population aged above 60 has risen to 14.98% in rural area and 11.68% in urban area, 4.07% and 2% higher in comparison with those in 2000, respectively. From the above, it is obvious that aging process in rural area is much faster than that in urban area.

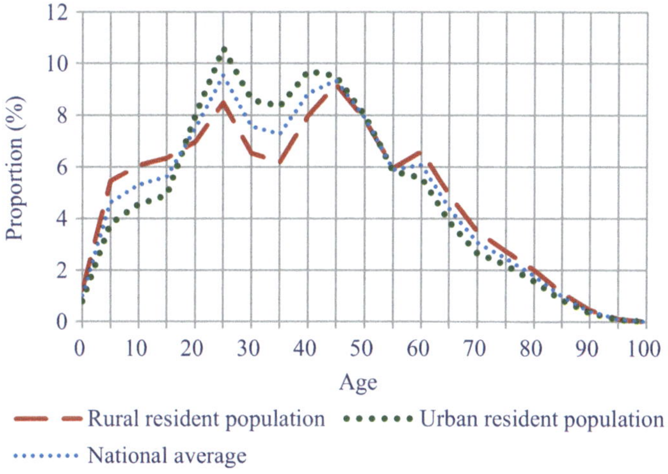

Fig. 6.15 The proportion of resident population in rural area and urban area in different age structure in China in 2010 (Long et al. 2016)

---

[1]Peasant laborer indicates a laborer has rural household registration, and engages in non-agricultural industry or migrates out for a job for six months or more in a year.

[2]Local peasant laborer indicates a laborer who is employed in the town or country where his/her household was registered at.

[3]Migrant peasant laborer indicates a laborer who is employed in a place except for the town or country where his/her household was registered at.

[4]Peasant Laborer Monitoring Report in 2014. (http://www.gov.cn/xinwen/2015-04/29/content_2854930.htm).

The migration has not only affected the age structure, but also changed the intellectual structure of the population in rural China. The migrant laborers have been led by better educated people, causing a "brain drain" that has seen rural communities lose their most active population segment (Table 6.11).

(2) **Changes of human resource deployment**

The outflow of rural population and the consequent influence on rural population structure have changed human resource deployment such as village cadres, health technician and agricultural labor force etc. in rural China.

Deployment of village cadres: Based on sampling survey on cadres of 124 villages in 13 provinces including Hunan, Jiangxi and Yunnan carried out by the author's research team in 2015, most village cadres are old and of the total, staff aged 40–60 takes up 75.82%, while aged 50–60 at 37.36% and above 60 at 10.99%. Old cadres are more familiar with rural work and agricultural activities, but they tend to be close-minded, blinkered and lack of management and innovation ability. As young men choose to work outside and don't have too much recognition on cadres' work, there is insufficient force reserve of primary organization in rural area with village cadres under 40 years old only accounting for 13.19%. In the view of educational attainment, there are 69.23% of cadres with junior and senior secondary school education, 29.67% with college and higher level. Most cadres with college and higher level education don't stay a long time and some of them are assigned by towns, hence

Table 6.11 The educational attainment of rural population and peasant laborers in China (%) (Long et al. 2016)

Educational attainment	Rural population in 2010	Peasant laborer		Migrant peasant laborer	Local peasant laborer
		In 2010	In 2014	In 2014	In 2014
No schooling	9.45	1.3	1.1	0.9	1.6
Primary school	33.17	12.3	14.8	11.5	18.1
Junior secondary school	46.25	61.2	60.3	61.6	58.9
Senior secondary school	8.77	15	16.5	16.7	16.2
College and higher level	2.36	10.2	7.3	9.3	5.2
Junior secondary school and higher level	57.38	86.40	84.10	87.60	80.30
Senior secondary school and higher level	11.13	25.20	23.80	26.00	21.40

there exist problems like lack of work experience, short tenure and discontinuity of work etc. The family income of village cadres is mainly from agriculture, and 58.6% of them have a per capita annual income under 5000 RMB yuan (exchange rate US$ to RMB yuan: 1–6.35), far behind the national rural per capita net income which is 8896 RMB yuan. This situation indicates that village cadres played a limited role in the village development and prosperity.

Deployment of rural health technicians: There were 9.79 million health workers nationwide in 2013. Rural and urban shared almost the same proportion, i.e. 54.15% in the rural and 45.85% in urban. However, rural health technicians are relatively insufficient and increase slowly (Fig. 6.16). In 2013, the number of health technicians was 3.64 per thousand people in rural area, equivalent of 69.07% of that in the whole country and 39.65% in the urban area. Among the technicians, the number of licensed (assistant) doctor was 1.48 per thousand people in the rural, equivalent of 88.62% of that in the whole country and 46.39% in the urban area. Due to inadequate attraction to talents, country doctor has been the main part of service in village clinics. In 2013, country doctor has taken up 70.9% in the village clinics while licensed doctor only 20.56%.

Deployment of agricultural labor force: With the development of industrialization and urbanization in China, rural labor force has transferred from agricultural to non-agricultural sector constantly. The agricultural labor force is declining while non-agricultural population is rapidly increasing. From 2000 to 2013, the proportion of population engaged in primary industry has declined from 50 to 31.4%. As opposed to this, non-agricultural population keeps on growing in rural area. Rural labor force engaged in non-agricultural industries has increased 48.53 million cumulatively since

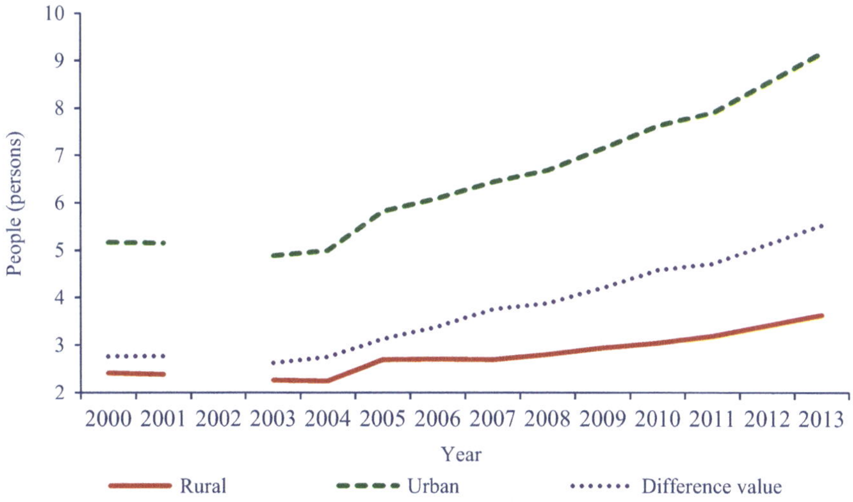

Fig. 6.16 The number of health technicians per thousand people in rural area and urban area of China during 2000–2013 (Long et al. 2016)

2008, among whom 20.73 million are local laborers.[5] The average annual growth of local non-agricultural employment is 2.96 million.

The flow of rural laborer to non-agricultural sectors has not only reduced the number of agricultural labor force, but also brought about significant changes to age structure of agricultural workers. In 2010, the proportion of "21–30 years", "31–40 years", "41–50 years" and "above 50 years" accounted for 16.7%, 19.3%, 26.4% and 34.6%, respectively, of the total agricultural labor force, while that of the total non-agricultural labor force was 35.9%, 23.5%, 21.2% and 12.9%, respectively. Therefore, with the rising of age structure, it presents the characteristics of increasing agricultural labor force and decreasing non-agricultural labor force (Fig. 6.17). The proportion of China's agricultural workers has reached to 32.5% for laborers aged above 50 and 11.2% for those above 60 in 2006. The proportion of laborers over 60 years old is 3% higher than that of the first survey in 1996.[6] However, the percentage has risen to 34.58% for laborers aged above 50 and 13.64% for those above 60 in 2010.

This phenomenon of aging of rural labor force has also been proved in our case surveys. For instance, there are 252 villagers in He village, Ganguyi town, Baota district, Yan'an city, Shaanxi province in 2015. Among them, 67 people are engaged in agricultural production with 76.12% over 50 years old, 37.31% over 60 years old and 17.91% over 70 years old. With the socio-economic development and technical progress, it is an inevitable trend to decrease the number of agricultural labor force.

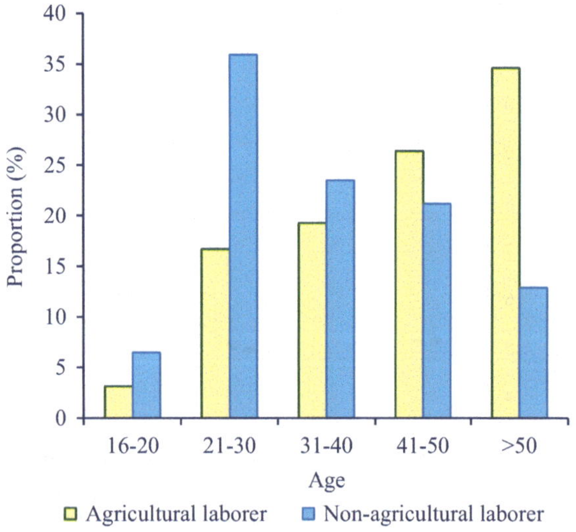

Fig. 6.17 Comparison of age structure between agricultural labor force and non-agricultural labor force (Long et al. 2016)

[5]Peasant Laborer Monitoring Report in 2009. (http://www.stats.gov.cn/ztjc/ztfx/fxbg/201003/t20100319_16135.html); Peasant Laborer Monitoring Report in 2014. (http://www.gov.cn/xinwen/2015-04/29/content_2854930.htm).

[6]The First National Agricultural Survey Data. (http://www.stats.gov.cn/tjsj/pcsj/nypc/dycnypc/200308/t20030826_39928.html).

However, problems like insufficient input of production means and manpower, great difficulty in popularizing agricultural science and technology, brought by aging rural labor force and low education level, have had a negative effect on the advance of production efficiency and development of modern agriculture.

To sum up, with the globalization, industrialization and urbanization, a large number of rural people have migrated to urban areas in China since the beginning of the new century. The outflow of young high-quality workforce has not only changed the population structure and human resource allocation in rural area, but also caused problems such as lack of talents, less-educated human resource, weakened mainstream of development. The change in human resource deployment has broken the traditional social structure and agricultural production mode, and will have profound impact on the utilization of land resource, the development of rural industry, the restructuring of rural social governance system, etc.

6.5.1.2 Land Resource

Driven by rapid industrialization and urbanization and related socio-economic changes and innovations, tremendous changes have occurred in the aspects of the structure, mode and efficiency of land use in rural China, as evidenced by the nonagriculturalization and non-grain preference of farmland use, farmland abandonment, transfer of land use right, rural housing land expansion and village hollowing.

(1) **Allocation of farmland resource**

Nonagriculturalization, non-grain preference and abandonment of farmland use: China's farmland resource keeps on decreasing with the expansion of the city and the increasing demand of industrial land. There were 121.72 million ha of farmland in China at the end of 2008. China's farmland has reduced by 11.48 million ha during 2000–2008.[7] After taking out the increased area from land consolidation, reclamation and development, there was a net decrease of 7.49 million ha of farmland in China. Besides disaster destroyed farmland, implementation of Grain-for-Green program (Long et al. 2006), agricultural structure adjustment etc., nonagriculturalization of farmland use, which refers to the process of changing from agricultural land with lower economic benefits into construction land with higher economic benefits, is an important reason for the decrease of farmland. During 2000–2008, the area of nonagriculturalized farmland has accumulated to 1.90 million ha, accounting for 25.31% of the net decreased area of farmland. The rate of nonagriculturalization of farmland[8] has accumulated to 1.52%, with an annual average of 0.16%. From the spatial

[7]In China, there are obvious differences in the data statistical caliber between the second land use survey finished in 2009 and the first land use survey finished in 1996. In addition, data of land use change survey of all provinces have been updated to 2008. So, the time span of land use data in this study is from 2000 to 2008.

[8]The rate of nonagriculturalization of farmland means the proportion of nonagriculturalized farmland in a region during a certain period in the total farmland area of the region in the beginning year of study period.

distribution of nonagriculturalization of China's farmland, during 2000–2008, nonagriculturalization of farmland was most rapidly advanced in developed regions like Shanghai, Beijing, Zhejiang and Tianjing with the rate of nonagriculturalization of farmland up to 14.1%, 12.84%, 8.54% and 7.3%, respectively; while the speed was slower in western and northeastern China such as Ningxia, Shaanxi, Yunnan, Tibet, Guizhou, Jilin, Inner Mongolia, Xinjiang, Hainan and Gansu etc. with the rate of nonagriculturalization of farmland all less than 1%.

The non-grain preference of farmland use refers to the phenomenon that farmland is used for planting cash crops instead of grain, which can be reflected by the fluctuation of grain sown area and yield. The grain sown area was 111.96 million ha in 2013 with an increase of 3.22% compared with that in 2000, and the proportion in agricultural crop sown area reduced from 69.39 to 68%. On the whole, there were not too many changes on grain sown area in China. However, thc phenomenon of non-grain preference of farmland use is prominent in some areas with the largest reduction of grain sown area occurred in Beijing, Zhejiang, Shanghai and Fujian, and the proportion has decreased by 48.46%, 45.50%, 34.89% and 34.26%, respectively. Corresponding to grain sown area, the grain yield in above mentioned four regions has declined by 33.34%, 39.73%, 34.40% and 22.27%, respectively (Fig. 6.18).

The non-grain preference of farmland use is intimately linked with the transfer of rural population, the change of the spatial pattern of farmland resource, the change of the structure of food consumption, the adjustment of agricultural structure etc., but its direct driving force lies in the relatively low benefits of grain-growing. The total cost of grain production has risen from 356.18 RMB yuan per mu in 2000 to 1026.19 RMB yuan per mu in 2013, with an average annual growth rate of 7.85%. The benefits of grain-growing have been continuously squeezed by the increasing cost, which caused low enthusiasm in planting grain among peasants.

The abandonment of farmland refers to the status of leaving arable land derelict or underutilized due to halted and reduced farming. Driven by the rising wages and incomes from non-agricultural employment and the jacking up opportunity cost of farming, the economic value of farmland has been reduced due to redistribution

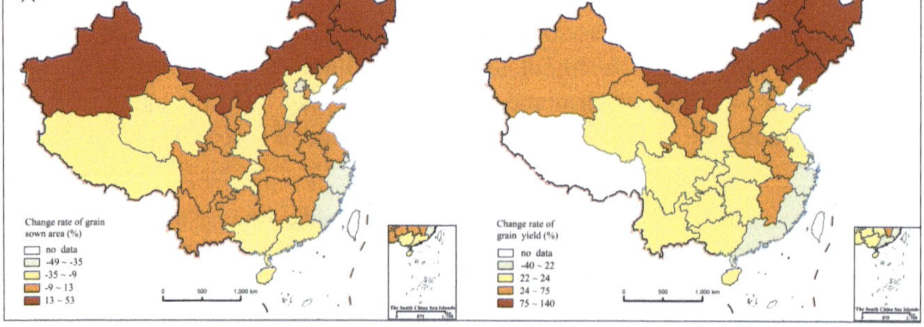

Fig. 6.18 Change rate of China's grain sown area and yield during 2000–2013 (Long et al. 2016)

and migration of rural labor force, which has consequently brought about extensive utilization of farmland such as "paddy field from double cropping to single cropping", "poplar expansion and cropland shrinkage", abandonment of farmland etc. This problem is particularly prominent in hilly and mountainous area with poor farming conditions and difficult in achieving laborer substitution. According to incomplete statistics, there were up to 161 research sites with such phenomenon, involving 107 counties (cities) covering 21 provinces (autonomous regions and municipalities) in China in the last two decades. For instance, in 2014, in Muyang town of Yunnan province, there were 749 ha of abandoned area with abandonment proportion taking up 25%; in Yulin city of Shaanxi province, there were 80.8 thousand ha accounting for 13.9%.

The issue of China's nonagriculturalization, non-grain preference and abandonment of farmland use, is a reflection of the changes of rural man-land interrelations on land use structure and mode under the background of rural transformation development, and is the change of allocation of land resource caused by the adjustments of land use structure and mode which are made by rural household from economic rationality based on scarcity degree and relative price of changed production elements. However, the phenomenon of regional nonagriculturalization and non-grain preference of farmland use, will ensure the pressure on national food security be passed to the main grain producing areas in the northern and central China. The center of gravity of China's grain production has taken on the changing trends of "heading northwards and moving to the central" and "heading northwards and marching westwards" (Liu et al. 2009). The original situation of "transporting grains from the South to the North" has been evolved into that of "sending grains from the North to the South (Liu et al. 2009), which has important influences on the regional social economy and ecology such as aggravating the contradiction of water and land resources in traditional agricultural areas in central China, potential eco-environment risk in the activity of "increasing grain yield through reclamation of arable land" in ecologically fragile areas of northern China, soil hardening and soil fertility decline caused by overloading operations of land resource in northeastern China.

The phenomenon of "Upside down" between grain production contributions and economic income level has intensified the imbalance of regional economic development between the East and the Midwest, the developed areas and underdeveloped areas. The abandonment of farmland is the equivalent of fallow cropping system which is conducive to the recovery of farmland soil to some extent, but it is not only a great waste of land resource, but also an expression of an urgent need to strengthen regional resource allocation and regulatory means under such a kind of man-land interrelation as imbalance between land supply and demand and serious shortage of land resource in China.

Land transfer: The migration of rural labor force and the introduction of new production tools have brought about changes of marginal productivity of different elements and requirements for re-allocation of agricultural production resources. Land transfer is a process of re-allocation and optimization of rural contracted land use right among different management mainstream. The transfer of rural contracted land use right began to appear in early 1980s and for a long period of time, the area of transfer has accounted for about 4.5% of the household contracted farmland in general. As proposed in the Notice on How to Do a Good Job in the Transfer of Rural Contracted Land Use Right of Peasant Household issued by the Central Committee of the Communist Party of China (CPC) in December 2001, it was allowed to transfer the contracted land use right on the basis of stabilizing the household contract management system. And since then, the transfer scale of farmland has been enlarged and the progress been advanced gradually. During 2000–2009, the average transferred farmland has been increased from 0.62 to 1.06 mu per household with its proportion in household contracted farmland increased from 8.34 to 14.89%. The total transferred area of household contracted farmland has reached to 22.74 million ha in 2013, with an increase of 7.54 million ha compared with 2012 and an increase rate of 22.5%, which accounted for 25.69% of the total area of household contracted farmland.

The speed of China's land transfer is closely related to the small-scale and fragmentation of farmland management under household contract responsibility system. The area of each field plot was averagely 1.26 mu in household contracted farmland in 2000. The intensive large-scale management has been strengthened in farmland running in recent years, but the fragmentation of farmland management has not been changed fundamentally. In 2009, the area of each field plot was averagely 1.74 mu in household contract management farmland, with the pieces of farmland "below one mu", "1–3 mu", "3–5 mu" and "above 5 mu" accounting for 58.78%, 27.8%, 7.32% and 6.1%, respectively, of the total pieces of average household management land (Fig. 6.19). The fragmentation of farmland was more serious in western and eastern China than that in central China, with the pieces of farmland "below one mu" in the eastern, central and western China taking up 66.14%, 49.08% and 61.89%, respectively, of the total pieces of average household management land. The small-scale management of farmland has increased the communication and transaction costs of land transfer, which has restricted the scale of land transfer and has resulted in the regional difference of land transfer.

The scale and speed of farmland transfer has presented obvious regional differences under the background of enlarged transfer scale of farmland with the proportion of transfer is higher in fertile areas, developed areas and regions with higher transfer of rural laborers and lower in the central and western areas. In 2006, the proportion of households participating in the transfer of farmland is up to 44.49%, 28.01%, 26.76%, 23.93%, 23.88% respectively in Shanghai, Heilongjiang, Zhejiang, Beijing, Fujian, while the proportion is only 3.68%, 3.81%, 4.06%, 4.68%, 5.03% respectively in Henan, Shandong, Hebei, Gansu, Shaanxi (Fig. 6.20). In the central and western China, peasant income is highly dependent on land output due to underdeveloped non-agricultural industries and lack of employment opportunities. In such situation, they are unwilling to rent out their farmland. In addition, there is such a situation in some regions as difficult transfer and abandonment of farmland because of high input cost of renting caused by low quality of farmland and underdeveloped irrigation and water conservancy facilities.

Moreover, it is unsustainable for the profit mode of renting land for grain growing on account of low benefit of grain growing and mounting leasing price. The non-grain

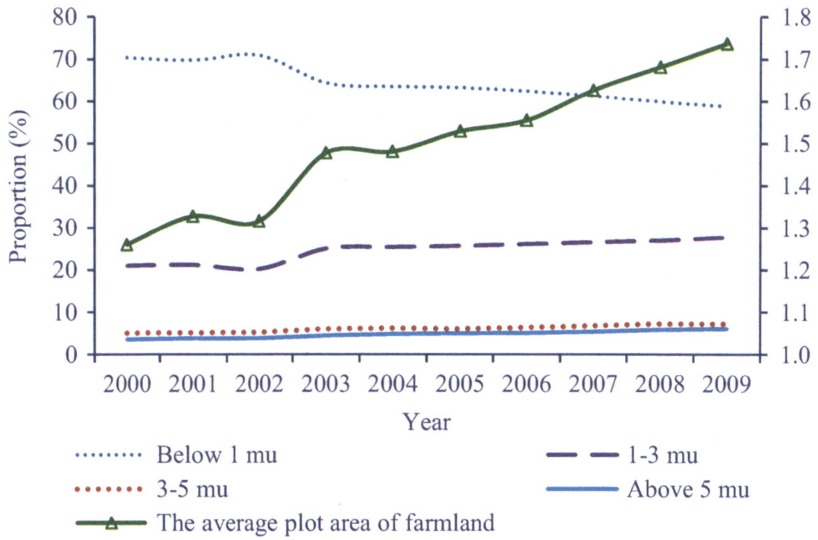

Fig. 6.19 The proportion of the pieces of farmland with different sizes in total pieces of average household management land and the average plot area of farmland during 2000–2009 (Long et al. 2016)

preference of farmland use is prominent in land transfer with the blinded introduction by some local government of industrial and commercial enterprises who has rent large area of farmland for planting economic crops or developing non-agricultural industry for a long time. In 2013, there were 12.85 million ha of transferred land for food crops only with a proportion of 56.52% in the total transferred farmland. Based on the survey on "non-grain" growing of transferred land, the rate of "non-grain" growing of transferred land has reached to 61.1% in the four major grain producing areas including Hebei, Henan, Shandong and Anhui provinces.

(2) **Allocation of rural construction land**

With a large number of rural labor force flow into the city, it is expected that the rural construction land in China will be decreased. However, on the contrary, it takes on a trend of rural construction land increase with rural population decline, subjecting to the constraints of current household registration system, rural social security system and rural land use policy (Long et al. 2012). From 2000 to 2007, China's average annual growth of rural housing land amounted to 16 thousand ha, with an annual growth rate of 0.1%. In terms of regional differences, rural housing land increased rapidly in the eastern and the western of China, with an annual growth rate of 0.19% and 0.15%, respectively. While the growth rate in northeastern China was only 0.07%, even a slight decrease trend in central China. During 2000–2007, there was an obvious increasing trend of rural housing land in some provinces such as Tibet, Zhejiang, Hebei, Xinjiang, Ningxia, Qinghai and Tianjin; while it was decreased in Shanghai, Jiangxi, Anhui, Gansu, Chongqing and Shaanxi.

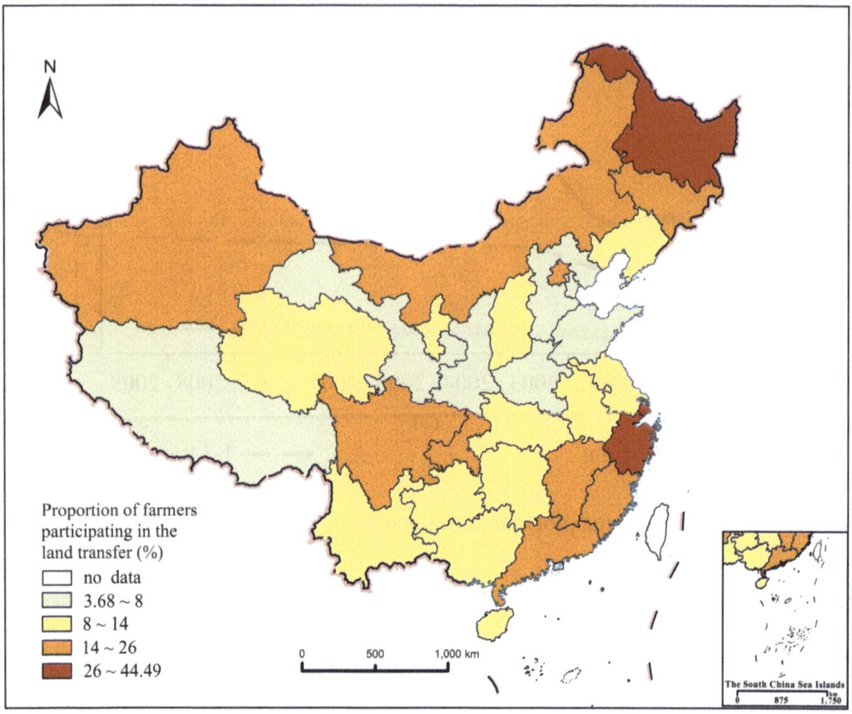

Fig. 6.20 The proportion of farmers participating in land transfer in the provinces of China in 2006 (Long et al. 2016)

The overall increasing trend of rural housing land, together with the decrease of both rural resident population and registered population, made the per capita rural housing land rapidly increased. During 2000–2007, per capita rural housing land in terms of rural resident population and rural registered population increased by 12.68% and 4.4%, which exceeded 50% and 25% of the national standard, respectively. The change of China's rural housing land per capita in terms of rural registered population took on a trend of rapid growth in eastern developed areas, slight decline in central and western areas, and little change in northeastern China (Long et al. 2013). The change of China's rural housing land per capita in terms of rural resident population showed a rapid increase in the eastern developed areas and the central and western areas from which a lot of rural laborers migrated out (Long et al. 2013). In this aspect, most provinces experienced a growth process except for Shanghai and Tibet. In the provinces with a growth trend, Chongqing, Hebei, Jiangsu, Hunan and Beijing had an annual growth rate of more than 3% (Fig. 6.21).

The status quo of coexisting rural population decline and rural construction land increase caused the low efficiency of rural housing land use. In 2007, there were 27 provinces whose rural housing land per capita in terms of rural resident population exceeded the national standard, and 23 provinces whose rural housing land per capita

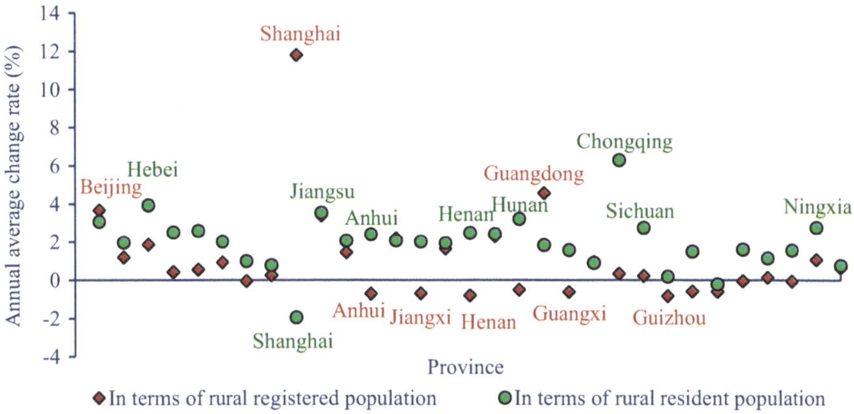

Fig. 6.21 The annual average change rate of rural housing land per capita in terms of rural resident and registered population during 2000–2007 (Long et al. 2016)

in terms of rural registered population exceeded the national standard. The author led a field investigation in May 2015 in Yangqiao village, Yucheng city, Shandong province, which showed that the phenomenon of village hollowing was shocking. Yangqiao village had 237 houses, 83 of which were idle or derelict, accounted for one-third of the total rural housing land area of this village, and 70% of the households in the village have more than one house. Under the background of rapid urbanization, the change of man-land interrelation and settlement spatial structure caused village hollowing and related socio-economic problems. The research of the Institute of Geographic Sciences and Natural Resources Research, CAS, indicated that the potential of increasing farmland via consolidating hollowing villages in China was 7.6 million ha under the background of pushing forward urbanization in batch (Liu et al. 2011b).

In order to accomplish the optimal allocation of land resource, the central government promulgated the land use policy of "Linking up Increased Urban Construction Land with Decreased Rural Construction Land" and "The National General Land Use Planning (2006–2020)", both emphasizing the consolidation of rural construction land, especially that in the hollowing villages. The central government tried to solve the contradiction between the spatial constraints for increasing urban construction land and the inefficient use of rural construction land through the innovation of land consolidation policy. For these, national and local governments have carried out considerable land consolidation projects at different scales. However, the implementation of these land consolidation projects has brought about some problems, such as excessively emphasizing on construction land quota, blindly pursuing land finance and land consolidation potential, engaging in violent demolition of rural houses and violating farmers' desire, depriving the development rights of the farmers, etc.

6.5.1.3 Capital

Capital plays an important role in promoting the development of rural economy and increasing the income of farmers. Since 2000, the national financial department has gradually increased the intensity of the capital supporting for rural development. The expenditure at national level for supporting rural development rose from 123.15 billion RMB yuan in 2000 to 1376.3 billion RMB yuan in 2013, accounting for the proportion of national fiscal spending by 7.75% to 9.82%, respectively. Recently, some agricultural leading enterprisers and industrial and commercial enterprisers have gradually increased their investment into the rural, which have led to a rapidly increase in rural investment. To some extent, the fund shortage problem impeding rural development has been alleviated. However, it is still problematic in the aspects of insufficient capital for developing rural public infrastructure and affairs, limited supply of rural financial capitals, and low efficiency of capital use due to the system obstacle concerning its allocation and management.

(1) **Allocation of capital for developing rural public infrastructure and affairs**

In recent years, the national financial expenditure has been gradually inclined to promoting the development rural social affairs, to achieve the equalization of basic public services. Compared with 2007, the spending of rural social affairs development increased by 3.27 times in 2013, accounting for 43.89% of the total expenditure for rural development. But compared with the city, the capital allocation for developing rural public infrastructure and affairs is still inadequate. The structural imbalances of capital allocation for developing public infrastructure and affairs between the urban and the rural as well as among different areas were too evident.

Allocation of public financial funds for infrastructure construction. In China, the disequilibrium phenomenon of the allocation of public financial funds for infrastructure construction between the urban and the rural is obvious, that of the rural is far lower than the urban. In 2001, per capita public financial investment for infrastructure construction at the levels of village, country, town, county and city was 20, 62, 169, 374 and 658 RMB yuan, respectively. The investment into the city was 33.2, 10.6, 3.9 times of that into the village, country and town, respectively. The investment into the county was 18.9, 6.0 and 2.2 times of that into the village, country and town, respectively. From 2001 to 2013, per capita public financial investment for infrastructure construction at different levels took on an increasing trend, with a growth of 223, 431, 885, 2424 and 3679 RMB yuan to village, country, town, county and city, respectively (Fig. 6.22). However, the investment growth of village, country and town was relatively slow, only accounting for 6.06% and 9.02%, 11.72% and 11.72%, 24.06% and 36.53% that of city and county, respectively. Inadequate investment into rural public infrastructure construction has brought about many negative effects on the improvement of the living standards of farmers, the development of the rural market system and the introduction of new elements in favor of agricultural production.

Allocation of educational finance. In 2000, the per student educational finance spending of local rural junior secondary school and primary school were 884.41 and

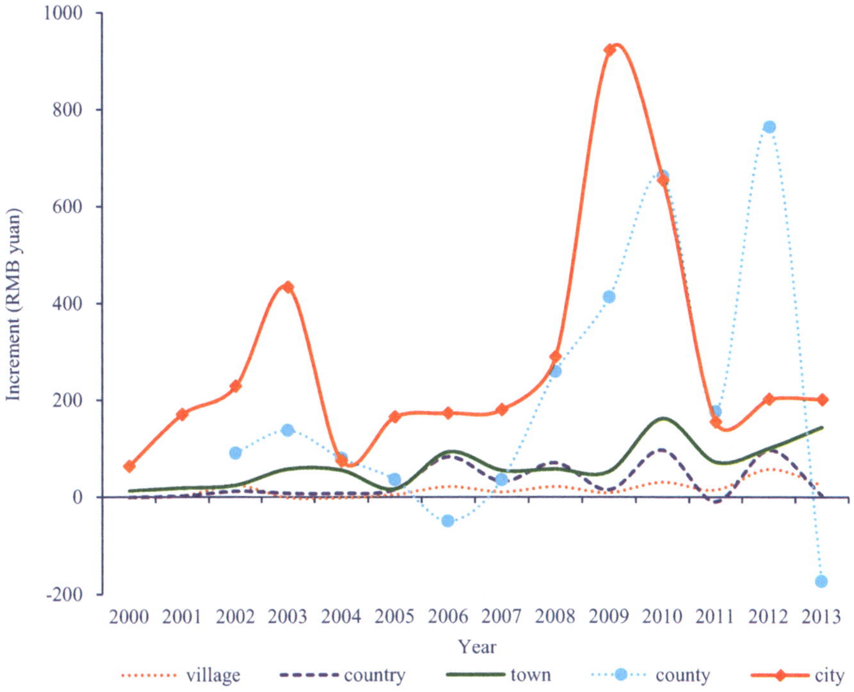

Fig. 6.22 The increment of per capita public financial investment for infrastructure construction at the levels of village, country, town, county and city during 2000–2013 (Long et al. 2016)

647.01 RMB yuan, accounting for 73.01% and 81.63% that of the local ordinary junior secondary school and primary school, respectively. From 2000 to 2011, the educational finance for local junior secondary school and primary school increased rapidly. Compared with the year 2000, per student educational finance spending of rural junior secondary school and primary school in 2011 has increased by 7.41 and 7.84 times, respectively. However, due to the imbalance education funding allocation between the urban and the rural, the difference between local ordinary primary and secondary schools and rural primary and secondary schools per student education spending was gradually expanded (Fig. 6.23). During 2000–2011, the difference of per student educational finance between local rural junior secondary school and primary school and local ordinary junior secondary school and primary school increased from 326.91 and 145.62 RMB yuan in 2000 to 739.64 and 398.53 RMB yuan in 2011, respectively.

From the perspective of the regional allocation of per student education finance in local rural primary and secondary schools, the gap between different provinces was obvious. In 2013, Beijing, Tianjin, Shanghai, Jiangsu, Zhejiang and other developed provinces had a higher per student education finance of rural primary and secondary schools, but some central and southwestern provinces, e.g., Henan, Hunan, Hubei,

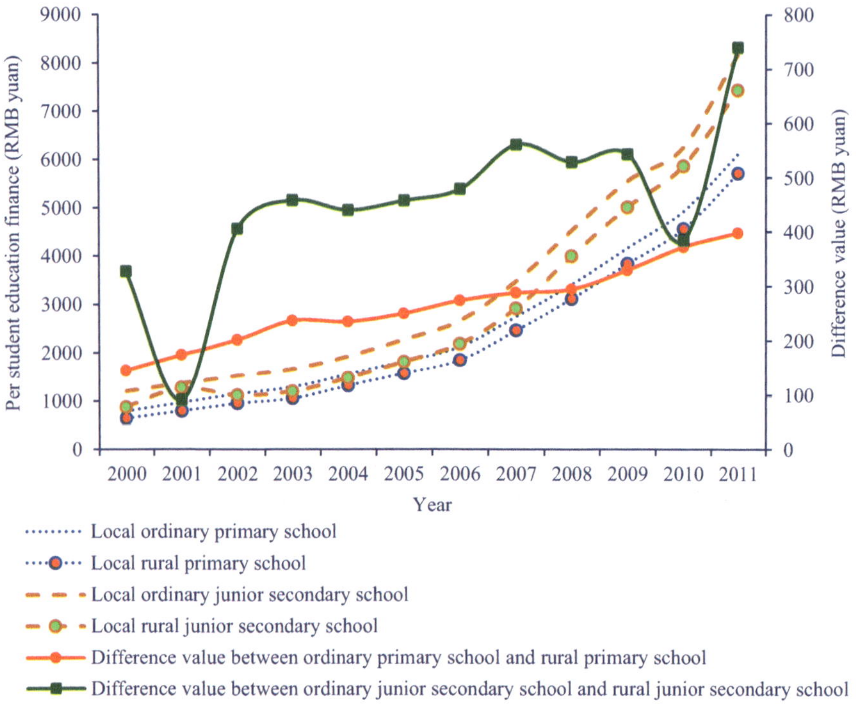

Fig. 6.23 The differences of per student educational finance between local rural and ordinary junior secondary school and primary school during 2000–2011 (Long et al. 2016)

Jiangxi, Guangxi and Guizhou had a less share. In recent years, with the implementation of preferential policy about national public affairs for the western China, the education investment of Tibet, Xinjiang, Inner Mongolia, Ningxia, Qinghai provinces have increased rapidly (Fig. 6.24). In general, per student education finance of China's local rural primary and secondary schools appears higher in the eastern, northwestern and northeastern areas, while that in the southwestern and central areas is relatively low.

(2) **Allocation of financial capital**

China's rural financial capital market has been developed slowly, and the difficulty of obtaining loans and the high cost of loans are common problems faced by rural households. Due to the characteristics of long-term investment and slow return of agricultural project, together with many uncertain factors and high risk make the financial institutions lack investment enthusiasm for rural area, and move their attentions to the large and medium-sized cities. As a result, rural areas get less financial support than the cities, which can be reflected by two indices of the difference of per capita loans between the urban and the rural and the ratio of per capita loans between the urban and the rural. From 2000 to 2008, the absolute size of the difference of per capita loans between the urban and the rural rose from 4121.39 yuan to 8106.72

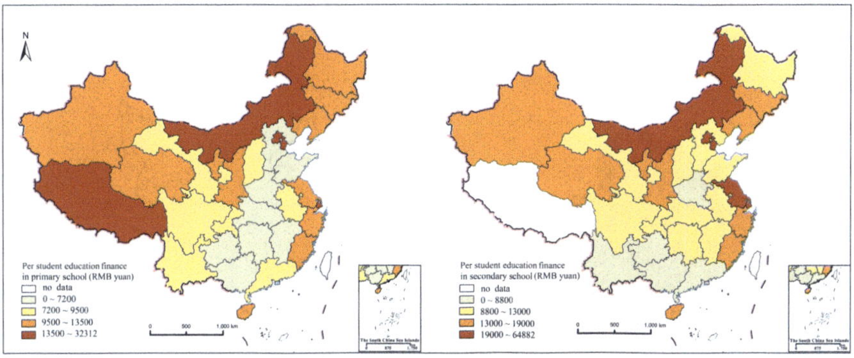

Fig. 6.24 The regional pattern of per student education finance of China's local rural primary (left) and secondary (right) schools in 2013 (Long et al. 2016)

yuan (in constant price of 2000), and the average value of the ratio of per capita loans between the urban and the rural was 13.17:1, which means that the urban can get the financial support 13 times more than that of the rural. However, the ratio of per capita loans between the urban and the rural in the central and western provinces was higher than that in the eastern provinces (Fig. 6.25), and it showed that the development of rural financial capital market in the central and western regions is more sluggish.

Due to the under-developed rural financial market and most households and rural enterprises lacking of effective mortgage goods, the difficulty for their obtaining loans from formal financial institutions is increased. As a result, loans from informal financial institutions and folk lending become the main channels of rural financing. The household loans from banks and credit cooperatives only accounted for 37.09% of the cumulative amount borrowed in 2009. Although this value increased by 7.56% compared with that in 2000, the amount of private loans still accounted for 61.68% of the cumulative amount borrowed in 2009. However, private finance exposed many problems such as high interest rate, disordered management, lacking of legal protection, and so on. Insufficient loans from formal financial institutions make rural development in a capital shortage state for a long time, and it has constrained the development of China's rural economy to some extent .

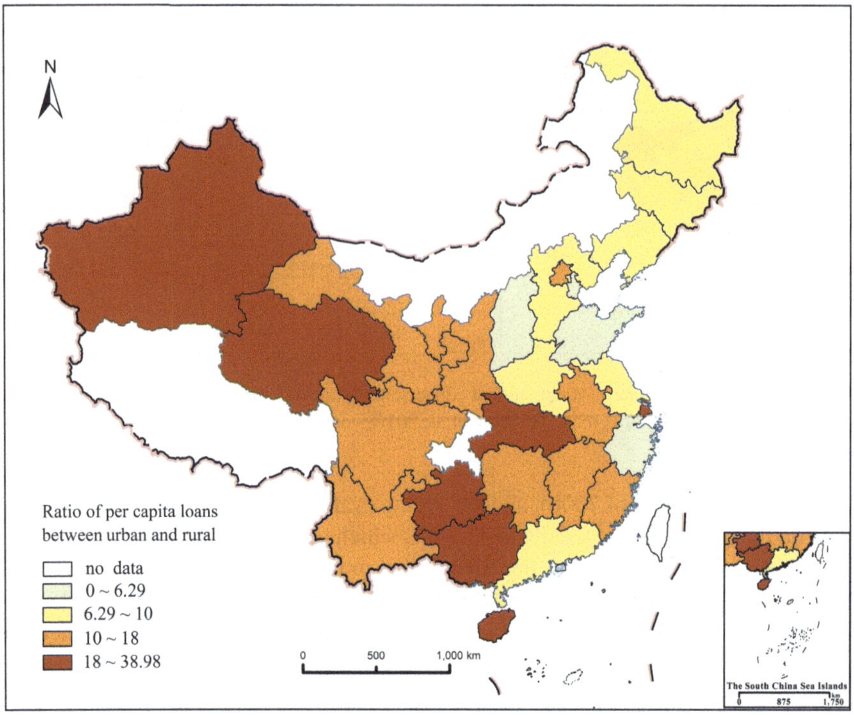

Fig. 6.25 The ratio of per capita loans between the urban and the rural at provincial level (Long et al. 2016)

(3) Industrial and commercial capital

The shortage of funds has long been the bottleneck afflicting China's rural development. Aiming at the status quo of funding shortage in China's rural development, some industrial and commercial capitals are continuously flowing into rural market taking the advantages of land transfer policy being implemented by the central government, which promotes the development of agricultural scale management and agricultural products processing. In 2013, 9.44% of transferred farmland in China was contracted by industrial and commercial enterprises. In Zhejiang province, the accumulated industrial and commercial capitals invested into modern agriculture have exceeded 20 billion RMB yuan during 2010–2013. To some extent, investing industrial and commercial capital into agriculture makes up for the shortage of financial capital. However, approximately 10% of the enterprises withdrew afterwards on account of the out-dated rural financial service system, the shortage of technician, and the unstable land circulation contract. In addition, due to the profit pursuing nature of capital, industrial and commercial capital involved in rural development will probably result in some new issues and contradictions such as the resources flow from rural to urban, the radical changes of land use, and so on.

As for the existing problems of rural capital allocation, they can be reflected in several aspects including the insufficient investment in public resources, the limited supply of rural financial capital and the imbalance allocation between the rural and the urban. Additionally, the system obstacle of fund allocation and management, which causes multiple management and repeated investments, imposes a great effect on the efficiency of capital utilization. As the No. 6 Auditing Results Announcement in 2008 based on 50 counties' audit investigation results of central specialized fund in supporting agriculture published by The Audit Bureau of China shows, the specialized fund was distributed into many departments step by step and it covered numerous projects involving various fields. Consequently, it was difficult to concentrate limited financial resources to solve critical issues. Among the 50 counties chosen to check, the total of central specialized fund is 2.387 billion RMB yuan and it was dispersedly invested into 45.5 thousand projects.[9] According to The State Council's 2013 Auditing Working Report Concerning Central Budgetary Implementation and Other Financial Revenue and Expenditure, 'farming, forestry and water' category includes 66 special projects and they are controlled by 50 departments and bureaus of nine competent departments at central administration level. At the local level, they are related to 20 provincial competent departments.[10] All kinds of problems, such as the function cross, disunited policy requirements, unsmooth information communication among departments, multiple management and multi-channel distribution of the specialized fund, contribute to the repeated projects setting and contents duplicating. Furthermore, capital input at national level has not effectively solved the subsistent obstacles in a real sense. Pilots of New Countryside Construction and stronger villages probably attain more projects and supporting funds depend on their higher popularity (Rogers 2014).

In conclusion, the insufficient investment in public resources, the limited supply of rural financial capital and the imbalance allocation between the rural and the urban and other problems are becoming increasingly serious. Considering the production element attribute and scarcity of capital, capital input from government fiscal expenditure, financial credit, and industrial and commercial capital play an important role in leading the flowing of land, technology, laborer and other production elements. Owing to the shortage of funds and the tremendous gap of capital allocation between the rural and the urban, productive resources, the high quality medical and educational resources will be centralized in cities and developed areas inevitably. Thus, polarization effect can also prohibit the development of rural areas and underdeveloped areas, so as to enlarge the imbalance resources allocation among different regions.

[9]The Auditing and Investigating Results of 50 Counties About the Use of Central Specialized Fund in Supporting Agriculture (No. 6 Document in 2015 issued by National Audit Bureau of China). (http://www.audit.gov.cn/n1992130/n1992150/n1992500/2302050.html).

[10]The State Council's 2013 Auditing Working Report Concerning Central Budgetary Implementation and Other Financial Revenue and Expenditure. (http://www.audit.gov.cn/n5/n26/c64269/content.html).

6.5.2 Prospects

Under the background of globalization, industrialization, urbanization, informatization and agricultural modernization, the interaction and allocation changes of human resource, land resource, financial capital and other rural critical development elements have brought about many challenges for China's rural development. Owing to the phenomenon that a growing number of prime rural laborers flow into cities, serious brain drain arises in rural area. Rural human resource tends to be low level, and the mainstream of rural development has been weakened continuously. Limited by the change of rural labor force, the increasingly grain production cost, the decrease of the economic value of agriculture land use and the existing institution defect, the phenomenon of nonagriculturalization, non-grain preference and abandonment of farmland use has become increasingly prominent. The changes of rural labor force and agricultural mechanization level give rise to the requirement of reallocating land contractual management right between different management mainbodies. The situation of land transfer allows of no optimism, and it is restricted by many factors such as the underdeveloped non-agricultural industry and land fragmentation. The unsound household registration system, imperfect rural social security system, the unsmooth rural homestead circulation and other related factors impose great restrictions on rural development. As a result, large-scale labor force transformation is not linked to the reduction of rural housing land. Instead, the existing obstacles embrace the expansion of rural housing land, as evidenced by widely spread vacant and derelict rural settlements. The unfairness and structural imbalance of resources allocation owing to rural-urban dual system cause rural development space further occupied by the urban and rural self-development ability further weakened. Integrating and restructuring rural space, industry and social organization is in great urgency to solving the problems and challenges resulted from the changes of the allocation of human resource, land resource and financial capital.

6.5.2.1 Spatial Restructuring

A majority of China's rural settlements are characterized by the spatial distribution features of 'small scale, scattered and chaos' and in a natural evolution state. The scattered distribution of rural settlements, rural-to-urban migration and idle and derelict rural construction land contribute to a series of problems. On the one hand, production elements will be depleted in the flow process due to chaos production space and the cost of economic development will be increased. On the other hand, it is difficult to put the optimal allocation of rural public infrastructure and socialization services into implementation. The shortage of scale benefit and output efficiency originated from public resources investment occurs in some areas. What appears to be in serious insufficiency is the allocation of educational, medical and public infrastructures in some parts of rural China. Hence, rural development placed under the contours of rural-urban territorial system is of vital necessity to improve town-village spatial system and restructure the rural spatial pattern.

Initially, town-village planning system should be established based on coordinating urban-rural development. It is crucial to build the spatial agglomeration axis and the structure network of town-village planning, taking into account the development scale, spatial distribution, territorial function and main transportation artery network, so as to forming a three-level rural settlements system composed of central town, major town and central village (or community). On the basis of the node effect of central town and major town, there should be a greater concern to giving full play to the radiation and driving function of central town and major town, so as to laying a material foundation for in situ urbanization and employment in industrial park by accelerating the construction of central town and major town, intensifying the cultivation of characteristic industry, improving public service facilities, enhancing comprehensive service function, and strengthening the agglomeration function of towns.

Then, it is essential to restructure the production, living and ecological space in the village at the micro level by carrying out land consolidation. Meanwhile, it is feasible to put the unified planning of residential district, industrial district, basic farmland protection region and ecological region into implementation according to the principle of relative centralization and resources optimal allocation. Aiming at the status quo of scattered, disordered and hollowing rural housing land, based on natural environment, socio-economic development, cultural customs and other various territorial differences, probing the institution innovations and regional model of pushing forward hollowed village consolidation and central village construction is helpful for rural spatial restructuring. While the implementation of village combination which guides peasants to centralized community or administrative village, has been recognized as an effective way to economizing administrative resources and improving the social service efficiency of public resources. Furthermore, the scientific assessment on the suitability of public facilities and infrastructures allocation and strengthening rural infrastructure construction have been favored to restructure a livable rural living space (Long 2014). However, as for the scale and spatial layout of central village and town, as well as construction and housing design of the central community, local government should give sufficient consideration to the factors such as natural environment carrying capacity, regional production mode, industry development orientation and living custom of local residents, and it is essential to probe the rural spatial restructuring models in different areas, e.g., plain agricultural area, mountainous and hilly area, urban and suburban area, under developed area and developed area. Eliminating the fragmentation of agricultural land management is a challenge that requires effective measures. At first, village combination, rural housing land consolidation as well as agricultural land consolidation should be carried out. And establishing large-scale high standard basic farmland is helpful for providing space for agricultural scale management and agricultural production base construction. In order to restructure intensive and efficient rural production space and promote nonagriculturalization industry agglomeration development, making existing scattered, disorderly and small-scale industries centralized in a place with perfect infrastructures is an effective way (Long 2014). The restructuring of environmental friendly and harmonious rural ecological space needs to push forward no public hazard agricultural production, establish ecological interception system, strengthen the comprehensive treatment of pollutants, and improve ecosystem corridor.

6.5.2.2 Economic Restructuring

Economic restructuring imposes great effect on widening farmers' employment channel, transforming land use mode, enhancing rural economic strength and self-developing ability, is becoming increasingly important in rural restructuring. The improvement of rural economic strength makes it possible for operating rural spatial restructuring and social restructuring. In the special environment of the decelerated growth of world economy, regional development mode transformation and industrial structure adjustment, great changes have taken place in rural socio-economic system. These changes have resulted in the reduction of laborer employment, subsequently a growing number of peasant-workers were returned to home to get or wait employed. The Chinese government has taken a series of corresponding measures to prevent the impacts of economic downturn on peasant's employment and income. For instance, proactive fiscal policy was implemented through expanding the central fiscal expenditures and the local government issuance of new bonds; the national strategy of "Mass Entrepreneurship and Innovation" was carried out,[11] and supporting the entrepreneur of the home-going peasant-workers[12]; reform of the fiscal and taxation system was accelerated in order to further reduce the cost of enterprise start-up and operating[13]; the right of rural land use was confirmed gradually which is beneficial to increase farmers' property income[14]; strategy of new-type urbanization was implemented to settle the former agricultural population who have migrated to the cities[15]; targeted measures were taken in poverty alleviation and the efforts in helping the vulnerable

[11]Opinions of the State Council on Some Policies and Measures to Promote Mass Entrepreneurship and Innovation (No. 32 Document in 2015 issued by the State Council of China). (http://www.gov.cn/zhengce/content/2015-06/16/content_9855.htm).

[12]Opinions of the State Council on Supporting the Entrepreneur of the Home-Going Peasant-Workers (No. 47 Document in 2015 issued by the State Council of China). (http://www.gov.cn/zhengce/content/2015-06/21/content_9960.htm).

[13]Opinions of the State Council on Further Promoting the Employment and Entrepreneurship in the New Situation (No. 23 Document in 2015 issued by the State Council of China). (http://www.gov.cn/zhengce/content/2015-05/01/content_9688.htm).

[14]Guidance of the State Council on Carrying Out the Mortgage Loans Pilot of the Management Rights of Rural Contracted Land and Farmer's Housing Property (No. 45 Document in 2015 issued by the State Council of China). (http://www.gov.cn/zhengce/content/2015-08/24/content_10121.htm).

[15]The Central Committee of the Communist Party of China and the State Council of China. National Plan on New-type Urbanization (2014–2020). (http://www.gov.cn/zhengce/2014-03/16/content_2640075.htm).

groups were strengthened.[16] However, fostering open and thriving industrial system is crucial to enhancing the ability of absorbing employment and increasing income. Hence, restructuring the rural economic engine with the core of industrial cultivation is one of the fundamental issues of rural restructuring.

To cope with the new trend of economic operation and meet the needs of rural-urban development transformation, the development model of rural industry should be established on the deep analysis of regional advantages and external market demand. Additionally, improving the development model based on the dynamic grasp of macro development situation is good for complying with the external development environment. Currently, rural ecological function and cultural function should be emphasized in the process of industrial reshaping in order to meet the demands of transforming personalized and diversified consumption market. Characteristic agricultural industry, industry providing services for retired and old people, industry preserving people's health, ecotourism industry and other rural economic development models should be encouraged to pour vigour into rural development. Taking advantages of the great potential for high quality, green and ecological agricultural product market, more attentions should be paid to support the breeding, industry base construction, postharvest processing, market circulation and brand cultivation of characteristic agricultural products. It is urgent to establish characteristic industry bases that connected with the related leading enterprises by supporting some leading enterprises with great ability in driving local industry development. Then, the benefit sharing mechanism that connects leading enterprises with farmer cooperation and peasants should be established to strengthen the organic combination of production, processing and sale, so as to realize the specialization, standardization, intensive and scale production. As for the policy dimension, supporting various market mainbodies and improving rural investment environment is expected to be emphasized to create the conditions for capital inflow. Meanwhile, there should be a great concern to formulating a series of preferential policies involving land use, tax deduction, getting loans, prime rate and infrastructure complement, to promote rural industrial development and reshaping.

6.5.2.3 Social Restructuring

Along with the advancing of globalization, urbanization and industrialization, as well as the implementation of village relocation and combination, the foundation maintaining Chinese traditional rural society consists of agricultural civilization, village culture and family clanship has been broken. Generally, traditional village community has been disassembled, and local rural development is facing some serious problems, such as outflow of rural population, the downfall of organization system and distortion of value system, which subsequently have caused some rural social

[16]Opinions of the State Council on Further Mobilizing All Aspects of Social Resources to Participate in Poverty Alleviation and Development (No. 58 Document in 2014 issued by the State Council of China). (http://www.gov.cn/zhengce/content/2014-12/04/content_9289.htm).

chaos concerning eroticism, gambling and drug. There is an urgent need to reconstruct rural social structure emphasizing particularly on the organization system and rural management mainstream.

(1) **Improving the governance system of rural social organization**

Enormous rural "brain drain" has brought about the function decline and talent shortage of autonomous organization in rural areas of China. Especially with the cancellation of the agricultural tax nationwide in 2006, the public affairs service function of rural grassroots administrative organization has been further weakened, which has caused the loss of channels and platforms for defending the internal order and attending external conversation in the name of collective action. The reconstruction of the governance system of rural social organization is one of the important means of rural social restructuring. The following two aspects are crucial to reconstruct the governance system of rural social organization. On the one hand, government should fully play the guiding role in macro decision-making and public resources allocation, so as to arouse the initiative spirit of peasants, enterprises and social organizations to participate in the rural planning and construction. On the other hand, in order to make up for the supervising vacuum of government administrative intervention, the autonomy function of primary mainstream should be strengthen through promoting the establishment of diverse rural organization system, e.g., village committee, economic organizations, social intermediary organizations, public service organizations at the grassroots level, mass organizations, and so on. In terms of autonomous organization construction of village committee, the leadership team construction needs to be strengthened by improving the mechanism of cadre selection and encouraging local elite to govern the village, so as to enhance the ability of village management. Meanwhile, the balance mechanism of rural democracy at the grassroots should be established through perfecting the villagers' congress system, which is an important measure to ensure that villagers can participate in public affairs and play a supervisory role well. In addition, developing various economic organizations including farmer cooperatives is an effective approach to make the farmers access to market, and to transform original rural community based on geographical relationship and kinship into more stable partnership one based on common goals and interests.

(2) **Cultivating new-type agricultural management mainstream**

Due to relatively low benefit from grain planting and outflow of rural workforce, China's agricultural laborers presented such characteristics as part-time, aging and low education attainment. Current allocation of human resource has induced a negative effect on improving the efficiency of agricultural production and promoting the development of modern agriculture. As opposed to this, with the implementation of rural land transfer, new agricultural management mainstream is appearing, such as professional investors, family farms, farmer cooperatives and so on. Statistical data from the Ministry of Agriculture indicated that there were 877 thousand family farms in China by the end of 2012, with operating farmland of 176 million mu, accounting for 13.4% of China's total farmland. The average operation scale of family farms amounts to 200.2 mu, which was nearly 27 times that of per peasant household contracted farmland area. By the end of 2013, professional large investors operating more than 50 mu exceeded 2.87 million. Aiming at the situation that traditional agricultural management mainstream is weakening, it is urgent to explore professional farmer registration system and cultivate new-type agricultural management mainstream (Zhang 2014). Besides, it is also important to foster other industry talents in rural areas. The local government should encourage civil organizations and economic groups to train rural entrepreneurs, and try to form a normalized training mechanism to cultivate rural talents. At the same time, in order to guarantee the legitimate rights of rural innovator and entrepreneurs, the related interest distribution mechanism should be perfected so as to fully display all kinds of rural talents' demonstration effect on innovation and entrepreneurship.

6.6 Conclusions

This Chapter establishes indicator systems of three dimensions used to measure China's rural transformation development (RTD): the rural development level (RDL), the rural transformation level (RTL), and the urban-rural coordination level (URCL). The spatio-temporal characteristics of China's RTD from 2000 to 2008 were analyzed in relation to these three indicators. Our results show that China's rural territorial system has universally experienced an intensive transformation since 2000. Constrained by physical conditions and general socio-economic development patterns, China's RDL in 2000 showed a gradient, declining spatial pattern from east to west. With the rapid development of China's economy from 2000 to 2008, regional RDL at a national level generally shows a clear increase, with different rates of development resulting from different regional development modes and policies. In general, due to these different regional development modes, China's RTL from 2000 to 2008 took on a complex spatial pattern, which caused a dynamic spatial pattern for regional URCL during the same period without an obvious spatial pattern. In general, China's urban-rural coordination development declined between 2000 and 2008. The regions that experienced a reduction in URCL were mainly concentrated in areas with rapid economic development but without a good balance between urban and rural areas.

By analyzing the territorial types and mathematical statistical features of China's RTD, the internal mechanism of regional RTD was identified. The results show that low initial RDL, together with defective or excessive RTL inevitably leads to uncoordinated urban-rural development. In the regions with certain initial RDL grades, advancing RTD will effectively coordinate development between the urban and rural areas. Appropriate RTD that corresponds to certain RDLs will lead to the effective development of the regional rural system and the improvement of urban-rural relationships. In the regions with high initial RDL grades, more powerful measures should be taken to advance regional RTD to ensure a coordinated urban-rural development pattern; otherwise, the URCL will continue a negative growth trend as the pace of regional RTD falls drastically behind the pace of rapid industrialization and urbanization.

Counties with 'very high' rurality index are mainly located in the hilly areas of inner China with socio-economic development seriously lagged behind, and Huang-Huai-Hai traditional agricultural areas which are China's major grain production areas. These counties distribute centralized and contiguously. The rurality index of most counties in other major agricultural areas such as Northeast Plain, Jianghan Plain, Sichuan Basin and Jiangxi province is relatively 'high'. But these areas are in the process of rapid urban-rural transformation development, their ruality is changing toward the 'average' type. Counties with low rurality index are mainly around the eastern coastal big cities, northeast national forest and northwest areas with rich mineral and energy resources experiencing large-scale exploitation. Although they are rural areas, the population property, settlement mode, and industry pattern have differed largely from the traditional agricultural areas. It was found that, in most cases, the rurality pattern reflects the impact of physical geography, resource endowment, traffic location, territorial culture, economic foundation and socio-economic policy on the rural system effectively. The rurality index could largely reflect the basic characteristics of the spatio-temporal pattern of China's rural development, and the pattern identified by rurality index significantly improved our knowledge on the recent development of China's rural areas.

Rural territory is an open system consists of diversified elements including natural endowments, geographical conditions, economic base, human resource, cultural customs, etc. The essence of rural restructuring is a process of optimizing the allocation and management of the material and non-material elements affecting the development of rural areas, including human resource, land resource, capital, etc., and accomplishing the structure optimization and the function maximum of rural development system. Since the turn of the new millennium, with the influence of globalization and implementation of the national strategies on industrialization, urbanization, informatization and agricultural modernization, the changes of human resource, land resource and capital allocation have brought about many problems and challenges for the development of rural China, due to the constraints from the maintained rural-urban dualism of land ownership and household registration. These problems and challenges are protruding in the following aspects: rural human resource allocation takes on the trend of low level and the mainstream of rural development is weakening; nonagriculturalization, non-grain preference and abandonment of farmland use together with the derelict and idle rural housing land result in low efficiency of rural land use; unfair urban–rural allocation of capital and its structural imbalance gradually weaken the self-development ability of rural China. Aiming at how to resolve the abovementioned problems and adapt to the challenges due to the changes of allocation of critical resources including human, land and capital, it is pivotal to restructure the rural development space, rural industry, and rural social organization and management mainstream.

References

Ali AM (2007) Population pressure, agricultural intensification and changes in rural systems in Bangladesh. Geoforum 38:720–738

Anselin L, Sridharan S, Gholston S (2007) Using exploratory spatial data analysis to leverage social indicator databases: the discovery of interesting patterns. Soc Indic Res 82(2):287–309

Blunden JR, Pryce WT, Dreyer P (1998) The classification of rural areas in the European context: an exploration of a typology using neural network applications. Reg Stud 32(2):149–160

Cai Y, Smit B (1994) Sustainability in Chinese agriculture: challenge and hope. Agr Ecosyst Environ 49:279–288

Cloke P (1977) An index of rurality for England and Wales. Reg Stud 11:31–46

Cloke P (2006) Conceptualizing rurality. In: Cloke P, Marsden T, Mooney P (eds) Handbook of rural studies. Sage, London, pp 18–28

Cloke P, Edwards G (1986) Rurality in England and Wales 1981: a replication of the 1971 index. Reg Stud 20:289–306

Dunford M, Li L (2010) Chinese spatial inequalities and spatial policies. Geogr Compass 4(8):1039–1054

Fan C (2005) Interprovincial migration, population redistribution, and regional development in China: 1990 and 2000 census comparisons. Prof Geogr 57(2):295–311

Fløysand A, Sjøholt P (2007) Rural development and embeddedness: the importance of human relations for industrial restructuring in rural areas. Sociol Ruralis 47(3):205–227

Frouws J (1998) The contested redefinition of the countryside: an analysis of rural discourses in the Netherlands. Sociol Ruralis 38(1):54–68

Fujita Y, Phanvilay K (2008) Land and forest allocation in Lao People's Democratic Republic: comparison of case studies from community-based natural resource management research. Soc Natur Resour 21(2):120–133

Goodman DS (2008) China's regional development, 5th edn. Royal Institute of International Affairs, London

Griffith D (2003) Spatial autocorrelation and spatial filtering. In: Gaining understanding through theory and scientific visualization. Springer, Berlin

Halfacree KH (1993) Locality and social representation: space, discourse and alternative definitions of the rural. J Rural Stud 9(1):23–37

Halfacree KH (1995) Talking about rurality: social representations of the rural as expressed by residents of six English Parishes. J Rural Stud 11(1):1–19

Heilig GK, Zhang M, Long H et al (2006) Poverty alleviation in China: a lesson for the developing world? Geogr Rundschau (International Edition) 2:4–13

Li X, Qiao J (2001) 20世纪90年代中国县际经济差异的空间分析 (County level economic disparities of China in the 1990s). Acta Geogr Sinica 68(2):136–145

Li Y, Liu Y, Long H et al (2013) Local responses to macro development policies and their effects on rural system in China's mountainous regions: the case of Shuanghe village in Sichuan province. J Mt Sci 10(4):588–608

Li Y, Liu Y, Long H et al (2014) Community-based rural residential land consolidation and allocation can help to revitalize hollowed villages in traditional agricultural areas of China: evidence from Dancheng county, Henan province. Land Use Pol 39:188–198

Li Y, Long H, Liu Y (2015) Spatio-temporal pattern of China's rural development: a rurality index perspective. J Rural Stud 38:12–26

Liu Y (2007) 中国东部沿海地区乡村转型发展与新农村建设 (Rural transformation development and new countryside construction in eastern coastal area of China). Acta Geogr Sinica 62(6):563–570

Liu Y (2011) 中国新农村建设地理论 (Geography of new countryside construction in China). Science Press, Beijing

Liu Y, Liu Y, Chen Y (2011a) Territorial multi-functionality evaluation and decision-making mechanism at county scale in China. Acta Geogr Sinica 66(10):1379–1389

Liu Y, Long H, Chen Y et al (2011b) 中国乡村发展研究报告——农村空心化及其整治策略 (Research report on rural development in China: Hollowed village and its renovation strategy). Science Press, Beijing

Liu Y, Wang J, Guo L (2009) 中国粮食生产与耕地变化的时空动态(The spatial-temporal changes of grain production and arable land in China). Sci Agr Sinica 42(12):4269–4274

Long H (2012) 中国乡村转型发展与土地利用 (Land use and rural transformation development in China). Science Press, Beijing

Long H (2014) Land consolidation: an indispensable way of spatial restructuring in rural China. J Geogr Sci 24(2):211–225

Long H, Heilig GK, Li X et al (2007a) Socio-economic development and land-use change: analysis of rural housing land transition in the transect of the Yangtse River, China. Land Use Pol 24:141–153

Long H, Tang G, Li X et al (2007b) Socio-economic driving forces of land-use change in Kunshan, the Yangtze River delta economic area of China. J Environ Manage 83(3):351–364

Long H, Heilig GK, Wang J et al (2006) Land use and soil erosion in the upper reaches of the Yangtze River: some socio-economic considerations on China's grain-for-green programme. Land Degrad Dev 17(6):589–603

Long H, Li Y, Liu Y et al (2012) Accelerated restructuring in rural China fueled by "increasing vs. decreasing balance" land-use policy for dealing with hollowed villages. Land Use Pol 29(1):11–22

Long H, Li Y, Liu Y et al (2013) Population and settlement change in China's countryside: causes and consequences. In: Cawley M (ed) The sustainability of rural systems: global and local challenges and opportunities. The Commission on the Sustainability of Rural Systems (CSRS) of the International Geographical Union (IGU) and the Whitaker Institute, Galway, pp 123–133

Long H, Liu Y, Li X et al (2010) Building new countryside in China: a geographical perspective. Land Use Pol 27:457–470

Long H, Tu S, Ge D et al (2016) The allocation and management of critical resources in rural China under restructuring: problems and prospects. J Rural Stud 47:392–412

Long H, Zou J, Liu Y (2009) Differentiation of rural development driven by industrialization and urbanization in eastern coastal China. Habitat Int 33(4):454–462

Long H, Zou J, Pykett J et al (2011) Analysis of rural transformation development in China since the turn of the new millennium. Appl Geogr 31:1094–1105

Lópezigelats F, Tàbara JD, Bartolomé J (2009) The rural in dispute: discourses of rurality in the Pyrenees. Geoforum 40(4):602–612

Madu I (2010) The structure and pattern of rurality in Nigeria. GeoJ 75(2):175–184

Mann S (2009) Institutional causes of urban and rural sprawl in Switzerland. Land Use Pol 26:919–924

Marsden T (2010) Mobilizing the regional eco-economy: evolving webs of agri-food and rural development in the UK. Camb J Reg Econ Soc 3(2):225–244

McGee TG (2008) Managing the rural-urban transformation in East Asia in the 21st century. Sustain Sci 3:155–167

Murdoch J (2000) Networks—a new paradigm of rural development? J Rural Stud 16:407–419

Nelson PB (2001) Rural restructuring in the American West: land use, family and class discourses. J Rural Stud 17:395–407

Ocañariola R, Sánchezcantalejo C (2005) Rurality index for small areas in Spain. Soc Indic Res 73(2):247–266

OECD (1996) Territorial indicators of employment: focusing on rural development. Organisation for Economic Co-operation and Development (OECD) Publications, Paris

OECD (2006) The new rural paradigm: policies and governance. In: OECD rural policy reviews. Organisation for Economic Co-operation and Development (OECD), Paris

Prietolara E, Ocañariola R (2010) Updating rurality index for small areas in Spain. Soc Indic Res 95(2):267–280

Putterman L (1997) On the past and future of China's township and village-owned enterprises. World Dev 25:1639–1655

Rogers S (2014) Betting on the strong: local government resource allocation in China's poverty counties. J Rural Stud 36:197–206

Terluin IJ (2003) Differences in economic development in rural regions of advanced countries: an overview and critical analysis of theories. J Rural Stud 19:327–344

Tolbert CM, Mencken FC, Riggs TL et al (2014) Restructuring of the financial industry: the disappearance of locally owned traditional financial services in rural America. Rural Sociol 79(3):355–379

Unger J (2002) The transformation of rural China. Routledge, London

Woods M (2007) Engaging the global countryside: globalization, hybridity and the reconstitution of rural place. Prog Hum Geog 31:485–507

Woods M (2010) Performing rurality and practising rural geography. Prog Hum Geog 34(6):835–846

Woods M (2011) Rural. Routledge, London

Woods M (2013) Regions engaging globalization: a typology of regional responses in rural Europe. J Rural Community Dev 8(3):113–126

Xu W, Tan K (2002) Impact of reform and economic restructuring on rural systems in China: a case study of Yuhang, Zhejiang. J Rural Stud 18:65–81

Yang X, Ma H (2009) Natural environment suitability of China and its relationship with population distributions. Int J Env Res Pub He 6:3025–3039

Zasada I, Reutter M, Piorr A et al (2015) Between capital investments and capacity building-development and application of a conceptual framework towards a place-based rural development policy. Land Use Pol 46:178–188

Zhang X (2014) 构建新型城乡关系 (Reconstruction of new-type urban-rural relationship). Social Science Academic Press, Beijing

Chapter 7
Rural Restructuring in Eastern Coastal China

Abstract The differentiation of rural development in eastern coastal China has been exaggerated by the rapid rural restructuring under globalization, since economic reforms and an open-door policy were initiated in 1978. There are both high-developed villages and very poor hamlets (mainly located in the remote and mountainous areas) in eastern coastal China, which can be considered a miniature of China's rural development. Firstly, this chapter makes the definitions of four rural development types, i.e. farming industry dominated rural development type, industry dominated rural development type, rural development type focusing on business, tourism and services industries, and balanced rural development type, and classifies the rural development types in eastern coastal China. Then, taking the social representation approach and basing on the major factors affecting the long-term rural development and the exertion of the functions of the countryside with regard to society, the assessment indicator system of rurality degree index was established to distinguish the rurality degree of different types. The problems associated with rural restructuring in China may in part be addressed by drawing on experiences and achievements from other countries, including Britain, which may have experienced similar developmental stages as part of their trajectory of rural progression. Secondly, this chapter examines the different trajectories of rural development and different policies and strategies for rural development in eastern coastal China and Wales, identifying points of convergence and divergence, and considering the possible lessons that China might take from the experience of rural development in Wales.

7.1 Introduction

Rural area is a spatio-territorial system, which includes all the sphere other than urban regions. Since the adoption of economic reform and open-door policy in 1978, China's rural area has experienced an unprecedented development, but the regional differentiation of which is obvious. Although the eastern coastal China has experienced rapid industrialization and urbanization, nearly half of the population still live in rural areas, which are affected to varying degrees by the industrialization and

H. Long, *Land Use Transitions and Rural Restructuring in China*,
https://doi.org/10.1007/978-981-15-4924-3_7

urbanization. There are both high-developed villages and very poor hamlets (mainly located in the remote and mountainous areas) in eastern coastal China, which can be considered a miniature of China's rural development. Rural development can be viewed from a multi-perspective including the political, economic, social, cultural, environmental and many other aspects. The regional differentiation of rural development in eastern coastal China is very obvious, due to the influences of physical conditions, the existing economic fundamentals, historical and cultural background, together with the location and traffic conditions and other factors.

Global integration has also brought convergence between the dynamics driving change in rural China, and those experienced in other parts of the world. Increased political, economic and cultural openness has exposed rural localities in China to globalization processes—such as global trade and economic competition, international migration, and cultural assimilation—that are already contributing to the reconstitution of rural places elsewhere (Woods 2007). Furthermore, there are some notable resonances between challenges facing China today and experiences of rural regions in Europe, North America, Australia and New Zealand during periods of urbanization and social and economic restructuring in the last century. From this global perspective, it is legitimate to consider how China's policies for rural development fit into the wider international context. Indeed, there has been notable interest from both government and academic circles in China over the last decade in examining rural development policies and practice in other countries, and in learning from international experience, through various conferences, exchanges and research visits. Globalization has increased the international transferability of knowledge and practice in economic development, and there are many successful examples of transferring international knowledge and practice and of providing international solutions to local strategic problems.

This chapter firstly defines the rural development types, then carries out a case study in eastern coastal China combining with the assessment of rurality degree, and discusses some of the major implications for promoting the rural development. In the second part, the similarities and differences between Chinese and European approaches to rural development are cautiously examined through the juxtaposition of two contrasting narratives of regional rural restructuring, in eastern coastal China, and in Wales, Great Britain in order to reveal similarities and contrasts, and to consider whether there are lessons or examples of good practice that could be translated from Europe to China to enhance future rural development.

7.2 Development and Differentiation of Rural Eastern Coastal China Under Rapid Restructuring

Rural development and industrialization and urbanization are closely interrelated in the aspects of migration, employment, land use and natural environments. With the process of urbanization, the rural population is increasingly marginalized and natural environments are increasingly destroyed. The reform-induced industrialization and urbanization have rapidly altered the physical and human landscapes in China's rural areas, as evidenced by the substantial rate of rural housing development, rural to urban migration, agricultural to non-agricultural land conversion, widening rural-urban income gap, and regional rural inequalities. Zhang (2008) indicated three paradoxical dynamics concerning China's current rural-urban development: rapid urbanization concurrent with the under-urbanized feature, the massive and sustained flow of rural-to-urban migration concomitant with the bulk of migrants without urban citizenship, and the proliferating addition of new cities with inclusion of ample rural attributes. Driven by industrialization and urbanization, rural landscapes have been transformed into urban like ones, the ecological functioning and spatial structure of which has been being changed.

To some extent, agriculture and the countryside have made a big contribution to (simultaneously, the huge sacrifices for) the development of industries and the cities in China in the process of industrialization and urbanization since 1978. As a result, a series of problems that hampers the social and economic development of China occurred, such as degrading eco-environment, decreasing cultivated land, widening rural-urban income gap, and so on (Cai and Smit 1994; Yang and Li 2000). With the implementation of an important long-term development strategy on "building a new countryside" (Long et al. 2010), a brand new rural-urban relationship that industries support agriculture and cities support the countryside is expected to be shaped.

7.2.1 Defining Rural Development Types

Rurality has proved very difficult to define in an all-embracing manner for three important reasons involving functions, dynamics and variation. As rural areas are undergoing considerable spatio-temporal change due to social, economic and technological developments, coupled with the interaction of various elements affecting rural development is difficult to quantify, to some extent, it increases the difficulty in our understanding the rural development types. In general, rural issues are closely related to the marginalization of rural economic, political, social and cultural development, which occurs differently at the intersection of material and experiential elements of rural lifestyles (Cloke et al. 1995). Compared with the city, the rural is always in a marginalized position. From the perspective of geographical location, the development of remote rural areas is even more marginalized. The marginalization and interaction of the material and non-material elements affecting rural production and lifestyles generates considerable regional differentiation at every scale.

It has generally been agreed that a combination of a pull factor of cities and a push factor of villages promotes the process of industrialization and urbanization (Soemarwoto 1976), which in turn drives the development of rural areas. Usually, the situation of industrial development determines the level of economic development and employment in an area. In developing countries, the industrial division of labor in rural areas is characterized by (Zhou and Fan 1988): (1) underdeveloped industrial division of labor and diversified part-time farmers; (2) to great extent, industrial division of labor in rural areas is in thrall to the neighboring urban economic development level; (3) the development level of industrial division of labor is directly depended on the linkages of the rural area with the outside world; (4) as a rule, the industrial division of rural labor begins with labor-intensive industries, then extends to other industries with the improvement of labors' education level; and (5) the limitation of intra-regional flow of production elements constrains the development of industrial division of labor in rural areas. Redfield (1941) put forward the concept of rural-urban continuum, which favoured a sliding scale rather than an abrupt halt between rural and urban poles. There have a thousand and one links between rural industry and urban industry. So, it is unrealistic to discuss the development of rural areas breaking away the city.

Rural–urban interdependence plays an important role in rural development. Using spatial economic data from four small English towns, Courtney et al. (2007) measured the strength of economic integration between town and hinterland and estimated the magnitude of town–hinterland spill-over effects, and indicated that the potential value of a town as a 'sub-pole' in local economic development is dependent on structural differences in the local economy. As for the production side of the economy, in the process of industrialization through rural–urban interdependence, that agricultural productivity was enhanced by utilizing non-agricultural inputs was proved by many case studies of India, Japan, the United States, Taiwan, West Pakistan, and so on. In addition, Rigg's (1998) case study of southeast Asian showed that the diversification of the household economy and the interpenetration of rural and urban have created multiple hybridities through individuals and households' shifting between agricultural and industrial pursuits and crossing between rural and urban areas.

To recognize some of the differences between degrees of rurality, Cloke and Edwards (1986) constructed an "index of rurality" for local government districts in England and Wales using a range of statistics from the 1971 and 1981 censuses. The indicators, including population, household amenities, occupational structure, commuting patterns and the distance to urban centers, were fed into a formula that placed districts into one of five categories—extreme rural, intermediate rural, intermediate non-rural, extreme non-rural and urban. Although the indices of rurality marked an improvement on simple dichotomous definitions, it still provokes a number of critical questions (Woods 2005). Concerning China's rural communities development types, Unger and Chan (1999) put forward a distinct four-type model based on relations of production through investigating rural south China, i.e., predominance of private industry, collectives, foreign industry, and no industry.

Before China's economic reforms launched in 1978, the rural China took on a look of balanced development. Three decades' reform has greatly changed the homogeneous status in rural coastal China. Rural development difference and industrial divergence were exaggerated by the rapid industrialization and urbanization and absorbing advanced technologies and foreign capitals. Recently, new change of the scale, structure and pattern of economic growth in rural coastal China has occurred due to the spatio-territorial recombination and industrial restructuring around the urban agglomeration, which were embodied in three primary aspects, i.e., transferring traditional agriculture to scale management agriculture, shifting dispersed rural housing to town or rural new-type communities, and developing industries in development zone or industrial park.

Therefore, in rural coastal China, industrial structure has been being changed by rural–urban interpenetration, which also resulted in the transformation of rural employment structure. It is a common phenomenon that farm is in thrall to non-farm, and industry is often relying on 'rural' labor. This kind of interaction between the material and non-material elements of rural production and lifestyles has shaped different rural development types relying on a carrier, which is composed of different industries. Accordingly, it is feasible to divide the rural development types basing on the forces of production. Usually, the rural areas close to urban agglomeration have the obvious advantages for fostering manufacturing industry. However, remote and mountainous rural areas may focus on developing farming industry or tourism if there were good tourism resources. Of course, there are some rural areas with no obvious individual advantages.

The author argues that there are four rural development types in China: (1) farming industry dominated rural development type (FIT), usually located in the remote and mountainous areas, with a slow process of industrialization and urbanization, where the local economic development depends on agriculture-oriented industries to some extent; (2) industry-dominated rural development type (IDT), where the rapid industrialization and urbanization exerts far-reaching impacts on local rural industrial structure and employment structure; (3) rural development type focusing on business, tourism and providing services (BTT), where the rural development mainly depends on tertiary industry such as trade, tourism and services industries, and so on; and (4) balanced rural development type (BDT), where the urban and rural development keep almost the same pace, and there is no obvious individual advantages in the development of primary industry, secondary industry and tertiary industry.

7.2.2 Study Area

The eastern coastal China, the study area, covers 10 provinces or municipalities: Beijing, Tianjin, Hebei, Shandong, Shanghai, Jiangsu, Zhejiang, Fujian, Guangdong and Hainan (Fig. 7.1). The eastern coastal China has a population of 469 million, and it is unique because of its high population density. In 2006, the population density in this region amounted to 512 persons/km^2, which was much higher than the average population density (137 persons/km^2) of China at the same period (Table 7.1).

Relying on the national policies of making use of the sea in connecting China to the global economy, the eastern coastal China is developing very rapidly in the last three decades since the adoption of economic reform and open-door policy in 1978, and plays a significant role in promoting China's rapid economic growth and enhancing the nation's international competitiveness. Concomitant to this development is the

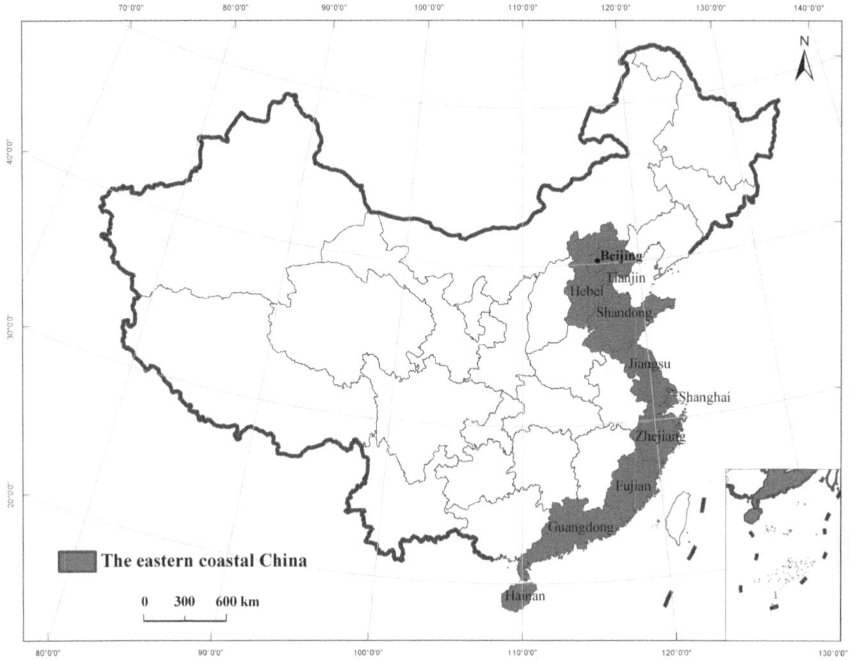

Fig. 7.1 Location of the study area (Long et al. 2009)

rapid change of landscape in both urban and rural areas, and the obvious enhancement of socio-economic development level (Table 7.1). In 2006, the GDP per capita of this region amounted to 27 567 RMB¥, which was much higher than that of China (16 084 RMB¥) at the same period. In 2006, the primary industry only provides 7% of the total GDP in eastern coastal China; however, 52 and 41% are provided by the secondary industry and tertiary industry, respectively.

During the last three decades, the industrialization and urbanization in coastal China has been accelerated and the economy of metropolitan areas such as "Beijing–Tianjin–Hebei", "Yangtze River Delta" and "Pearl River Delta" has grown vigorously, and the rapid urbanization and centralization of population and industries have been the obvious characteristics of urban-rural integrated development in the region (Liu et al. 2008). However, the rapid industrialization and urbanization in eastern coastal China increased the demand for construction land, which diminished cultivated land. In addition, the industrialization pushed forward the development of TVEs, which is helpful for improving rural employment but caused the pollution of local environment. Due to loose and neglected protection of the environment, the consequences of uncontrolled rural industrialization in this region are particularly serious. Rapid loss of farmland and ongoing deterioration of local environment are major concerns in eastern coastal China.

Table 7.1 The situation of rural-urban development in eastern coastal China (Long et al. 2009)

Indicator	National total	Eastern coastal China	% of national total
Area of land (10 000 km^2)	960.0	91.6	9.5
Area of cultivated land (km^2)	1217760	263960	21.7
Per capita cultivated land (ha)	0.0926	0.0563	
Population at the year-end (10 000 persons)	131448	46906	36.3
Rural population (10 000 persons)	73742	22217	30.1
Employment of TVEs at the year-end (10 000 persons)	14680	7262	49.5
Population density (persons/km^2)	137	512	
Gross Domestic Product (GDP) (100 million RMB¥)	210871	128593	55.7
Primary industry	24737	9343	37.8
Secondary industry	103162	66798	57.8
Tertiary industry	82972	52452	57.9
The ratio of primary, secondary and tertiary industries	12:49:39	7:52:41	
Per capita GDP (RMB¥)	16084	27567	
Economic density (10 000 RMB¥/km^2)[a]	220	1404	
Grain output (10 000 tons)	49748	13316	26.8
Per capita disposable income of urban households (RMB¥)	11759	14967	
Per capita net income of rural households (RMB¥)	3587	5188	
The ratio of urban-rural residents' income	3.28:1	2.88:1	

Notes (1) All the data are the value of 2006, the data of cultivated land are from the Ministry of Land and Resources of China, and other data are from National Bureau of Statistics of China; (2) In 2006, exchange rate US$ to RMB¥: 1–7.8

[a]Economic density was calculated by dividing the gross domestic product (GDP) of country or eastern coastal China by the total area of country or eastern coastal China

7.2.3 Materials and Methods

7.2.3.1 Data Source and Processing

Rural development is usually depending on various industries. Therefore, the status quo of rural industrial development can be used to diagnose the rural development type, to which a region belongs. In this section, the output value data of total GDP

Table 7.2 The mean and standard deviation of indicators (Long et al. 2009)

Indicator (%)	Mean	Standard deviation
GDP1	19.19	12.76
GDP2	47.13	13.36
GDP3	33.66	8.13

as well as its primary, secondary and tertiary industries of 615 counties (or cities at county level) in eastern coastal China was collected from the abovementioned ten coastal provinces' statistical yearbook in 2006. The proportions of the output value of primary (GDP1%), secondary (GDP2%) and tertiary (GDP3%) industries in the total GDP, respectively, were chosen as the indicators to classify the rural development types. Then, the mean and standard deviation of indicators GDP1, GDP2 and GDP3% was counted (Table 7.2).

The degree of rural development will be different even in the same rural development type because of the difference of regional physical conditions and socio-economic development level. Rurality degree index (RDI) was established to diagnose the difference of rural development types. Defining the degree of rural development, we have employed the idea of the functions of the countryside with regard to society. In general, the major factors affecting the long-term rural development and the exertion of the functions of the countryside with regard to society involve: (1) coverage of society's food and agricultural raw material production; (2) the change of rural population; (3) rural economic development level; and (4) rational spatial organization of the activities of the rural population.

The social representation approach can produce a more robust and flexible way of defining rurality, through accommodating the effects of social and economic change in rural environ-ments (Woods 2005). Usually, the effects of social and economic change in rural environ-ments can be primarily showed by the change and output value of cultivated land, change and employment of rural population, and the productivity of rural labor. Accordingly, the assessment indicators of RDI were constructed (Table 7.3). It seems that there is a strong correlationship between the change of cultivated land and its output value. But the actual situation in coastal China is that many farmland are idle and even the utilized farmland are not in an intensive way because of the low incentive for raising cultivated land-use intensity (Li and Wang 2003). Similarly, there is not a strong correlationship between the change of rural population and the percentage of employed population in primary industry and the productivity of rural labor, for existing lots of surplus rural labors in the statistic rural population and most of them are rural-urban migrants. So, the five indices used to synthetically measure the rurality degree in Table 7.3 are relatively independent, and the variables also passed the tests for independence in the software of SPSS 11.5 for Windows.

Table 7.3 Assessment indicators of rurality degree index (RDI) (Long et al. 2009)

Indicator (Weight)	Definition	Explanation
Change rate of cultivated land (0.262)	$\frac{CA_l - CA_e}{CA_e}$	CA_l = cultivated land area of the later period; CA_e = cultivated land area of the early period; positive indicator, the higher the value, the stronger the rurality degree
Change rate of rural population (0.225)	$\frac{RP_l - RP_e}{RP_e}$	RP_l = Rural population of the later period; RP_e = Rural population of the early period; positive indicator, the higher the value, the stronger the rurality degree
The percentage of employed population in primary industry (0.154)	$\frac{ELP}{TL}$	ELP = the employed labors in primary industry; TL = the total labors; positive indicator, the higher the value, the stronger the rurality degree
Output value of cultivated land (0.167)	$\frac{AOV}{CA}$	AOV = agricultural output value; CA = cultivated land area; negative indicator, the lower the value, the stronger the rurality degree
Productivity of rural labor (0.192)	$\frac{OVFFAF}{LFFAF}$	$OVFFAF$ = gross output value of farming, forestry, animal husbandry and fishery; $LFFAF$ = labors employed in farming, forestry, animal husbandry and fishery; negative indicator, the lower the value, the stronger the rurality degree

Notes (1) The weights for the indicators were determined by conducting semi-structured interviews with relative experts; (2) The data used to calculate the change rate are the value of 1996 and 2004, other data are the value of 2004; (3) All the data are from the socio-economic statistical yearbooks of abovementioned 10 coastal provinces

7.2.3.2 Methods

To classify the rural development types, we regard that, if a certain indicator value of a county exceeds the sum of the indicator's mean and standard deviation, it can be determined that rural development in this county is dominated by the factors relative to the indicator. Accordingly, the indicator system classifying rural development type was constructed (Table 7.4).

Because the socio-economic data for the various indicators in Table 7.3 are in many different dimensions, it was impossible to directly compare the rurality variation in the various measures. Therefore, they were transformed into common units by normalizing all measurements, according to the formula (7.1):

Table 7.4 The indicator system classifying rural development type (Long et al. 2009)

Rural development type	Indicator and its value
FIT	GDP1% ≥ 31.95
IDT	GDP2% ≥ 60.49
BTT	GDP3% ≥ 41.79
BDT	Not conform to the above three conditions

$$S_i = \begin{cases} \dfrac{x_i - x_{\min}}{x_{\max} - x_{\min}} \\ \dfrac{x_{\max} - x_i}{x_{\max} - x_{\min}} \end{cases} \quad (7.1)$$

(adapted to positive indicator)
(adapted to negative indicator)
where S_i is the standardized value of indicators; x_i is the value of indicator *i*; x_{min} is the minimum value of indicator *i*; x_{max} is the maximum value of indicator *i*; and *i* is the number of indicators.

We used the normalized values to calculate the RDI according to formula (7.2):

$$RDI = \sum_{i=1}^{5} w_i S_i \quad (7.2)$$

where RDI is the rurality degree index, the higher the index value, the stronger the rurality degree; S_i is the standardized value of indicators; w_i is the weight for indicator *i*.

Geographic information system (GIS) analysis was adopted for the research. In particular, the ESRI's ArcGIS spatial analysis module was used to aggregate, synthesize and analyze the databases, and to identify spatial relationships.

7.2.4 Results

What kind of rural development types the 615 counties (or cities at county level) in 10 provinces of eastern coastal China belong to was determined by diagnosing their rural development status, and the RDI of each counties were calculated according to formula (7.2).

Relying on the advantages of location and national lean policies, 57.6% of the total territory of eastern coastal China belongs to BDT type, and 18.64, 11.99 and 11.75% of the total area is shared by FIT type, IDT type, and BTT type, respectively (Fig. 7.2 and Table 7.5).

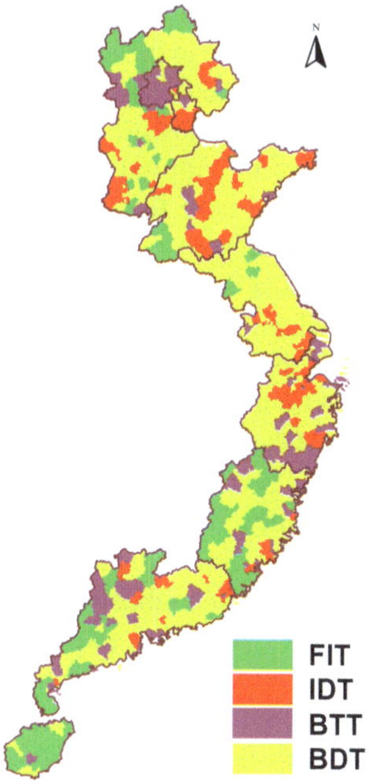

Fig. 7.2 Rural development types in eastern coastal China (Long et al. 2009)

Table 7.5 Results of classification and assessment of rural development types in eastern coastal China (Long et al. 2009)

Rural development type	Area (10 000 km^2)	Proportion of type area in the total area (%)	RDI value	Number of counties	Area (10 000 km^2)	Proportion of area of certain type with different RDI value in the total area of the type (%)
FIT	16.28	18.64	0.35–0.4	4	0.5973	3.67
			0.4–0.5	32	6.0655	37.26
			0.5–0.751	53	9.6174	59.07
IDT	10.48	11.99	0.2–0.4	30	3.4963	33.37
			0.4–0.5	37	4.4883	42.84
			0.5–0.646	19	2.4931	23.79
BTT	10.26	11.75	0.2–0.4	33	4.3847	42.73
			0.4–0.5	16	3.0738	29.96
			0.5–0.725	17	2.8025	27.31
BDT	50.34	57.63	0.2–0.4	83	8.6783	17.24
			0.4–0.5	164	23.0625	45.81
			0.5–0.87	127	18.6032	36.95

Among the FIT types, 59.1% of the type area shows a strong rurality (RDI > 0.5), mainly locate in the northern Hebei province, southwestern Shandong, the border of Hebei and Shandong, parts of Hainan province, western Fujian, and northern Jiangsu. These areas, mostly locate in the remote and mountainous areas, are the plant areas of primary agricultural products, with more rural labors employed in primary industry. There are only four counties with a weak rurality (RDI < 0.4), characterized by rapid decrease of cultivated land and rural population, and a higher output value of cultivated land. The type with a moderate rurality (0.4 < RDI < 0.5), in parts of Fujian and Guangdong, accounts for 37.3% of the type area (Fig. 7.3 and Table 7.5).

In the type of IDT, the area with a weak rurality and a moderate rurality takes up 33.4 and 42.8%, respectively, mainly located at the junction of Beijing, Tianjin and Hebei, the Yangtze River Delta, central and southern Shandong, and along the coastal belt of Shandong, Guangdong and Fujian (Fig. 7.3 and Table 7.5). In these areas, the industries are characterized by high added-value products, and high-grade, high-precision, advanced technology. A lot of cultivated land was occupied due to rapid industrialization and urbanization, and most rural labors were employed in factories. Area with a strong rurality accounts for 23.8% of the IDT type, located in the counties of southern Hebei and northern Guangdong, where the level of urbanization is not high, with a relatively laggard economic development, although the secondary industry is in a leading position, but most products is low value-added primary industrial products.

The area with weak, strong and moderate rurality shares 42.7, 27.3 and 30.0% of the total area of BTT type, respectively (Table 7.5). The regions with a weak rurality mainly located in Beijing, Shanghai, Jinan, Qingdao, Nanjing, Guangzhou, Zhanjiang, Shenzhen, and some counties of Zhejiang and Fujian (Fig. 7.3), where the highly developed business, tourism and other services industries, have played an important role in promoting local rural development both in the aspects of industry and employment. The rest areas of this type with a strong and moderate rurality mainly distributed in western Hebei, northwestern Guangdong and the border of Zhejiang and Fujian, where is characterized by the development of primary tourism products and low level business and services industries, the role of which to drive local rural development has much room for improvement.

Although slightly more than half area of the eastern coastal China gets a balanced development among the primary, secondary and tertiary industries, the level is different. There is only 17.2% of the BDT type with a weak rurality, mainly located in Huairou, Zhangjiakou, Sanya, Yantai and Rizhao, as well as the Yangtze River Delta, the Pearl River Delta and coastal Fujian, with the characteristics of high added-value agricultural products, developed high-technology, tourism and leisure agriculture, and relative high-level business and tourism industries. Area with a strong rurality accounts for 37% of the BDT type, mainly in some counties of Hebei, central and western Shandong, western Zhejiang and Fujian, and northern Guangdong, where belong to low-level balanced development (Fig. 7.3 and Table 7.5).

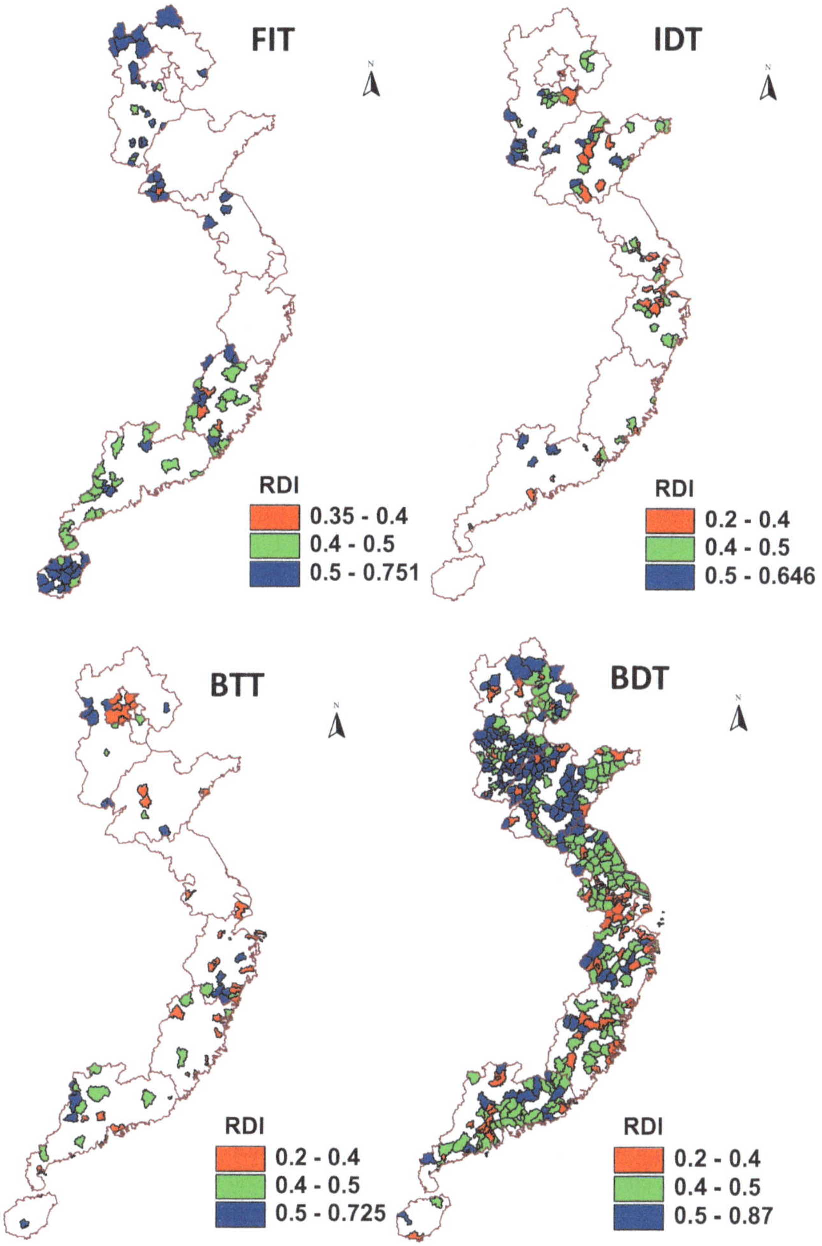

Fig. 7.3 Rurality degree index (RDI) of different types in eastern coastal China (Long et al. 2009)

7.2.5 Discussion

The interaction between the material and non-material elements affecting rural production and lifestyles shaped different rural development types depending on a carrier, which is composed of different industries. Accordingly, the above studies make the definitions of four rural development types, i.e. FIT type, IDT type, BTT type and BDT type, and classifies the rural development types in eastern coastal China. Then, basing on the major factors affecting the long-term rural development and the exertion of the functions of the countryside with regard to society, the assessment indicator system of rurality degree index (RDI) was established to distinguish the rurality degree of different types.

There are 18.64, 11.99, 11.75 and 57.6% of the total territory of eastern coastal China shared by FIT type, IDT type, BTT and BDT type, respectively. In the type of FIT, 59.1% of the type area shows a strong rurality (RDI > 0.5). In the type of IDT, the area with a weak rurality (RDI < 0.4) and a moderate rurality (0.4 < RDI < 0.5) takes up 33.4 and 42.8%, respectively. The area with weak, strong and moderate rurality shares 42.7, 27.3 and 30.0% of the total area of BTT type, respectively. Almost half (45.8%) of the BDT type has a moderate rurality. To some extent, the RDI may accurately reflect the status quo of rural development and the exertion of the functions of the countryside with regard to society. The RDI can also reflect the different stage in what the same rural development type in different region stays. Combining with regional physical conditions and socio-economic development level, it will be helpful for us to analyze the regional differentiation of rural development and promote rural economic development by taking measures suited to local conditions.

Currently, both rural development and urban development in China are experiencing a transition period—the transformation of a traditional agricultural society into a modern industrial and urban society, and the economy is changing from a traditional planned economy to a modern market system. With the accelerated rural industrialization and urbanization process, rapid population growth and development of the market economy, the industrial structure, employment structure and land-use pattern in rural coastal China has been transformed tremendously. The functions and management mode of agriculture also come in multiple forms, step by step with the characteristics of changing from paying more attentions to the quantity turns to the quality and economic returns, and springing up of agriculture parks, urban agriculture, and tourism and leisure agriculture. With the economic and social transformation, regional development factors recombination and followed industrial restructuring have changed the rural areas in eastern coastal China deeply.

China's accession to the World Trade Organization (WTO) will bring about obvious impacts on the agricultural production and its market. So, to confront with this challenge, rural associations will play an important role especially in the counties belong to FIT type. The government's decision on agricultural products prices and food security will influence the farmers' enthusiasm for pursuing agricultural production. Correspondingly, the state corporatist rural associations what Unger and Chan (2008) argued need to help farmers improve traditional cultivation techniques and popularize agricultural standardization, to promote the strategic adjustment of agriculture structure by optimizing the industrial structure, the product structure and regional distribution of agriculture, and to improve rural (agricultural products) circulation system and strengthen the building of agricultural service system. However, civil rural organizations will be more flexible in finding domestic information on specific agricultural production and its market.

Rapid industrialization and urbanization have greatly changed the rural areas in the aspects of cultivated land loss for factory workshop, and rural labors transformation for workers. Although the counties belong to IDT type have the advantages in absorbing surplus rural labors, the monotonous industry mix is fragile, to some extent. In China's Pearl River Delta, seriously influenced by global economic crisis in 2008, many export-oriented enterprises were closed down, which resulted in a lot of rural migrators unemployed.[1] However, in the Yangtze River Delta, the situation causing unemployed rural labors is largely mitigated because of its diversified employed modes. So, in the counties belong to IDT type, the adjustment of industrial structure needs to be paid more attention so as to evade this kind of risk.

Rural development depends on the change of its affecting factors, including material and non-material aspects of two major groups involved in natural resources, human resources, industry and employment, information, science and technology, customs and so on. The change of these elements, even their inter-regional flow results in the occurrence of rural development change from quantitative to qualitative. So, the rural development types classified in this section cannot always keep the same situation, at least the quantitative change occurs momently. The study was to make clear the differentiation pattern and degree of rural development in eastern coastal China, and to provide scientific basis for understanding its further development trend and setting down corresponding developmental strategy.

In addition to the shape of a rural development type, another manifestation of the interaction of these elements is the socio-economic development mode of a specific rural area. The author argues that the study on the interaction of rural development factors in the process of economic and social transformation and the subsequent rural development mode is very important to deeply understand the rural development and to smoothly achieve coordinated and balanced rural-urban development in developing countries, which are experiencing rapid industrialization and urbanization.

[1]Slowing Economy Spurs Disquiet in China (http://chinadigitaltimes.net/2008/12/slowing-economy-spurs-disquiet-in-china/).

7.3 What Can Be Learned from Wales?

7.3.1 Globalization and Rural Development in Europe and China

For most of the period since the early twentieth century, rural development has been primarily framed by the discourse of modernization. Based on the conflation of development and modernity, and the assumption that rural areas lag behind urban areas in both respects, the modernization paradigm holds that rural development requires the spatial diffusion of modernity from cities to the countryside. This is understood in social, technological and cultural terms, as well as economic terms, such that rural modernization commonly involves four parallel processes: agricultural modernization, involving the mechanization and industrialization of farming; economic modernization, diversifying rural economies away from agriculture to embrace 'modern' industries; infrastructure modernization, expanding transport and communications infrastructure, installing electricity and improving housing; and social modernization, challenged perceived 'backward' traditional rural social structures (Woods 2010).

In Europe, the modernization paradigm reached its zenith in rural development after the Second World War. The continent faced major challenges in securing food supplies for a booming urban population, and in maintaining a viable economic and social structure in rural communities experiencing rapid depopulation. The response involved all four dimensions of modernization. Agriculture was mechanized and industrialized with the assistance of subsidies and price supports provided by national governments and, after 1957, through the Common Agricultural Policy of the European Economic Community (later the European Union). As the agricultural workforce contracted even further as a result, the state intervened to promote economic diversification, particularly by promoting the development of manufacturing industry. Support was provided through grant-aid, but also by infrastructure projects, including new roads, bridges, tunnels and airports, power stations and reservoirs supplying mains electricity and water, and new factories and housing. In turn, these infrastructure projects facilitated social modernization and cultural change, as rural society was integrated into modern consumer society.

Modernization programs in Europe helped to stabilize rural population decline, diversify rural economies and improve accessibility. In Britain, in particular, rural modernization initiatives created the conditions that facilitated a reversal in the migration trend, from urbanization to counter-urbanization. Modernization was also closely associated with globalization in rural areas. Improvements to the transport and communications infrastructure increased access to rural areas for trade, tourism and migration; agricultural modernization promoted specialization of production within an increasingly international market; economic diversification was achieved through inward investment by trans-national corporations establishing branch plants in rural localities; and social modernization opened up rural culture to global media. These globalization processes stimulated the economic development of many rural districts, but they also introduced new risks, including, for example, the vulnerability of rural branch plants to distant decision-making by transnational corporations (Woods 2010). At the same time, the principles of modernization came to be challenged by in-migrants pursuing the anti-modern ideal of the rural idyll, by environmentalists attacking the ecological impacts of modernization projects, and by political activists claiming that modernization had failed to benefit endogenous rural populations (Woods 2010).

From the 1980s onwards, the critique of the modernization paradigm gathered force in European rural development policy and practice, with van der Ploeg et al. (2000) observing that, "rural development is on the agenda precisely because the modernization paradigm has reached its intellectual and practical limits" (p. 395). In its place emerged a new approach that van der Ploeg et al. (2000) label 'the new rural development paradigm', and which included a shift in emphasis from inward investment to endogenous development, from top-down planning to bottom-up initiative, and from sectoral modernization to territorially-based integrated rural development (Woods 2010). The principles of the new paradigm were articulated for European rural policy in the Cork Declaration issued by the European Conference on Rural Development in 1996, which has informed subsequent reforms of the EU Common Agricultural Policy and implementation of EU Structural Fund programs for rural and regional development. Agricultural policy now emphasizes diversification and agri-environmental schemes over increasing production; whilst rural development schemes are delivered through a partnership approach engaging local people in projects that tend to focus on exploiting endogenous natural and cultural resources.

Although rural development policies are primarily the domain of national governments, or supra-national regional bodies such as the European Union, ideas about approaches to rural development have always circulated internationally. Nation states copied each other in the race for modernization, and the emerging academic disciplines of agricultural economics and rural sociology helped to disseminate the modernization discourse. The modernization paradigm also framed approaches to rural development in imperial colonies, and was continued by post-independence governments in Africa and South East Asia. Initially, international bodies such as the World Bank reinforced the modernization approach, but more recently they have become champions of a new 'community-centred' approach to rural development that has mirrored the new rural development paradigm in the global north.

China's political and economic independence has enabled it to follow a more distinctive trajectory, and to diverge from global trends in rural development. Rural modernization was first promoted in the 1930s, but significantly through models that were marked out from western approaches by emphasizing and preserving aspects of traditional Chinese rural culture. Rural modernization was further de-westernized during the Cultural Revolution, with the assertion of a new policy of collectivization of agriculture, which critics have argued led to at least two decades of agricultural stagnation, with the result that China's rural economy remained relatively undeveloped. It was only following the introduction of Deng Xiaoping's economic reforms in 1978 that rural development returned to the political agenda, as it was realized that objectives for economic growth could only be achieved by modernizing social and economic structures in rural China.

Since the 1970s, rural development in China has been advanced around three key policy strategies: the implementation of a household responsibility system; the development of township and village enterprises (TVEs); and the goal of 'building a new countryside'. Each of these policy objectives has had a substantial impact across the social and economic canvass of the Chinese countryside, involving changes to the rural population structure, rural lifestyles, employment structure, industrial structure, community organization, accessibility of rural areas, and rural cultures.

The introduction of the 'household responsibility system' between 1979 and 1984 has been an epoch-making reform in rural China, transforming the agricultural sector from collective management to private production. The household responsibility system contracted land to individual households for a period of 15 years,[2] with relative autonomy over land use decisions and crop selection. After fulfilling procurement quota obligations, farmers are entitled to sell their surplus on the market or retain it for their own use. By linking rewards directly to effort, the contracting system enhanced incentives and promoted efficient production. As such, the introduction of the household responsibility system has encouraged entrepreneurialism and aroused enthusiasm in Chinese peasants for private enterprise, helping to shift the Chinese countryside from self-sufficiency towards a market economy, hence creating opportunities for development of TVEs. The number of TVEs increased from 1.5 million in 1978 to over 23 million in 1993 (Liang et al. 2002), with the current figure likely to be significantly higher still.

China's TVEs have contributed significantly to the increase in rural income levels and employment by making full use of the local resources, utilizing the capital scattered in the peasant's hands, developing the expertise of skilled craft-workers, and improving incomes for peasant households. Almost half of TVEs are manufacturing enterprises, with significant numbers also involved in construction, commerce and services. As such, TVEs have facilitated the diversification of the rural labour market, with non-agricultural employed in rural China increasing from 5% in 1978 to 25% in 2000 (ibid.). However, the rapid rate of development has also generated problems, such as shortages of materials, energy and capital due to the growth of rural enterprises outstripping supply, and worsening environmental quality because of a lack of treatment for the disposal of polluted water, poisonous gas and wastes (Long et al. 2009).

Furthermore, whilst the implementation of the household responsibility system and the development of TVEs have promoted rural modernization, they have been progressed in a wider context of economic liberalization in China that has continued to be driven by urban expansion and industrialization. As such, economic growth has been greater in cities than in rural areas, and the income gap between urban and rural

[2]Initially, the term of land contracts was 15 years. In 1995, it was extended to 30 years and although an extension to 70 years has been discussed, the continuation of 15-year term was confirmed in government documents in early 2009 (http://www.bjreview.com.cn/document/txt/2009-01/05/content_173368_3.htm).

regions has widened. The disparities in wealth have contributed to large-scale rural-to-urban migration in China, producing an estimated 100 million internal migrants within the country by the late 1990s (Liang et al. 2002).

In response to these pressures, the Chinese government recognized the importance of stabilizing rural regions and adopting a more integrated approach to rural development. From 2004 to 2009, the top-priority annual 'number one policy document' of the Chinese government was devoted to rural issues. These documents, which set out the key policy task of the central government for the coming year, successively addressed increasing farmers' incomes (2004); improving agricultural production capacity (2005); advancing the 'building a new countryside' scheme (2006); developing modern agriculture (2007); resolving rural problems (2008); and stabilizing agricultural prices and increasing farmers' incomes (2009). Together, these stated actions were encompassed in the overarching agenda of 'building a new countryside', which targeted five major objectives (Long et al. 2010):

(1) Advanced production: Developing modern agriculture and strengthening the productive forces of the countryside.
(2) Improved livelihoods: Increasing the living standards and incomes of farmers.
(3) A civilized social atmosphere: Building a more civilized and harmonious society.
(4) Clean and tidy villages: Expanding the use of clean fuels such as marsh gas and solar energies and improving the standard of rural housing according to a scientific village plan.
(5) Efficient management: Promoting the establishment of new primary organizations, strengthening their service function, and implementing democratic management and open government.

As such, the 'building a new countryside' agenda has shifted the emphasis of rural policy from the concern of earlier strategies with structural adjustments aimed at stimulating private enterprise, to a more holistic concern with broad social, economic and environmental improvements, involving collective action. Although 'building a new countryside' remains an overtly modernizing program, it contains elements that resonate with aspects of the 'new rural development paradigm' in Europe and elsewhere. It is in this context that there is growing interest from Chinese scientists and rural development practitioners in learning from international experiences.

The remainder of this chapter examines further the differences and similarities in rural development in Europe and China, and lessons to be learned, by focusing on rural development policy and implementation in two case study regions: Wales in the United Kingdom, and first, Eastern Coastal China.

7.3.2 Rural Restructuring in Coastal China

7.3.2.1 Economic Reform and Rural Restructuring

The eastern coastal region of China is a geographical and cultural territory that includes the administrative provinces of Hebei, Shandong, Jiangsu, Zhejiang, Fujian, Guangdong and Hainan, as well as the municipal regions of Beijing, Tianjin and Shanghai (Long et al. 2009). Its population of over 469 million represents 36.3% of the total Chinese population in 2006, and incorporates major cities including Shanghai, Hangzhou, Nanjing, Guangzhou and Qingdao. However, the region also includes an extensive rural area, embracing fertile plains in Jiangsu and Shandong provinces and the mountainous districts of Fujian, and Zhejiang provinces, and varying in degrees of acccssibility and infrastructure development, agriculture type and natural resources. The eastern coastal region was the first area of China to be opened to international trade and thus has the most developed degree of engagement with the global economy (Liu et al. 2008; Long et al. 2009). As early as 1986 the region was producing three-quarters of China's exports (Goodman 2008), and its coastal proximity has enabled it to benefit from investment from Japan and South Korea, as well as from developing economic relations with Taiwan.

Whilst the scale and pace of rural restructuring in eastern coastal China has been strongly influenced by the region's geographical position and context, the processes and conditions of change have been steered by the evolution of macroeconomic policy in China. From the foundation of the People's Republic of China in 1949 to the initiation of Deng Xiaoping's reforms in 1978, economic policy in China followed a central planning model. The development of agriculture and industry was unbalanced and a "dual track" structure was formalized in the national economy, with industrialization rapidly progressed at the expense of agriculture and the peasant economy. As described above, it was only following the liberalization of China's economy after 1978 that significant rural development and restructuring commenced, including the implementation of the household responsibility system and the development of township and village enterprises (TVEs).

The geographical position of eastern coastal China and its early engagement with the global economy aided market reforms and economic development, and the region quickly started to experience pronounced changes in its rural economy and society. In particular, these included a dramatic re-alignment of its economy away from agriculture. Between 1978 and 2005, the contribution of primary industries to the region's GDP decreased from 23.3 to 7.9%, whilst that of tertiary industries increased from 19.8 to 40.5% (Liu 2007). Over the same period, the proportion of the regional labor force employed in agriculture fell from 90.8 to 47.9% (ibid.). The amount of farmland in the region has similarly declined steadily, with 1.71 million hectares lost between 1996 and 2005, largely due to industrialization, urbanization, the expansion of rural housing, and the adjustment of agricultural structures and household withdrawal from farming (ibid.). Meanwhile, the fragmentation and structural complexity of agricultural holdings has increased, with consequences for economies of scale in the management of farming (Long et al. 2009). Rampant urbanization has also encroached on rural areas, and according to analysis by Long et al. (2009), only 38.4% of the total territory of eastern coastal China can be classified as strongly rural (with a rurality degree index of above 0.5).

The effects of political-economic reforms and growing prosperity can also be observed in aspects of social restructuring in eastern coastal China, notably changing household structures and the consequential demand for new housing. Before 1978, growth in the rural housing stock in China was limited for cultural, political and economic reasons. The Chinese tradition of 'Si Shi Tong Tang', in which three or four generations would commonly live together in one house, restrained demand for new housing, as did social and political suspicion of multi-property ownership or aspirations for a more capacious house as expressions of capitalism (so-called 'capitalism's tail') (Long et al. 2007). Additionally, the low incomes obtained from work on collective farms meant that most rural households lacked the resources to improve their housing conditions. However, the increasing affluence of many farmers after 1978 created the opportunity for rural households to improve their property, or to move to new houses, with growing preferences for multi-functional, more comfortable or more spacious houses. At the same time, cultural changes have meant that the model of 'Si Shi Tong Tang' is no longer popular, and many young couples prefer to have their own housing. Accordingly, as Sargeson (2002) argues, the combination of the expanded social and demographic aspirations of families and the reconfiguration of rural households' economic activities has stimulated a major boom in rural house-building in eastern coastal China. Thus, although the rural population in eastern coastal China decreased from 306 million in 1978 to 219 million in 2005, the amount of land per capita used for rural housing increased (Liu 2007). This has been reflected in a change in the topography of rural settlement in the region, as farmers have demonstrated a preference for building their houses in valleys, close to roads and other infrastructure, which in turn has contributed to the loss of agricultural land (Yang and Li 2000; Long et al. 2007).

Restructuring has also further integrated rural and urban economies in eastern coastal China, including through the dynamics of the labor market. The processes of economic restructuring and modernization have reinforced the division between the unskilled and manual labor of primary industries and the skilled labor of the secondary and tertiary sectors, which has a distinctive spatial expression. Limitations in the intra-regional flow of production elements have constrained the development of demand for skilled technical and professional labor in rural regions, thus consolidating an urban-rural divide (Zhou and Fan 1988; Long et al. 2009). Accordingly, although the initial impact of liberalization reforms was to reduce the income gap between urban and rural residents, since the mid-1980s the gap widened considerably in line with the acceleration of industrial development in urban areas, from 1.86:1 in 1985 to 3.31:1 in 2008. The income gap is further reinforced by the operation of 'price scissors' in which farmers are disadvantaged by the dissonance between artificially low prices for agricultural products and the high prices of exchange for industrial goods. Collectively, these factors have fuelled mass economic migration from the Chinese countryside to towns and cities, including intra-regional migration to eastern coastal China's booming cities.

China's rural-to-urban labor migration has delivered some benefits to rural districts through remittances and through the physical and human capital brought back by return migrants, however these positive effects have been more than outweighed by the problems generated by the depopulation by the countryside. In particular, out-migration has been led by younger and better educated rural residents, producing a 'brain drain' that has seen rural communities lose their most active population segment. In many cases, adult migrants have left children with family members, creating social issues as families are divided and children raised without parental care and support. Rural migrants arriving in cities are vulnerable to exploitation and economic volatility, and discriminated against politically and socially. Rural migrants are not granted permanent household registration in cities and are not eligible for many social welfare and economic opportunities that are reserved for permanent urban residents.

The migration out-flow has also impacted on the fabric of the countryside in eastern coastal China. In many villages houses have become vacant as owners are absent as migrant workers in cities, or due to abandonment in depopulation or for newer housing (Long et al. 2007). Similarly, some high quality farmland has been abandoned as farmers have become migrant workers, and as the agricultural labor force has been depleted (Li and Wang 2003; Long et al. 2007, 2009). These social, economic and environmental impacts of rural out-migrant have intensified with restructuring, yet the speed of restructuring left policy lagging and its only recently that the Chinese government has acted to respond to the rural problem with the adoption of the 'building a new countryside' agenda.

7.3.2.2 Building a New Countryside

The implementation of the 'building a new countryside' policy has had far-reaching impacts in the rural districts of eastern coastal China, with actions focused on transforming traditional farming into modern large-scale agriculture; developing industries in new development zones and industrial parks; and consolidating rural housing in new town communities (Long et al. 2009). The last dimension, shifting from a pattern of dispersed rural housing to new nuclear communities, has been the most prominent aspect of the policy in the region. Aimed primarily at identifying releasing land for industrial and urban development, the consolidation of rural housing has been advanced through an innovative land management policy balancing the increase in land for urban construction with a decrease in construction on rural land. In practice, this has involved a county-level land use planning and development control system that involves the identification of land for construction in concentrated settlement pockets, and the corresponding identification of existing dispersed rural properties for demolition and return to cultivated land.

Significantly, the process of land displacement and centralized settlement planning has been implemented at the local level, involving the collective action of local actors.

However, Long et al. (2009) have argued that this mode of operation is problematic. There is fundamentally a disjuncture between the aspirations of developers and the skills and expectations of the rural population. The policy is aimed at creating the infrastructure for industrial development, but the upgrading of the skills base of the peasants has not kept up with the demands of the industrial revolution. Similarly, whilst the construction of high-rise apartment buildings has been favored in order to maximize land for industrial development, this style of housing is often not desired by the displaced rural residents used to living in individual houses in the rural landscape. Thus, whilst the 'building a new countryside' policy has been implemented through local actors, it has arguably not been responsive to local priorities and opinions.

Xu (1999) argued that China's economic policy relies too much on a top-down approach in monitoring, control and supervision. The style has been continued in the implementation of the 'building a new countryside' policy in eastern coastal China. A lack of local embeddedness and naturalization in the development of policies has created difficulties in the implementation of strategies such as housing consolidation, which have encountered reluctance and resistance on the part of peasants. Regeneration policies have similarly failed to take account of rural circumstances and needs. For example, some development zones and industrial parks are established in small towns with poor infrastructure that have difficulty in attracting outward investment. Local cadres have simply imitated examples from other towns, without taking local differences into account, and hoped for a miracle. As such, local advantages in endogenous resources, technologies and production traditions are frequently neglected in rural development strategies by decision-makers who have prioritized finding funds for immediate actions and benefits over long-term planning for sustainable development. Similarly, the interdependence between agriculture and the rural economy, society and environment has arguably received limited attention in eastern coastal China, with, for example, the environmental cost of industrialized agriculture and the impact on the rural labor market ignored.

As China consequently wrestles with the search for appropriate and effective mechanisms for rural development, it may be instructive to turn to the experience of rural development in Europe, where policy-makers have tried and tested a number of approaches over the last fifty years, which will be illustrated in the next section by focusing on the example of Wales, in the United Kingdom.

7.3.3 Experiences and Lessons of Rural Restructuring in Wales

7.3.3.1 Rural Restructuring in Wales

Wales has been a predominantly urban and industrialized country since the late nineteenth century, yet the majority of its land area continues to be rural, and its rural heritage is a powerful element in its national identity. Around 960,000 people, out of Wales's total population of 3 million, live in rural areas, but rural Wales has a relatively low population density by British standards, no towns of over 20,000 inhabitants, relatively poor transport networks, and a difficult upland topography. Most of the region is classified as a 'less favorable area' for agriculture, but whilst less than 2% of the population of Wales are now employed in farming, this is still the highest proportion of the constituent nations of Great Britain.

The trajectory of rural restructuring in Wales can be traced to the late nineteenth century and the development of the South Wales coalfield and the associated urbanization and industrialization of the South Wales valleys and coastal plains, as well as urbanization and industrialization in North East Wales. Towns and cities such as Cardiff, Swansea, Newport, Wrexham, Merthyr Tydfil and the Rhondda, became industrial powerhouses connected to global economic networks, and acted as magnets for migrants from the struggling rural areas of Wales. The rural counties of Mid Wales lost a quarter of their population between 1871 and 1961, a greater rate of population decline than any other part of Britain. The scale of depopulation in rural Wales, and the implications for the fabric of rural society and Welsh culture, was such that the new Welsh nationalist party, Plaid Cymru, unsuccessfully campaigned on a platform of rural resettlement during the 1930s and 1940s.

By the 1940s and 1950s, rural Wales has confronted by a range of problems that have resonance with the contemporary situation of rural China. Agriculture was in dire need of modernization, farm incomes were precarious, and opportunities for employment outside primary industries were limited. The infrastructure of the region was weak and standards of living were low. At least a quarter of houses in Mid Wales were without a piped water supply and over 3,000 farms did not have mains electricity (Hooson and Jenkins 1965). Depopulation not only continued, but accelerated to around 4% a year between 1951 and 1961. Rural Wales was increasingly portrayed as a region in crisis, and in response the British government established a committee of inquiry to examine the reasons for the region's decline and propose action to stimulate development.

The priorities of the British government in responding to the problems of restructuring in rural Wales during this period were again remarkably similar to the objectives of rural policy in contemporary China: stabilizing agricultural incomes; stimulating enterprise; and modernizing the rural economy and society. The problem of farm incomes was addressed through the system of agricultural subsidies and price supports put in place by the 1947 Agriculture Act, which offered farmers a guaranteed price for their produce, but which also encouraged mechanization and intensification, thus reducing the agricultural labor force. Moreover, the small hill farms of rural Wales were poorly suited to the type of intensive, industrial agriculture that the productivist system promoted, and farm incomes in Wales remained relatively low, with many Welsh farmers dependent on subsidies.

The problems of diversifying the economy, stimulating enterprise and modernizing the infrastructure of rural Wales required a more targeted approach. The Mid Wales Industrial Association was set up in 1957 to encourage industrialization in the region, and following the recommendations of the Beacham Committee, the Mid Wales Development Corporation was established in 1965 to oversee the expansion

Newtown in Powys as a centre for industrial and population growth. In 1976, the functions of these two bodies were taken over by the Development Board for Rural Wales (DBRW). These government agencies sought to develop rural Wales through a strategy that included encouraging small enterprises, training the labor force with new skills, building industrial estates and factory units, and attracting inward investment. In particular, an emphasis was placed on developing manufacturing industry. At the same time, efforts were made to improve infrastructure, with plans for new roads and airports, and an ambitious scheme to restructure the traditionally dispersed rural settlement pattern around a number of 'growth centres'. In many of these respects the post-war strategy for rural development in Wales has parallels in the contemporary experiences of eastern coastal China.

The strategy produced mixed results. The ambitions for infrastructure development and settlement rationalization were never fully realized, but economic development achieved more success. Between 1977 and 1985, over 200 new factories were established in the DBRW area, increasing manufacturing employment by 61% from 3,921 to 6,330 (Edwards 1985). Together with the more general process of counter urbanization, which gathered momentum in Britain during the 1970s and 1980s, this industrialization helped to reverse the trend of depopulation and improve income levels and standards of living.

However, the essentially top-down strategy of rural development, driven by a non-elected central government agency, was increasingly criticized for failing to engage with local communities, failing to reflect local needs, and for creating jobs for in-migrants rather than for endogenous residents. In a particularly strident critique, focused on the county of Gwynedd but relevant for rural Wales as a whole, Lovering (1983) attacked the development strategy for subordinating the economy of rural Wales to economic conditions and decision-making elsewhere:

> So what was post-war development all about? Did it really herald a 'new industrial era' as local MPs promised? It was becoming obvious by the mid-1970s that this was far from the truth. Post-war economic change created a new structure of economic dependence—the economy of North West Wales being dependent in the literal sense that its different segments depended on economic conditions elsewhere, and did not sustain each other. Working people in Gwynedd had been given a new role as providers of cheap unskilled short-term labour. Those who wanted more either had to leave the area or, if they were fortunate, enter the professions. Modern construction projects recreated a large temporary, rootless workforce reminiscent of the 'navvies' of the previous century. Modern manufacturing and service industries, largely owned outside, created a lop-sided labour market, which in turn generated pools of real poverty. The state, through a variety of channels, softened the effect by pumping money in, and by directly creating a large proportion of local jobs, but at the same time it encouraged the process through regional aid. (Lovering 1983, p 53)

Moreover, by the late 1980s the economic stimulation of industrialization was beginning to falter. Faced with increasing global competition, manufacturing in rural Wales went into decline as factories were closed and production relocated to cheaper locations elsewhere, including China. In contrast, the more sustainable elements of growth in rural Wales appeared to come from the in-migration of ex-urban residents, often for amenity or lifestyle change reasons, the expansion of the service sector, and the increasing population of rural tourism. Yet, the methods of DBRW's development strategy were arguably ill-suited to capitalizing on these opportunities. As such, for both political and economic reasons, the approach of rural development in Wales was fundamentally realigned during the 1980s and 1990s to a new emphasis on bottom-up, community-led regeneration.

7.3.3.2 Community Action: A Process of Embeddedness and Naturalization

The first steps towards community-led rural regeneration in Wales can be traced back to the late 1960s (Edwards 1998). Following the success of an experimental community-based strategy adopted in Glyncorrwg, a deprived community suffering from depopulation, economic decay and social disadvantage, a series of initiatives were introduced in Wales in the 1970s involving community participation as the planning solution for tackling rural disadvantage (Edwards 1998). Examples included the Antur Llanaelhaearn initiative on the Llŷn peninsula in North Wales launched in 1974, which ran economic aims in parallel with an objective of restoring social vitality to rural communities, and Antur Teifi in southern Ceredigion, which emerged in 1978. These two pilot initiatives served as role models for small-scale, area-based community action in Wales, involving the key elements of employment of a local animateur, wider community involvement, and the development of bids by local residents for external funding and support from local authorities or development agencies (Edwards 1998).

In the late 1970s, Development Board for Rural Wales (DBRW) launched its Social Development Grant Fund with the aim of pump-priming community projects throughout rural mid-Wales. It encouraged local problem/need identification, local participation and self-help in delivery and partnership guidance and funding from DBRW, and led to considerable investment in social and environmental improvement projects targeted at both the total community and various age-specific subgroups throughout a large area of rural Wales over the next 15 years. Edwards (1998) argued that such involvement with community action initiated, and subsequently embedded, the idea of rural development as a partnership engagement that was agency-led but resident-driven. During the 1980s, the Welsh Development Agency (WDA) initiated a related strategy focusing on economic regeneration in rural districts in north and south Wales, which were seen as pilot initiatives involving a partnership between statutory bodies, local authorities, the private sector and communities. The schemes were to have ownership by the communities, integration between various agencies, and appropriate delivery mechanisms for the effective provision of services to the community. Similarly, in 1987 a partnership of the Countryside Commission in Wales, the Prince of Wales's Committee and the British Trust for Conservation Volunteers established the 'Local Jigsaw/Jigso' scheme to encourage local participation in community development (Edwards 1998). The success of these initiatives with their clear focus on economic regeneration led to the expansion of this mode of engagement after 1990, including through Welsh participants in the European

Union's LEADER program of community initiatives, and was enshrined in the 1996 'rural white paper' policy document for Wales.

As such, responsibility for rural development and regeneration in Wales has been increasingly decentralized. Local authorities are required to take a leading role in delivering national policies whilst addressing local needs and priorities. They have a statutory duty to prepare, in partnership with community stakeholders, community strategies for promoting the well-being of their areas. Local authorities covering rural areas are also encouraged to prepare integrated rural development strategies. Following the establishment of the devolved National Assembly for Wales in 1999, the Welsh Assembly Government (WAG) and the Welsh Local Government Association (WLGA) have created a joint set of policy priorities which outline shared priorities to achieve change at the local level, including better opportunities for learning, a better stronger economy, better health and wellbeing, better quality of life, and better simpler government.

Additionally, there is growing recognition of the potential for non-statutory organizations to contribute to rural community regeneration, which can be showed by an increasingly strong emphasis on community empowerment. The Welsh Assembly Government and its sponsored bodies have committed themselves to providing the opportunities and support for communities to participate more fully in their own development and regeneration. At the same time, communities are urged to take responsibility for identifying and addressing their own needs, problems and challenges. Increasingly, regeneration initiatives require the inclusion of a range of local stakeholders, with partnership between the public, private and voluntary sectors seen as integral to sustainable rural development in Wales. As the Wales Council for Voluntary Action has argued, "a strong relationship between local government and the voluntary sector is the basis for a vibrant local democracy and active, inclusive communities".

The engagement of local communities and the principle of cross-sectoral and trans-scalar partnership working between different agencies are supposed to make rural development more responsive to local needs and facilitate the capitalization of endogenous resources. In principle, this 'bottom-up' approach is argued to provide a flexible and efficient way of addressing the needs of rural areas, whilst the decentralization of support from established statutory agencies is contended to maximize a region's regeneration capacity. Ray (2001) refers to this approach as a neo-endogenous approach to rural development which reflects the balance between extra-local funding, resources and policy frameworks with local endogenous decision-making and responsibility. Yet, these strategies can also produce fragmentation and duplication in the process of rural development. In Wales, the adoption of the adoption has led to a proliferation of partnerships active in rural development, with overlapping objectives and territorial remits (Edwards et al. 2001).

Moreover, Edwards et al. (2001) and others have suggested that regeneration is still in many ways heavily 'top-down' because of the funding rules set by central government and the leading role played by local authorities and government agencies. Many programs in Wales are dependent on grass-roots delivery, but the structure of programs and conditions of funding are often imposed from the top-down by the sponsoring agencies. The criteria that local partnerships must meet even to be eligible to bid for regeneration funding are centrally defined, as are the audit paths groups must follow if they do receive funding. A typical case is the Communities First program, which aims to build leadership capacity and encourage empowerment within the most deprived communities in Wales. To some extent, this degree of steerage may undermine the endogenous nature of these schemes, and there are also reservations over the extent to which local partnerships are actually representative of, and democratically accountable to, local communities.

7.3.3.3 Integrated Rural Development Policy and Agricultural Restructuring

The shift in strategy for rural development in Wales represented not only a change in the style of governance and policy implementation, but also a change in the focus and objectives of rural development policy. Whereas the emphasis during the 1960s and 1970s had been on attracting inward investment, the emphasis is now on utilizing and valorizing the endogenous resources of the region, including its people and its environment. Inevitably this has brought agriculture back into the picture for rural development in Wales. During the post-war period the policy concerns of agriculture were largely separated from other aspects of the rural economy. The imperative to maintain farm incomes and modernize agriculture was delivered through the mechanisms of price supports and subsidies put in place by the 1947 Agriculture Act, quite independently of other economic development measures. Rural development, per se, was about developing the non-agricultural rural economy, and agencies such as the Development Board for Rural Wales had no remit over agriculture.

Since the early 1990s, however, the dominant rural policy discourse in Europe has been that of integrated rural policy, recognizing the inter-connection and interdependence not only of different economic sectors, but also of economic, social and environmental processes in the countryside. Integrated rural development has become a core feature of European Union rural policy, but has also been specifically promoted in the Welsh context from the 1996 Rural White Paper onwards and implemented through subsequent rural development initiatives (Bristow 2000). Significantly, whilst in England and Scotland integrated rural policy has been associated with the weakening of agriculture's privileged place in rural policy and the assertion of a new vision of the countryside in which farming plays a minor part (Woods 2008), in Wales agriculture remains at the heart of the new integrated rural policy. This was evident in the positioning of the Welsh Assembly Government's key policy document on rural policy, *Farming for the Future*.

Thus, the farming sector is seen as a key contributor to endogenous rural development that revolves around new premium agri-food products—in some cases produced largely for export—sustainable tourism, and other elements in a new 'eco-economy' (Kitchen and Marsden 2009). Equally, integrated rural development is also now seen as the preferred method for stabilizing farm incomes, both by rewarding farmers for good environmental stewardship and improvements to footpaths and recreational infrastructure—particularly through schemes such as Tir Gofal and Tir Mynydd—and by encouraging the diversification of farm enterprises into tourism or specialist food production. As such, individual farmers are coaxed through these policies into making decisions about the future development of their business, that essentially concern five possible options: (1) extending the agro-industrial model of farming through further development of conventional products; (2) redeploying resources into the development on-farm of 'new agricultural products or services'; (3) redeploying resources into non-food/fibre products on-farm; (4) redeploying human resources into off-farm employment; (5) continuing with conventional production and accept lower business income; or (6) moving to hobby farming or semi-retirement (Evans 2009).

Integrated rural policy also means recognizing that the economic well-being of the countryside is affected by political decisions taken about social policies or about public services. The greater awareness of these connections has been promoted by the implementation of 'rural proofing' in Wales and England, whereby new policies developed by any government department are now supposed to be assessed for their impact on rural communities (DEFRA 2000).

7.3.3.4 Longevity and Sustainability of Rural Development Initiatives

Achieving sustainable rural development is as much dependent on the continuity of support over time as it is on the type and level of support within a locality. This can be seen in the case of Pembrokeshire, in south west Wales, which has been cited as an excellent example of a region which has had a long-standing commitment to building community capacity. Pembrokeshire initiated a progressive approach to community development in the 1970s, when a newly-created local group assumed responsibility from the local authority for developing day-care, child care, information technology and some aspects of sports provision. The scope of community action in the locality was extended in the 1980s when the group, SPARC (South Pembrokeshire Action for Rural Communities) (leader renamed PLANED—Pembrokeshire Local Action Network for Enterprise and Development), achieved LEADER status in the first round of the program operated by the then European Community, becoming involved in a range of bottom-up rural regeneration projects. It maintained this status through LEADER II and LEADER+, becoming the only Local Action Group in Wales to be involved in all three rounds of LEADER, as well as gaining funding from the EU Objective 1 program after 2000. As a result, Pembrokeshire has been one of the main beneficiaries of financial support from the European Union in Wales. Thus, the continuity of organization and funding has been significant in creating strong institutional foundations in Pembrokeshire, facilitating the area's capacity to benefit from new programs.

As such, there is increasing recognition that programs targeted at the sustainable development and regeneration of communities in Wales must themselves be sustained. However, many of the early community regeneration programs targeted at communities in Wales often funded 'one-off' projects that were limited in scope, short-term in focus and frequently unsustainable once funding from external sources was withdrawn. For example, Edwards et al. (2000) found that almost half of the partnerships operating in Mid Wales in the late 1990s had been constituted for a fixed limited life, often of less than five years, usually because funding has only been secured for a limited period, or because the partnership forms part of a limited-life program. The voluntary sector, in particular, is critical of the current trend for project funding and ad hoc, short-term grants, and has called for more secure arrangements (Collis 2003). In response to these concerns, recent funding programs, such as Communities First, have begun to place more emphasis on a longer-term approach.

7.3.4 Discussion

Eastern coastal China and Wales are vastly different regions. They differ in size, population, natural environment, economic structure and level of development. They also represent different stages in the trajectory of rural restructuring—whilst Eastern coastal China struggles with problems of rural depopulation, agricultural modernization and alleviating severe rural poverty, the major challenges confronting rural Wales include coping with in-migration, gentrification, a precarious remnant agricultural sector and a collapsing manufacturing sector. However, there are more parallels between the two regions then might be expected. In particular, many of the issues facing rural eastern coastal China today are similar to those found in rural Wales after the Second World War: significant out-migration, especially of younger and more skilled residents; limited employment opportunities; an out-dated agricultural sector; household incomes below those in urban areas; poor housing conditions and an inadequate infrastructure. Moreover, the rural development strategy adopted in eastern coastal China has resonance with that followed in post-war Wales. As such, it becomes possible to suggest that rural development in China might be improved by learning lessons from the Welsh experience.

Table 7.6 summarizes the different approaches to rural development in eastern coastal China and in Wales, and highlights potential lessons that could be learned. Essentially, these can be reduced to a contrast between a top-down approach focused on modernization and overcoming geographical peripherality, as practised in eastern coastal China today and as previously practised in rural Wales, and an alternative bottom-up approach emphasizing endogenous development and place-based regeneration, as has become dominant in Wales over the last two decades. Both approaches have their merits and disadvantages. Bristow (2000) argued that a top-down emphasis on inward investment and job creation is important for generating employment and income growth but may fail to make the links with other rural activities that would establish resilience to the increasingly competitive pressures evident; however, a bottom-up emphasis based on local empowerment is likely to lead to more diverse and embedded activities, but may not be sufficient on its own to buttress the rural economy against the pressures of globalization. The experience of Wales has indicated that community-led initiatives are more likely to succeed in the long

Table 7.6 Comparison of rural development approaches between Wales and eastern coastal China (Long and Woods 2011)

Measures	Wales	Eastern coastal China	What China can learn
Community action	A process of embeddedness and naturalization	Lack of embeddedness and naturalization	To foster and pay more attention to embeddedness and naturalization
Approach to rural community regeneration	'Bottom-up' accompanied by 'top-down' elements	'Top-down' accompanied by few 'bottom-up' elements	More 'bottom-up' and less 'top-down'
Rural development policy and agricultural restructuring	Integrated rural development policy emphasizing the interdependence between agriculture and the rural economy, society and environment	Waiting to be improved in this aspect	Paying more attention to the interdependence between agriculture and the rural economy, society and environment
Implementing rural Proofing	Yes	No	Introducing rural proofing
Towards limited-life Funding	Lessons: it will restrict the potential for long-term planning for community and economic development	Refusing nobody's offer	Drawing the lessons of rural Wales and paying close to long-term sustainable development
Rural development mode responding to globalization	Towards place-based rural development	Not clear	Adopting space-based or place-based rural development mode according to different local socio-economic conditions

term than 'top-down' approach (Greenlees 1998). However, in eastern coastal China, the approach to rural community regeneration is mainly top-down accompanied by few bottom-up elements. Accordingly, an approach to rural community regeneration with more 'bottom-up' and less 'top-down' will be helpful for resolving the exposed issues in the process of rural restructuring in eastern coastal China.

The objective of community action in rural Wales has been to target material economic and social needs, through local specification and engagement. In so doing the initiatives built on natural co-operation, but allowed expression of local dissatisfaction with central delivery and a degree of self-determination (Edwards 1998). As practised in rural Wales, bottom-up development encompasses different strategies for local participation in community action that exhibit multiple perspectives on its role and purpose; these reflect the interests of the actors and agencies involved.

As such, whilst the bottom-up approach has become increasingly embedded in policy in Wales, it has also been constantly re-examined and re-made, fine-tuned to fit particular circumstances (Edwards 1998).

Moseley (2003) points out that it is important to note the danger of what might be called 'project-ism'—the assumption that promoting discrete one-off projects comprises all of what is possible or necessary to achieve one's goals. Sometimes, long-term and necessarily wide-ranging strategic plans need to be implemented through short-term and more precise 'operational plans', which stress the role of individual 'tasks' linked to strategic objectives, each with their own targets, costing and allocation of responsibility. Under this condition, 'local' actors may need to seek to persuade the key partners in rural development—the transport providers, the land-use planners, the agricultural agencies, the housing providers—to respect 'sustainability' in their day-to-day dealings with the local area in question (Moseley 2003, p. 24). Sustainable local development must be based on a vision relating more to long-term human welfare than to maximizing the production of goods and services or the crude creation of as many jobs as possible.

Turning to eastern coastal China, the emphasis has been on top-down large-scale projects and strategies, but in line with Moseley's observation, these have frequently required implementation at the local scale. The attempt to balance the increase in urban construction land with a decrease in rural construction land is a case in point. It is a centrally-directed policy, but in theory should be implemented by village collectives through village plans. Yet, in practice local discrimination and input has been limited and there has been resistance from rural residents who resent being compelled into new housing projects. The short-comings of the program can therefore be identified as resulting from a lack of embeddedness and naturalization in rural communities. However, the fostering process of embeddedness and naturalization will take a long period, which also depends on many factors such as local socio-economic conditions and residents' educational level. The experience of rural Wales shows that at least ten years is needed.

The lessons of rural Wales also show that the limited-life funding in rural regeneration restricts the potential for long-term planning for community and economic development. However, in eastern coastal China short-termism is reinforced by the eagerness of local cadres to bid for every available funding source. In order to actively respond to the 'building a new countryside' policy, there are special funds earmarked for investing in rural development in China's local governments and different central ministries, such as the Ministry of Civil Affairs, the Ministry of Water Resources, the Ministry of Land and Resources, and the Ministry of Construction. Usually, pilot villages or counties are chosen to receive these funds. In general, these pilot villages or counties have a good location and a relatively strong economic base, which means that they are the easiest to develop, and successes in these communities may not translate to other more peripheral or economically deprived areas. Furthermore, as these funds are normally allocated for a specific 'project' defined by the sponsoring ministry or administration, and as each agency and scheme will have its own conditions, in order to obtain the funds villages are required cater to the requirements of the departments who fun them, even changing their existing plans to fit. As such, long-term strategic planning is trumped by the need to meet short-term objectives dictated by limited life funding. In the process of rural restructuring in eastern coastal China, the lessons of rural Wales need to be drawn on, with longer-term initiatives introduced and projects unfavorable for local long-term sustainable development avoided.

The shift from top-down to bottom-up rural development in Wales has meant not only a change in the administrative structure and processes, but also in the focus of development strategies. For much of the post-war period, rural development strategy generally focused on measures to combat rural Wales's disadvantaged position in the twentieth-century space-economy, helping it to participate in the emerging global economy though improved infrastructure, support for export-orientated industries and inward investment. However, these measures require large-scale top-down intervention and planning. Individual communities, supposedly empowered through bottom-up approaches, do not have the resources or the scope to engage in these activities and therefore instead focus on utilizing local resources and place-based regeneration. Place-based regeneration has hence become a hallmark of neoliberal rural development not only in Wales, but across Europe, North America, Australia and New Zealand, and is championed by advocates as the way forward (Halseth et al. 2010). Yet, place cannot be divorced from the space-economy, and place-based endogenous development will generally only be successful if geographical conditions are appropriate to allow access to external markets.

This paradox is again evident in the contemporary situation in eastern coastal China. Over the last three decades since the start of Deng Xiaoping's economic reforms, policies for rural development have focused on overcoming the backward and marginalized position of rural regions, resulting in rapid industrialization and urbanization (Long et al. 2009). This strategy has delivered benefits, but it increasingly seems that a new approach is required for the next phase of rural development, which could emphasize place-based regeneration. The major requisite for such an approach to be successful is the presence of a large and affluent urban middle class with the money and time to spend on consuming countryside commodities. This includes not only the consumption of speciality craft and food products from specific rural localities, but also rural tourism and recreation and even migration to rural communities for lifestyle purposes. It is the presence of such a class in western society that has enabled the switch to place-based rural development, and the nature of the growth of the Chinese economy suggests that these conditions are likely to come into existence in China very soon. Indeed, there are already some isolated examples of place-based rural development for tourism and amenity, such as the 'water town' of Luzhi in Jiangsu province (Fan et al. 2008). However, these opportunities are likely to be largely limited to those rural areas closest to urban centers, at least initially. More remote rural communities may exhibit a strong sense of solidarity and social capital, which are essential requirements for successful bottom-up development, but their potential for attracting tourists, capturing passing trade or pulling-in lifestyle migrants from towns and cities, will be restricted by distance and poor accessibility. As such, a combination of place-based and space-focused rural development will continue to be needed in eastern coastal China.

Achieving successful rural development in the globalizing countryside is a difficult challenge that requires sensitivity to both economic and political factors. Top-down, infrastructure-focused strategies, as currently dominate in eastern coastal China, can deliver economic benefits but can also be politically unpopular. The pursuit of such an approach in rural Wales during the 1960s and 1970s was successful in helping to halt depopulation, diversify employment opportunities and improve living standards, but the lack of engagement of local communities and the perceived bias towards creating jobs for in-migrants also contributed to the rise of Welsh nationalism as a political force in the region. If China wishes to marry economic growth and political stability, some move towards a more inclusive and participatory form of rural development would be advisable, but with the qualifying understanding that not all rural district will benefit equally from endogenous place-based regeneration. In negotiating this precarious path, China has the opportunity to learn from the experiences of Wales and other western countries, but any international transfer of models and strategies for rural development will need to be modified for the particular cultural, economic and political traditions and circumstances of China.

7.4 Conclusions

The differentiation of rural development in eastern coastal China has been exaggerated by the rapid rural restructuring under globalization, since economic reforms and an open-door policy were initiated in 1978. Firstly, this chapter makes the definitions of four rural development types, i.e. farming industry dominated rural development type (FIT), industry dominated rural development type (IDT), rural development type focusing on business, tourism and services industries (BTT), and balanced rural development type (BDT), and classifies the rural development types in eastern coastal China. Then, taking the social representation approach and basing on the major factors affecting the long-term rural development and the exertion of the functions of the countryside with regard to society, the assessment indicator system of rurality degree index (RDI) was established to distinguish the rurality degree of different types. The results indicated that, to some extent, the RDI may accurately reflect the status quo of rural development and the exertion of the functions of the countryside with regard to society, and can also reflect the different stages in what the same rural development type in different region stays. The problems associated with rural restructuring in China may in part be addressed by drawing on experiences and achievements from other countries, including Britain, which may have experienced similar developmental stages as part of their trajectory of rural progression. Secondly, this chapter examines the different trajectories of rural development and different policies and strategies for rural development in eastern coastal China and Wales, identifying points of convergence and divergence, and considering the possible lessons that China might take from the experience of rural development in Wales.

References

Bristow G (2000) Structure, strategy and space: issues of progressing integrated rural development in Wales. European Urban Reg Stud 7:19–33

Cai Y, Smit B (1994) Sustainability in Chinese agriculture: challenge and hope. Agr Ecosyst Environ 49:279–288

Cloke P, Edwards G (1986) Rurality in England and Wales 1981: a replication of the 1971 index. Reg Stud 20:289–306

Cloke P, Goodwin M, Milbourne P et al (1995) Deprivation, poverty and marginalization in rural lifestyles in England and Wales. J Rural Stud 11:351–365

Collis B (2003) Wales voluntary sector almanac 2003: A new Era?. Wales Council for Voluntary Action, Caerphilly

Courtney P, Mayfield L, Tranter R et al (2007) Small towns as 'sub-poles' in English rural development: investigating rural–urban linkages using sub-regional social accounting matrices. Geoforum 38:1219–1232

Department for Environment, Food and Rural Affairs (DEFRA) (2000) Our countryside: the future (Rural White Paper). Defra, London

Edwards JA (1985) Manufacturing in Wales: a spatial and sectoral analysis of recent changes in structure, 1975–1985. Cambria: A Welsh Geogr Rev 12:89–115

Edwards B (1998) Charting the discourse of community action: perspectives from practice in rural Wales. J Rural Stud 14:63–77

Edwards B, Goodwin M, Pemberton S et al (2000) Partnership working in rural regeneration: governance and empowerment?. The Policy Press, Bristol

Edwards B, Goodwin M, Pemberton S et al (2001) Partnerships, power and scale in rural governance. Environ Plann C 19:289–310

Evans N (2009) Adjustment strategies revisited: agricultural change in the Welsh Marches. J Rural Stud 25:217–230

Fan N, Wall G, Mitchell CJ (2008) Creative destruction and the water town of Luzhi, China. Tourism Manage 29:648–660

Goodman DS (2008) China's regional development, 5th edn. Royal Institute of International Affairs, London

Greenlees A (1998) Building communities: a guide for local government working with community-led initiatives. Welsh Council for Voluntary Action, Caerphilly

Halseth G, Markey S, Bruce D (eds) (2010) The next rural economies: constructing rural place in a global economy. CABI, Wallingford

Hooson E, Jenkins G (1965) The heartland: a plan for Mid-Wales. Liberal Publications Department, London

Kitchen L, Marsden T (2009) Creating sustainable rural development through stimulating the eco-economy: beyond the eco-economic paradox? Sociol Ruralis 49:237–294

Li X, Wang X (2003) Changes in agricultural land use in China: 1981–2000. Asian Geogr 22:27–42

Liang Z, Chen Y, Gu Y (2002) Rural industralisation and internal migration in China. Urban Stud 39:2175–2187

Liu Y (2007) 中国东部沿海地区乡村转型发展与新农村建设 (Rural transformation development and new countryside construction in eastern coastal China). Acta Geogr Sinica 62:563–570

Liu Y, Wang L, Long H (2008) Spatio-temporal analysis of land-use conversion in the eastern coastal China during 1996–2005. J Geogr Sci 18:274–282

Long H, Woods M (2011) Rural restructuring under globalization in eastern coastal China: what can be learned from Wales? J Rural Community Dev 6(1):70–94

Long H, Tang G, Li X et al (2007) Socio-economic driving forces of land-use change in Kunshan, the Yangtze River delta economic area of China. J Environ Manage 83(3):351–364

Long H, Zou J, Liu Y (2009) Differentiation of rural development driven by industrialization and urbanization in eastern coastal China. Habitat Int 33(4):454–462

Long H, Liu Y, Li X et al (2010) Building new countryside in China: a geographical perspective. Land Use Pol 27(2):457–470
Lovering J (1983) Gwynedd—a county in crisis—Coleg Harlech occasional papers in Welsh studies, No. 2. Coleg Harlech, Harlech
Moseley MJ (2003) Rural development: principles and practice. Sage, London
Ploeg JD, Renting H, Brunori G et al (2000) Rural development: from practice and policies to theory. Sociol Ruralis 40:391–408
Redfield R (1941) The folk culture of Yucatan. Chicago University Press, Chicago
Rigg J (1998) Rural-urban interactions, agriculture and wealth: a southeast Asian perspective. Prog Hum Geogr 22:497–522
Sargeson S (2002) Subduing "the rural house-building craze": attitudes towards housing construction and land-use controls in four Zhejiang villages. China Quart 172:927–955
Soemarwoto O (1976) Rural-urban relationships in Indonesia. Habitat Int 1:247–250
Unger J, Chan A (1999) Inheritors of the boom: private enterprise and the role of local government in a rural south China township. China J 42:45–74
Unger J, Chan A (2008) Associations in a bind: the emergence of political corporatism. In: Unger J (ed) Associations and the Chinese state: contested spaces. M.E. Sharpe, Armonk, pp 48–68
Woods M (2005) Rural geography: processes, responses and experiences in rural restructuring. Sage, London
Woods M (2007) Engaging the global countryside: globalization, hybridity and the reconstitution of rural place. Prog Hum Geog 31:485–507
Woods M (ed) (2008) New labour's countryside: rural policy in Britain since 1997. Policy Press, Bristol
Woods M (2010) Rural. Routledge, London
Xu H (1999) Environmental policy and rural industrial development in China. Hum Ecol Rev 6:72–80
Yang H, Li X (2000) Cultivated land and food supply in China. Land Use Pol 17:73–88
Zhang L (2008) Conceptualizing China's urbanization under reforms. Habitat Int 32:452–470
Zhou Z, Fan J (1988) 农村发展经济学 (Rural development economics). China Renmin University Press, Beijing

Chapter 8
Rural Restructuring in Typical Villages of China

Abstract Rural restructuring is an important means of implementing and pushing forward the strategy of rural vitalization, and villages are the basic unit of rural social and economic activities in China. Under the background of rapid urbanization, the rural territory continuously exchanges the material, energy and information with external urban system in China, and it is not a homogenous space, but a differentiated patchwork of hybrid places affected by rural restructuring. While most villages are still in the initial stage of social and economic restructuring, some villages have gradually embarked on the road of prosperity through integrating space and traditional industries, and the settlements, employment structure, regional space and urban-rural relations have changed gradually. Based on the theoretical research of rural restructuring, this chapter takes three typical villages dominated by tourism industry in metropolitan suburb areas, by traditional agriculture in plain farming areas and by e-commerce in central China, respectively, to carry out empirical studies on the process of rural restructuring through adopting method of the participatory rural assessment and geographic information system technology. The aims of this chapter are to reveal the process of and to explore the interactions of economic restructuring, spatial restructuring and social restructuring, and to put forward some suggestions on promoting local rural restructuring. The results show that, among the typical villages, the function of traditional agricultural production is declining gradually, and the industrial production, ecological culture and other multi-functional values of the rural territory have successively appeared.

8.1 Introduction

Rural restructuring is an important means of implementing and pushing forward the strategy of rural vitalization, and villages are the basic unit of rural social and economic activities in China. Under the background of rapid urbanization, the rural territory continuously exchanges the material, energy and information with external urban system in China (Tu and Long 2017), and the functions of rural territory include the aspects of living, production, ecology and culture, have evolved inevitably. Especially in the metropolitan suburbs of China, as the flowing of material, energy and

H. Long, *Land Use Transitions and Rural Restructuring in China*,
https://doi.org/10.1007/978-981-15-4924-3_8

information among the nodes of weaving and interlinking urban-rural territory, the "elements-structure-function" of rural territory have changed prominently accompanied with continual boost of urbanization, and the socio-economic morphology and spatial pattern are undergoing dramatic restructuring due to the emergence of leisure tourism, cultural and creative industries and other new forms of rural economy. Meanwhile, the agriculture-dominated villages in the plain farming areas are becoming differentiated influenced by the factors such as the flow of urban and rural development factors, market economy value-oriented, innovation of production tools. While most villages are still in the initial stage of social and economic restructuring, some villages have gradually embarked on the road of prosperity through integrating space and traditional industries, and the settlements, employment structure, regional space and urban-rural relations have changed gradually.

The continuous infiltration of e-commerce is of great significance for vitalizing the world's countryside (Cui et al. 2017; Liang 2010). The number of Taobao village[1] in China rose from 3 in 2009 to 2118 in 2017, with particular growth in southeast coastal China and slightly diffusion in northwest inland regions (Ali Research 2017). According to the data from the Ministry of Commerce of the People's Republic of China and Ali Research (Ali Research 2017), the retail sales of rural e-commerce in 2016 amounted to 894.5 billion RMB yuan (exchange rate US$ to RMB yuan: 1–6.35), directly creating more than 840,000 jobs, which not only means the endogenous motivation of traditional rural society to seek growth and modernity, but also implies that internet technology is restructuring the rural socio-economic structure with unprecedented force (Chen 2016; Jalali et al. 2011). In recent years, China's rural areas have undergone intense restructuring motivated by various element flows derived from e-commerce, which triggered a new wave of rural rejuvenation.

Based on the theoretical research of rural restructuring (Long et al. 2016; Tu and Long 2017), this chapter takes three typical villages dominated by tourism industry in metropolitan suburb areas, by traditional agriculture in plain farming areas and by e-commerce in central China, respectively, to carry out empirical studies on the process of rural restructuring through adopting method of the participatory rural assessment (PRA) and geographic information system (GIS) technology. The aims of the research are as follows: (1) to reveal the process of and to explore the interactions of economic restructuring, spatial restructuring and social restructuring; and (2) to put forward some suggestions on promoting local rural restructuring.

[1]Taobao village is initiated by Alibaba Group, and it refers to an aggregation of workshops that using a digital commercial platform provided by Taobao to conduct internet-based retail. A Taobao village requires the following conditions: (1) the basic unit of trading venues is administrative village; (2) the annual turnover exceeds 10 million RMB yuan (exchange rate US$ to RMB yuan: 1–6.35); and (3) the number of active online stores is over 100 or the proportion of active online stores in local households is more than 10%.

8.2 Typical Village Driven by Tourism Industry in Metropolitan Suburb Areas

8.2.1 *Study Area*

The Huangshandian village is located in the Zhoukoudian town, Fangshan district of Beijing, with a total area of 20.2 km^2, southwest of the metropolitan area of Beijing and 10 km away from the world cultural heritage-Zhoukoudian site (Fig. 8.1).

Owning to being situated in hilly area, the village is rich in mineral resources and vegetation coverage. Based on the advantages of resources and location, the spatial patterns, economic forms and social relations in Huangshandian village have undergone major changes since 2000, the multi-values of rural areas have become increasingly prominent, the living space has been gradually intensive and the production space has been further expanding, as well as self-government organizations, infrastructure and public services have been successively improved. There were 565 households, a population of 1600, and its farmers' per capita net income amounted to 16,000 RMB¥ in 2015, and it was awarded the titles of "National Civilized Village" and "Demonstration Village in Beijing". In the process of rapid urbanization, with the changes in population, land, industry as the core, parts of rural areas in metropolitan suburb have been universally facing with insufficient motivation in industrial development, increasingly empty rural housing land, decline grassroots self-governance organization, weak rural development mainstream, the shortage of

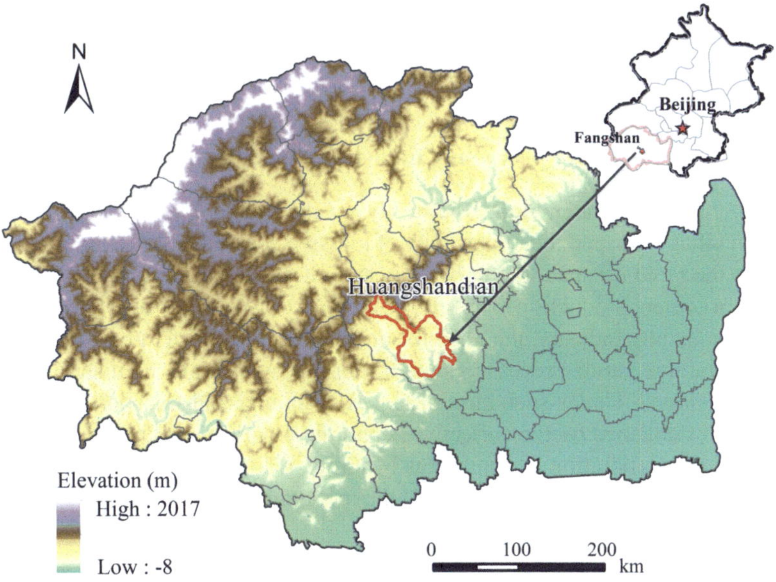

Fig. 8.1 Location of the Huangshandian village in Beijing (Tu et al. 2018)

public facilities and other issues. Under the above background, the paths and modes of rural restructuring in Huangshandian Village may provide a good reference for the rural development in metropolitan suburbs of China, and it is of great significance to research the process of rural restructuring and the enlightenment of land use policy to promote spatial restructuring in this kind of village. Therefore, we regard it as a typical case study of rural restructuring at village level under rapid urbanization in metropolitan suburbs of China.

8.2.2 Process of Rural Restructuring

8.2.2.1 Economic Restructuring

Since 2000, relying on mineral resources, tourism resources and regional advantages being adjacent to the capital, the industrial development in the Huangshandian village has experienced different transformations from traditional agriculture to primary industrial processing and eco-tourism. At present, the function of traditional agricultural production is declining gradually, and industrial production, ecological culture and other multi-functional value have successively appeared.

(1) **Phase one: Traditional agricultural production (before 2000)**

Before 2000, the industry in the Huangshandian village was dominated by the planting of traditional crops such as corn, wheat, soybean, sorghum and millet. Restricted by high mountains, steep slopes, thin soil layer, soil erosion and other physical factors, the agricultural cultivation could not completely meet villagers' basic demands for food and clothing. In 1990, the farmers' per capita net income was 1200 RMB¥, and it was a typical poor village.

(2) **Phase two: Primary processing industry (from 2000 to 2008)**

In 2002, under the leadership of the newly set up village committee, the Huangshandian village grasped the opportunity of massive demands for building materials owing to the rapid urbanization in Beijing, and introduced an investment enterprise engaged in cement production to develop primary processing industry. Subsequently, around the leading industry, the need for upstream and downstream products, such as the ore mining, stone processing, cargo transport and commercial services were further developed. Under such circumstances, a mining and transportation corporation was established by the collective economic organization via integrating idle funds, and the corporation has enormously stimulated the economic thriving in the Huangshandian village, and once shaped the predominant industrial structure based on the ore mining and primary processing.

Fig. 8.2 Pofengling scenic spot in the Huangshandian village (Photos taken by Wenzhi Gao)

(3) **Phase three: Eco-tourism industry (since 2009)**

With the negative impacts on the local environment resulting from the ore mining activities and the introduction of national policy of closing small-scaled mine, the Huangshandian village has stopped the resource-mining industry gradually and initiated exploring the way of eco-tourism industry via taking advantages of vigorous tourism market demand around the metropolis and rich mountainous landscape resources. In 2009, marked as the construction of Pofengling scenic spot, the village commenced the second industrial transformation to develop the collective tourism. As of 2016, with a total scenic area of 133 ha and gross investment of more than 40 million RMB¥, Pofengling scenic spot has held five Red Leaves Festivals, and has gradually become a popular tour site of climbing mountain and viewing red leaves in autumn in the suburbs of Beijing (Fig. 8.2). Simultaneously, the tourism industry has brought about satisfied social and economic benefits, the comprehensive income of tourism amounted to 20 million RMB¥ in 2016, and more than 100 villagers engaged in ticketing sales, parking management, environmental maintenance and other related jobs in the village.

8.2.2.2 Spatial Restructuring

Accompanied with the evolution and restructuring of village socio-economic morphology, the rural space of the Huangshandian village has accordingly taken place tremendous changes in the aspects of quantity, structure and pattern, and the ratio of living, production, ecological space was changed from 1.5:6.9:91.6 in 2000, respectively, to 1.1:16.0:82.9 in 2016. It shows that the rural space is gradually undergoing a process of restructuring (Fig. 8.3).

(1) **Living space restructuring**

Due to the special terrain condition, the Huangshandian village suffered repeated attacks of floods and waterlogging in history. Combined with the inefficient use of homestead caused by urbanization, the village began to restructure the living

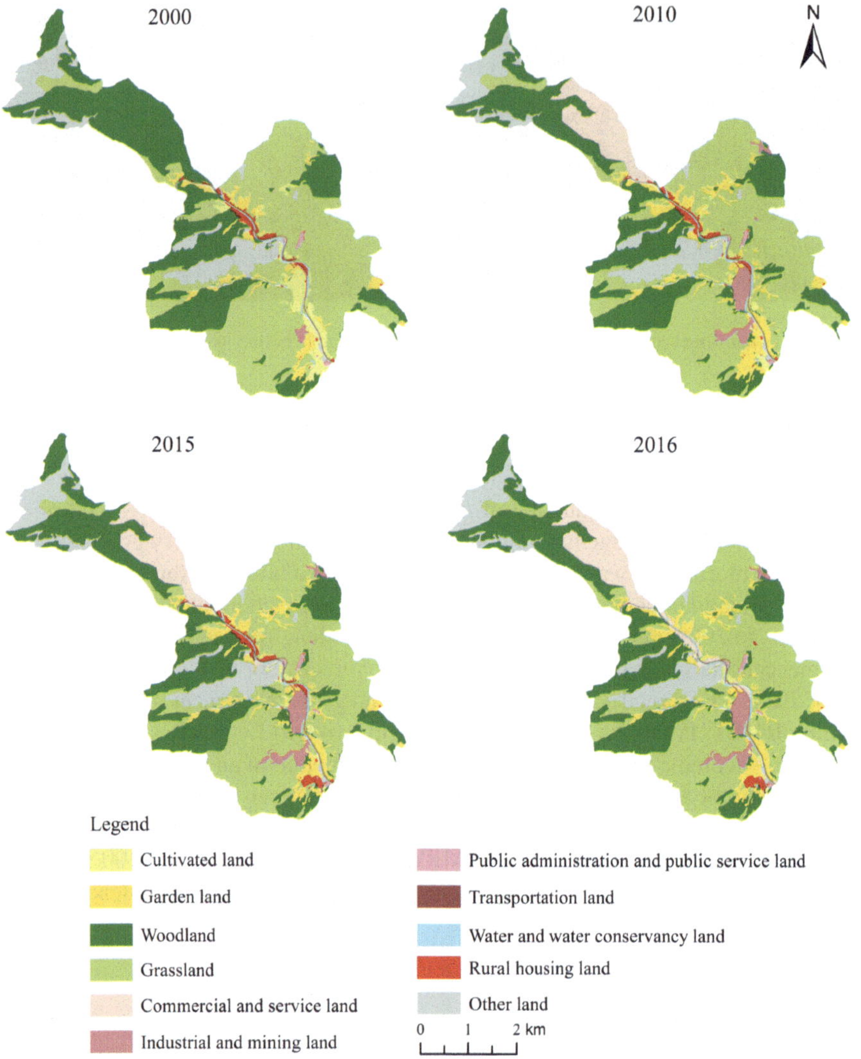

Fig. 8.3 The evolution of land use patterns in the Huangshandian village during 2000–2016 (Tu et al. 2018)

space through building a new community outside the village in 2013. By the end of 2016, the new community occupied 9.8 ha, included 35 residential buildings, and simultaneously was equipped with water, electricity, gas, roads, heating and other supporting infrastructure (Fig. 8.4), which has improved the living environment of farmers and promoted the complete separation between the rural production space and living space, and rural lifestyle has become increasingly urbanized. Meanwhile, in pace of moving into the new community in 2016, the old homesteads as well as

Fig. 8.4 The new (left) and old (right) living environment in the Huangshandian village (Tu et al. 2018)

streets and lanes in the village could be used to develop tourism industry, and the living space reduced by 33.9%.

(2) **Production space restructuring**

The restructuring of production space is closely related to local industrial development. With the industrial transformation and subsequent functional evolutions of rural territory in the Huangshandian village, land use types are increasingly complicated among production space, and their spatial distributions have undergone three stages, i.e., the outer circle around the rural settlements dominated by traditional agriculture, external decentralization around the rural settlements dominated by industry, and the overall expansion and the partial spatial agglomeration at the same time dominated by eco-tourism industry.

Compared with the slight decrease of agricultural production space (composed of cultivated land and garden land), agricultural land management is gradually becoming large-scale and centralized. Since 2000, the amount of cultivated land has been largely cut down. The decrease of cultivated land is directly bound up with the expansion of industrial land, and also affected by the national policy of returning farmland to forest. With the construction of tourism projects such as cherry picking garden, rose plantation and others, garden land have rose by 57.54% during 2000–2015, which to some extent inhibited the decreasing trend of agricultural production space. In addition, owing to the cultivated land and garden land being transferred to the collective organization in the village for unified operation, agricultural land management is moving towards large-scale and centralized development from fragmentation, and agricultural production function has been gradually strengthened with the characteristics of standardization and modernization.

With the development of cement and quarrying industries, industrial and mining land expanded quickly. From 2000 to 2005, the area of industrial and mining land has increased by 46.8 ha, which was mainly derived from cultivated land, garden land, woodland and idle land. In the wake of the shutting down of resource-based enterprises, there are large quantities of abandoned or idle industrial and mining land needed to be further reclaimed.

Parts of the ecological and living spaces have evolved into tourism service space. Along with the construction of Pofengling scenic spot and the corresponding supporting facilities e.g., catering and parking, a large quantity of ecological space such as woodland and other ecological land were transformed into commercial and service land, which accounted for 7.3% of the total land area in 2016. After the overall relocation of the villagers, more than 400 old homesteads covering an area of 21.9 ha would be transformed into guesthouse with the function of tourism, leisure and vacation (Fig. 8.5), which is expected to increase the production space by 12.1 ha subsequently.

(3) **Ecological space restructuring**

With the expansion of production space, the area of ecological space has relatively reduced in the Huangshandian village, but the ecological function has been strengthened to some extent due to the national policy and local measures concerning ecological environment construction.

The area of ecological space has reduced sharply. From 2000 to 2016, the proportion of ecological space (made up of grassland, woodland, water and water conservancy, idle land in other land use types) have decreased by 8.67% on account of the conversion to industrial and mining land, commercial and service land, garden land, protected agricultural land. Separated by dot-like or linear production spaces, the morphology of ecological space tends tobe broken and fragmented.

The production-ecological composite space has massively risen. Along with the transformation of rural industrial coordinated development among agriculture, tourism service and other functional diversification of rural territory, some of commercial and service land converted from idle land, cultivated land and garden land, have both tourism and ecological functions, and the production-ecological composite space has rapidly increased by 155.1 ha from 2000 to 2016. Besides, the implementation of returning farmland to forest in China as well as barren hills blasting and afforestation projects covers an area of 159 ha, part of the free land have been transformed into woodland. The adjustment of land use structure within ecological space together with local ecological construction programs, e.g., cleaning river course, ecological function of the village has presented a situation of getting better recently.

Fig. 8.5 The guesthouse transformed from old homesteads in the Huangshandian village (Tu et al. 2018)

8.2.2.3 Social Restructuring

To cope with the social issues accompanied with the urbanization and industrialization, such as the downfall of organization system and lack of infrastructure and public services, the Huangshandian village has carried out rural social restructuring in recent years through establishing the autonomous organizations and implementing a series of projects to improve infrastructure and public services.

(1) Restructuring rural autonomous organizations as the core of the village committee. Generally, the village committee plays an important role of leader, organizer and demonstrator in the process of rural restructuring. In 2001, the new village committee and the leadership team have been established by improving the cadre selection system and encouraging local elites to govern the village. After that, in order to improve the governance ability and the credibility of the village committee, a series of working mechanisms have been set up gradually, e.g., improving the villagers' congress system, carrying out public bidding for key programs and democratic decision-making system for major matters, which were conductive to playing the participation and supervision roles of villagers in public affairs and administrative management. In addition, for the purpose of enhancing the cohesion of villagers and developing more stable community-based common interests, diversified economic organizations in the fields of agricultural corporation and tourism management, and some building firm as well as minerals and transportation companies were established one after another.

(2) Improving the infrastructure and public services. With the promotion of economic strength, the village has implemented a whole string of projects benefitting people's livelihood, such as voltage rectification, road hardening, river course cleaning, construction of centers for pension and disabled villagers, as well as sanitation, tap water and cable television installation, which effectively improved the infrastructure and public service facilities. Moreover, village collective has provided an annual grant of 360 RMB¥ for 60 years old or above since 2006.

To summarize, along with the advancement of rapid urbanization, as well as the guidance of national macro policy and the driving of market demand, the "elements-structure-function" in the rural territory of Huangshandian village has experienced tremendous changes since 2000. Under the intervention of local social behavior mainstream by effectively integrating the elements (i.e. land, capital and labor), recombining the structure (i.e. industry, employment and land use) and optimizing the function, its socioeconomic forms and spatial structure are undergone rapid restructuring (Fig. 8.6).

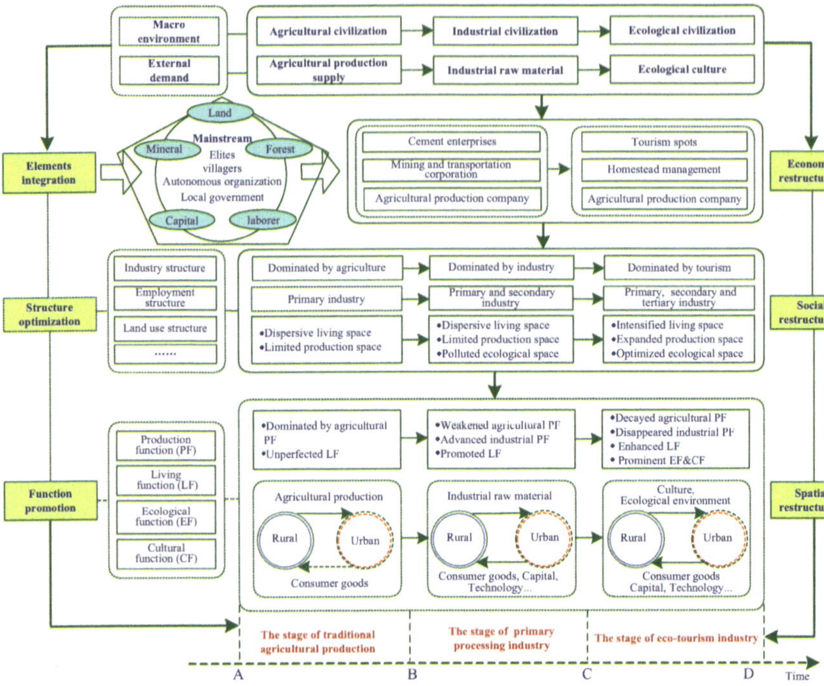

Fig. 8.6 The process of rural restructuring under rapid urbanization in the Huangshandian village (Tu et al. 2018)

8.2.3 *Discussion*

8.2.3.1 The Relationships of Economic-Spatial-Social Restructuring

The rural restructuring involves spatial restructuring, economic restructuring and social restructuring, three mutually reinforced and restrictive dimensions (Fig. 8.7), and it occurs in an interaction framework of external system and kernel system composed of natural subsystem, economic subsystem, social subsystem and ecological subsystem at any time. Under the human intervening, the interactions of economic restructuring, spatial restructuring and social restructuring drive jointly the systematic development of the "natural-ecological-economic-social" subsystems and the comprehensive promotion of the "production-living-ecological-cultural" functions in the rural territory (Tu and Long 2017).

(1) Economic restructuring plays a leading role in the process of spatial and social restructuring. Usually, rural restructuring begins with economic development, which not only provides the necessity of spatial integration due to the changes of economic factors, but also generates the feasibility of spatial integration and social restructuring based on the promotion of material foundation. Just as

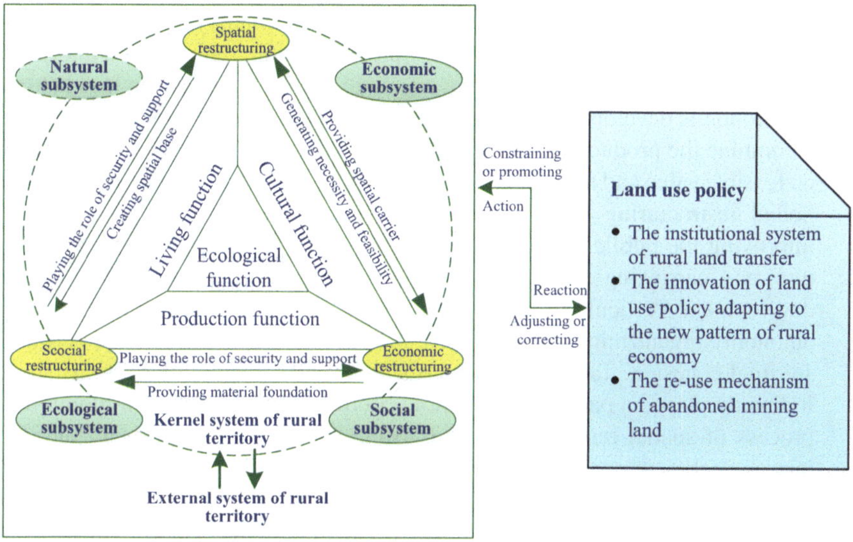

Fig. 8.7 Land use policy and institutional innovations pushing forward rural restructuring (Tu et al. 2018)

the Huangshandian village shows, the industrial development directly brought about the non-agricultural transfer of the labor force and subsequent transformation of the employment structure, which created the important prerequisite for the farmland circulation, thus further promoted the agricultural large-scale management and the production spatial restructuring. At the same time, the increase of the villagers' income, changes of employment structure, the transformation of land use patterns and the improvement of the self-development ability resulting from developing industry and tourism, accelerated the pace of the residential centralization and the construction of new community, which led to the living spatial restructuring as well as social restructuring in aspects of infrastructure, education, culture and health. Therefore, rural restructuring in practice should take industrial cultivation as the core to remodel the vitality of rural economy, and drive the linkage development of the subsystems as well as the comprehensive promotion of functions in the rural territory.

(2) The spatial restructuring is of particular significance to provide spaces for economic and social restructuring. Like the Huangshandian village, the production spatial restructuring has played an important role in the process of economic development. Initially, it was the farmland transfer and the introduction of external finance based on this that promoted industrial transformation from traditional agriculture to industry. Subsequently, the exploitation of barren mountains as well as the old homesteads converted into guesthouse re-excavated the multiple values of rural resources, further expanded production space and boosted the thriving economy. Additionally, the spatial restructuring based on farmland circulation not only promoted large-scale management and modern agricultural

development, but also stimulated the prosperity of secondary and tertiary industries to some extent by transferring surplus agricultural labor force into non-agricultural industry. Generally, the course of spatial optimization lags behind the economic development. It is necessary to take human intervention timely to recombine the production, living and ecological spaces in rural territory based on feasible value and target orientation as well as scientific planning technology.

(3) Social restructuring such as cultivating the social behavioral mainstream, improving the public service and social security system, inheriting local culture, and narrowing the urban-rural gap is the support system for the rural development. Particularly as the core of the social behavioral mainstream, the autonomy organizations usually exert an influence on intellectual support and spiritual cohesion. Taking the Huangshandian village committee for instance, it directly played a part of organization, coordination and demonstration in the process of spatial restructuring, and also constantly pushed forward the village to seek new economic growth point and smoothly realize the economic transformation. Moreover, as the improvement of village committee credibility, villagers gradually formed the common value of building beautiful hometown, and actively participated in the land circulation, industrial development, environmental conservation and new community construction with high enthusiasm and cohesion, which enormously stimulated the simultaneous improvement of production, living and ecological functions in the Huangshandian village.

(4) Economic restructuring, spatial restructuring and social restructuring, the three dimensions of rural restructuring are mutually reinforced and restrictive. Initially, rural economic restructuring produces crucial influence on increasing the farmers' income, realizing the non-agricultural transfer of surplus labour and improving the value of land use, which further enhance the necessity and feasibility of production, living and ecological spatial restructuring, and offer the material foundation for the rural social restructuring (e.g. infrastructure and public services); Secondly, rural spatial restructuring by optimizing production, living and ecological land use patterns and improving land use efficiency, is not only conductive to provide space for developing scale farm management and non-agricultural industries, which would further exert important impacts on industrial transformation and economic capability, but also create the spatial base for social restructuring such as the living conditions and public service facilities; As far as social restructuring is concerned, as the most dynamic factor, behavioral mainstream involving villagers, autonomous organization and rural elites usually provide intellectual support for rural economic restructuring and play a role of organization, coordination and demonstration in the process of rural spatial restructuring, while the perfect public service and social security system is of particular significance to provide guarantee for the effective organization of spatial and economic restructuring.

8.2.3.2 Suggestions for Innovations of Land Use Policy Aiming at Promoting Rural Spatial Restructuring

Recently, interactively influenced by natural resources, national macro strategies, market demand, technological progress, macro-regional policies as well as the flow of urban-rural developments elements under the rapid urbanization, the economic forms, industrial structure, employment model and land use efficiency have changed greatly in metropolitan suburb of rural China, and it is urgent to restructure the rural socio-economic morphology and spatial pattern. As the carrier of socioeconomic development, land and its utilization play an important role in the process of rural economic and social restructuring. While the effective implementation of spatial restructuring is closely related to a series of land policies and institutions, which not only hinder or promote the practice of rural restructuring via influencing the allocation of production factors, but also have been constantly adjusted and corrected along with the practice of rural restructuring. To cope with the changes of "elements-structure-function" in metropolitan suburb of rural China, the following proposals on land use policy and institutional innovations should be paid more attentions for the purpose of optimally utilizing the land resources and promoting the rural spatial restructuring (Fig. 8.7).

(1) Accelerating the institutional framework design of rural land transfer. With the advancement of urbanization and farmers' livelihood diversification in the metropolitan suburbs, the resident population in rural areas has declined sharply and the inefficient use of housing land is increasingly prominent in China. At the same time, the changes of the marginal productivity of different factors due to the migration of rural labor force and the introduction of new production tools have resulted in the re-allocation of farmland between different management streams. As a consequence, there is an urgent need to accelerate the institutional framework design of rural land transfer in order to promote the optimal allocation of land resources and regulate the behavior of land use in the process of living and production spatial restructuring. First of all, it is essential to set up the important premise to promote land transfer through pushing forward the reform of rural land property rights system and establishing rural property value and profit assessment mechanism to make a comprehensive evaluation of the productive value, ecological value and cultural value of the land (Long et al. 2016). Meanwhile, it is suggested to explore the incentive and restraint mechanism of residential land use right circulation as the core of regulating the basic procedure of homestead withdrawal and innovating the model of the resettlement in the context of urban-rural integration development, and pay more attention to the institution of property rights transfer and transaction involved in the process of homestead circulation and the agglomeration of rural settlements. Besides, in view of the de-agriculturalization and non-grain preference of farmland use after land transfer, the informatization construction monitoring the behaviors of farmland operation rights transfer and rural housing land

mortgage should be strengthened so as to accurately assess the risk and make an early warning timely to ensure the healthy operation of the land market.

(2) Exploring the tourism land management system adapting to the new pattern of rural economy. In recent years, some rural areas in metropolitan suburbs take advantage of the ecological natural landscape, various festival activities and the location conditions to develop tourism, as the new pattern of rural economy, which is playing an increasingly important role in boosting the industrial transformation and promoting economic restructuring. However, due to the blurry classification of tourism land, the absence of supporting regulations on tourism land supply together with the weak cohesion and convergence among tourism planning with other related departments' planning in the aspects of urban construction, forestry, environmental protection and cultural protection, it is very difficult to obtain legal tourism land, which has become a bottleneck to development rural tourism. For the sake of adapting to the new pattern of rural economy, exploring the tourism land management system from the following aspects is of great significance to expand rural production space and boost rural economic restructuring. Firstly, on the basis of establishing the definite classification standard of tourism land, it is suggested to gradually develop the classified supply and manage system of tourism land. Secondly, there is a tremendous need to implement the incentive polices and measures encouraging the inefficient and abandoned construction land to be transformed into the tourism land, and the supporting institutions related to the procedures of property right alteration as well as the revenue sharing of land transfer should be set up timely in order to effectively protect the legitimate rights and interest of land owners, users and operators. Last but not the least, it is necessary to clearly define the applicable scope of tourism land and establish the monitoring mechanism to ensure that the property of tourism land is unchanged instead of being transformed into residential or commercial housing and others.

(3) Pushing forward the re-use mechanism of abandoned industrial and mining land. Affected by the national macro policy of industrial transformation, the shutdown of resource-based industries have resulted in a large number of abandoned industrial and mining land in some areas. However, restricted by the characteristics of small-scale and scattered distribution and the high reclamation cost, the pace of the abandoned industrial and mining land reclamation is seriously lagging. Consequently, in respond to the booming demands of construction land in metropolitan suburb, exploring a complete set of re-use mechanism of abandoned land is an effective approach to excavate the land potential, optimize the allocation of land resources and restructure rural development space. First of all, it is critical to set up the information management system in order to systematically grasp the amount of exploitable abandoned industrial and mining land. Secondly, for providing technical support for the practice of abandoned land reclamation, a series of key reclamation and re-use standards and operating specification should be laid down through carrying out multidisciplinary demonstration research involving environment, chemistry, physic, geography,

management, economics and other disciplines. In particularly, in order to promote the rural plural spatial reconstruction, it should be paid more attention to set up the linkage and aggregation mechanism of actors at different levels of government and local multi-subjects, and ultimately stimulate the forming the diversified market operation system of the reclamation and revitalization of abandoned industrial and mining land e.g. giving full play to the role of government in planning, coordinating and supervising the utilization of abandoned land, encouraging the introduction of supporting policies that guide social funds and civil economic entities to participate in reclamation of the industrial and mining land, and increasing the proportion of local collective organizations benefiting from the land consolidation so as to arouse the enthusiasm of local behavior mainstream.

8.3 Typical Village Driven by Traditional Agriculture in Plain Farming Areas

8.3.1 Study Area

Yangqiao Village is located in the southwest of Yucheng City, Shandong Province, about 24 km away from Yucheng District (Fig. 8.8). The village has a flat terrain and a semi-arid and semi-humid climate with an average annual temperature of about 13 °C and an average annual precipitation of 500 mm. In 2015, there were 131 households and 519 people in the village. The land area of the village was 1.3 km^2, of which 94.34 ha were cultivated land. Since the 1980s, the Yangqiao Village has gone through the stage of traditional agricultural leading and concurrent farming production. At present, the main production function of rural areas is traditional agriculture, and the social and economic morphology has no significantly change. Rapid industrialization and urbanization have triggered serious population loss in villages, and the village is facing such prominent problems as weakening the main body of rural development and extensive land use. In order to cope with the problem of extensive use of residential land and rural hollowing caused by rural population loss, Yangqiao Village started a new community construction process in 2013 with the support of the local government. At present, the human settlement environment in the village area has been improved obviously, but the social and economic development are still relatively backward due to the insufficient potential of industrial development. The development status and restructuring stage of the village are quite representative in the agricultural-dominated village areas in the plain agricultural areas.

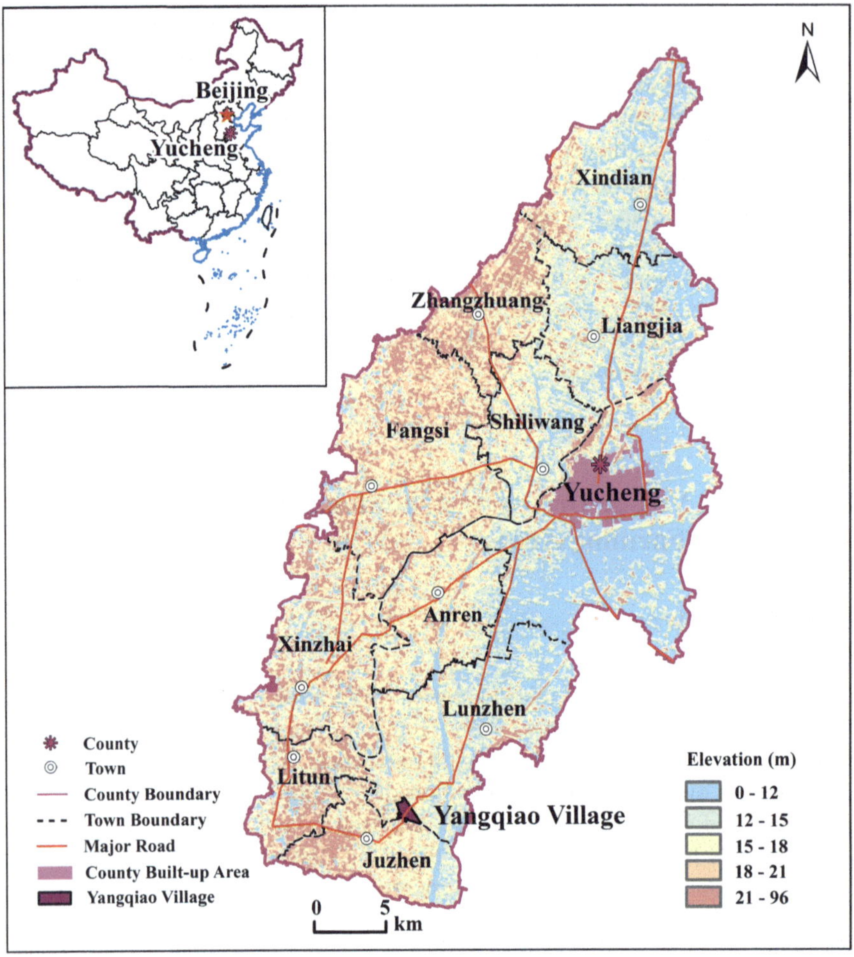

Fig. 8.8 Location of the Yangqiao village in Yucheng city (*Source* Drawing by Dazhuan Ge)

8.3.2 Process of Rural Restructuring

8.3.2.1 Economic Restructuring

(1) **Phase 1: The dominant stage of traditional agriculture (1980–2000)**

Yangqiao Village's traditional agricultural cultivation is mainly based on traditional bulk grain crops such as Winter wheat and summer corn, supplemented by peanut, soybean, sesame, cotton and other cash crops. However, due to without adequate income sources for farmers besides agricultural cultivation, the economic development stays at the stage of meeting basic food and clothing. After 1990, with the progress of urbanization in China, the phenomenon of cross-provincial migrant workers began to appear sporadically, mainly flowing into developed areas such as Beijing and Guangzhou. They mainly engage in non-agricultural industries such as construction and transportation, and their income level increased slowly.

(2) **Phase 2: Agricultural concurrent production (2000–present)**

In the wake of rapid improvement of agricultural mechanization and irrigation facilities, grain production has increased significantly again. However, due to the increase of price of agricultural production materials, land rent, labor costs and other factors of production, the trend of low income of grain planting is becoming obvious. Since 2000, a large number of non-agricultural employed people have emerged in the village, and the phenomenon of part-time farming is widespread. The income sources of farmers tend to be diversified. Concurrently, the development of villages is facing such prominent problems as the serious loss of young and middle-aged labor force, the weakening of the main body of rural development, and the insufficient development potential of villages.

From 1990 to 2015, the non-agricultural proportion in the village labor force increased from 10 to 76%. Among the non-agricultural employment population, the proportion of migrant workers increased steadily year round. From 1990 to 2015, the proportion of migrant workers increased from 7 to 35%. The majority of migrant workers were young and middle-aged workers, mainly to developed areas such as Beijing, Jiangsu, Zhejiang, Shanghai, and Qingdao and Yantai in the province. In addition to migrant workers, some peasants are engaged in construction odd jobs in the town area or employed by large land transfer households/enterprises in the surrounding villages and towns for seasonal crop cultivation. The phenomenon of farmers' concurrent business is widespread. From 1990 to 2010, the concurrent proportion increased from 3 to 46%, while the proportion of pure farmers decreased from 90% in 1990 to 24% in 2015. The age structure of migrant workers, concurrent peasants and pure peasant households shows gradient decreasing.

8.3.2.2 Spatial Restructuring

The industrial structure of Yangqiao village is relatively single, so are the functions of life, production and ecological space. The evolution of space is mainly manifested in the expansion of living space and the erosion of production and ecological space. The proportion of living, production and ecological space in villages ranged from 9.12:73.36:17.52 in 1980 to 12.85:73.53:13.62 in 2015 and to 8.80:86.43:4.77 in 2016 (Fig. 8.9).

(1) Restructuring of living space

The expansion of living space in Yangqiao Village has gone through three stages: enclave expansion to the outer edge of the settlement, internal infilling and filling development and external agglomeration development. In the 1980s, due to the long-term absence of village planning and management and the absence of strict land management, the new housing construction expanding outside the village, and the area of

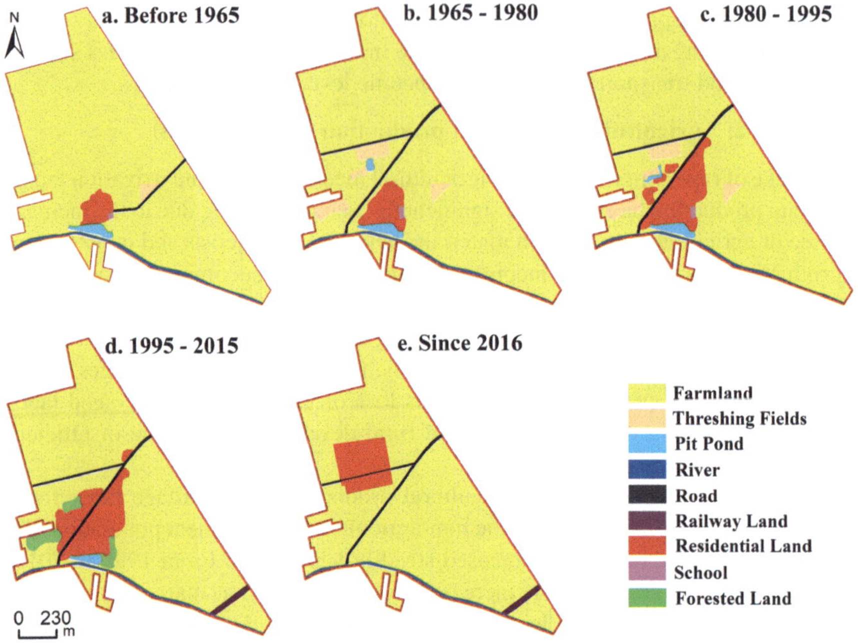

Fig. 8.9 Evolution of land use patterns in Yangqiao village (1965–2016) (*Source* Drawing by Dazhuan Ge)

village homestead continued to expand (Fig. 8.10). From 1980 to 2015, the number of residential land increased from 149 to 269, and the area increased from 612,000 to 1088,000 m^2. Homestead expansion is mainly manifested in the erosion and occupation of cultivated land, idle land and woodland. In 2013, in order to improve the human settlements environment, Yangcunqiao carried out comprehensive renovation of hollow villages based on the project of "increasing vs. decreasing balance", and started construction of new communities in Yangqiao village (Fig. 8.11). The

Fig. 8.10 The appearance of the village before the demolition of Yangqiao village (Long et al. 2018)

Fig. 8.11 New community restructuring in Yangqiao village (Long et al. 2018)

new community involved the relocation of 11 villages, of which the first phase of the project resettled 5 villages of Hekouzhao, Hekouli, Zhu wang, Yangqiao and Qinzhuang. By November 2016, 120,000 m^2 (including residential buildings, greening and internal transportation roads) have been built, 26 residential buildings and 352 sets have been built. In the former Yangqiao Village, 146 households and 578 people concentrated on resettling new communities. The residential land was equivalent to 467,000 m^2, and the new residential land was equivalent to 467,000 m^2.

(2) Restructuring of production space

At present, the industrial structure of Yangqiao Village is still dominated by traditional crop cultivation, and the production land is mainly cultivated land. From 1980 to 2015, the production space showed that the area of cultivated land was decreasing gradually, from 954,000 to 943,400 m^2. It is closely related to the expansion of living space in the context of specific social and economic development stage in China. With the expansion of residential land as the driving force, the change of production space structure morphology is characterized by eroded constantly along the road, along the outskirts of settlements. After the reclamation and cultivation of old villages, 63,100 m^2 of cultivated land added from the old residential land. In addition, the idle land, street land, scattered forest land and pond water surface land in villages have been transformed into cultivated land, and the production space increased by 215,600 m^2. From the perspective of spatial morphology, the production space of Yangqiao Village has gone through three stages: primitive inner circle stratification, settlement edge erosion under the expansion of residential land, and centralized fragmentation under the construction of new communities. Along with the concurrent industrialization and non-agricultural transfer of population, there is a small amount of land transfer phenomenon of alternative cultivation in villages. However, because of tits small scope and scale, the number and area of agricultural land have not changed significantly, and the spatial structure characteristics of arable land fragmentation have not changed significantly. The fragmented, fragmented and small-scale land management situation limits the improvement of agricultural production efficiency, as well as promotes the concurrent production to a certain extent.

(3) Restructuring of ecological space

With the expansion of living space, the total amount of ecological space in villages decreases, and the structure of ecological space is gradually fragmented. From 1980 to 2015, the total amount of traditional ecological space of villages consisting of woodland, water area and idle land decreased by 507,000 m^2. From the perspective of the internal structure of ecological space, the area of woodland and idle land continued to decrease, while the area of water area was relatively stable. Among them, forest land decreased by 411,000 m^2, and the reduction of ecological space mainly manifested in the transformation to residential land and cultivated land.

8.3.2.3 Social Restructuring

(1) **Phase 1: Extreme lack of infrastructure and public services (1980–2015)**

In the 1980s, the level of social development in villages was still low, and infrastructure and public services were extremely scarce. It was not until 1987 that Yangqiao Village was electrified, but because of the problems of circuit and voltage, normal electricity use is still difficult to guarantee. In 1995, Yangqiaocun Primary School was abolished with some school-age children following their parents. Under the guidance of the national inclusive policy, the whole village launched a new rural cooperative medical insurance in 2004; in 2007, the village realized its own deep well water supply; villagers began to pay and enjoy the basic old-age insurance in 2010; fixed garbage collection points were set up in the main streets of the village, and domestic garbage was collected and transported in a sealed way in 2011. However, due to the low level of economic development, rural social development is faced with such problems as the imperfect democratic system at the grass-roots level, the deterioration of the human settlement environment in villages, the inadequate allocation of public facilities, and the decline of traditional village culture.

(2) **Phase 2: Spatial restructuring promotes social restructuring (since 2016)**

The newly-built community has realized the construction of water, electricity, gas, ground hardening, environmental sanitation, night lighting and other supporting facilities simultaneously, and has allocated community convenience service center, Party member activity center, multimedia activity room, culture room, kindergarten, community health service station, apartment for the elderly, cultural fitness square, standardized village health service rooms and so on, the living environment of peasant households has been significantly improved (Fig. 8.12). In order to solve the inconvenience of the elderly living upstairs and downstairs, 20 old people's houses were built in the community. However, with the increase of the proportion of peasant

Fig. 8.12 Yangqiao community convenient service center (left) and reading room (right) (Long et al. 2018)

households going out to work, the attention of villagers to the development of village affairs has declined, the development of village social organizations is weak, and the cohesion within the village is seriously insufficient.

8.3.3 Discussion

The social-economic development drives the agricultural-dominated villages in plain agricultural areas to gradually diverge. Some villages have gradually embarked on the road of specialization and industrialization of agricultural development through the transformation of traditional agriculture, and the production and living space has been reconstructed. However, most villages in plain agricultural areas in China, such as Yangqiao Village, are generally faced with such prominent problems as single village economic structure, low comparative efficiency of grain planting, the weakening behavior mainstream, fragmentation of cultivated land management, idle and abandoned residential land, etc. Rural population loss, land waste, industrial vacuity and lack of socialized services are intertwined, which together constitute the main obstacles to the development of traditional rural areas and villages. How to optimize the internal structure of agriculture and establish a modern agricultural production and operation system, effectively integrate the factors of rural development, transform and activate traditional industries to achieve agricultural transformation and upgrading, strengthen policy support and guide to improve the system of agricultural support and protection, and properly handle the population-land-industry relationship are the keys to realize rural revitalization in this type of villages.

(1) Integrating rural land elements and promoting the restructuring of village production space. Agricultural-dominated villages generally face the problem of low efficiency of grain planting. Behind the low efficiency of grain planting is scattered, fragmented and small-scale land management, which is not conducive to mechanized farming, but also hinders specialized management and affects the further improvement of labor productivity. The key to solve this

problem lies in promoting the restructuring of production space and moderate large-scale operation, taking the road of specialization and industrialization of agricultural production and operation, to save labor costs and input of other production elements. However, in reality, due to many factors such as imperfect policy system, long agricultural production cycle and lagging construction of agricultural supporting service facilities, the difficulty of land scale transfer and the unstable relationship between land transfer and lease have become obstacles to the sustainable promotion of large-scale agricultural management. In view of the problems existing in land transfer, it is urgent to establish a stable and standardized market order from outside the village area, so as to provide system guarantee and market platform for the smooth land transfer. In addition, land transfer is a progressive process. Land resources not only provide a place for production demand and its corresponding economic benefits, but also maintain the livelihood of the population in a certain area and ensure social stability through a small amount of land management. In the process of promoting land transfer, we should focus on the non-agricultural employment opportunities provided by the local government and the livelihood of the elderly, the weak and the disabled after losing their land.

(2) Improving the organizational system of agricultural production and promoting the restructuring of village economy. At present, the dominant rural areas in the case area are mostly traditional bulk grain planting, such as wheat and maize, with a single product structure and low economic benefits. In some villages, by adjusting the internal structure of agriculture and relying on local resources, large-scale cultivation of vegetables, fruits, flowers and other specialized farming, such as cows, pigs and fisheries, has been carried out. But the agricultural product structure assimilation is prominent in economic development, and influenced by natural risks, market fluctuations and agricultural policies, the production of single agricultural products faces marketing and market risks. The reconstruction of village economy urgently needs to take the improvement of agricultural production organization system as the core, and build an integrated industrial system of agricultural production, processing and marketing through the demonstration and promotion of leading enterprises, specialized cooperative organizations and capable people, so as to increase the added value of agricultural production. In the form of market-led, base-led and base-linked peasant households, diversified agricultural specialization and industrialization modes have been formed to gradually realize the integration of trade- industry-agriculture, and the one-stop production and operation system of production, supply and marketing.

(3) Properly handling the population-land-industry relationship. Under the background of "upside-down" relationship between grain production and economic development, the development of economy and the improvement of agricultural production efficiency depend on the specialized and industrialized management based on the adjustment of traditional agricultural structure; specialized and industrialized management depend on the appropriate scale of land transfer;

while the smooth promotion of land transfer depends on the liberation of peasants from the traditional mode of farming land and the non-agricultural transfer of surplus labor; the non-agricultural transfer of surplus labor force needs to be supported by a large number of employment posts relying on the development of non-agricultural industries. Population migration, land transfer and industrial development are complementary and mutually restrictive. It is the key for rural restructuring in agricultural-dominated villages to correctly handling the relationship between population, land and industry, and attach great importance to land and industry. Taking land renovation and land transfer as the starting point, we should promote the restructuring of living and production space to provide space for the development of modern agriculture and the cultivation of non-agricultural industries. Taking industrial cultivation as the core, we should enhance the self-development ability of villages, and provide operational feasibility for population non-agricultural transfer, land use mode transformation and rural spatial restructuring.

8.4 Typical Village Driven by e-Commerce in Central China

8.4.1 Study Area

Xiaying village, located in Yunxi County, Shiyan City of Hubei Province (Fig. 8.13),

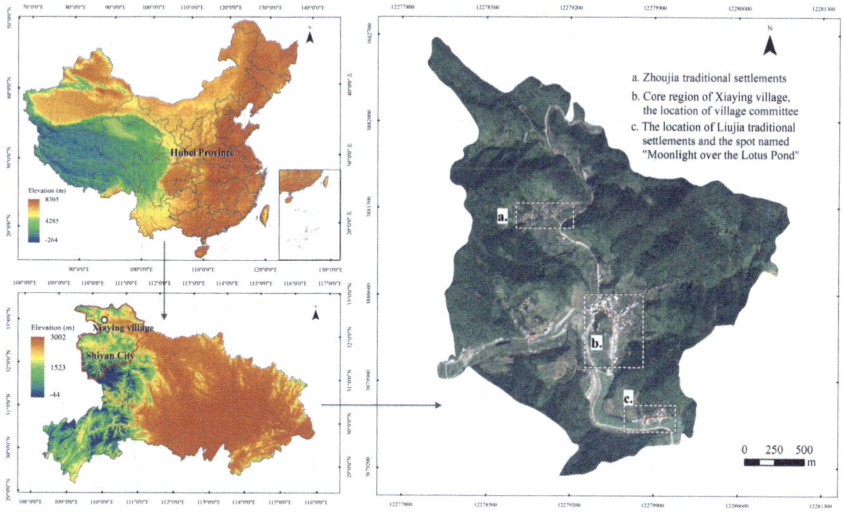

Fig. 8.13 Profile of Xiaying village (Zhang et al. 2018)

is renowned for abundant kallaite resources. It is one of the existing 25 Taobao villages in central and western China. More than 70% of the villagers get involved in the manufacturing of kallaite and rural e-commerce, which makes Xiaying village is of significant representativeness.

The whole village occupies an area of 4.5 km^2 and has 339 families who live locally. Its total population is 1413, whose per capita farmland area is merely about 500 m^2. Prior to 2008, an overwhelming majority of villagers migrated for work and the small group ran physical stores outside selling kallaite, which finished primitive accumulation of capital for rural e-commerce. Inspired by returning college students, e-commerce is gradually accepted and adopted by villagers, and is rapidly spread through the "acquaintance society" uniquely linked by kinship and relationships of proximity in native society. Since 2016, villagers have been exploring a sustainable development path that combines e-commerce, characteristic agricultural products selling and rural tourism. Xiaying village commissioned the China Rural Construction Academy team to conduct strategic consulting and spatial planning to transform and upgrade rural e-commerce in 2016. Generally, the village can be divided into three parts in space (Fig. 8.13). Zhoujia traditional settlements situates in the north, represented by contiguous old houses. The middle part is the most densely populated region, where the village committee sits and dramatic changes have taken place. The third part consists of the small-scale Liujia Old Settlements and the spot named "Moonlight over the Lotus Pond" (so called "He Tang Yue Se" in Chinese) as the main body.

Like other Taobao villages in eastern China, these villages follow the path of "bottom-up" initiative initially, and then local government intervened to provide relevant support. The majority of Taobao villages have greatly dependence on their special resources or products, and nearly all rural e-businesses are conducted by taking household as the basic unit. Moreover, most Taobao villages situate in the outskirts of urban areas or have superior traffic conditions. However, Xiaying village is unique because it is located in the remote Qinling-Bashan Mountain Area, which is one of the most impoverished regions in China. As a result, selecting Xiaying village as an example is of great significance for poverty alleviation and vitalizing remote countryside in China, or even the worldwide.

8.4.2 Process of Rural Restructuring

8.4.2.1 Economic Restructuring

(1) **Industry transformation and upgrading**

Before e-commerce was introduced to Xiaying village, scarce land resources and low-efficiency traditional agriculture are difficult to meet the demands of the growing population. Relying on the abundant kallaite resources, some villagers had left to work for enterprises selling accessories made from kallaite since 1980s, which are mainly concentrated in Chengdu and Lhasa. The majority of young people chose to migrate to other cities to find employment. A small number of young people and nearly all the older residents who stayed in the village relied on cultivating wheat and tea to make a living. Traditional agriculture was still the dominant industry.

In 2008, initiated by returning college students, e-commerce and the unique kallaite resources were combined. E-commerce has gradually been accepted and adopted by the villagers, and it has spread rapidly through the "acquaintance society", which is based on kinship relation and geographic proximity. The number of online shops has experienced a rampant increase since more and more villagers became involved in 2013. By the end of 2016, it has reached 139 with the annual revenue breaking through 70 million RMB yuan, and the fast-growing tendency has proven irresistible (Fig. 8.14). The long tail effect initiated by the internet enables the extensive flows of online transaction information and offline product transport to be formed in Xiaying village. The vigorous information flows involving product quality, price and after-sale service, as well as the tremendous capital flow attached, continues to be strongly formed by instant communication between network salesman and customers. The information flow and product flow jointly constitute the predominance of commodity trading pattern and contribute greatly to the establishment of long-distance spatial networking. For analysis purposes, we collected 1500 shipment orders that randomly extracted from 30 households and produced from January 2017 to June 2017 to map out the spatial pattern of product flows (Fig. 8.15). The number of orders was classified into five levels by Natural Breaks Method to intuitively present the connecting intensity between Xiaying village and different cities. In terms of the outcomes, the economic linkage extended drastically, and nearly covers 70% of prefecture-level cities in China. In particular, frequent trading activities intensively occur between this remote village and economically prosperous areas, such as southeast coastal cities, the provincial capital, thus attracting tremendous hidden capital flows to converge at this node. Deemed as the fundamental premise of element flows, the demand from urban areas determines whether the flow is strong or weak. For the

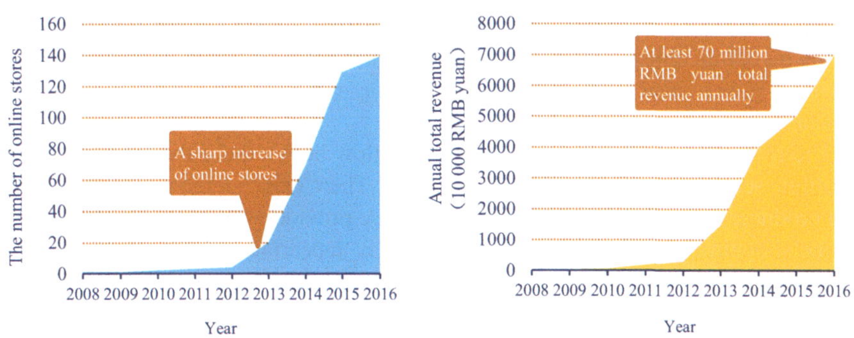

Fig. 8.14 The number and annual total revenue of online stores in Xiaying village during 2008–2016 (Zhang et al. 2018)

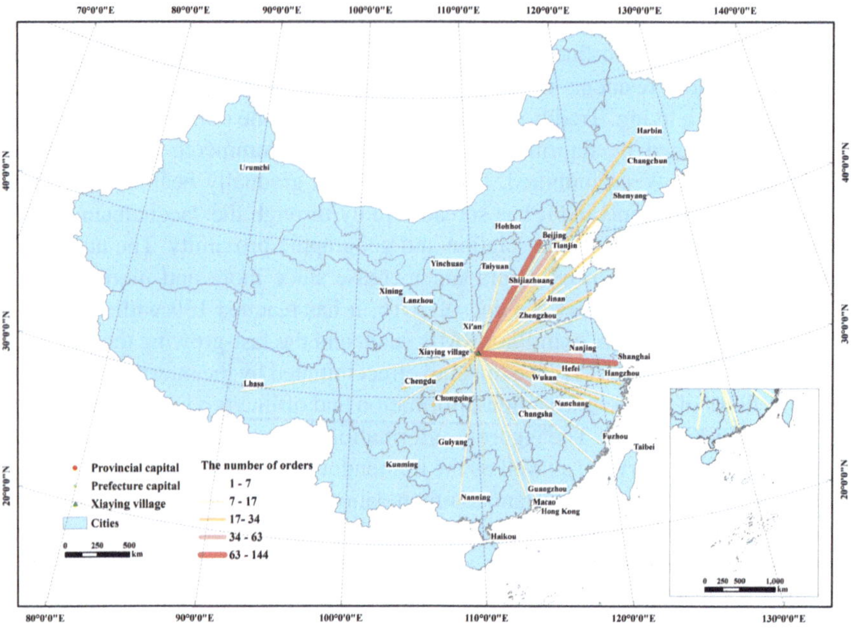

Fig. 8.15 The number of delivery orders to each destination from Xiaying village (Zhang et al. 2018)

purpose of ensuring the effective operation of information and product flow, the government has given tremendous financial and technical support in the improvement of transportation and communication infrastructure. In view of the necessary network infrastructure, the original optical fiber has been upgraded under the coordination of superior government, achieving full coverage of 100-megabit fiber connectivity and 4G signals. The original roads were difficult to carry the increasing transportation flow. Since 2016, the roads have been widened and the roadbed has been reinforced. As the online transaction increases, the high fees for interbank transfer and the limitation on online transfer restrict the kallaite transaction. To tackle this problem, the township government held a symposium aiming at providing convenient financial assistance.

The formation and components of product flow both require the agglomeration of offline service providers and online retailers. There exist a large number of physical producers, distributors and derivative service providers behind the fast-growing online shopping platform. When scale benefit is appearing in logistics transportation, all kinds of logistics enterprises successively settle in. As one online shop owner selling kallaite in his 41s said:

> Previously, the products can merely sell in surrounding cities and the farthest is Wuhan. In order to make a living, some villagers migrate to Sichuan Province or Tibet to sell ornaments or decorations made from kallaite at the scenic spots. Currently, I can sell my products to customers all over the country at home. My customers mainly come from economic

prosperous cities, such as Beijing, Hangzhou, Shanghai, and Guangzhou. Moreover, they often buy high-end products. In recent years, a large number of households have purchased professional machines for carving kallaite due to the increased sales, and more and more express service companies have settled in.

The extension of the industrial chain has evolved progressively, with the accompanying process of regional labor division. Xiaying village mainly engages in the industries at both ends of the industrial chain with highest added value, i.e., raw material selection and selling (online store operation and customer communication). Majority of merchants invite relevant specialists for various training containing photography, picture processing, websites design so that attracting customers' attention through distinctive users' interface design and absorbing special offers. Then, the service industry has flourished with the special and demanding needs of customers. In nutshell, a complete industrial chain consisting of kallaite purchase, processing, selling, and logistics has been established. Induced by growing population mobility and migrants, service business successively appeared and expanded, such as supermarkets, restaurants, and so on.

Considering the fact that the mineral resources are nonrenewable, villagers gradually launch into industrial transformation and upgrading with the enhancement of crisis awareness in recent years. Some of them take full advantage of a network platform to peddle local characteristic agricultural products and exquisite rural scenery, so that attracting urban population and their attached capital flows, and reducing the cost of product promoting. Besides, this pattern not only overwhelmingly expands marketing channels, but also organically combines resources, capital, and market. In 2016, two professional cooperatives were established, one target for integrating lotus planting and sightseeing, another is characterized by developing experience type of agriculture and organic farming through creating high-quality tea brand. Oolong tea planting experts were invited by informed local people to guide the cultivating, processing and package, and to improve the quality of local tea. Currently, one tea processing base with first class equipment introduced from Taiwan has been established. To some extent, kallaite industry accumulates enough capital for the development of characteristic agricultural undertakings. Meanwhile, the villagers began to develop characteristic rural tourism resorting to the beautiful natural scenery, traditional residences and local culture (Fig. 8.16). In response, industries solving the intractable rural issues related to farmers, agriculture and rural areas (so-called "San Nong Wen Ti" in Chinese), as a supplement to kallaite industry, improving the rural economic structure, and continuously sustain the sound development of remote rural areas. As one village official shouldering the responsibility of economic development in his 36s said:

> There is no doubt that kallaite will run out eventually. At present, the cost of raw materials is becoming higher and higher as it is getting scarcer. What should we do if the whole area has been mined out? So we have no choice but to seek for other channels of obtaining income. But we can't follow the other villages of developing industry and then destroying the environment. We will take full use of our organic vegetables, clean water, and rural scenery to attract urban citizens to come here.

Fig. 8.16 The newly built scenic spot named 'Moonlight over the lotus pond' in the Xiaying village (Zhang et al. 2018)

There is an obvious synergy relation between the flow of economic production elements such as information, technology, capital and labor, and the industrial transformation in Xiaying village. Originally, under the leadership of a few returning elites, the e-commerce platform was combined with the rural characteristic resources to form the initial flows of information, product, and capital. With the proliferation of technology, the number of online stores has gradually increased, and product and capital flows have also expanded rapidly. Subsequently, Xiaying village has an active output as a basis for its differentiated positioning depending on the unique products and regional influences. This has attracted the in flux of tourists, while also allowing a large amount of capital to flow into rural areas. Therefore, the directions of various elements flows are different. The information flow is bidirectional, which is conducive to the interaction and feedback between urban and rural areas. Population, capital, and technology mainly flow from cities to rural areas, while the product flow is the opposite. External funds, technology, information, and population enter into rural areas through different channels, forming the fundamental engine for industry restructuring.

(2) **Changes in the household economy**

The increasing employment opportunities generated by the formation of the industry chain and the industry transformation logically makes it possible for villagers to obtain more benefits. Ordinary villagers make full use of their leisure time by doing small-scale business. Prior to the rise of e-commerce, the villagers' income is mainly originated from working outside (41%) and traditional agriculture (26%) (Fig. 8.17). By 2016, the income structure has undergone tremendous changes compared to 2008. E-commerce and other business activities have become the main source of villagers' income, accounting for 43% and 13% of the total, respectively. Rural residents can effectively arrange their own time to engage in diversified business so that rural human resources are properly allocated. As one online shop owner in his 43s said:

> In the past, my family merely depended on the slender income that earned by working at other places. Since the opening of the online shop, the income from online store nearly constitutes the whole income of my family. As long as I take care of my online store every

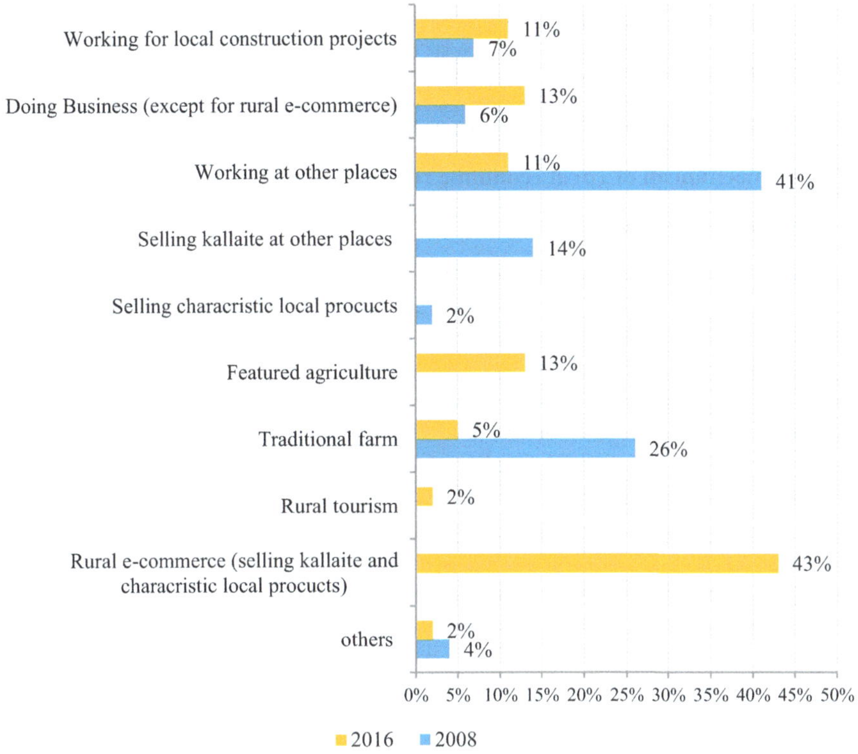

Fig. 8.17 Changes in villagers' income sources in Xiaying village (Zhang et al. 2018)

> day, I can meet the daily expenses and have a surplus. The original cultivated land is now only planted with some vegetables and crops to meet daily needs.

One manager of the logistic branch in his 28s said:

> My parents planted wheat at home and can merely make ends meet before I returned to start a logistic branch. My wife and I were working outside, and our monthly income was about 5,000 RMB yuan. At present, the demand for delivery service is increasing, and we are so busy every day. Of course, the income is also considerable.

The combination of virtual and traditional markets has greatly expanded the income sources and income levels. Daily sales are subject to large fluctuations in the number of products sold and the quality of the products. Therefore, the annual income levels of the online store operators are quite different and it is difficult to investigate. But their income level can directly reflect through their behavior. Most online shop owners have built two or three floors of well-decorated buildings, and they have purchased new furniture at home. According to our interviews, nearly half of the respondents purchased real estate in the county to receive a better education for their children. As one online shop owner in her 47s said:

> I only received very little education, and I even did not graduate from primary school. I used to use the mobile phone only to make a phone call and was unable to send a text message. With the help of my neighbors, I learned to type and communicate with customers through We Chat (a sort of social software in China) and opened an online store in 2015. The annual income of my online store reached 200,000 RMB yuan in 2016.

(3) The mechanism of rural economic restructuring driven by e-commerce

1. Rural elites

Rural elites are identified as the fundamental motivation for activating the entrepreneurial activities (Fig. 8.18). The rural elites mentioned in this section refer to the rural economic elites, who are the pioneer of achieving prosperity and always generate widespread repercussions. They have the ability to mobilize social resources in order to solve the tough issues in industry development. At the early stage, the returning graduate college students successfully made full use of e-commerce platform to expand the demand for local products. Commodity transaction needs to be supported by tremendous information flow, capital flow, and products flow. In other words, rural elites play a decisive role at the early stage of the development of the space of flows, who gradually awaken the villagers' entrepreneur awareness through the effect of "neighborhood demonstration". Technology flow in social networks gradually spreads with the help of kinship and geographic proximity in the rural areas. The spread of technology has led migrant workers to return to their hometowns, thus making rural development entities more vigorous. Meanwhile, advanced concepts are brought into promote industry transformation and upgrading. Combining technology, information and population will produce more economic benefits. Therefore, the positive externalities of population influx will also eventually be precipitated as human capital, improving the structure of the elements.

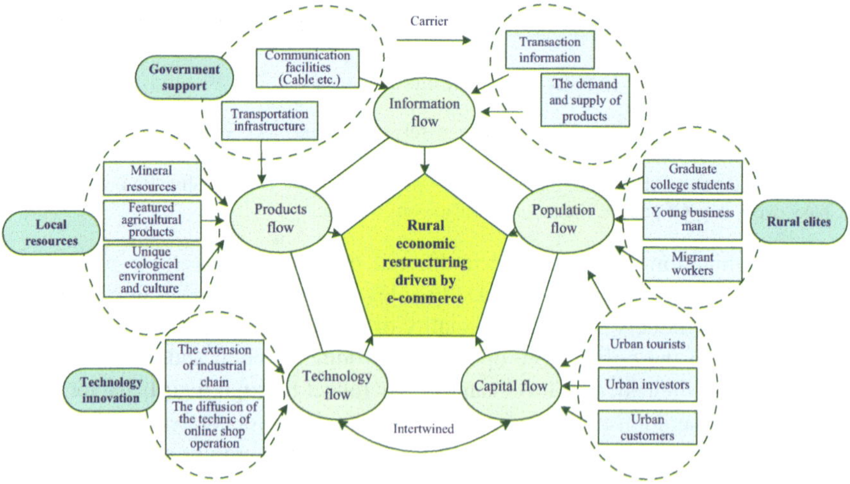

Fig. 8.18 The mechanism of rural economic restructuring driven by e-commerce (Zhang et al. 2018)

2. **Technology innovation (e-commerce platform)**

Information technology plays a key role in improving the resource allocation process. The impact of technology innovation on the economic transformation of a Taobao village is mainly reflected in two aspects, one is the diffusion of e-commerce, and another is that the extension of technology flow promotes the formation of industrial chain. One important reason for the flourishing of online shops is the inclusiveness of the e-commerce platform, which provides equal opportunities for rural entrepreneurs. This platform reduces the material cost, risk cost and learning cost of starting up business, which exactly corresponds to the cost-oriented nature of grassroots' entrepreneurship. From the perspective of the consumer, online consumption is more attractive than offline consumption. At first, consumers can easily compare the basic conditions such as prices, previous evaluations, etc. depending on the advantage of information aggregation. Secondly, it is convenient for consumers to enjoy the fast delivery service without leaving their houses. The third is the advantage of borderless consumption. The network enables the consumers' behaviors are not restricted by geographical location. Therefore, this technology innovation expands the demand for commodities by breaking through information asymmetry. The platform produces a large amount of transaction information by effectively coordinating the supply and demand of commodities, which is a prerequisite for the formation of products flow. As e-commerce gradually develops, its demand for related services will increase, which is also the initial driving force for related technology innovation. The emergence of new technology is considered as the premise of new industries, e.g., the online shop's design technology, commodity packaging technology, and so on. As the related technology improves, the upstream and downstream industry chains are gradually formed, thus making the rural industry more diversified.

3. **Resource endowments**

Featured resources are the guarantee for the formation of effective elements output, which contributes greatly to sustaining the commodity flow. The kallaite and featured agricultural products successfully stimulate the market potential, also support the industry transformation and the establishment of brand value. Rural culture and ecological environment are very attractive resource endowments for urban residents. Relying on these special resources, the village can seek a new foothold in the differentiated urban-rural framework, so as to finding a new discourse in the market system, taking the advantage of the value and charm of the countryside. The output elements from rural areas can reflect the positive externalities of the agricultural industry and rural society on account of their scarcity. In the framework of the urban-rural circulation, local resources fundamentally sustain the elements exchange. E-commerce and rural tourism are unable to be developed without these regional indigenous factors.

4. **Government support**

In the process of economic restructuring, the government has become a catalyst of industry agglomeration. The support for e-commerce mainly comes from the village committee and superior governments (township government and county-level government). As for village committee, the village leaders are composed of rural elites, who are mainly responsible for coordinating and managing the issues about funds, technology, and regulations regarding e-commerce development. The intervention of the village committee will coordinate the interests of different villagers to support the development of e-commerce to a certain extent and regulates the behavior of each business entity through formulating relevant norms. The village committee also entrusts relevant corporations to jointly make rural development plans, which will point out the direction of industrial transformation and upgrading. The superior governments always begin to get involved after the scale of e-commerce has reached a certain level. On the one hand, the government attempts to facilitate the elements flow by improving infrastructure and investment environment. On the other hand, the government can actively coordinate capital loans for entrepreneurs and provide necessary entrepreneurial skills training. In general, the government's involvement and support contribute to the improvement of the entrepreneurial system and support the formation of various elements flows.

8.4.2.2 Social Restructuring

Villagers' activities and behaviors constitute rural social space. Rural E-commerce has changed the behavior logic of villagers' employment options, political participation, and lifestyle. Rapid information and population flow make urban civilization and local civilization continues to blend, and rural residents' lifestyle to become modernized, as well as the traditional settlement to gradually disappear.

(1) **Employment de-agriculturalization**

Space of flows is not the single one in our society, and it demonstrates dominant spatial logic and needs to be mobilized, conceived, determined and implemented by social actors. The increase of product flows and transaction information flows promoted by industrial transformation and upgrading require more administrators and executants to ensure their effective operation. Upon their own subjective initiative, rural elites play a dynamic role in activating the market demand and entrepreneurial motive of rural areas, and this enthusiasm for entrepreneurship gradually spreads due to the ripple effect of rural social networks. Meanwhile, a great many migrant workers have returned to start up their own businesses or obtain employment in response to the economic rationality manifesting as far higher earnings. Based on online sale platform, villagers make the most of local competitive and featured resources to participate in e-commerce oriented commercial activities in the forms of self-sale or commission sale. In this way, a new path of nonagricultural employment was created locally, which was mainly embodied in associating rural households with network merchants. Villagers perform multiple and heterogeneous roles, such as material processing workers, customer service staff, logistic senders, network technician and salesman, in the collaborative space composed of knowledge and information that connecting purchasers and sellers. Specifically, kallaite ornament processing demands massive labor for sculpturing. Consequently, numerous carving workers from Henan, Shaanxi, and Jiangsu gathered here for manufacturing high-end decorations and intricate artware. By the end of 2016, more than 700 people have already engaged in online store management, and nearly 100 outsiders have come here for jobs. However, social strata differentiation is increasingly prominent as the income disparity continuously enlarges. Minority group who are difficult to operate online stores tend to be marginalized owing to their age, disease or other troubles.

One online shop owner selling kallaite and local tea in his 26s said:

> When I graduated from college, it is difficult to find an ideal job facing the fierce competition. So, I wanted to start my own business. Coincidentally, I heard that many young people sold kallaite by using Taobao platform and earned a lot of money. And the village leader invited me to return to expand this emerging industry and contribute to my hometown. My business is getting better and better, and I hired two sculptors to deal with the raw material. Staying with my family, I feel really happy every day.

Another shop owner selling kallaite in her 38s commented as follows:

> Before I returned home, I and my husband worked in Wuhan together, and our child was taken care of by my parents. Although the income is acceptable, the work is very hard. Besides, we don't have the sense of belonging because we are separated with my parents and child. Now, we can do business at home and the working time is flexible. Meanwhile, we can obtain more income than working outside.

(2) **Changes in rural governance**

The actors who manipulate the networks have become more diverse as a result of the return of former migrants. Interconnected issues comprising gaining widespread trust, favorable conflict coordination mechanisms, and satisfying villagers' public service demand should be always emphasized in rural governance (Wang and Yao 2007). E-commerce has intervened in the traditional social interest pattern and produced far-reaching implications for the structure of rural governance. To satisfy the diversified demands and counter economic relations, a transition in rural governance was advanced by means of leading by elites, accepting multi-participants and improving public undertakings (Fig. 8.19).

Undoubtedly, rural elites play a decisive role in the construct of the modern rural governance system. Village secretary organizes capable young people and persuades villagers working outside to return to participate in the decision making, planning and enforcement of important affairs. Besides, network businessman association was actively propelled and established for the purpose of setting up the platform for product processing, sample exhibition and experience exchange. Thus, rural elites successfully got involved in rural governance and turned out to be political transition by neighborhood demonstration and public affairs participation. It is obvious that e-commerce has enhanced the interest linking among villagers and awoken the subject consciousness of interests and rights. As a result, an evident positive transition of villagers' attitude towards social affairs occurred compared to previous political indifference. Some villagers donated to build dams by the riverbank or sacrificed their own land to construct a square. We made a comparative analysis of the frequency

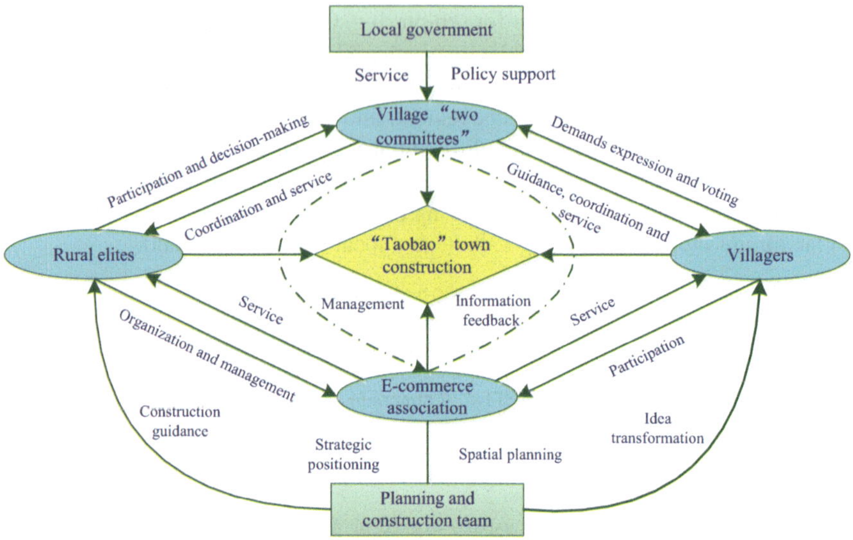

Fig. 8.19 The relationship of rural governance mainstream in Xiaying village (Adapted from Zhang et al. 2019)

of villagers taking part in significant issues before and after online store operation. The results revealed that the proportion of people who always get involved in vital affairs increase from 16 to 59%.

One returned young man selling kallaite and agricultural products in his 29s said:

> I was invited by the village leader to return home and help him to expand the rural e-commerce and push forward the transitional development. The village leader, other rural elites and I often gather to discuss the issues of promoting rural e-commerce, developing ecological agriculture and establishing a local brand.

Another online shop owner selling Kallaite and agricultural products in his 40s said:

> I never ask about the administrative affairs of my village before I started the online store. What is worth caring about? Whether they make what decisions, it makes no difference to me. But it is totally different now. The development plan or some regulations that they make are tightly associated with my earnings. I attend all relevant meetings at present. Besides, the village officials also consult villagers for advices.

This bottom-up initiative explored by villagers swiftly gave rise to great repercussions locally and drew the attention of local government. In consequence, a serious policy and service support have been implemented since 2014. With the help of village committee and superior governments, more logistics companies were introduced. To address the limitation of backward network facilities, the original optical fiber was upgraded to realize more than 100 M optical fiber connecting. Against the high service charge of interbank transfer and the quota of online transaction, local governments coordinated relevant banks to provide financial service for online store operators. As income increases, high-quality education and medical service, as well as recreation and leisure activities are increasingly demanded. Therefore, a standardized kindergarten, a clinic and a square with modern fitness equipment have been successively established in recent years. Moreover, the leaders vigorously pursued scientific spatial planning and commissioned the team from China New Rural Planning and Design Academy to undertake a comprehensive layout of industrial transformation and space management in 2016.

(3) **Changes in social communication and lifestyles**

The space of flows enlarges geospatial scales, and the behavior mainstream will respond to space demands through a series of spatial decision-making changes. The instantaneity of information flows tends to compress the space-time cost of long-distance contact so that the villagers' communication space tends to be distant and virtualized. The intertwined social networks in this village were shaped by the original kinship and occupational relationship, as well as the burgeoning online social activities. The closely connected links incorporating production, selling and logistics creates more opportunities for villagers to exchange with others. The majority of respondents said they exchanges with their neighbors more frequently after running online stores, and the main constituents of exchange are the experiences and interesting things taking place in online store management. It indicates that e-commerce enables villagers to have more common discourse and become closely connected. All online store operators use social software for along time every day to communicate instantly with customers, families, and friends. In addition, 72% of the interviewees added into group chats related to antique, jewelry and ornaments to sell their products. In general, the internet has strengthened the "actual communication space" between villagers and their originally acquainted groups such as relatives, friends, classmates, and colleagues. On the other hand, getting acquainted with customers through the "group chat" and other channels has expanded the "virtual interaction space" (Fig. 8.20). As one online store owner selling kallaite in her 29s said:

> Since the network is connected, I usually chat with my nonlocal relatives. By using the instant communication software, it is convenient for me to contact with my friends. In order to peddle my products and provide after-sale service, I spend most of my time with my cellphone or laptop. So, don't be surprised if you find I stare at my cellphone at every minute.

New-type spaces were derived under the influence of various element flows. Online store operators always follow a unique routine. They often get up at about ten o'clock in the morning and rest at about twelve o'clock at night because their working time is closely related to the customer visiting time, while evening is the peak. In terms of consumption space, sophisticated logistics and increasingly diverse and high-level shopping needs make "online shopping" has become the first choice for villagers. Generally speaking, the mainstream lifestyle of Xiaying village has already been affected by the unique production mode of e-commerce. The nonagricultural tendency collides with the traditional rural lifestyles creates a unique "Taobao Style" that significantly different from the routines of traditional rural and modern city. As one online store owner selling kallaite in her 36s said:

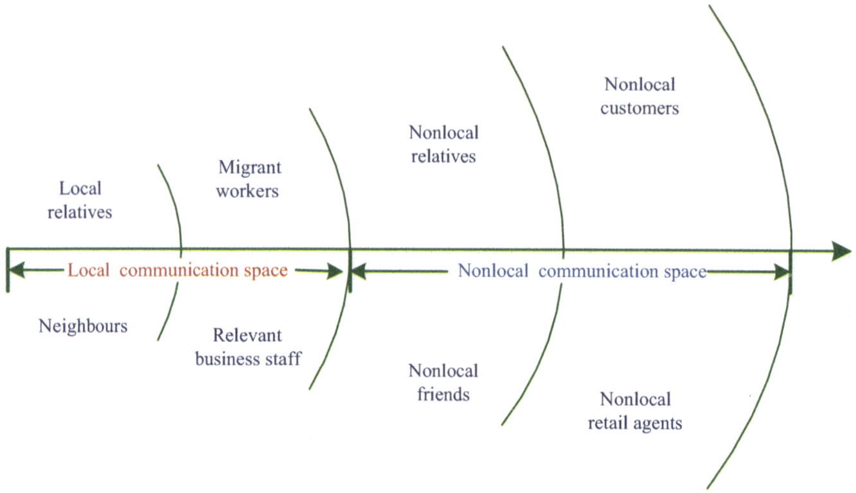

Fig. 8.20 The evolution of villagers' communication space in Xiaying village (*Source* Drawing by Yingnan Zhang)

> My working time depends on the customers. If the sound of message reminding rises, I realize I need to chat with the customer. You know that most urban people are accustomed to sleeping late. It's common for me to work late. Naturally, I often get up late.

8.4.2.3 Spatial Restructuring

From the spatial morphology of substance composition shaped by elements flows, inner element flows and their physical movement process require the support of materialized space. The stretching of technology flow and the expansion of products flows, as well as the returning population, flows jointly attributed to the restructuring of living, production and ecology, which is mainly manifested in the optimized settlements space structure, promoted land use efficiency and improved residential environment.

(1) **Living space restructuring**

Significant changes have taken place in the central part where village committee situated, which covers the largest area and is the most densely populated among the three settlements. While the other two merely present small variations. Generally, increasing demands for production space engendered by e-commerce have made the living space tend to be multifunctional, intensification and vertically utilization. Moreover, the public service space has experienced an extensive expansion.

1. For the convenience of production and management, an alley mainly used for e-commerce was uniformly planned and built up, which nearly occupied 2.2 ha

farmland. About 30 households moved there together for aliving and operating online stores. Others who have certain economic basis also use their own existing homesteads for reconstruction. The average height of newly built houses is 2.5 floors (Fig. 8.21). Most villagers put up a simple shack on the rooftop using metal sheet so that they store chores.

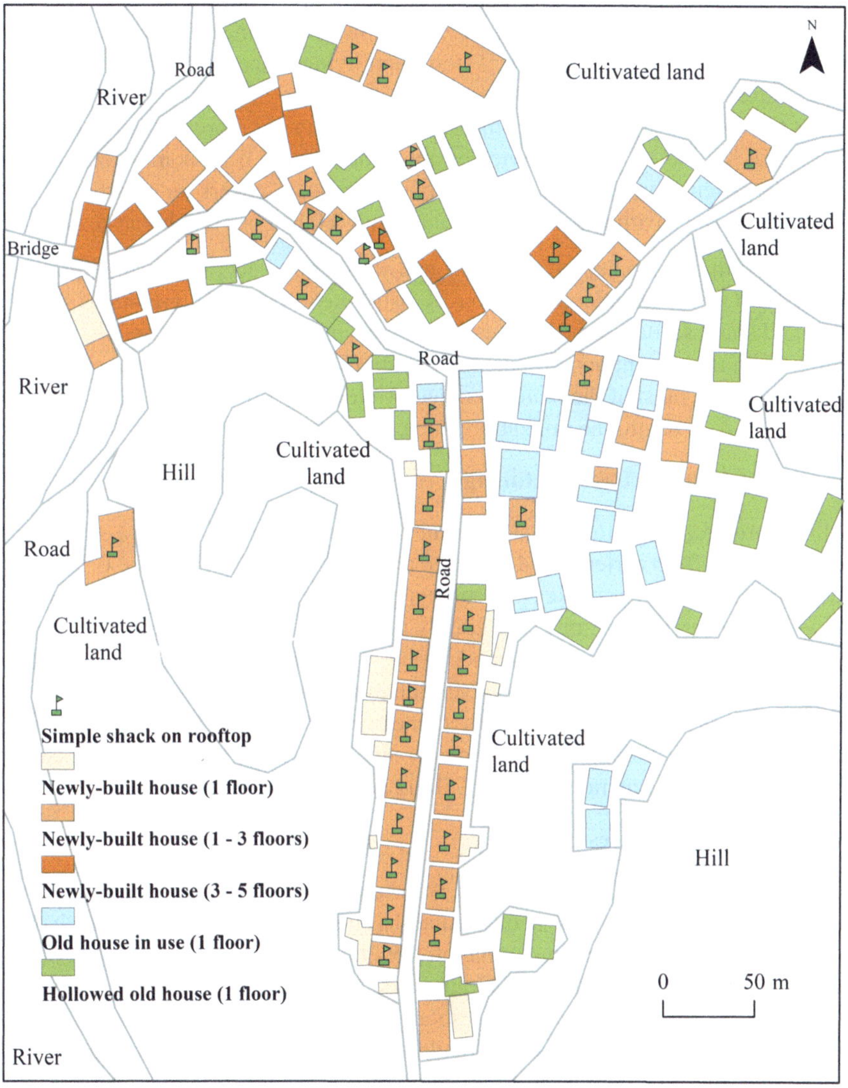

Fig. 8.21 Status of rural housing in the central part of Xiaying village (Adapted from Zhang et al. 2019)

2. The rising demand for production space prompted villagers to make the utmost of their homesteads in view of the scarce land suiting for construction. Newly-built houses tend to combine production and residential function, and mainly present three modes. (1) Integrating stores and manufacturing sites: The ground floor was used for physical stores, the second or a certain upper floor is used for kallaite processing and storage, the other space is for daily life. (2) The store in front and a plant at the back: A simple plant is built behind the house for processing and warehousing. (3) The mode of stores and manufacturing sites separation: The new houses are used for physical stores and living space, and the old houses are used for processing and warehousing.
3. As the income rises, village committee began to invest more in social undertakings. A serious of public service establishments have been completed, such as kindergarten, clinic and square. All of these totally covers an area of 1.29 ha including approximately 1.2 ha farmland and 0.09 ha original homestead.

Notably, the phenomenon of hollow and idle houses gets highlighted progressively. Most of the abandoned old houses distributed in the high ground of sloping filed, and some scatter in the settlement. At present, the idle houses nearly account for 62.1% of the current households, with an area of 0.9 ha. In spite of the old houses currently in use, most of them are occupied by elderly people used to store debris or production. As a result, the utilization efficiency is extremely low.

(2) **Production space restructuring**

With regard to production space, its function mainly manifested in a substantial transformation from agriculture dominated to non-agriculture dominated (Fig. 8.22). The appearing of supermarket, restaurants and other service sites, as well as the introduced logistics company, have jointly resulted in the loss of farmland and the renovating of original homestead since 2009. Thereinto, farmland was diverted to service providing. On account of the bankruptcy of previous state-running mine, the newly-added production space chiefly comes from the newly-built tea processing plant. It should be noted that the relatively small space for production workshops and storage mostly generated by inner filling and nearby expansion.

(3) **Ecological space restructuring**

Since 2008, ecological land (forest land, water area except for newly-developed lotus pond and idle land included in another land) has increased almost 13 ha, which originated from the large scale transformation from farmland to forest land promoted by the continuous implementation of Grain for Green Policy. Furthermore, more and more villagers choose to discard or desolate their only remaining farmland in view of the farmland is so steep that is difficult to cultivate and their increasing income. Thus, some of them has evolved into the forest. Nonetheless, this transformation just scatteredly occurs at the margin of the forest with a small scale. As the industry transforms, the composite space that synchronously functions as a production and ecological conservation has gradually extended. For instance, the lotus pond that occupying 7.6 ha farmland was dredged up in 2015. Therefore, the successful

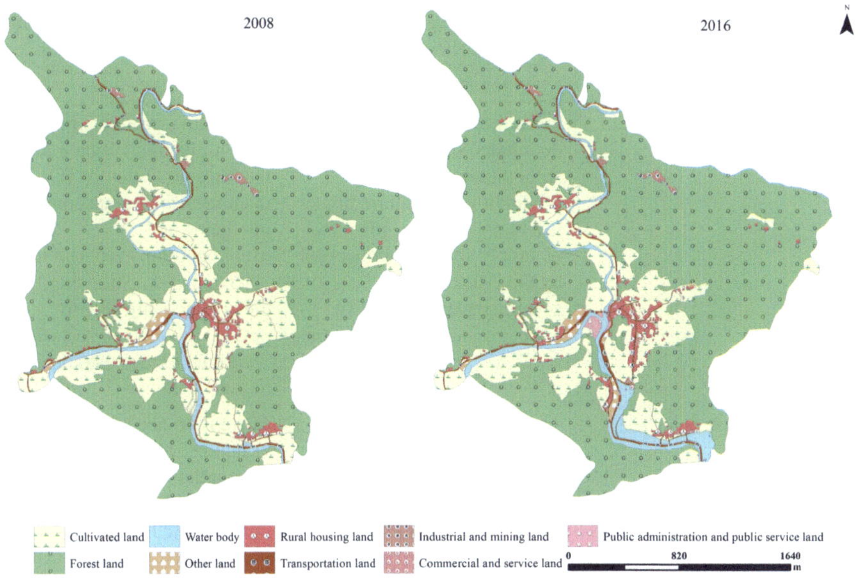

Fig. 8.22 Land use map of Xiaying village in 2008 and 2016 (*Source* Drawing by Yingnan Zhang)

practice of fully exploiting the multiple functions through characteristic and leisure agriculture has been demonstrated to a superior choice.

8.4.3 Discussion

(1) **Comparative advantages of Taobao villages**

Actually, a large number of Taobao villages were converted from traditional specialized villages, and have become a special morphology of specialized villages. Embedding in an original production system, internet exerts impacts on both supply and demand sides of the commodity market. Compared to traditional specialized villages, the dominant advantage of rural e-commerce is that it can compress the cost through reducing intermediate links and circulation expense at the supply side and avoiding information asymmetry and breaking information bulwark at the demand side (Table 8.1). Exposed to national or global market, marketing channels of Taobao villages are extremely extended.

Furthermore, the income-added effect of the e-commercialization was analyzed from the perspective of the economy. We hypothesize that the Taobao village was a specialized village previously, which faced a completely competitive market. In this market, all producers can enter or exit it freely and every enterprise performs the same cost curve. When the saving produced by eliminating middlemen's profit, inventory cost and promotion cost is larger than the new expense of online store

Table 8.1 Analysis of the comparative advantages between the common specialized village and Taobao village (Adapted from Zhang et al. 2019)

Influential factors	Common specialized village	Taobao village
Production mode	Some specialized villages dominated by manufacture still conduct large scale production and require higher inventory cost. The mismatch of supply and demand caused by the information asymmetry has a negative effect on the farmers' income	Just-in-time production and just-in-time marketing make the Make to Order (MTO) manufacturing mode come into being and overwhelmingly reduced the inventory cost. Additionally, the operators are able to provide products or service according to the customers' requirements
Transaction cost	The profit-taking brought by the middlemen will infringe on the farmers' interest. The separation of production and marketing requires more promotion costs to extend marketing channels	There is no need for the virtualized transaction to rely on middlemen to sell products, which reduces the intermediate links. Increasing propaganda by using internet greatly cuts down the cost of advertising. The integration of production, supply, and marketing lowers the mobility cost
Time and space of transaction	Stable time and location of the transaction are necessary. The long distance between sellers and buyers raises the communication cost	The transaction time is at the discretion of the consumers. The limitation of space and time between sellers and buyers is solved by the internet
Market	The market is always limited within the city or province that the village belongs to	The limitation of long distance is broken so as to extend the market to the whole nation or even the world. The increasing market demand and the saving of the intermediate cost both contribute to more profits for online store managers

management and logistic, the supply curve of the market will move to the right ($S_1 \rightarrow S_2$); in view of the demand side, an increasing number of consumers are attracted to purchase the products, so as to push the demand curve to move to the right ($D_1 \rightarrow D_2$) and create a higher equilibrium price and output (Fig. 8.23).

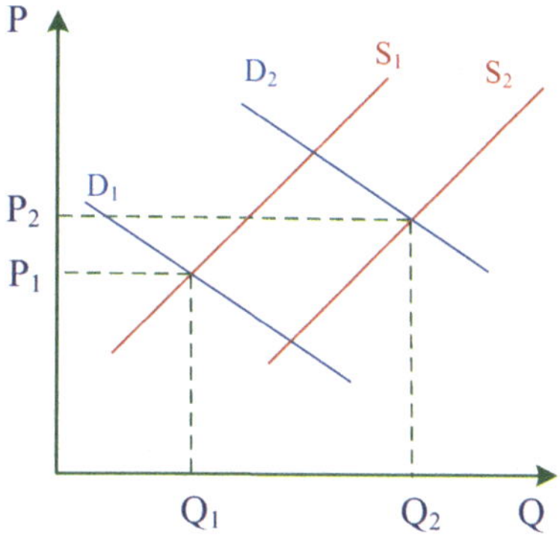

Fig. 8.23 The changes of market supply and demand curve after introducing e-commerce (*Source* Drawing by Yingnan Zhang) *Note P* represents the price; *Q* represents quantity; *S* represents supply; *D* represents demand

(2) **What are the critical forces driving e-commerce-oriented rural restructuring?**

Space of flows is essentially a kind of social space that is contingent on the behavior space of actor network (Woods 1998). Elements mobility creates differentiated opportunities and provides diverse information or other pivotal resources for disparate actors. As for the inner mechanism of rural restructuring in Xiaying village, what counts is the initiative of the organically integration of kallaite, featured agriculture products and e-commerce platform. Additionally, the actor network with aspecific target, explicit function, and feasible technic, resorting to external forces, has made a significant contribution to this pioneering undertaking (Fig. 8.24). It shares

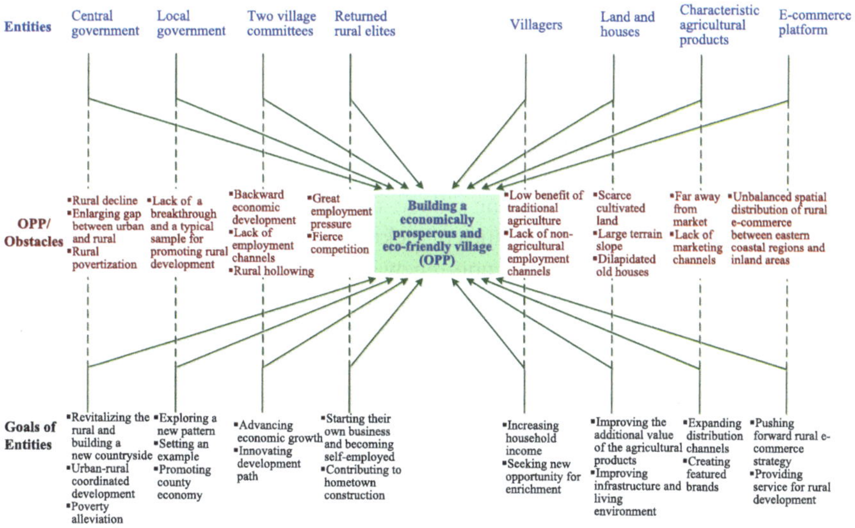

Fig. 8.24 The entities and their goals, OPP or obstacles of rural restructuring driven by rural e-commerce in Xiaying village (*Source* Drawing by Yingnan Zhang)

the similar generating mechanism with other specialized villages because the inducement of technology progress, the promotion of relevant regulation arrangements, the guidance of market demand, the driving of rural elites and the support of resources endowment are all essential for the transformation development of Taobao villages. As a matter of fact, it is the immediate cause of the restructuring brought by rural e-commerce that the network may change the pattern of elements exchange between urban and rural as it can eliminate the friction of distance. The networking logic can make up for the shortage of mutual interaction among different objects in the physical geographic space, and it may reform the traditional cognition of accessibility. The vigorous mobility of elements can effectively ease the segmentation of physical space and make the information amount of urban and rural to be symmetrical, which naturally generates the demand for elements exchange between urban and rural. Importantly, the network-generated restructuring corresponding with the progress of "difference-demand-flowing-integration" may break through the dichotomy of urban and rural development.

The corridor of elements flow is the foundation of constructing "space of flows", which means that superior transportation condition is crucial. At present, central and western regions in China are not able to provide sufficient and necessary support for rural e-commerce due to the weak commercial basis and unsound infrastructure. Therefore, it is still impractical for rural e-commerce to make a mass diffusion in underdeveloped areas. With the implementation of rural vitalization, some specialized villages with sound infrastructure will be introduced e-commerce platform and attract urban capital. However, it needs to be vigilant that the less developed rural area may be exposed to increasing external fierce competition, thus causing the rural areas with unsound infrastructure tend to be marginalized (Ding and Liu 2009).

(3) **Obstacles and countermeasures of network-driving rural restructuring from the perspective of stimulating element flows**

The effective demand between urban and rural is the prerequisite for stimulating the flowing of elements. Moreover, the technical infrastructures for transportation and information constitute the essential transmission media for material and non-material elements. Currently, Xiaying village, as well as the numerous remote rural areas, is confronted with the difficulties of lacking sustainable attractiveness and the inferior location accessibility. The non-renewable kallaite resources are incapable of continuously sustain rural development. Additionally, some agricultural products are challenged by the exacerbated homogenous competition from surrounding areas. Although its development plan attempts to promote economic transition by launching ecological tourism, resources and environment carrying capacity are neglected in this village. Profit comes ahead of the environment and resources, which will definitely not creates sustainable core competitiveness.

Consequently, establishing the featured brand locally, resorting to diverse products, and conducting differential operations for the households can satisfy the multiple demands of urban residents. Excavating local distinctive culture and combining it with local environment and resources should be encouraged so as to create local characterized products. In order to construct convenient transportation network, appropriate plans for roads may be formulated by considering the feasibility and suitability of the existing and planned transportation infrastructure under the support of local government. At the same time, providing convenient public transport service is necessary for enhancing the accessibility.

8.5 Conclusions

In this chapter, three typical villages were taken to carry out an empirical study on the process of rural restructuring. The results show that: (1) Since 2000, the Huangshandian village has experienced different industrial transformations from traditional agriculture to the industries of primary processing and eco-tourism. The function of traditional agricultural production is declining gradually, and the industrial production, ecological culture and other multi-functional values of the rural territory have successively appeared. With the evolution and restructuring of socio-economic morphology, there are significant changes in the quantity, structure and pattern of rural living, production and ecological spaces, accordingly. (2) The industrial development in Yangqiao village has experienced the stages of traditional agricultural leading and concurrent farming production. Since 2000, it has taken on a sign of socio-economic restructuring. Recently, the living space has been restructured under the promotion of local governments, but the economic morphology has not changed significantly. At present, the village is still at a low development level, to a great extent. (3) E-commerce directly serves as the dramatic change of economic morphology, specifically, the combination of primary, secondary and tertiary industries in Xiaying village. Induced by the complete industry chain, numerous non-agricultural employment opportunities are generated for villagers. This development also makes the relationship of interest among villagers be increasingly complicated, which further diversifies the rural governance structure, modernizes the villagers' lifestyle. Prompted by the growing elements flows, living space is becoming multifunctional and intensive, and the production space has dramatically expanded.

The practice of rural restructuring in the typical villages show that, economic restructuring, spatial restructuring and social restructuring are mutually reinforced and restrictive. Among which the economic restructuring plays a leading role in the process of rural restructuring, the spatial restructuring is of particular significance to provide a carrier for socio-economic development, and the social restructuring is the support system for rural development. The interactions of the three dimensions have driven jointly the systematic development of the "natural-ecological-economic-social" systems and the comprehensive promotion of the "production-life-ecology-culture" functions in rural territory.

References

Ali Research (2017) China Taobao village research report (2016). http://www.aliresearch.com. Accessed 15 Jan 2018

Chen R (2016) 地方自觉与乡土重构: "淘宝村"现象的社会学分析 (Self-consciousness and rebuilding of rural areas: social analysis of the phenomenon of Taobao village). J Huazhong Agr Univ (Soc Sci Edn) 03:74–81 + 134–135

Cui M, Pan S, Newell S et al (2017) Strategy, resource orchestration and E-commerce enabled social innovation in rural China. J Strat Inf Syst 26(1):3–21

Ding J, Liu W (2009) 农村信息技术应用的地理学研究进展与评析 (Progress in geographical studies of rural informationalization). Prog Geogr 28:759–766

Jalali AA, Okhovvat MR, Okhovvat M (2011) A new applicable model of Iran rural e-commerce development. Procedia Comput Sci 3:1157–1163

Liang C (2010) The present situation of and prospects for e-business in China. J Interact Advert 11:74–81

Long H, Li Y, Tu S et al (2018) 平原农区空心村整治关键技术研究与示范 (Key technologies research and demonstration of Hollowed village consolidation in plain farming areas of China). Science Press, Beijing

Long H, Tu S, Ge D et al (2016) The allocation and management of critical resources in rural China under restructuring: problems and prospects. J Rural Stud 47:392–412

Tu S, Long H (2017) Rural restructuring in China: theory, approaches and research prospect. J Geogr Sci 27(10):1169–1184

Tu S, Long H, Zhang Y et al (2018) Rural restructuring at village level under rapid urbanization in metropolitan suburbs of China and its implications for innovations in land use policy. Habitat Int 77:143–152

Wang S, Yao Y (2007) Grassroots democracy and local governance: evidence from rural China. World Dev 35:1635–1649

Woods M (1998) Researching rural conflicts: hunting, local politics and actor-networks. J Rural Stud 14:321–340

Zhang Y, Long H, Ma L et al (2018) Analysis of rural economic restructuring driven by e-commerce based on the space of flows: the case of Xiaying village in central China. J Rural Stud. https://doi.org/10.1016/j.jrurstud.2018.12.001

Zhang Y, Long H, Tu S et al (2019) 电子商务影响下的"淘宝村"乡村重构多维度分析: 以湖北省十堰市郧西县下营村为例 (A multidimensional analysis of rural restructuring driven by e-commerce: a case of Xiaying village in central China). Sci Geogr Sinica 39(6):947–956

Part III
Rural Restructuring Promoted by Land Use Transitions

Chapter 9
Land Use Transitions and Rural Restructuring: A Theoretical Framework

Abstract Different economic and social development stages correspond to different land use patterns and land use transitions stages. Socio-economic development inevitably results in a specific land-use transition process. Rural restructuring is an important means to implement the strategy of rural vitalization, also the link between land use transitions and rural vitalization. Population, land and industry are closely interlinked, and economy, space and society are intertwined. It is a complex and systematic project to realize rural vitalization relying on land use transitions and rural restructuring. Based on expounding the relationship between land use transitions and rural transformation development, as well as explaining rural vitalization as a new-type rural restructuring, this chapter explores the relationship among land use transitions, rural transformation development and restructuring, and rural vitalization to enrich and expand the research field of rural geography. The recessive morphology of land use and its changes should be the focus of research on rural vitalization through controlling land use transitions. Thriving business features heavily in rural vitalization. Linking the revitalization of land resources with the rural industry is of great significance for the sustainable development of rural economy and the efficient use of resources. Therefore, it is an urgent need to innovate land management policies and regulations by controlling recessive morphology changes of land use, and to explore a new land use transition mode combined with multifunctional agriculture to promote rural vitalization.

9.1 Introduction

Against the background of long-term urban-rural dual structure, the lack of an effective method and platform to assist rural transformation development has led to the restricted urban-rural integration development and a large number of rural issues have arisen and been exposed. Land is the carrier of human's major socio-economic activities. Yi-Fu Tuan, a renowned geographer, believes that land use is a mirror of society (Tuan 1971). All kinds of socio-economic problems exposed in the process of rural transformation development can be reflected in land use patterns. These issues also can be alleviated and even solved through restructuring the development

H. Long, *Land Use Transitions and Rural Restructuring in China*,
https://doi.org/10.1007/978-981-15-4924-3_9

pattern of rural territory, which includes social restructuring, spatial restructuring, and economic restructuring (Tu and Long 2017). Rural vitalization, as a new type of rural restructuring, focuses on solving the key problems of rural development and aims at improving the sustainable development ability and competitiveness of rural territorial system, so as to promote the rural transformation development. Studying the changes, problems and corresponding solutions of land use patterns and conflicts in the process of rural transformation development is a new way to study the issues arising from rural transformation development. Based on expounding the relationship between land use transitions and rural transformation development, as well as explaining rural vitalization as a new-type rural restructuring, this chapter explores the relationship among land use transitions, rural transformation development and restructuring, and rural vitalization to enrich and expand the research field of rural geography.

9.2 Land Use Transitions and Rural Transformation Development and Restructuring

9.2.1 Background

Rural transformation development refers to the reconstruction of socio-economic morphology and spatial patterns of rural areas as a result of the reorganization and interaction between urban-rural population flow and socio-economic development elements, which is also a reconstruction held by local participants in response to such changes. It mainly involves the transformation of the spatial organization structure of town-village, development model of rural industry, employment mode, consumption structure, the relationship between the industry and agriculture, urban-rural relationship, and urban-rural differences (Liu 2007; Long 2012). In essence, the term "rural transformation" has a similar conceptual connotation with an internationally popular term "rural restructuring" (Woods 2005; Long and Woods 2011).

Land use transitions have never occurred outside the framework formed by the interactions of the natural system, economic system and institutional system (Barlowe 1986; Li 2002). Among them, the exhaustion of natural resources such as the reduction of cultivated land in quality and quantity, as well as environmental degradation such as land pollution or salinization, often manifest changes in natural systems. As long as these problems are serious enough to draw public attention, the institutional system is able to adjust the land use system through resources and environmental management methods such as laws, regulations, and policies. For example, when the problems like farmland occupation, labor transfer, and the rural environmental issues caused by industrialization are serious to affect the development of urban-rural transformation and, the government will adjust the land use system by some policies and regulations such as farmland protection, building new countryside, rural vitalization, rural land consolidation, and so on.

There is no doubt that rural transformation development is also restricted by this framework that includes the natural system, economic system, and institutional system. On the one hand, the driving factors of land use transitions, such as the investment of capital and labor force, the development of industry, the employment and migration of rural population, will affect the development of rural transformation to some extent. On the other hand, all kinds of socio-economic problems exposed in the process can be reflected in the way of land use. The complex socio-economic evolution leads to diversified land use patterns, which conversely have impact on the process of regional natural, ecological and social development. To some extent, there is an interaction, and a coupling relationship between land use transitions and rural transformation development.

9.2.2 Land Use Transitions Under Rural Transformation Development and Restructuring

With the rapid development of Chinese economy and society, the demand for construction land continues to increase. The development of industrialization and urbanization has continuously eroded the rural areas, which has put tremendous pressure on ensuring food safety, adhering to the cultivated land red line of 1.8 billion mu, and maintaining stable social and economic development. During the process of rural transformation development, the contradictions and problems, which mainly involve the utilization and protection of farmland, rural hollowing, and land contamination, caused by land use transitions are increasingly prominent. The changes in land use patterns in a region can be reflected basically from the changes in cultivated land and urban and rural construction land that are most closely related to human production activities (Long 2006). Cultivated land and rural housing land are two kinds of land use types that play an important role in the process of rural transformation development and land use transitions. Their changes in form and nature can reflect land use transitions in the process of rural transformation development as a whole.

9.2.2.1 The Transition of Farmland Use in the Process of Rural Transformation Development and Restructuring

The rapid expansion of urban and rural construction land has led to the large occupation of cultivated land, which has further stimulated the contradiction of relations between human and land. According to the statistics of the Ministry of Natural Resources, in 2017, although the cultivated land area increased by 26 million hectares through land consolidation and agricultural structural adjustment, the cultivated land area reduced by 32 million hectares due to construction occupation, disaster damage, ecological restoration, and agricultural structural adjustment. Actually, the cultivated land continues to decrease and the pressure to stick to the "red line" of cultivated land is increasing. In addition, Chinese total population has exceeded 1.4 billion. In order to ensure national food security, a sufficient amount of arable land must be maintained. Meanwhile, to ensure national ecological security, it is also necessary to strengthen the protection of agricultural land with ecological functions, especially arable land. With the increase of cultivated land protection and ecological conservation, only limited land resources are available for the development of construction land that is still eager for a large number of land supplements. This phenomenon results in the unprecedented pressure on the supply of construction land, which may further affect the stable development of the socio-economic aspect. Because the rapid advancement of industrialization and urbanization will still inevitably occupy a part of the cultivated land, while China has limited land reserves available for development into arable land. The mutual restriction and contradiction between land supply and cultivated land protection will have a profound impact on the transition of farmland protection and utilization in the process of rapid urban-rural transformation in China.

9.2.2.2 The Transition of Rural Housing Land Utilization in the Process of Rural Transformation Development and Restructuring

Since the reform and opening up, the rapid development of China's industrialization and urbanization has led to the acceleration of the non-agricultural transfer of the population and the gradual decrease of the permanent resident population in the countryside. During this process, great changes have taken place in the relationship between people and land in rural areas, among which the most prominent issue is the phenomenon of "hollowing houses". This phenomenon gradually expanded from individuals to a certain area of the village or even the entire village, resulting in a large-scale hollowing of rural settlements. Furthermore, this has gradually evolved from the hollowing of the population to the hollowing of rural areas including population, land, industry, and infrastructure, leading to the emergence of a large number of "hollowing villages" (Liu et al. 2011). Although the rural population decreased from 1996 to 2007, rural residential land use increased by 18.28%, with an average household size of more than 667 m^2 and an average of 229 m^2 per person, which means that the potential of rural land consolidation is considerable.

9.3 Rural Vitalization: A New-Type Rural Restructuring

9.3.1 Connotations of Rural Vitalization

The rural territorial system, an interactive formation of various key elements, is an open, dynamic system with comprehensive multidimensionality. Population, land, and industry are the core elements that affect rural socio-economic development (Ge et al. 2018a, b; Long et al. 2018; Tu and Long 2017). Among them, the population as the mainstream of development in rural areas creates economic, social, and ecological values for rural areas through the exploitation and utilization of resources, business management and other activities related to production and daily living. The population is the momentum driving rural vitalization. Land resources shoulder the basic role of providing resources support in rural vitalization, it not only has multifunctional value in meeting the needs of production, living and ecological space but also spawns different rural industrial development modes. Industrial development forms the main driving force for rural vitalization. Population and land are combined organically through land resources utilization and employment of the population, becoming the crucial power of rural economic development. How appropriate the industrial structure is will affect farmers' incomes and even determines the rural socio-economic sustainable development. Boosting the integration of "resources, capital, and assets" by coordinating the elements of "population, land, and industry" is the key to fostering development momentum and enhancing vitality in rural areas.

Rural vitalization is the opposite of rural decline (Liu 2018). The large-scale outflow of the rural population during urbanization and industrialization is the origin of rural decline (Long et al. 2012; Tu et al. 2018), which has led to "aging or weakening" management mainstream and eventually caused the following issues: (1) lag in industrial development, such as low agricultural efficiency and undeveloped non-agricultural industries; (2) inefficient use of land resources, such as the abandoned farmland and vacant homesteads; and (3) a series of social problems, such as landscape devastation, disordered governance, dilapidated infrastructure, and weak local culture (Long et al. 2016).

To cope with the loss and decline of the intrinsic elements in rural areas, rural vitalization is a development strategy aimed at realizing economic, political, cultural and ecological rejuvenation in rural areas by reshaping socio-economic morphology and spatial pattern in rural territory (Tu and Long 2017; Zhang 2018), which is intended to address the conflicts between the people's ever-growing need for a better life and unbalanced, inadequate development during the mid-term urbanization in China. Besides, the value orientation of rural vitalization is to narrow the gap between urban and rural development, initiate the sharing of development achievements and build a balanced urban-rural development pattern to create equal development opportunities. The complexity of the rural territorial system determines the implementation process of rural vitalization strategy with systematized, hierarchical, and regional characteristics.

(1) Systematized: Rural vitalization is the coordinated symbiosis of various elements that are interconnected and mutually constrained. "Thriving industry, pleasant living environment, refined rural civilization, effective governance, and prosperous life" are the basic principles for rural vitalization, covering economic, social, ecological, and cultural aspects. As a comprehensive humanistic process, on the one hand, rural vitalization requires the multidisciplinary integration of various subjects such as economics, sociology, management, engineering technology, and geography in order to make scientific development

planning with clear objectives, reasonable position, and functional optimization. On the other hand, rural vitalization requires multi-departmental coordination to promote developing a mutually complementary governance system with clear responsibilities and standardized management.

(2) Hierarchical: Considering the different rules and characteristics of rural development at different spatial scales, rural vitalization should adopt differentiated goal orientation and strategic positioning. At the regional level, rural vitalization should cover whole regions and form an overall vitalization pattern. While at the village level, it is crucial to selectively support according to the law of economic development. For some villages in unfavorable natural environments, poor geographical locations or difficult supplies of basic public services, it is necessary to guide villagers to relocate and allow the original villages to dissipate naturally. On the contrary, the villages with excellent resource endowments, strong industrial bases, and better locations should be given priority support.

(3) Regional: Influenced by various factors such as resource endowments, location conditions, industrial bases, policies, urban-rural relationships, market environments, central-city leading forces, and cultural heritages, rural socio-economic development differs significantly among regions of China. Therefore, excavating local rural values, cultivating diverse industrial structures locally, and scientifically formulating strategies and paths for rural transformation are the keys to promoting rural sustainable vitalization.

9.3.2 Rural Vitalization and Rural Restructuring

According to the overall goals of rural vitalization that are "thriving industry, pleasant living environment, refined rural civilization, effective governance, and prosperous life", the relationship between rural vitalization and rural restructuring is reflected in following three aspects, restructuring rural production space to achieve industrial revitalization, restructuring rural ecological space to achieve ecological revitalization, and restructuring rural living space to achieve organizational and cultural revitalization.

9.3.2.1 Restructuring Rural Production Space to Achieve Industrial Vitalization

As the space for the development of rural enterprises, the restructuring of rural production space is the inevitable course to vitalize the rural industrial economic system (Long and Tu 2017). In the process of restructuring rural production space, the industrialization and modernization of agriculture, and the popularity of the industrial park, should be accelerated. This can be achieved by the following aspects. First, through the preparation and implementation of the plan of urban-rural integration spatial layout, the industries that are highly dependent on agricultural products and labor-intensive will be appropriately concentrated in rural areas. Secondly, the high-end industries of technology intensive and capital intensive will be distributed in urban areas to achieve structural coordination and functional complementarity between urban and rural territorial systems (Long and Tu 2017; Long 2014). Further, based on reconstruction of intensive and efficient rural production space and innovation of rural industrial management mode, it is necessary to transform and revitalize rural traditional industries, actively foster rural economic forms and other reconstruction measures of rural economy, so as to finally form an industrial pattern that can ensure the steady improvement of agricultural comprehensive production capacity and the integration of rural primary, secondary and tertiary industries.

9.3.2.2 Restructuring Rural Ecological Space to Achieve Ecological Vitalization

Rural ecological space reconstruction emphasizes on natural ecological laws and principles of landscape ecology, generally attaching importance to ecological network and green infrastructure construction on a large scale, and improving habitat quality and landscape diversity on a small scale, thereby improving land productivity, ecological landscape service capability, as well as carbon sequestration capacity. Moreover, in order to make the rural areas an important green space and ecological barrier for urban and rural development, it is required to maintain and repair natural ecological processes and biological chains from different scales, to build a good rural ecological spatial pattern, and to improve the elasticity of rural ecosystems and ecological service functions.

9.3.2.3 Restructuring Rural Living Space to Achieve Organizational and Cultural Vitalization

Under the premise of promoting urbanization, effective planning and regulation should be adopted to achieve reasonable integration of rural settlements and the moderate concentration of rural population. On this basis, the quality of rural human settlements can be effectively improved and the rural living space that is conducive to the coordinated interaction between urban and rural areas can be formed through the rational allocation of rural infrastructure. The land use layout of rural residential areas in China is characterized as "scattered, disordered and empty" determining that the renovation of "hollowing village" is the key to reconstruct rural living space (Long 2014; Long et al. 2009). The core of it is to promote the integration of rural organizations and establish a town-village space system to coordinate urban and rural areas. This is also a necessary condition for achieving the goal of revitalizing rural areas with a pleasant living environment and effective governance. Therefore, during the process of living space restructuring, attention should be paid to strengthening the effective connection and mutual support between spaces and functions of the rural territorial system. In particular, based on the node role of central towns and key towns in the energy transmission chain of urban and rural territorial systems, the construction of central town and key town should be accelerated with the core of strengthening the cultivation of characteristic industries, improving public service facilities, and enhancing the comprehensive service function. At the same time, protection of characteristic rural cultural landscapes and enhancement of rural cultural functions also need to be given priority. The specific approach is to encourage distinctive settlements, based on the protection of traditional features, explore local cultural characteristics, develop tourism, cultural and creative industries, and seek economic driving forces for traditional cultural protection, so as to achieve the benign interaction between industrial development and the protection and revival of characteristic culture (Long and Tu 2017). Indeed, this is also an important content of rural social restructuring and cultural vitalization. In a word, the spatial pattern, which serves the goals such as pleasant living environment, refined rural civilization, and effective governance of rural vitalization, can be finally formed by restructuring rural living space.

9.4 Theoretical Framework

The advancement of urbanization and industrialization has triggered changes in the rural industrial structure and the transfer of key elements of rural development (population, capital, land, etc.), and the functions of rural areas have changed accordingly. This series of changes and the involvement of new factors such as economic globalization and international competition have made China's rural development enter a new stage of transformation, upgrading, and restructuring. Rural restructuring is such a process in which actors optimize the allocation of material and non-material factors that affect rural development and manage them effectively, then reconstruct the rural socio-economic morphology and optimize the regional spatial pattern. Its purpose is to adapt to changes in rural internal factors and external regulation, so as to achieve the optimization of the internal structure and function of the rural territorial system, and to achieve the structural coordination and functional complementarity between the urban and rural systems (Long and Tu 2017; Long and Tu 2018a).

The realization path of rural restructuring includes three dimensions: spatial restructuring, economic restructuring, and social restructuring. Rural spatial restructuring mainly focuses on the establishment of a village-town spatial system that can be conducive to coordinate urban and rural areas, and then, on the basis, to reconstruct the internal production, living and ecological spatial pattern of the village following the principle of relatively concentrated space and optimal allocation of resources. As the most active part of rural restructuring, the core of economic restructuring is industrial cultivation, whose purposes is to promote the transformation of moderate-scale agricultural operations to revitalize traditional rural industries. Besides, it also aims to develop and cultivate new forms of rural economy by using the multi-functional value of rural areas and modern technological means (Long and Tu 2018a).

In order to adapt to changes of population, land and other rural development elements during the process of urbanization and a series of challenges for sustainable development, the report of the 19th National Congress of the Communist Party of China in 2017 proposed several measures and means to realize rural vitalization, for instance, "accelerating the modernization of agricultural and rural areas", "deepening the reform of rural land system", "developing various forms of moderate scale operation", and "promoting the integration of rural primary, secondary and tertiary industries".

The key to rural vitalization lies in stimulating internal power and absorbing external resources through the means of economic, political and cultural construction to reasonably reallocate development elements (rural population, land, and industry) and make efficient use of them, so as to stimulate rural vitality, optimize elemental structure, improve regional function, as well as reshape rural morphology. What is more, a new pattern of comprehensive economic, social and ecological revival in rural areas and urban-rural integration will be achieved. In addition, based on the connotation interpretation of the rural vitalization strategy by the 19th National Congress of the Communist Party of China (Liu 2018), it is found that the rural restructuring is an important approach to implement the strategy of rural vitalization (Long 2018). Moreover, the activation of multi-factors of rural development such as population, land, and industry is the key to fostering growth drivers and solving the "Three Rural Issues" (peasant, agriculture and countryside). The essence of rural reconstruction and rural vitalization, therefore, is to adjust the evolution direction of the three key development elements of population, industry, and land, so as to form a coupling and coordinated development pattern among the various elements and promote the multi-dimensional revival of rural areas (Liu 2018; Long 2014).

Land use transitions usually happened under the framework formed by the interaction of the natural system, economic system and institutional (management) system (Li 2002). Land use morphologies of a country or region often change with the stage of economic and social development of the country or region. The essence of rural restructuring is the process of restructuring rural socio-economic morphology and territorial spatial patterns. In order to promote this process, governments, rural elites, industrial and commercial enterprises, peasants, rural autonomous organizations and other actors intervene in the development of key elements of rural development, such as population, land, and industry, in response to the current changes in urban and rural development factors.

The land is the spatial carrier of human social and economic activities. On the one hand, the flow of production factors and the subsequent responses of actors in the process of rural socio-economic reconstruction will inevitably cause changes in regional production, living, and ecological spatial structure, and land use patterns. For example, the land-use mainstream adjusts the land-use structure by using economic means such as land rent and land price, controls the rate of non-agriculturalization of farmland, and intervenes into changes in dominant morphology of land use (Long 2014). In the process of reconstruction of production and living space, a lot of measures have been used to promote the transformation of dominant and recessive morphology of land use.

These measures mainly include the implementation of engineering and technical means such as land consolidation and village planning, and promulgation of related policies and systems to promote the reuse of vacant, abandoned homestead and the transfer of contracted management right of agricultural land. On the other hand, the transformation of land use morphologies, in turn, directly or indirectly affects the practice of rural restructuring and plays an important role in rural vitalization. For example, promoting land transfer through consolidating the land in inefficient use can provide space for the development of modern agriculture and non-agricultural industries, and create the necessary conditions for the improvement of living environment and infrastructure, and the efficient allocation of public services (Tu and Long 2017). At the same timc, it develops local industries on these lands to exert a positive influence on the non-agriculturalization of the rural population and the improvement of rural self-development capabilities (Fig. 9.1).

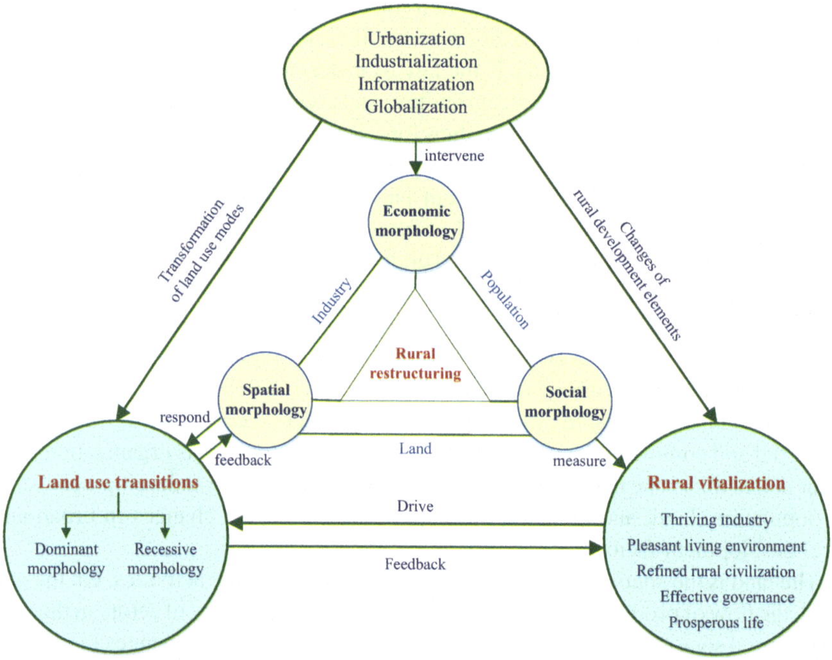

Fig. 9.1 The relationship among land use transitions, rural vitalization and rural restructuring (Adapted from Long and Tu 2018b)

9.5 Conclusions

Different economic and social development stages correspond to different land use patterns and land use transitions stages. Socio-economic development inevitably results in a specific land-use transition process. Rural restructuring is an important means to implement the strategy of rural vitalization, also the link between land use transitions and rural vitalization. Population, land and industry are closely interlinked, and economy, space and society are intertwined. It is a complex and systematic project to realize rural vitalization relying on land use transitions and rural restructuring. The practice of rural restructuring and vitalization directly or indirectly implicates the matching and arrangement of a series of resource allocation systems and macro policy environment, such as land, finance, social security, etc. (Long and Tu 2018a). This goal can be achieved through restructuring rural production space to realize industrial vitalization, restructuring rural ecological space to realize ecological vitalization, and restructuring rural living space to realize organizational and cultural vitalization. In this process, mainstream of land use adjusts the allocation of land resources and optimize the land use morphology by means of spatial restructuring such as economic leverage, engineering technology and policy system. The transformation of land use morphology in turn affects the practice of rural restructuring and rural vitalization.

The first three chapters have introduced and explained land use transition in detail, and pointed out that land use morphology includes dominant morphology and recessive morphology. The changes in rural labor factors, the dominance of land economic value, and the impact of existing systems and policy systems during the process of urbanization have driven major changes of these two land use morphologies. In order to cope with the new situation and problems arising from changes in land use patterns in the context of economic and social development and transformation, decision-making departments need to comprehensively use economic, legal and engineering technologies and other means to timely adjust relevant measures for land resource management. The formulation of land management regulations and policies should be based not only on a thorough understanding of the stage of land use transitions in the areas where rural vitalization are implemented, but also on the forward thinking about the stage of land use transitions that is coming with the transformation of socio-economic development. This contributes to the scientificity of land resource management decisions. The recessive morphology of land use and its changes should be the focus of research on rural vitalization through controlling land use transitions. Thriving business features heavily in rural vitalization. Linking the revitalization of land resources with the rural industry is of great significance for the sustainable development of rural economy and the efficient use of resources. Therefore, it is an urgent need to innovate land management policies and regulations by controlling recessive morphology changes of land use, and to explore a new land use transition mode combined with multifunctional agriculture to promote rural vitalization.

References

Barlowe R (1986) Land resource economics: the economics of real estate, 4th edn. Prentice-Hall, Englewood Cliffs

Ge D, Long H, Zhang Y et al (2018a) Analysis of the coupled relationship between grain yields and agricultural labor changes in China. J Geogr Sci 28:93–108

Ge D, Long H, Zhang Y et al (2018b) Farmland transition and its influences on grain production in China. Land Use Pol 70:94–105

Li X (2002) 土地利用变化的解释 (Explanation of land use changes). Prog Geogr 21(3):195–203

Liu Y (2007) 中国东部沿海地区乡村转型发展与新农村建设 (Rural transformation development and new countryside construction in eastern coastal China). Acta Geogr Sinica 62:563–570

Liu Y (2018) 新时代城乡融合与乡村振兴 (Research on the urban-rural integration and rural revitalization in the new era in China). Acta Geogr Sinica 73(4):637–650

Liu Y, Long H, Chen Y et al (2011) 中国乡村发展研究报告: 农村空心化及其整治策略 (Research report on rural development in China: Hollowed village and its renovation strategy). Science Press, Beijing

Long H (2006) 中国农村宅基地转型的理论与证实 (Rural housing land transition in China: theory and verification). Acta Geogr Sinica 61(10):1093–1100

Long H (2012) 中国乡村转型发展与土地利用 (Land use and rural transformation development in China). Science Press, Beijing

Long H (2014) Land consolidation: an indispensable way of spatial restructuring in rural China. J Geogr Sci 24(2):211–225

Long H (2018) 乡村重构专辑序言Preface to the special issue of rural restructuring. Prog Geogr 37(5):579–580

Long H, Ge D, Zhang Y et al (2018) Changing man-land interrelations in China's farming area under urbanization and its implications for food security. J Environ Manag 209:440–451

Long H, Li Y, Liu Y (2009) 中国空心化村庄演化特征及其动力机制 (Analysis of evolutive characteristics and their driving mechanism of hollowing villages in China). Acta Geogr Sinica 64(10):1203–1213

Long H, Li Y, Liu Y et al (2012) Accelerated restructuring in rural China fueled by 'increasing vs. decreasing balance' land-use policy for dealing with hollowed villages. Land Use Pol 29(1):11–22

Long H, Tu S (2017) 论乡村重构 (On rural restructuring). Acta Geogr Sin 72(4):563–576

Long H, Tu S (2018a) 乡村重构的理论认知 (Theoretical thinking of rural restructuring). Prog Geogr 37(5):581–590

Long H, Tu S (2018b) 土地利用转型与乡村振兴 (Land use transition and rural vitalization). China Land Sci 32(7):1–6

Long H, Tu S, Ge D et al (2016) The allocation and management of critical resources in rural China under restructuring: problems and prospects. J Rural Stud 47:392–412

Long H, Woods M (2011) Rural restructuring under globalization in eastern coastal China: what can be learned from Wales? J Rural Community Dev 6(1):70–94

Tu S, Long H (2017) Rural restructuring in China: theory, approaches and research prospect. J Geogr Sci 27(10):1169–1184

Tu S, Long H, Zhang Y et al (2018) Rural restructuring at village level under rapid urbanization in metropolitan suburbs of China and its implications for innovations in land use policy. Habitat Int 77:143–152

Tuan Y (1971) Geography, phenomenology, and the study of human nature. Canadian Geogr 15(3):181–192

Woods M (2005) Rural geography: processes, responses and experiences in rural restructuring. Sage, London

Zhang J (2018) 乡村价值定位与乡村振兴 (Village value orientation and rural revitalization). Chinese Rural Econ 1:2–10

Chapter 10
Land Consolidation: A Way of Rural Restructuring and Vitalization

Abstract Land consolidation is a spatial problem-solving land management instrument that attempts to eliminate certain types of land fragmentation, to enhance land productivity, and to improve rural production and living conditions for the purpose of coordinating urban-rural development, through a process of concentration of plots or rejuvenation of failing and aging rural settlements and abandoned industrial and mining land, which is usually accompanied by the construction of new roads, irrigation facilities and other auxiliary services. Rural spatial restructuring mainly includes the restructuring of rural production, living and ecological spaces. Land consolidation can provide space and supporting platform for rural restructuring and vitalization and realizing urban-rural integrated development in China. This chapter puts forward that land consolidation, as an indispensable way of rural restructuring and vitalization promoted by land use transitions, includes three types, i.e., agricultural land consolidation, hollowed villages consolidation and industrial and mining land consolidation. The optimization and restructuring of suitable living space with amenity, intensive and high-efficient production space and the ecological space with green mountains and clean water in rural areas will rely on the innovations of regional engineering technology, policy and mechanism, and mode of rural land consolidation. As a crucial means of promoting rural vitalization, land consolidation is endowed with new connotations such as activating the key elements of rural development, coordinating the vitalization of the physical space and the promotion of the spiritual core, integrating the restructuring of the actual physical space and the rural governance system.

10.1 Introduction

Currently, the implementation of new-type industrialization and urbanization and agricultural modernization strategies lacks of a major hand grip and spatial supporting platform, due to long-term existed "dual-track" structure of rural-urban development in China as well as unstable rural development institution and mechanism. Meanwhile, major changes have taken place in the principal contradiction of Chinese society. In response to the conflict between the people's ever-growing need for

H. Long, *Land Use Transitions and Rural Restructuring in China*,
https://doi.org/10.1007/978-981-15-4924-3_10

a better life and current unbalanced and insufficient development, the 19th National Congress of the Communist Party of China (CPC) put forward the "rural vitalization" strategy to alleviate issues of the uncoordinated urban-rural development. Rural vitalization is the construction of economy, politics, culture, ecology, and people's well-being, aiming at realizing the comprehensive rejuvenation in rural areas by systematically allocating and efficiently managing various developmental elements, such as population, land, and industry (Hillman 2004; Li et al. 2016; Liu 2018). It is necessary to restructure rural production, living and ecological spaces by carrying out land consolidation, so as to establish anew platform for realizing rural vitalization and urban-rural integrated development in China.

This chapter probes the connotations of land consolidation and rural restructuring and vitalization as well as their relationships. Based on the effects analysis of industrialization and urbanization on rural production, living and ecological spaces, the mechanism of pushing forward rural spatial restructuring by carrying out land consolidation is analyzed. A conceptualization of the modes of rural production, living and ecological spatial restructuring is analyzed combining with agricultural land consolidation, hollowed villages consolidation and industrial and mining land consolidation. Finally, prospects on rural restructuring in China based on resources optimal allocation are put forward.

10.2 Land Consolidation and Rural Restructuring and Vitalization

10.2.1 Connotations of Land Consolidation

Traditionally, rural land consolidation has been limited to engineering and technical aspects. Its core objectives are mainly to expand the scale of farmland, improve the quality of cultivated land, and optimize the village's layout, and even in some areas, land consolidation is considered merely a means of providing more space for urban development. Currently, rural vitalization is reflected not only in the upgrading of the living environment and public services and other material aspects but also in the in-depth revival of a vibrant industry, a unique culture, and an orderly governance system. Comprehensive rural vitalization should be based on the distinctions between urban and rural territorial system and the multifunctional values of rural areas. Land consolidation is endowed with the following new connotation and multi-functions from the perspectives of rural restructuring and vitalization:

(1) Activating the key elements of rural development such as population, land, and industry. Beginning with comprehensive land consolidation, it is better to restructure production, living, and ecological spaces, strengthen the integration of modern agriculture and multi-industrial forms such as Bed and Breakfast

(B&B) industry, pension industry and ecological tourism etc., and it is necessary to promote the non-agricultural transfer of the rural population and the transformation of land use patterns.

(2) Coordinating the vitalization of the physical space and the promotion of the spiritual core. Comprehensive land consolidation aims at changing the quantity and quality of the cultivated land and the rural construction land as well as vitalizing rural land resources. Meanwhile, it should take into account the inheritance of rural characteristic culture, the maintenance of the architectural appearance and settlement texture in order to preserve the unique charm of the countryside and enhance the rural ecological and cultural functions.

(3) Integrating the restructuring of the actual physical space and the rural governance system. With the aggregating of rural living spaces, it is of vital important to establish multilayer rural governance modes through combining grassroots autonomous organizations and communities, as well as cultivating skilled manpower capable of appropriately scaled operations for the purpose of realizing a smooth linkage between the rural spatial system and the governance system.

10.2.2 Reciprocity via Interlinking "Population-Land-Industry"

10.2.2.1 Coupling Relationship Between Land Consolidation Evolution and Rural Transformation Development in China

The evolution process of land consolidation in China can be roughly divided into three stages (Fig. 10.1).

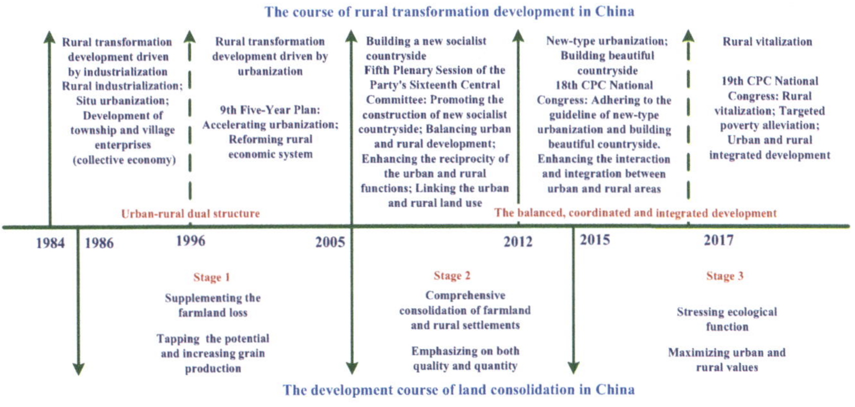

Fig. 10.1 Phase characteristics of land consolidation and rural transformation development in China (Long et al. 2019)

(1) **The stage of tapping the quantitative potential**

The development of township and village enterprises and the rapid expansion of urban construction land resulted in a sharp shrinkage of cultivated land and increasing pressure on food security (Long et al. 2018). To tackle these issues, land consolidation aimed at supplementing farmland loss was widely carried out in China's rural areas (Long 2014), among which some successful modes emerged including the "three concentrated" consolidation mode in Shanghai (the concentration of farmers' houses to central villages and small towns, township and village enterprises to industrial zones, and farmland to large-scale operations). From 1998 to 2005, China witnessed a rapid increase in the number of funds, projects and scale for land consolidation, and more than 4 million mu of cultivated land was replenished annually through land consolidation. During this period, land consolidation focused on increasing the amount of cultivated land and providing space for urbanization and industrialization, which has guaranteed agricultural production, increased farmers' income and ensured food security to a certain extent.

(2) **The stage of emphasizing on both quantity and quality**

In 2006, the establishment of 116 national prime farmland protection demonstration zones marked the transformation of China's land consolidation from emphasis on a quantitative increase to a comprehensive consideration of both quantity and quality of newly-added cultivated land. In the same year, the former Ministry of Land and Resources deployed the first batch of pilot projects of "Linking up Increased Urban Construction Land with Decreased Rural Construction Land", which indicated that rural construction land consolidation has been an important part of land consolidation (Long et al. 2012). In 2005, the State proposed the strategy of "Building New Countryside" aiming to form a new pattern of coordinated urban-rural development (Long et al. 2010), and land consolidation laid a solid material foundation for the building of the "new countryside" through improving the production and living conditions as well as ecological environments in rural areas.

(3) **The stage of stressing ecological functions and maximizing urban-rural values**

The "13th Five-Year Plan" of land consolidation highlights the construction, utilization, and protection of ecologically beneficial fields and advocates "green" land consolidation, which shows that China's land consolidation is shifting to paying more attention to ecological function of land use. In 2017, the 19th CPC National Congress proposed the strategy of rural vitalization to promote the integration of urban and rural development. Meanwhile, land management departments at all levels have successively issued a series of comprehensive rural land consolidation policies to accelerate the supply-side reform in the agricultural sector, optimize the space of urban and rural development and lay a solid foundation for rural vitalization. Different socio-economic development stages have different requirements for land use morphology patterns. The process of rural transformation development prompts land users to optimize land use morphology patterns through land consolidation projects, and the transition of land use morphology in turn acts on rural development and vitalization (Long 2012; Long and Tu 2018). Chronologically, the modernization of a region or a county usually undergoes the periods of the "urban bred by countryside, urban-rural separation, urban-rural opposition, and urban-rural integration", as is the case with China's rural transformation development (Fig. 10.1). With the advancement of socio-economic development in different periods, the objectives of land consolidation in China have changed from the emphasis on exploitation of unused land, the increase of cultivated land and the provision of space for urbanization to focusing on the comprehensive improvement of farmland and villages, the equal emphasis on "quantity" and "quality", and eventually the multi-functionality of land use and the maximization of urban-rural values. Apparently, the orientation and effectiveness of land consolidation have been highly consistent and intrinsically coupled with the needs of socio-economic development and transition in China.

10.2.2.2 Reciprocity Between Land Consolidation and Rural Vitalization from the Perspective of "Population-Land-Industry"

The advanced or lagging transformation of the urban-rural socio-economic development elements could cause different feedbacks and responses to the rural territorial system, which could affect the sustainable development of regional agriculture and rural areas. The common point of rural land consolidation and rural vitalization is to adjust key development elements and promote coupling and coordination among various elements. Rural land consolidation mainly includes farmland and rural construction land consolidation, both of which are land use behaviors interfering with rural production, living, and ecological spaces through engineering and technical means (Long 2014) (Fig. 10.2).

As the space carrier of major socio-economic activities, the optimization of inner structure and the improvement of quality for land resources directly affect the economic benefits of agricultural production and the cultivation of non-agricultural industries, thus exert an important influences on the income of local residents, the transformation of industrial structure and the allocation of human resources. Compared with the objectives of agricultural land consolidation to increase the amount of cultivated land, improve agricultural production conditions and promote the scale of agricultural management (Jiang et al. 2017), the goal of rural construction land consolidation is to achieve a positive interaction between urban capital and rural idle land by guiding rural population agglomeration and industrial concentration, and eventually build up the channels of elements flows between urban and rural areas.

The essence of land consolidation is to adjust land ownership and supervise land use, and the multi-functionality of land resources also extends their derivative values such as coordinating urban and rural development, preserving the cultural landscape, and maintaining social stability. Confronted with various problems, such as the inefficient utilization of land resources, the poor infrastructure, the rapid loss of development elements, the weakening of the manpower etc., in rural areas, land consolidation

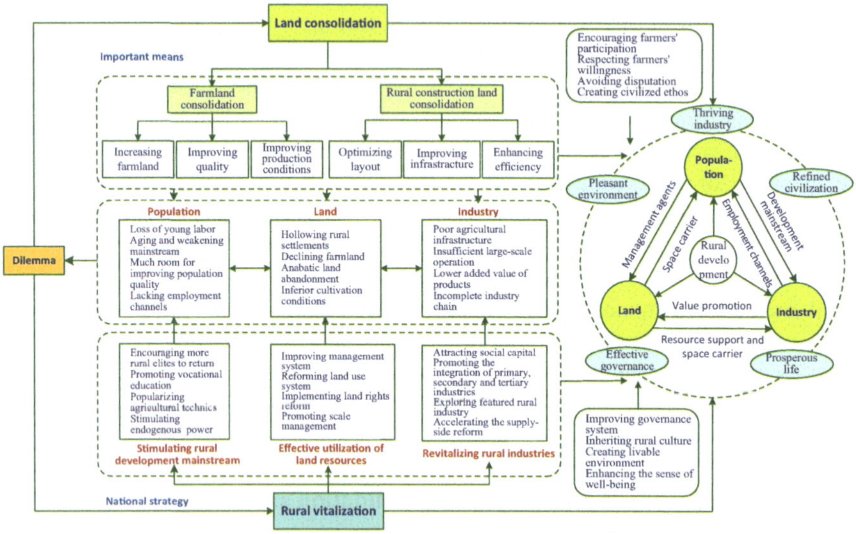

Fig. 10.2 The relationship between land consolidation and rural vitalization in China (Long et al. 2019)

within the context of rural vitalization should be gradually transformed from merely increasing land to prospering rural economy, improving farmers' income, and protecting the ecological environment etc., as well as from the single-factor regulation of "land-based" to the comprehensive regulation towards coordinating the multiple elements of "population, land and industry" through the organic integration of new enterprises, advanced technologies, and skilled manpower.

10.3 Rural Spatial Restructuring Based on Land Consolidation

10.3.1 Land Consolidation and Rural Spatial Restructuring

Land consolidation is a spatial problem-solving land management instrument that attempts to eliminate certain types of land fragmentation, to enhance land productivity, and to improve rural production and living conditions for the purpose of coordinating urban-rural development, through a process of concentration of plots or rejuvenation of failing and ageing rural settlements and abandoned industrial and mining land, which is usually accompanied by the construction of new roads, irrigation facilities and other auxiliary services (Coelho et al. 1996; Huang et al. 2011; Long et al. 2010, 2012). Land Consolidation and Rehabilitation Center of the Ministry of

Land and Resources of China was established in 1998, and some land consolidation projects were initiated at national level in the same year. At first, the national objective of land consolidation in China emphasized on increasing the quantity of cultivated land by land (e.g. unused land, pasture land and wetland) reclamation and consolidation, to compensate the decrease of that due to rapid urbanization and industrialization.

China's vast rural areas have experienced much notable restructuring process since the economic reforms of 1978 (Cai 2001; Long et al. 2011, 2012; Long and Woods 2011). Seeing from the practice of rural development in China, rural restructuring is a rural development strategy with economic, social and spatial connotations, aiming at fostering the spirit of equity, harmony, and coordinated development in the relationships between industry and agriculture, and between the urban and the rural by altering the dichotomy of socio-economic structures owing to long-term urban-rural segmentation, so as to achieve a virtuous interaction of urban-rural development in the end. And it is accomplished by promoting the function and position of the countryside in the urban-rural system, continuously developing rural economy and society, improving the standards of material and spiritual civilizations, and advisably organizing the spatial distribution. More specifically, one major form of rural restructuring is the change of its spatial pattern, i.e. rural spatial restructuring.

Rural spatial restructuring is an optimization and adjustment even utterly changing process of rural production, living and ecological space accompanied by the reshaping of socio-economic structure in rural areas owing to an integrated result of endogenous development need and exogenous driving forces of rural development, under the pressure of rapid industrialization and urbanization. Besides, rural spatial restructuring is an integrated approach aiming to optimizing urban-rural space organization and promoting coordinated urban-rural development. Accordingly, the connotations of rural spatial restructuring involve the following three aspects: (1) assembling industrial development, i.e. facilitating the cluster of industrial production in industrial parks, and the transformation of large-scale managerial agricultural production; (2) centralizing farmers' living, namely the migration of rural population from villages to townships and new rural communities, so as to grapple with problems resulting from living dispersedly like huge investment needs for public infrastructure and its low utilization efficiencies, as well as to effectively keep the land use area of rural settlements within bounds and preserve traditional rural cultural landscape; and (3) the intensive utilization of resources, that is, to achieve the coordinated development of population, resources, and the environment, and to solve problems like low efficiency of resources utilization and environmental pollution in rural production process as well as rural daily living, by agglomerating industries and concentrating rural settlements. To summarize, rural spatial restructuring mainly includes the restructuring of rural production, living and ecological space.

Both the selection of land consolidation types and the mechanism analysis of facilitating rural spatial restructuring via land consolidation need an overall analysis of the effects of industrialization and urbanization on rural production, living and ecological space in China. Furthermore, the understanding of the status quo of rural production, living and ecological space as well as the mechanism of rural spatial restructuring promoted by land consolidation is the important base and premise of adopting suitable modes and approaches for rural spatial restructuring.

10.3.2 Mechanism

10.3.2.1 The Effects of Industrialization and Urbanization on Rural Production, Living and Ecological Space

With the continual boost of industrialization and urbanization, the two key elements of rural development, population and land, have undergone dramatic changes, which brought about far-reaching effects on rural production, living and ecological space. On the one hand, the nonagriculturalization and concurrent occupation of rural population plays a role in shaping rural living and production space; on the other hand, the nonagriculturalization and non-grain preference of rural land use exerts effects on rural production and ecological space.

As industrialization quickens its pace, rural population continuously floods into cities. This fast out-draining leads to serious depopulation in rural areas, and issues like the abandoned and unused construction land (mainly rural housing land) in village centers also come to surface. Because of the flaws in them, rural homestead withdrawal mechanism and property rights system fail to act as a brake on the hollowing process in the countryside, thus "hollowed villages" sprouting and thriving all over China (Long et al. 2012). "Hollowed villages" have two major characteristics: one is that many homesteads inside the village is getting increasingly dilapidated and even ruined, with the central area taking on an obsolescent landscape look; on the other hand, due to building new houses but leaving them unused as well as not dismantling old ones, the spatial morphology differentiation of peripheral extensive development and internal declining was formed. In China, the majority of villages have features like "scattered, small, and messy", due to the combined influences of natural conditions, traditional local customs, and the lack of rural planning (Long et al. 2012). Up to 2009, China still had 567.5 thousand administrative villages, while the natural ones rocketed to 2.71 million, among which 1.31 million had a population less than 200 (MHURDC 2010). Village hollowing in China have led to several immediate consequences, such as the weakening function of critical rural organizations, the fragility of structures and networks, the chaotic flow of rural development elements, and the lack of economies of scale and output efficiency, resulting in lagged effects like slow rural economic development, woeful inadequacy of rural infrastructure, and ultimately the extreme inconvenience of rural living space. Besides, the extensive management of cultivated land is common in rural areas, and some even abandon farming totally and just leave their farmland uncultivated. Under the circumstance of defective social security and undefined collective property rights, it is impossible for farmers to completely cut off their connection with the countryside and leave, bringing about the major feature of concurrent occupation among farmers in Chinese agricultural production, which caused limited rural production space to some extent.

The bulky reduction of farmland takes place due to the nonagriculturalization and non-grain preference of rural land use in the process of industrialization and urbanization. On the one hand, the innate pursuit of profits makes farmland changed to orchard, rangeland, aquiculture and other usages in the wake of agricultural structural adjustments. On the other hand, remarkable yield gap between urban and rural areas in terms of land use leads to an irresistible trend of land transfer from the countryside to the city. Besides, rural settlements tend to locate in places with subdued topography and good facilities, and where usually high-quality farmland is. With the continuing encroachment of urban and rural construction land, farmland is getting increasingly fragmentized, plus the fact that there is a large population with relatively little (arable) land in China, which makes it very difficult to carry out large-scale farmland management and consequently confines rural production space (Long et al. 2009). The disordered arrangement of industrialization in its early stage, and the loose control of pollutant emissions lead directly to severe contamination of soil, surface water and groundwater in rural areas. As with environmental regulation, its reinforcement will increase the cost of the pollution treatment for enterprises, so they might relocate in rural areas with relatively less strict environmental regulation and cause serious resources and environmental problems as a consequence. For instance, relevant studies have proved that countryside is the safe haven for pollutions from joint venture enterprises in China (Dean et al. 2009). Therefore, with the continuing of urbanization and industrial structural transformation, industrial pollution from the city is looking for shelter in rural areas, plus the waste of agricultural production and rural life, and deterioration of rural ecological space seems to be inevitable. To some extent, the nonagriculturalization of rural land use has exerted its negative influence on the consecutiveness and completeness of rural landscape ecological processes and patterns.

10.3.2.2 Land Consolidation Types Facilitating the Restructuring of Rural Production, Living and Ecological Space

Basing on the analysis of the effects of industrialization and urbanization on rural production, living and ecological space, as well as the characteristics of rural land consolidation (Yun 2011), the land consolidation types facilitating rural spatial restructuring and their corresponding features are concluded as follows:

(1) **Agricultural land consolidation**

It means to increase the area and the quality of farmland, improve agricultural production conditions and local ecological environment, by land leveling, irrigation and drainage facilities upgrading, field road paving, farmland protecting, ecological environment conserving and other engineering measures. Its primary aim is to enhance the building of farmland infrastructure, strengthen the mass producing of high-standard and disaster-resistant prime farmland, optimize the layout of farmland with multi-functions, and effectively facilitate the concentrating and linking of farmland pieces so as to carry out large-scale management.

(2) **Hollowed villages consolidation**

It refers to rejuvenating disperse, abandoned, idle and low-efficient used rural construction land (mainly rural housing land), so as to improve rural infrastructures and public service facilities, better rural producing and living conditions, raise the intensive and economical utilization level of rural construction land, promote the building of new countryside, and ultimately achieve the overall allocation of urban-rural land. The major goal of hollowed villages consolidation is to improve rural infrastructures and public service facilities by rejuvenating vacant rural homesteads, and other abandoned and idle rural construction land.

(3) **Industrial and mining land consolidation**

It means to remediate the industrial land of township and village enterprises (TVEs), and mining land such as ore, quarry, sand and salt mining, as well as brick/tile kilns, so as to improve supporting facilities, enhance land saving, expand urban-rural development space, increase land values, better human residential environment, and promote the intensive and economical utilization of land resource. It prioritizes facilitating the concentration of TVEs in the industrial park, remediating polluted rural land, promoting land use efficiency, and improving rural production, living and ecological environment.

10.3.2.3 Land Consolidation and the Restructuring of Rural Production, Living and Ecological Space

The analysis of rural land consolidation types and features proves that, via rural land consolidation and rural spatial restructuring, we can solve problems brought about by industrialization and urbanization, such as inconvenient living space, limited production space and deteriorated ecological space. Currently, major extrinsic spurs for rural spatial restructuring are from rapid expansion of the urban, frequent urban-rural migration, and urban-rural industrial spatial restructuring. Meanwhile, main intrinsic incentives are from nonagriculturalization and low-efficient use of rural land, vast outflow of rural population, disordered layout of rural industries, undersupply of rural infrastructures, and increasingly deteriorating of rural ecological environment. Taking all of these extrinsic and intrinsic factors into account, the restructuring of rural production, living and ecological space may be promoted by the innovations of land consolidation engineering technology, policy and mechanism, and mode (Fig. 10.3).

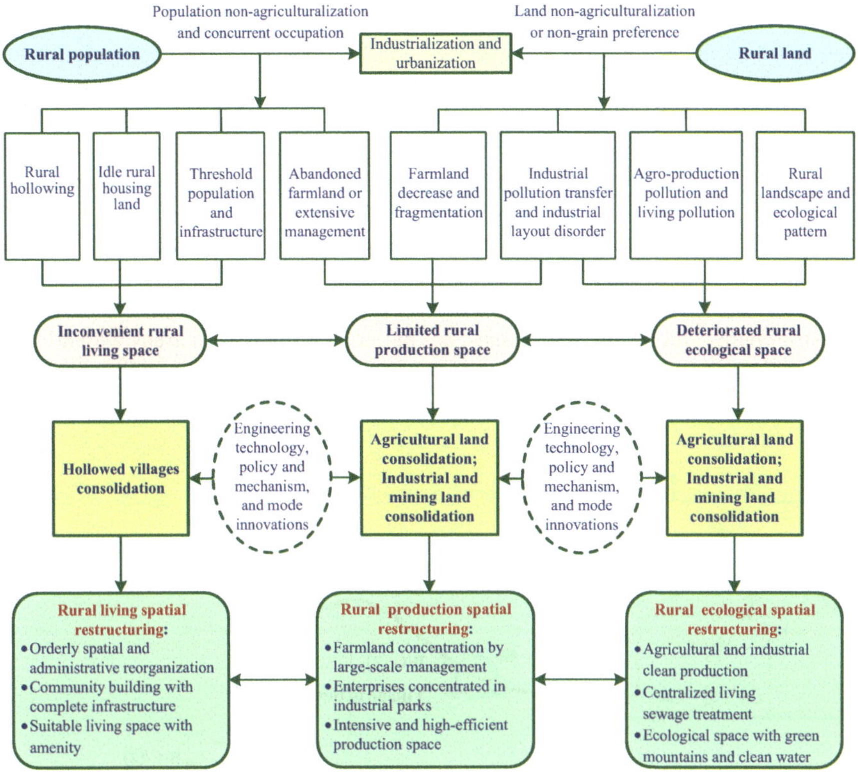

Fig. 10.3 The mechanism of promoting rural production, living and ecological spatial restructuring by carrying out land consolidation (Long 2014)

10.3.3 Modes and Approaches

10.3.3.1 Rural Production Spatial Restructuring

The prosperity of rural industrial economic system rests on the restructuring of rural production space. In this process, it is better to speed up the implementation of agricultural industrialization and the building of industrial parks, promote the extensive utilization of appropriate production technologies, improve the socialization level of production organization and the scientific level of its management, and ultimately achieve the unification of land output rate, labor productivity and resource utilization rate. The layout of urban-rural productivity needs to be balanced according to the guideline of planning and market mechanism, and the coordinated development of urban-rural economy can be achieved based on the principle of clustered development and intensive management. More specifically, it needs to map out and implement the urban-rural integration spatial layout plan, and to relocate the relatively low-end, labor-intensive, and resource-dependent (agricultural products) industries in rural areas, and the relatively high-end, technology-intensive, and capital-intensive industries in urban areas, so as to achieve the orderly division of labor cooperation in both urban and rural areas. With reference to the standard of "farmland regularly structured, trees in rows, roads interlinked, and canals adjacent and equipped to irrigate and drain", the large-scale construction and management of high-standard prime farmland and the agricultural production base may be promoted accordingly (Fig. 10.4).

Through carrying out industrial and mining land consolidation, manufactural industries are concentrated in industrial parks. Meanwhile, rural service industries are clustered in business subzones. These non-agricultural production bases, namely industrial parks and business subzones, are better to be located in areas with advantageous locations, good transportation and communication facility conditions, while agricultural production should be befittingly far away from cities, towns and central

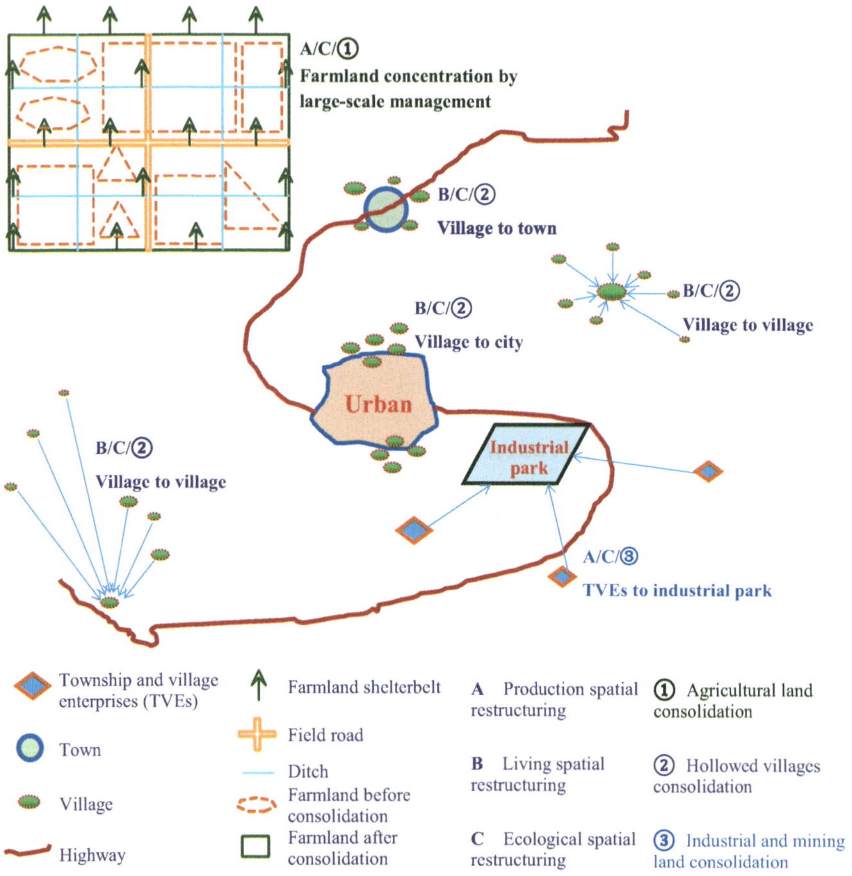

Fig. 10.4 A conceptualization of the modes of rural production, living and ecological spatial restructuring (Long 2014)

villages, for fear of nonagriculturalization of high-quality farmland. It is reckoned that rural social organizations like economic cooperation organizations, specialized cooperatives and industry associations should take the lead in promoting farmers' participation in social division of labor and industrial cooperation, in order to improve and restructure intensive and high-efficient rural production space, by innovating the allocation mode of essential rural production elements like land, labor and capital.

10.3.3.2 Rural Living Spatial Restructuring

During the process of rapid urbanization, it is better to promote the proper concentration of rural population through effective planning and regulation of rural settlements. After that, it needs to improve the rural infrastructures and human residential environment, so as to facilitate the rise of rural living space with beneficial and harmonious interactions with the urban. The notorious reputation of China's rural settlements layout, namely "being scattered, massy, and hollowed", calls for some emergency measures like the hollowed villages consolidation, whose top priority is to promote the spatial-territorial reorganization, which is accompanied in rural restructuring by administrative reorganization, aimed at reducing the number of administrative villages. There are three modes of administrative reorganization: (1) village to city, the incorporation 'urban villages' and peri-urban villages into larger urban areas; (2) village to town, to resettle residents of villages into nearby small towns; and (3) village to village, to reorganize remote villages into centralized villages or new rural communities (Long et al. 2012). In brief, this kind of restructuring encourages farmers to relocate in communities and incorporates villages by implementing hollowed villages consolidation and rural administrative reorganization, so as to solve problems like the disperse layout of rural settlements and meet the requirements for the threshold population of socialized service industry, to improve the effectiveness and efficiency of local governance in rural communities, to enhance farmers' sense of organization belonging, and to optimize and restructure suitable living space with amenity.

10.3.3.3 Rural Ecological Spatial Restructuring

In the process of restructuring rural production and living space, more attention is needed tobe paid to the restructuring of rural ecological space, which is inseparable from rural production and living spatial restructuring (Fig. 10.4). When restructuring rural ecological space, it needs to follow the law of the nature and the principles of landscape ecology, pay attention to the construction of ecological network and green infrastructure on the large scale, and improve habitat quality and landscape diversity on the small scale, whether it is agricultural land consolidation, hollowed villages consolidation or industrial and mining land consolidation.

It also needs to promote land productivity, landscape ecological service ability, and carbon sequestration capacity, so as to restore natural ecological processes and food chain, to improve the flexibility and service ability of rural ecological system on all the scales, and to make countryside the critical strategic space for establishing ecological barrier and green space for urban-rural development, ensuring food safety, conserving biological diversity, developing low carbon economy, and replying climate change (Yun and Yu 2011). Some specific matters needing attention in the restructuring process are: strict restrictions of chemical fertilizer and pesticide usage, and spurs of clean, green and pollution-free production by contrast, given the innate features of rural ecological system; construction of source-reduction and flow cut-off pollution control system, constituted by eco-systems (the ditch, canal, river, pond and reservoir) and other treatment projects to absorb and clean non-point source pollutants, so as to ensure the safety of rural ecological environment; renovation of rural ecological corridors and habitats to better protect the biodiversity in rural areas. To summarize, rural ecological spatial restructuring aims to provide clean production space for rural industrial development, and healthy and beautiful living space for rural residents, and restructure rural ecological space with green mountains and clean water, by carrying out land consolidation and rural ecosystem protection.

10.4 Prospects

10.4.1 Reshaping Value Orientation

Traditional land consolidation measures mainly focused on supplementing cultivated land, improving land quality, and increasing transactional construction land for the purpose of the maximization of economic benefits. However, from the perspective of rural vitalization, the value orientation of land consolidation must be reshaped based on a scientific understanding of urban-rural relations and rural territorial functions. Firstly, urban and rural areas constitute an integrated mutually beneficial system, in which the city's development is inseparable from the countryside's support because the newly added population, land, and agricultural products all come from the countryside. The development of the countryside is also inseparable from the leading influential force of the city because the city's capital, technology, and management system are crucial for rural development. Therefore, based on the principle of urban-rural structural coordination and complementary functions, rural restructuring and vitalization should make full use of urban capital to innovate the financing mechanism of land consolidation, absorb advanced technology and talents to participate in the planning of land consolidation, which is an effective way to realize the vitalization of rural resources while providing space for urban development.

Besides, rural territorial system have multiple functions such as economy, society, ecology and culture etc., and ecological value and cultural value are the unique charms that are different from city. Accordingly, guided by the concept of harmonious coexistence with nature, it needs to actively probe the organic integration of land consolidation and ecological construction, and create multi-level and multi-functional pastoral organisms via establishing ecological networks and green infrastructure. At the same time, it is necessary to coordinate "material vitalization" and "humanistic rejuvenation" in the process of land consolidation, on the basis of rational layout of rural development physical space. More attention should be paid to the inheritance of rural traditional culture, and make the countryside become a carrier with local culture and a homesickness.

10.4.2 Innovating Technologies and Modes

Land consolidation is a new demand of socio-economic development, as well as the inevitable result of deepening contradiction between land use status quo, socio-economic development and its land demand. In addition, land consolidation is an important approach for coordinating the relationship between the status quo and the goals of land use, and its contents and tasks will change with socio-economic development. Also, the differentiation of regional land use morphology under different socio-economic backgrounds calls for different modes and contents of land consolidation (Long and Li 2012). In other words, the gradually transformation development and restructuring of the rural is bound to need new modes of rural land use and rural land consolidation. Therefore, more attentions should be paid to the regional features and phase characteristics of rural spatial restructuring, and carrying out innovation research on regional engineering technology, policy and mechanism, and mode of rural land consolidation suiting local conditions, and systematic study on dynamic monitoring, early warning and forecast, and scientific control of rural spatial restructuring, so as to build favorable rural production, living and ecological space, and to facilitate urban-rural integrated development.

At present, China's agriculture is still dominated by traditional large-scale crop cultivation. Affected by price increases in production factors such as labor, agricultural materials, and land leases, the benefits of grain production are relatively low, and it further leads to a series of problems ensues such as the lack of labor force, insufficient agricultural production materials, extensive agricultural production, and difficulties in promulgating agricultural science and technology, which has become the keys and difficulties to coordinating rural man-land interrelations. Agriculture is the foundation of rural development, which bears multiple functions such as agricultural production, landscape aesthetics, recreation, maintenance of biodiversity, and inheritance of local culture. The connection between land resources and agriculture is significant for the sustainable development of the rural economy and the efficient use of resources. Therefore, in the process of rural production, living and ecological spatial restructuring, land consolidation should be combined with developing new agricultural production and operation forms such as organic agriculture, ecological agriculture, energy agriculture, tourism agriculture, cultural agriculture, so as to increase the added value of agricultural production, solve the shortcomings of low output of traditional crop production and effectively absorb the rural surplus labor force.

10.4.3 Coordinating Planning

The rural territorial system is a natural, ecological, social, and economic complex composed of multiple elements, and the spatial heterogeneity and the diversified development goals of rural territory require that rural vitalization planning should be scientifically formulated to make forward-looking arrangement for spatial morphology, infrastructure, industry etc., based on the identification of regional characteristics. Generally speaking, land consolidation planning makes overall layout for the priority and key areas carrying out land consolidation according to the natural and human factors. Currently, China's land consolidation planning has been implemented at four levels including national, provincial, municipal, and county, and rural vitalization planning at national level has just been introduced. In the future, based on a top-level design and an institutional guarantee, it is necessary to coordinate land consolidation and rural vitalization to ensure the positive interactions of land, industry and infrastructure under the unified spatial planning system including socio-economic development planning, urban-rural planning, land use planning, and ecological environmental protection planning. In addition, the village-level land consolidation planning has not yet been included in the overall planning system. Village is the basic unit of rural vitalization, and the lack of village planning could inevitably hinder the coordinated development of rural elements. Therefore, it is equally important to explore the synergy between village-level land consolidation planning and rural vitalization planning.

10.5 Conclusions

Land consolidation is a spatial problem-solving land management instrument that attempts to eliminate certain types of land fragmentation, to enhance land productivity, and to improve rural production and living conditions for the purpose of coordinating urban-rural development, through a process of concentration of plots or rejuvenation of failing and ageing rural settlements and abandoned industrial and mining land, which is usually accompanied by the construction of new roads, irrigation facilities and other auxiliary services.

Rural spatial restructuring is an optimization and adjustment even utterly changing process of rural production, living and ecological spaces accompanied by the reshaping of socio-economic structure in rural areas owing to an integrated result of endogenous development need and exogenous driving forces of rural development, under the pressure of rapid industrialization and urbanization. Rural spatial restructuring is an integrated approach aiming at optimizing urban-rural space organization and promoting coordinated urban-rural development, and its connotations involving three aspects: assembling industrial development, centralizing farmers' living, and the intensive utilization of resources. Rural spatial restructuring mainly includes the restructuring of rural production, living and ecological spaces. Land consolidation can provide space and supporting platform for rural restructuring and vitalization and realizing urban-rural integrated development in China.

Basing on the analysis of the characteristics of rural land consolidation and the effects of industrialization and urbanization on rural production, living and ecological spaces, this chapter puts forward that land consolidation aiming at pushing forward rural spatial restructuring includes three types: agricultural land consolidation, hollowed villages consolidation and industrial and mining land consolidation. The optimization and restructuring of suitable living space with amenity, intensive and high-efficient production space and the ecological space with green mountains and clean water in rural areas will rely on the innovations of regional engineering technology, policy and mechanism, and mode of rural land consolidation.

Comprehensive rural vitalization should be based on the distinctions between urban and rural territorial system and the multi-functional values of rural areas. As a crucial means of promoting rural vitalization, land consolidation is endowed with new connotations such as activating the key elements of rural development, coordinating the vitalization of the physical space and the promotion of the spiritual core, integrating the restructuring of the actual physical space and the rural governance system.

References

Cai Y (2001)中国农村转型与耕地保护机制 (The mechanisms of cropland conservation in Chinese rural transformation). Sci Geogr Sinica 21(1):1–6

Coelho JC, Portela J, Pinto PA (1996) A social approach to land consolidation schemes: a Portuguese case study: the Valença project. Land Use Pol 13(2):129–147

Dean JM, Lovely ME, Wang H (2009) Are foreign investors attracted to weak environmental regulations? Evaluating the evidence from China. JDev Econ 90(1):1–13

Hillman B (2004) The rise of the community in rural China: village politics, cultural identity and religious revival in a Hui hamlet. China J 51:53–73

Huang Q, Li M, Chen Z et al (2011) Land consolidation: an approach for sustainable development in rural China. Ambio 40:93–95

Jiang G, Zhang R, Ma W et al (2017) Cultivated land productivity potential improvement in land consolidation schemes in Shenyang, China: assessment and policy implications. Land Use Pol 68:80–88

Li Y, Hans W, Zheng X et al (2016) Bottom-up initiatives and revival in the face of rural decline: case studies from China and Sweden. J Rural Stud 47:506–513

Liu Y (2018) 新时代城乡融合与乡村振兴 (Research on the urban-rural integration and rural revitalization in the new era in China). Acta Geogr Sinica 73(4):637–650

Long H (2012) 论土地利用转型与乡村转型发展 (Land use transition and rural transformation development). Prog Geogr 31(2):131–138

Long H (2014) Land consolidation: an indispensable way of spatial restructuring in rural China. J Geogr Sci 24(2):211–225

Long H, Ge D, Zhang Y et al (2018) Changing man-land interrelations in China's farming area under urbanization and its implications for food security. J Environ Manage 209:440–451

Long H, Li T (2012) The coupling characteristics and mechanism of farmland and rural housing land transition in China. J Geogr Sci 22:548–562

Long H, Li Y, Liu Y et al (2012) Accelerated restructuring in rural China fueled by 'increasing vs. decreasing balance' land-use policy for dealing with hollowed villages. Land Use Pol 29(1):11–22

Long H, Liu Y, Li X et al (2010) Building new countryside in China: a geographical perspective. Land Use Pol 27(2):457–470

Long H, Liu Y, Wu X et al (2009) Spatio-temporal dynamic patterns of farmland and rural settlements in Su–Xi–Chang region: implications for building a new countryside in coastal China. Land Use Pol 26:322–333

Long H, Tu S (2018) 土地利用转型与乡村振兴 (Land use transition and rural vitalization). China Land Sci 32(7):1–6

Long H, Woods M (2011) Rural restructuring under globalization in eastern coastal China: what can be learned from Wales? J Rural Community Dev 6(1):70–94

Long H, Zhang Y, Tu S (2019) Rural vitalization in China: a perspective of land consolidation. J Geogr Sci 29(4):517–530

Long H, Zou J, Pykett J et al (2011) Analysis of rural transformation development in China since the turn of the new millennium. Appl Geogr 31:1094–1105

Ministry of Housing and Urban–Rural Development of China (MHURDC) (2010) 中国城乡建设统计年鉴 (2009 年) (China urban and rural construction statistical yearbook (2009)). China Planning Press, Beijing

Yun W (2011) 关于加快土地整治创新的思考 (Thoughts on accelerating land consolidation innovation). J Huazhong Agr Univ (Soc Sci Edn) 31(6):1–5

Yun W, Yu Z (2011) 中国农村土地整治生态景观建设策略 (Ecological landscaping strategy of rural land consolidation in China). TCSAE 27(4):1–6

Chapter 11
Land Use Transitions and Rural Restructuring Based on Land Consolidation

Abstract China has a vast territory, in which the eastern and central regions are the main agricultural areas with vast plain. Although they are both rapidly urbanizing and densely populated areas, significant regional differences exist, such as economy, social, culture, policy, and soil texture. Scientifically evaluating the potential of land consolidation in the process of land use transitions in the eastern and central agricultural areas at a micro-scale (village-scale) from the perspective of land consolidation in hollowed villages, depicting the effect of land consolidation in hollowed villages on rural restructuring, as well as examining the effectiveness of land consolidation in optimizing land use transitions and rural restructuring are of great significance to the optimization of China's rural territorial system and the integrated development of urban and rural areas. Based on the research needs and the foundation of preliminary work, this chapter selects Yangqiao community in Yucheng city of Shandong province and Chicun and Wangcun villages in Dancheng county of Henan province as typical demonstration areas to study the processes and characteristics of land use transitions and rural restructuring in traditional plain agricultural areas from the micro-scale based on the perspective of land consolidation. The results show that these typical villages have experienced a rapid process of rural hollowing, thus present a challenge for catch-up development. Meanwhile, they also show tremendous potential for rural development transformation that would assist in achieving an optimal allocation of urban-rural land resources as well as striking a balance between the demands of food security and urbanization.

11.1 Introduction

In the process of rapid industrialization and urbanization, various contradictions and problems have been exposed, accumulated, and intensified in the rural China under the background of long-term urban-rural dual structure, resulting in the "rural hollowing" (Liu et al. 2011). "Rural hollowing" is the main sign of the contradictions between urban and rural areas and the loss of rural value. The non-agriculturalization of the rural population has caused the phenomenon "people to leave their houses leaving empty houses" (ren zou wu kong), and the phenomenon "people build new houses

H. Long, *Land Use Transitions and Rural Restructuring in China*,
https://doi.org/10.1007/978-981-15-4924-3_11

without demolishing old ones on original homesteads" (jian xin bu chai jiu) (Long et al. 2009), resulting in damage to the value of rural land, lack of space production, destruction of the living environment and obstruction of rural development. Land consolidation, which has proven effective in specific situations in creating better conditions of life in rural areas and improving the sustainable use of resources and public facilities, as well as slowing rural depopulation (Vitikainen 2004; Miranda et al. 2006; Pasakarnis et al. 2013), has been employed as a particular kind of rural development instrument (Miranda et al. 2006). Therefore, there is an urgent need to activate the land elements with the help of the land consolidation as the platform, so as to realize the recombination of rural development elements, structural reorganization, and functional optimization. Employing engineering technology and management of land use transitions, land consolidation directly affects the reconstruction of rural space and then reconstructs the multidimensional rural production, living, and ecological spaces. Meanwhile, land consolidation stimulates the vitality of land elements to drive the optimization and reorganization of the core elements of human-land-industry in rural development, further promoting the structural optimization of rural development and the improvement of rural overall functions.

China has a vast territory, in which the eastern region and central region are the main agricultural areas with vast plain. Although they are both rapidly urbanizing and densely populated areas, significant regional differences exist, such as economy, social, culture, policy, and soil texture. Scientifically evaluating the potential of land consolidation in the process of land use transitions in the eastern and central agricultural areas at a micro-scale (village-scale) from the perspective of land consolidation in hollowing villages, depicting the effect of land consolidation in hollowing villages on rural restructuring, as well as examining the effectiveness of land consolidation in optimizing land use transitions and rural restructuring are of great significance to the optimization of China's rural territorial system and the integrated development of urban and rural areas. Based on the research needs and the foundation of preliminary work, Yangqiao community in Yucheng city of Shandong province and Chicun and Wangcun villages in Dancheng county of Henan province were selected as typical demonstration areas to study the processes and characteristics of land use transitions and rural restructuring in traditional plain agricultural areas (TPAA) of the eastern coastal and central agricultural areas of China from a micro-scale and the perspective of land consolidation.

11.2 Land Use Transitions and Rural Restructuring in Eastern Coastal China

11.2.1 Study Area

The typical research area was selected in Shandong province, an eastern coastal province, to carry out the typical village study of land use transitions and rural restructuring from the perspective of land consolidation according to the differential characteristics of the evolution of hollowing villages in plain agricultural areas and their comprehensive consolidation planning. Taking into account the natural conditions, the degree of hollowing, the potential for agricultural land increase, farmers' willingness on land consolidation, the support of local governments and the original work basis, as well as the principle of typicality and feasibility, Yangqiao community, a village in Lun town, Yucheng city of Shandong province, was selected as the study area based on field surveys of hollowing villages in Heze, Linyi, Qingzhou, and other counties and cities in Shandong province.

Yucheng city, located in the northwest of Shandong province, currently has 10 towns and 1 sub-district office, with a total of 997 natural villages. With animal husbandry, melon and vegetable, grain and cotton as the leading agricultural industries, this city has 26 leading agricultural enterprises above designated size. Lun town, one of the five central towns in Yucheng city, is located 18 km south of the urban areas of Yucheng city and 40 km from Jinan. It covers an area of 120 km^2, including 89,000 mu (around 5933 ha) of arable land. In 2009, the original 83 administrative villages were merged into 31 communities (villages) with a total population of 42,000. National Road 308, Jihan railway and the main canal leading to the Yellow River pass through the town and bring convenient traffic conditions to Lun town. In recent years, Lun town has experienced rapid economic development. In 2014, the tax revenue was 74.99 million RMB¥.

Since 2012, Lun town has started the construction of four communities: Lunzhen community, Chezhan community, Boji Wang community and Yangqiao community. Among them, the Yangqiao community is located in the southwest of Lun Town and involves the relocation of 11 villages including Yuanying, Juntun, Dingzhuang, Shihu Song, Wasun, Hebei Zhang, Yangqiao, Hekou Zhao, Zhuwang, Qinzhuang, and Hekou Li, where 4536 people are planned to be resettled. The first-phase project of Yangqiaoc community plans to resettle 5 villages including Zhu Wang, Hekou Zhao, Hekou Li, Yangqiao, and Qinzhuang, involving 489 households and 1,500 people. At the end of 2017, Hekou Zhao, Zhuwang, Yangqiao, and Qinzhuang have all been relocated.

11.2.2 Land Use Transitions and Land Consolidation Potential

The demonstration area for comprehensive land consolidation in hollowing villages in the plain agricultural area established by the Institute of Geographic Sciences and Natural Resources Research (IGSNRR), Chinese Academy of Sciences, is located in Yangqiao community, a typical village with plain agricultural production as the core industry in the eastern TPAA. The rural land consolidation (RLC) potential of the Yangqiao community was measured to provide the background resource support for rural restructuring from the perspective of land consolidation. Based on remote sensing images of villages and images information taken by Unmanned Aerial Vehicle

Table 11.1 Utilization of village homestead involved in the phase I Project of Yangqiao community (%) (Adapted from Long et al. 2018)

Category	In use		Vacant		Abandoned	
	% of total homesteads number	% of total homesteads area	% of total homesteads number	% of total homesteads area	% of total homesteads number	% of total homesteads area
Yanqiao	64.98	66.71	23.63	21.80	11.39	11.49
Zhu Wang	67.44	67.35	4.65	5.26	27.91	27.37
Hekou Zhao	73.85	75.25	20.77	19.94	5.38	4.82
Hekou Li	72.73	76.43	24.24	21.91	3.03	1.66
Qinzhuang	74.14	74.51	21.84	22.38	4.02	3.11
Mean	70.28	71.55	21.52	20.63	8.20	7.82

(UAV), field investigations were conducted in the five villages involved in the resettlement of the first phase of the comprehensive consolidation of hollowing villages in the Yangqiao community, including Zhu Wang, Hekou Zhao, Hekou Li, Yangqiao, and Qinzhuang. Through the homestead survey, the information such as the status, area, registered population, and the resident population of homesteads were obtained, and land-use status data such as land-use efficiency, hollowing degree, and homestead use patterns in typical villages were also got (Table 11.1). The land consolidation potential of typical villages was calculated based on the assessment technology of hollowing village consolidation potential.

According to the survey, Yangqiao community is a typical hollowing village in TPAA, with a low utilization rate of part homesteads. In 2015, there were 248 homesteads in the village, among which "one family with multiple homesteads" accounted for 70% of the total number of households, and 91 abandoned and vacant homesteads accounting for 35.02% of the total number of homesteads as well as 33.29% of the total area. In addition, 32.56% of the homesteads in Zhuwang were abandoned or vacant. Furthermore, the area of vacant and abandoned homesteads in the five villages amounts 32.8 mu and 24.2 mu, respectively; and the number and area of abandoned and vacant homesteads in five villages accounted for 29.72% and 28.45% of the total number and area of homesteads, respectively (Tables 11.1 and 11.2).

Based on the visual interpretation of high-resolution remote sensing images, household surveys, and field investigation, maps of the current status of land use in each administrative village of Yangqiao community were drawn, and the area of each type was classified and summarized by using the statistical table of village land use classification. Thereafter, according to the survey of the current situation of the use of each homestead in villages and consolidation scenarios of different land use ways, the near- and long-term RLC potential of hollowing villages in Yangqiao community were calculated (Table 11.3).

It is found that these hollowing villages have great potential for RLC, and the potential for consolidation of abandoned/vacant homesteads, vacant land, threshing grounds, and forested land has reached 100%.

Table 11.2 Status of land use in Yangqiao community (Adapted from Long et al. 2018)

Land-use code		Land-use type	Area (mu)	Land-use code		Land-use type	Area (mu)
01		**Farmland**	1157.7	Includes	102	Avenue	11.2
03		**Forested land**	152.8		103	Street and alley	23
06		**Industrial and mining land**	4.2		104	Rural road	37
07		**Housing land**	164.88	**11**		**Waterbody**	59.3
Includes	071	In use homestead	107.9	Includes	114	Pit-pond	31.9
	072	Vacant homestead	32.8		117	Canal and ditch	27.4
	073	Abandoned homestead	24.2	**12**		**Other land**	68.8
08		**Public administration and service land**	1.2	Includes	121	Vacant land	27.4
10		**Transport facilities land**	86		122	Agricultural facility land	53.8
Includes	101	Railway	14.7	**Total**			1953.7

Table 11.3 Calculation standard of land consolidation potential of the hollowing villages in Yangqiao community (Adapted from Long et al. 2018)

Types		Consolidation rate (%)	
		Near-term	Long-term
Residential land	Abandoned homestead	100	100
	Vacant homestead	60	100
	In use homestead	/	20
	Street and alley	30	30
	Vacant land	100	100
Non-residential land	Pit-pond	60	60
	Threshing ground	70	100
	Forested land	100	100

11.2.3 Rural Restructuring Based on Land Consolidation

Yangqiao villages is a typical village with grain production as the core in the plain agricultural area. The village is far away from the core development area of the town, and there are no large-scale industrial and mining enterprises around, which means that agricultural characteristics in the village is obvious. It is a common village among many villages in TPAA.

Ordinary villages are selected as the case villages for hollowing village consolidation in plain agricultural areas because they can reflect the common characteristics of this region. Based on the changes of rural man-land interrelations that Yangqiao village reflected in the process of urban-rural transformation development, as well as the evolution of rural space and the changes in farmers' production and living during the process of hollowing village consolidation, the problems exposed in the process of transformation development and restructuring of hollowing villages in plain agricultural area can be better explained, and the better experience presented in the process of hollowing village land consolidation can be summarized. In 2013, Yangqiao village implemented the project of "linking up increased urban construction land with decreased rural construction land". At the end of 2016, the old village was completely demolished and reclaimed into farmland.

11.2.3.1 Socio-Economic Transformation Development Process of Yangqiao Village

(1) **Phase 1: Dominant stage of traditional agriculture (1980–2000)**

The traditional agricultural planting in Yangqiao village is dominated by the traditional bulk grain cultivation such as winter wheat and summer corn, supplemented by the cultivation of cash crops such as peanuts, soybeans, sesame, and cotton. In the 1980s, with the implementation of the national rural land management and contract policy, farmers' enthusiasm was greatly mobilized. Thereafter, grain output has increased significantly, with wheat and corn yielding about 900 kg per mu. However, due to the lack of income sources for farmers except for agricultural cultivation, economic development entered a stagnant stage during which farmers can only be guaranteed basic food and clothing. In 1990, the per capita net income of farmers was about 1,200 RMB¥. After 1990, along with the advancement of urbanization throughout the country, the phenomenon of migrant workers in developed areas such as Beijing and Guangzhou began to appear sporadically. Migrant workers mainly engaged in construction, transportation and other non-agricultural industries, and their income level increased slowly. Affected by the stage of macro-economic development, infrastructures and public services in rural areas were extremely scarce at that time.

(2) **Phase 2: Multiple occupations of farmers (2000–2015)**

Since 2000, with the rapid improvement of agricultural mechanization and the improvement of farmland water conservancy facilities, grain output has increased significantly again. However, due to the rise in the cost of agricultural production factors such as agricultural capital prices, land rents and labor costs, grain cultivation has gradually shown a low-income trend. In addition, it was difficult to reverse the situation of atomized and fragmented agricultural land use at that time, which caused the coming of the period of bottleneck of the village economic development in TPAA. The influence of the social development process, the progress of agricultural production technology, the low efficiency of grain planting and other factors jointly drive the change of agricultural production mode. After 2000, a large number of non-agricultural employment populations appeared in the village, which means that the phenomenon of multi-professionalization of farmers was widespread, the income source of farmers was diversified, and their income level increased significantly. At the same time, the village development faced noticeable problems such as the serious loss of young and middle-aged laborers, the weakening of the mainstream of rural development, and the deficiency of village development potential. From 1990 to 2015 (Fig. 11.1), the proportion of non-agricultural employment in the village's labor force rose from 10 to 76%.

The reduction of the agricultural labor force is an inevitable trend accompanied by economic and social development and technological progress. However, the weakening of the mainstream of rural development, which because of the older agricultural labor force and multi-professionalization of farmers, has resulted in some problems such as the shortage of agricultural means of production and manpower inputs, and the difficulty of the popularization of agricultural science and technology. This has also affected the improvement of agricultural productivity and the development of modern agriculture. The existence and intertwining of factors such as population loss, land waste, industry hollowness, lack of public infrastructure and socialized

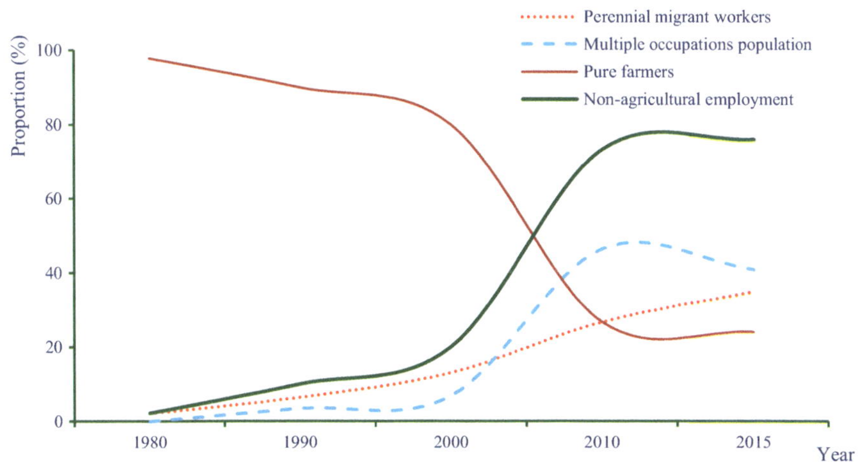

Fig. 11.1 Employment situation of Yangqiao village from 1980 to 2015 (Adapted from Long et al. 2018). *Note* The data in the figure are estimated by village officials. Perennial migrant workers refer to farmers who work outside the home for more than 10 months each year; pure farmers refer to the farmers who take planting and breeding as a single source of income; multiple occupations population refer to farmers who are engaged in both agricultural production and non-agricultural industries, except for "perennial migrant workers" and "pure farmers"; non-agricultural employment is equal to the total number of permanent migrant workers and multiple occupations population farmers

service industries have become the main obstacles to the development of traditional farming areas.

(3) **Phase 3: Hollowing village consolidation promotes rural spatial restructuring (since 2015)**

In order to improve the living environment in rural areas, Dezhou municipal government, Yucheng municipal government and Lun town government in 2013 made joint efforts to promote the comprehensive consolidation of hollowing villages based on the project of "linking up increased urban construction land with decreased rural construction land". The residential buildings that have been built in Yangqiao community consisted of residential buildings with "4 bedrooms plus one living room" and "3 bedrooms plus one living room" (total house area has three sizes, i.e., 80, 120 and 132 m^2), as well as business-living buildings along the street with "2 bedrooms plus one living room" (total house area has two sizes, i.e., 210 and 320 m^2). As of November 2016, 26 residential buildings with 352 units have been completed, supporting with community facilities including service center, kindergartens, leisure squares, medical offices and other public facilities. At present, 146 rural households have relocated from the original Yangqiao village with the newly-built residential area of 46,700 m^2. After the demolition, the original rural housing land of the whole village will be reclaimed as farmland, which is still in progress.

The new community has simultaneously realized the construction of supporting facilities such as water, electricity, gas, ground hardening, environmental sanitation, and night lighting. And a 900 m^2 community convenience service center has been built with a party member activity center, a multimedia activity room, a cultural room, a kindergarten, a community health service stations, and elderly apartments. In addition, one standardized village-level health service station (room) and one culture and fitness square have been reconstructed and/or expanded. Moreover, in order to solve the inconvenience of the elderly going up and down the stairs, 20 suits that suitable for the elderly have been built in the community. Furthermore, demolishing the old village and building the new community has contributed to a modest concentration of living and activity space, promoting the restructuring of the village space. Meanwhile, the production and living standards of farmers have been significantly improved.

11.2.3.2 Hollowing Process of Yangqiao Village

After the implementation of the household contract responsibility system in 1980, the demand for rural residential land in Yangqiao village was strong, which led to a large amount of farmland was encroached. This partly because the family structure has changed from a traditional big family to a small family, and farmers preferentially chosen to apply for a new homestead and build a new house in different places in order to maximize their benefits. Although the cost of building a new house elsewhere include both the cost of building the house and the potential cost of farmers applying for a new homestead, it is lower than that of demolishing an old house and building a new house on the original homestead. At that time, the houses in the old village were mainly rebuilt in situ between 1930s and 1940s, and the structure was still dominated by adobe structure.

From 1995 to 2015, the ongoing expansion of the village boundary and the hollowing of the interior coexisted. Significant changes have taken place in the production methods and employment patterns of peasants, and the degree of non-agriculturalization and multiple occupations was increasing. The production and operation mode of farmers remained at the stage of independent operation, but the promotion of socialized services for agricultural production has greatly improved the productivity of farmers.

The expansion of the village boundary and the internal infill development coexisted, which promoted the hollowing of the old village. From 1995 to 2015, the east side of Yujing Road was mainly renovated internally, but the newer houses were mainly brick-concrete houses that built after the 1980s. Due to the narrow roads, low-lying terrain and poor drainage in old villages, the adobe houses built in 1960, which was the main source of hollowing of the homestead in old village, were rarely renewed (Fig. 11.2). In 2015, "one family with more houses" accounted for 70% of the total number of households.

Throughout 1980–2015, before the relocation of Yangqiao village, the number of homesteads increased from 149 to 248, and the total area increased from 61,200 to 108,800 m^2. The 1980s witnessed the continuous expansion of the area of rural homestead caused by the explosive growth of newly-built housing. And 1990–2000 was the period when the size and intensity of the expansion of the homestead were greatest. Although people's enthusiasm for building houses gradually receded after 2000, the area of homesteads has increased by 37,600 m^2 during two decades from

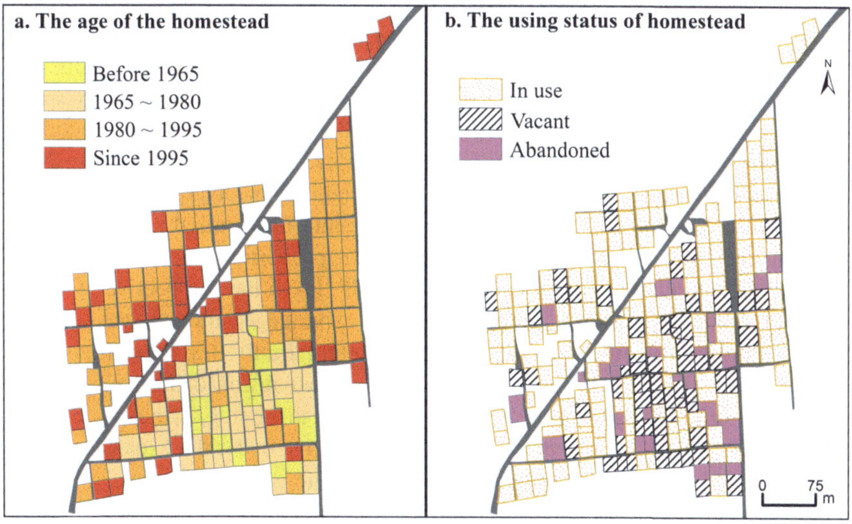

Fig. 11.2 Construction time and use status of houses in Yangqiao village in 2015 (Adapted from Long et al. 2018)

1980 to 2000, accounting for 77.4% of the total incremental area from 1980 to 2015. The expansion of the homestead was mainly achieved through the occupation of cultivated land, vacant land, and forested land. From 1980 to 2015, the per capita housing site area increased from 137.62 to 211.54 m^2, during which 1990–2000 witnessed the fastest growth of the per capita residential land use area, with an increase of 47.7 m^2 in the decade.

11.2.3.3 Spatial Restructuring in Yangqiao Village

Relying on the project of "linking up increased urban construction land with decreased rural construction land", Yangqiao village started the comprehensive consolidation of hollowing villages under the lead of the local government since 2013. In the process of Yangqiao village consolidation, the assessment technology, planning and technology system and rapid soil forming technology created by the research team of IGSNRR for the consolidation of hollowing villages in plain farming areas have been applied and practiced, and have obtained good demonstration results. By August 2017, the consolidation of Yangqiao community had achieved initial success. Firstly, the village living space was optimized. At the same time, the construction of supporting facilities such as water, electricity, gas, ground hardening, environmental sanitation, and night lighting was completed, so that the living environment of farmers was significantly improved. In addition, the overall relocation of farmers and the vacation of the original living space have increased the agricultural production space. Furthermore, the old village was demolished for reclamation to increase the area of farmland, and this area has become a demonstration base for high-standard farmland construction technology.

The internal structure of production, living and ecological spaces of Yangqiao village continuously evolved during the hollowing process and its subsequent consolidation process. From 1980 to 2016, the evolution and restructuring of production, living and ecological spatial morphology of Yangqiao village experienced two stages (Table 11.4). From 1980 to 2015, the share among the living, production, and ecological spaces changed from 9.12:73.36:17.52 to 12.85:73.53:13.62. However, in 2016, with the transformation of the old residential land, vacant land, street and alley, scattered forested land and pit-pond in the village into farmland, the area for living space and ecological space decreased, as well as the proportion between production, living and ecological land became 8.80:86.43:4.77. From the perspective of spatial morphology, the expansion of living space has experienced three stages: enclave expansion toward the outer edge of the village, internal infill development, and the agglomeration of peripheral areas. Along with the expansion of living space, the morphology of production space has also undergone three stages: the circle development in the old village, the erosion of the village edge under the expansion of the homestead, and the centralization and continuity of the new community.

In April 1981, the State Council issued the "Urgent Notice on Stopping Building Rural Housing on Encroached Farmland", after which a series of farmland protection policies and measures were successively issued. For example, the "Regulations on

Table 11.4 The evolution of living, production, ecological spaces in Yangqiao village (Adapted from Long et al. 2018)

Year	Share of living space (%)	Share of production space (%)	Share of ecological space (%)
1980	9.12	73.36	17.52
1990	9.99	73.61	16.40
2000	12.00	73.56	14.43
2010	12.69	73.66	13.66
2015	12.85	73.53	13.62
2016	8.80	86.43	4.77

Note Living space includes residential land, public management and service land, and transportation land; production space includes farmland, garden land, commercial land, industrial and mining storage land, and agricultural facility land in other land; ecological space includes forested land, grassland, waterbody, and vacant land in other land

the Protection of Basic Farmland" issued in 1994, the "Notice on Further Strengthening Land Management and Effectively Protecting Farmland" issued in 1997, and in 1999 the "Regulations on Implementing the Land Management Law of the People's Republic of China" has proposed the "system of land use control and the balance between the occupation and compensation of farmland". The promulgation of farmland protection policies has restricted the behavior of farmers occupying farmland to build houses. As a typical representative of the spatial evolution of rural villages in plain areas, the hollowing process of Yangqiao village and its consolidation are the reflections of changes in social and economic factors in rural areas under the background of urbanization and industrialization. This is the result of a combination of various factors. It is not only affected by the exogenous factors such as the national policies on homestead use and management, "linking up increased urban construction land with decreased rural construction land" policy, local government guidance, and transportation location, but also by the endogenous factors such as the topographic conditions, farmers' behavior, the economic basis, and the industrial development.

At present, while Yangqiao village is restructuring its living space, the village industrial structure that dominated by small-scale peasant economy has not been changed significantly. The problem of decentralized and fragmented production space also needs to be further improved. Therefore, it is necessary to properly handle the relationship among population, land, and industry for sustainability in rural territorial system. In this process, the rural restructuring should take improving economic efficiency, maintaining social equity, protecting the ecological environment, and achieving sustainable use of resources as its value and goals. And then to coordinate the industrial cultivation and the construction of human settlements, the equalization of services, and the improvement of the capabilities of social development mainstream, as well as the ecological conservation and cultural inheritance, so as to promote the comprehensive development of the "natural-ecological-economic-social" system and the comprehensive enhancement of the "production-living-ecological-cultural" functions in rural areas (Long and Tu 2017, 2018).

In the context of urban-rural integration, how to effectively improve the living conditions of farmers in plain farming areas, the overall appearance of villages, the state of rural habitats, and the rural space system, have put forward higher requirements for the development of the technology systems of hollowed village consolidation. The experience of hollowed village consolidation in Yangqiao village is of great significance to the scientific implementation of village agglomeration and the promotion of urban-rural transformation development. Indeed, the hollowing degree of different types of villages in plain farming areas is different, and the internal and external consolidation environments of the villages are significantly different. This will inevitably lead to the diversification of the hollowed village consolidation mode and the diversification of the participants. The consolidation of hollowed villages in plain farming areas, therefore, must be based on comprehensive scientific assessments, reasonable village consolidation programs, and scientific measures tailored to local conditions.

11.3 Land Use Transitions and Rural Restructuring in Central China

11.3.1 Study Area and Data

11.3.1.1 Overview of Dancheng County

Dancheng county, located in the hinterland of the Huang-Huai-Hai Plain Region, has been chosen for the case study (Fig. 11.3). The Huang-Huai-Hai Plain Region is the representative of China's farming culture and one of the most important agricultural areas of China. Currently, both the scale and the number of land consolidation projects in China are mainly concentrated in the Huang-Huai-Hai Plain Region. Further, a latest evaluation showed that the Huang-Huai-Hai Plain Region has medium hollowed village degree and big potential of rural land consolidation, and thus it is of importance and necessity to carry out rural land consolidation (RLC) practices in this region (Liu et al. 2013). Dancheng, under the governance of Zhoukou city, Henan province, is located in the core area of the Huang-Huai-Hai Plain Region, and the problems of rural development and potential of RLC are typical in this region and China's central agricultural areas. Thus, when concerning RLC in traditional farming areas in central areas, Dancheng is regionally and even nationally representative. Dancheng has eight towns, 11 townships, covering an area of 1471 km^2. In 2009, the county had a population of 1.32 million, with per capita farmland of 0.083 ha. The

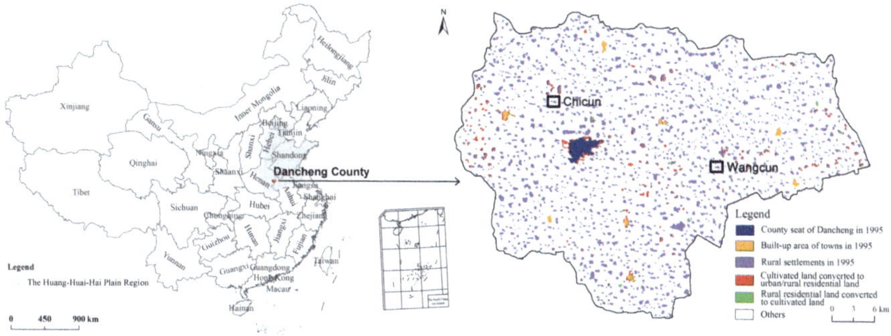

Fig. 11.3 Location of study areas and spatial pattern of rural residential land change in Dancheng county (1995–2000) (Li et al. 2014)

main agricultural products of Dancheng include wheat, corn, soybeans, cotton, pigs, goats, and cattle, etc. With per capita grain output up to 605 kg, Dancheng is one of the national commercial grain base counties. Incommensurate with its huge contributions to national grain security, socio-economic development level in Dancheng is rather backward, which reflects the basic characteristics of China's central TPAA (Li et al. 2011). In 2009, the county's per capita net income of farmers was 3915 RMB¥, 81.44% and 75.97% the provincial and the national average, respectively. The county's per capita fiscal revenue was 191 RMB¥, 16.88% and 3.71% of the provincial and the national average, respectively, and its GDP per capita was 9781 RMB¥, 47.49% and 38.24% of the provincial and the national average, respectively. More than 300 thousand rural population migrated out of Dancheng year round and thus rural hollowing in this county accelerated in the mid-1990s. A series of RLC projects have been carried out since then, and a large number of villages experienced RLC successfully,[1] which may have reference values for RLC Ricinin future.

11.3.1.2 Data Source

We collected socio-economic statistic data and village-level census data from local governments and village-level survey data from field investigations to analyze the hollowing state and the potential of RLC in this county. Moreover, vector data of land use/cover change in Dancheng in 1995 and 2000 were obtained through detection analysis of historical Landsat TM images from the Institute of Geographic Sciences and Natural Resources Research, Chinese Academy of Sciences. In order to probe deeply into the process, effects, experiences, adaptabilities and limitations of

---

[1]When concerning the definition of "success" of RLC, in our opinion, a successful RLC project should have the following characteristics: (1) construction scheme and investment plan are scientific and reasonable, (2) the process of implementation is approximately as scheduled, (3) the results are in line with expectations, (4) the later utilization of land resources are high efficient, and (5) most importantly, the majority of participants and stakeholders are satisfied with the outcomes.

typical RLC initiatives, further field investigations had been carried out twice in the selected villages (Chicun and Wangcun) since May 2010. We conducted semi-structured group interviews with villagers, village cadres and local government officers to obtain general information about RLC in the two villages. Finally, 24 and 22 valid questionnaires in Chicun and Wangcun were obtained, respectively. With these questionnaires, we can get more information on villagers' participation and effects of RLC on village system.

11.3.2 *Status Quo of Rural Hollowing*

11.3.2.1 Basic Characteristics of Rural Hollowing

As a typical county in the Huang-Huai-Hai Plain, Dancheng experienced a rapid process of rural hollowing characterized as the hollowing of rural industries, infrastructure, and residential population and settlements.

(1) **The depression of rural industries**

TPAA in inland China have rather weak industrial base and usually fall into the trap of lacking of skilled workers, entrepreneurship and investment, and thus local industries has been seriously lagging behind. In the selected 63 villages, only 16 villages have township/village enterprises or private/individual enterprises. Most of these 49 enterprises were in the field of agricultural products processing and building materials production. They were relative small and with low economic benefits and have been gradually marginalized for lack of advantages in comparison with the scale enterprises of same trade in cities. Moreover, though the central government of China has carried out a series of agricultural support policies, agricultural production is still lack of economic benefit when compares to off-farm work. From 2001 to 2008, the total sown area in Dancheng decreased by 9.45%, and the multiple-cropping index (dividing the crop area by the area of farmland) reduced from 202 to 183%.

(2) **The shortage of rural infrastructures**

Under the urban–rural dual structure system, urban-biased development mode caused a serious shortage of rural public goods. Taking infrastructure construction for example, the average per capita investment of infrastructure construction in villages of China was less than 100 RMB¥ in 2008, and was only 4.96%, 11.23%, 22.21% and 36.05% that of urban, county town, town and townships, respectively. Combined with the scattered, small and chaotic distribution of settlements, they have a serious negative impact on the accessibility of rural infrastructures. According to the census data of 475 administrative villages, irrigation is mainly depended on pumping groundwater; the road in 28.84% of these villages is still stone road or dirt road; only 6.28% of the villages have running water; 98.52% of the villages did not realize centralized garbage treatment; and nearly all the villages have no cultural and sports infrastructures. Inadequate supply of rural infrastructures largely deteriorated the rural living conditions and contributed a lot to the excessive outmigration of rural labors.

(3) **The massive out-migration of rural labors**

The total number of outmigration workers in Dancheng is more than 300 thousand in recent years, accounting for one third of the agricultural registered population. According to our investigation in 63 villages, 37.95% of rural labor forces year-round work outside Dancheng, accounting for 21.69% of the total rural population. In the 157 selected households, only 26 households have no family member work outside the county, and 43.39% of the workforces are migrant workers, accounting for 30.35% of the total population. While the average age of left-behind labors is almost 50 years old and 87.10% of them have experienced only junior high school education or below, migrant workers have the lowest average age and obtained the highest level of education among rural Dancheng. Obviously, the capability of rural actors has been seriously weakened due to the massive out-migration and thus largely restrained the rural development.

(4) **The hollowing of rural settlements**

The area of rural residential land in Dancheng accounted for 13.50% of the total land area in 2000, and this ratio was 1.19 times that of the average of Henan Province (11.39%). 1806 rural settlements patches were identified in the vector map of Dancheng in 2000 (Fig. 11.3), and the average patch density was 1.21 per square kilometers. From 1995 to 2000, we found an increase of 3.92% in rural residential land at the expense of 1297 ha farmland, and thus led to a 1.04% decrement of farmland. Based on the village-level survey data collected in 63 villages, we further found that 29,707 households had 38,413 homesteads (houses), with an average of 1.29 homesteads per household. Among these homesteads, the abandoned ones unsuitable for living accounted for 5.28% of the total; the idle ones used for less than one month in two consecutive years accounted for 7.64% of the total. Among the surveyed households, the ones that had two or more homesteads were up to 6906, accounting for 23.25% of the total. According to household survey data collected from 157 households, 33.33% of the rural households had two or more homesteads. For lack of unified planning and effective enforcement, villagers usually built their houses randomly or even on farmland (Fig. 11.4). The majority of the investigated villages have reached the stage of peripheral sprawl and internal hollowing.

11.3.2.2 The Potential of Increasing Farmland via RLC and Local Response

The process of rural hollowing led to the wasteful use of farmland resources, thus threatens the farmland protection and food security, and imposed huge obstacles on the optimization of land use efficiency and coordinated urban–rural development. However, from another point of view, hollowed villages in Dancheng show

Fig. 11.4 Newly built houses on high-quality farmland in Zhulou village, Dancheng county (Li et al. 2014)

tremendous potential for transformation development that would assist in achieving an optimal allocation of rural land resources, releasing and protecting agricultural land and helping to strike a balance between the demands of food security and urbanization. We calculate the increase potential of farmland via RLC based on Liu et al. (2011). According to our investigation in selected 63 villages, the proportion of the abandoned dwellings and vacant dwellings in residential land area in each towns/townships varies from 2.9% (Qiuqu) to 7.0% (Shicao) and from 4.7% (Huji) to 14.7% (Zhangwanji), respectively (Table 11.5). All of the abandoned dwellings and vacant dwellings and 60% of the ponds and 40% of the forested land could be consolidated to farmland averagely. Therefore, the potential of RLC in Dancheng is about 5649 ha, which was 37 times of the average annual quota of newly added construction land. Thus, Dancheng gave an effective response to rural hollowing, and a RLC project known as the 'consolidation of hollowed village land, brick kilns and abandoned lands of industry and mining' (tudi sanxiang zhengzhi) has been carried out since 2006. Until now, over 100 villages have implemented this project and more than 1200 ha residential land has been converted to farmland.

Table 11.5 The hollowing status quo and main sources of potential of increasing farmland by carrying out RLC in Dancheng (Li et al. 2014)

Town/township	Area of main land use types (ha)			Level of hollowing (%)		The main sources of potential (ha)				
	Residential land	Ponds	Forested land	Proportion of abandoned dwellings	Proportion of vacant dwellings	Abandoned residential land	Vacant residential land	Ponds	Forested land	Total
Wutai	1074.5	153.6	248.0	5.8	11.2	62.2	120.3	92.2	99.2	373.8
Nanfeng	1080.6	88.9	132.7	4.4	7.4	47.3	79.5	53.3	53.1	233.2
Baima	1469.8	81.1	233.6	4.4	8.2	65.0	121.1	48.7	93.4	328.2
Ningping	1208.5	118.3	160.8	4.5	7.3	54.9	87.7	71.0	64.3	277.9
Yilu	1157.5	84.8	202.7	4.1	7.6	47.1	88.0	50.9	81.1	267.0
Qiandian	1312.8	138.9	244.1	5.5	10.6	72.7	139.5	83.4	97.6	393.2
Jizhong	1217.6	226.6	147.3	5.8	10.3	70.5	125.0	136.0	58.9	390.4
Shicao	932.8	109.2	172.5	7.0	11.0	65.3	102.8	65.5	69.0	302.6
Chengjiao	743.4	74.3	116.8	4.0	5.2	29.6	38.7	44.6	46.7	159.6
Hutougang	1208.1	160.1	340.4	3.7	7.1	45.1	85.8	96.1	136.1	363.1
Jishui	1222.2	107.3	269.0	6.8	6.1	82.8	74.8	64.4	107.6	329.6
Zhangwanji	1229.8	45.7	182.8	4.5	14.7	55.2	181.1	27.4	73.1	336.8
Dingcun	1072.9	92.3	184.3	3.8	13.4	40.9	143.8	55.4	73.7	313.8
Shuanglou	812.4	100.5	104.3	3.7	6.1	30.1	49.7	60.3	41.7	181.8
Qiuqu	1128.3	76.0	257.4	2.9	9.5	32.6	107.2	45.6	103.0	288.3
Dongfeng	746.8	58.2	180.8	5.3	10.7	39.2	79.6	34.9	72.3	226.1
Baji	1100.0	137.2	144.6	5.6	12.0	62.0	131.7	82.3	57.8	333.9
Lilou	1199.5	176.0	221.9	4.2	9.7	50.9	116.7	105.6	88.7	361.9
Huji	954.4	105.0	127.9	3.0	4.7	29.1	44.4	63.0	51.1	187.7
Sum/average	20871.6	2134.2	3671.6	4.7	9.1	982.5	1917.5	1280.5	1468.6	5649.1

11.3.3 Rural Residential Land Consolidation and Allocation

11.3.3.1 RLC in Chicun: To Improve the Living Conditions

(1) **Problematization**

Chicun is an administrative village under the jurisdiction of Huji Town, with a population of 1220 in 2009, and its economy is dominated by agricultural production. During the new round of reform and opening up in 1990s, villagers started to seek off-farm work outside the county. The number of out-migrants increased from about 200 to 430 during 1992–2006, and the income per capita rose from 460 RMB¥ to 3850 RMB¥. As such, villagers had more intention to improve housing conditions. However, without scientific and longstanding guidance and restriction, villagers used to build their houses randomly, leading to the disordered distribution of new houses, and emptiness and discard of old houses in the village center. Severe hollowing displayed as 'empty inside but extended outside', increasingly bad living conditions, and thus more than five ha farmland had been occupied. In 2006, residential land area of Chicun was 31.67 ha, and per capita residential land area went up to 260 m^2, which was much larger than the maximum level set down by the central government (150 m^2 per capita).

(2) **Strategic decision and planning**

Given the accelerated village hollowing and consequent deterioration of village living conditions, Shi Tian, the village head of Chicun put forward a suggestion of settlement reconstruction and residential land reconsolidation in 1998. However, the villagers lacked of economic strength at that time, and the local government focused on grain production and agricultural structure adjustment and thus they were too busy to formulate correspondent policies to support RLC. Therefore, local government paid little attention to her suggestion at that time. However, Shi Tian still took advantage of all opportunities to clarify her thought of RLC. In October 2005, when the national strategy of building a new countryside was put forward in the Fifth Plenary Session of 16th CPC Central Committee, Shi Tian, who had been elected as new village party branch secretary, put forward her suggestion of RLC again. Her suggestion has been approved by the village party branch (hereafter VPB) and villager committee (hereafter VC) because the hollowing state and residential environment of the village got worse, villagers' wishes and abilities to improve living conditions increased and they might get support from local government.

A conventional two-story house would cost more than 100 thousand RMB¥ at that time, and it was still difficult for the villagers with per capita annual net income of 4000 RMB¥ to build. In order to get approval by all the villagers and take collective reconstruction activities, a comprehensive planning scheme was essential and advantages generated by the reconstruction need to be fully recognized by villagers. Therefore, the VPB and VC embarked on a series of preparatory activities in the early period. In order to make the villagers have an intuitive understanding of RLC, Chicun's VPB and VC organized several times of self-learning. They invited and

Table 11.6 The basic characteristics of villagers' participation in Chicun's RLC (Li et al. 2014)

Questions	Sampled households' answers
How did you participate in RLC planning?	Participated in making important decisions—**5** households; put forward some advices—**3** households; indirectly participated in it—**16** households
Did the planners seek your advice?	Yes—**24** households; no—**0** household
How did you know about the RLC plan?	Completely know it—**18** households; know a lot—**3** households; know a little—**3** households

Source Random sampling investigation of 24 households in Chicun

organized villagers to visit successful villages, which had experienced village reconstruction. This way of visiting and learning not only made villagers know more about village reconstruction and RLC, but also raised their inner desire to launch RLC.

Planning is essential for RLC. Under the guidance of Shi Tian, members of VPB and VC visited each household and collected the opinions and suggestions of majority of villagers on RLC planning. Then they worked out an initial planning scheme concerning how to allocate residential land, construct infrastructure, choose house style, consolidate and allocate the released residential, etc. The initial planning scheme was announced to all the villagers to collect their comments. From 2005 to 2007, Chicun held more than ten times of VPB and VC meetings to seek revise opinions widely. Finally, the planning scheme has been largely improved. After the autumn harvest of 2006, Chicun held the villagers' meeting about RLC. More than 600 villagers participated in secret ballot and 94% of them agreed on and adopted the planning scheme. Our investigation dada shows high levels of participation in Chicun'sRLC. All of the selected households had been asked for advice during the course of planning and the majority of them are familiar with their RLC plan (Table 11.6).

(3) **Settlement reconstruction, residential land consolidation and allocation**

Chicun's RLC started from March 2007. Because of the wide publicity and active mobilization in the early period, villagers had high enthusiasm. Moreover, Township government attached great importance to Chicun's initiatives and designated a deputy township head to coordinate the work. Several professionals selected from land management and rural construction department formed a work group to provide assistant in RLC. The area of residential land was measured. Villagers built their houses according to the plan in which the building type and location were designated. Fund supports from local government were as follows: firstly, Chicun became the hel ping objective of the secretary of CPC Dancheng County Committee in 2008, and the county government dispatched a work group led by Bureau of Land and Resources, who collected more than 600 thousand RMB¥ for land consolidation. Secondly, Chicun became the aid station of the secretary of the Discipline Inspection Commission of Zhoukou City in 2009, and more than 1 million RMB¥ has been collected under the coordination of agricultural bureau, electric power bureau, and land and resources bureau of the county, and thus infrastructures including road, electric power, internet, water supply and drainage were completed timely. Overall, under the combine efforts of various actors, including local government, the VPB and VC of the village, the representatives of the villagers, Party members and villagers and so on, RLC in Chicun was implemented smoothly.

RLC of Chicun achieved initial success in June 2010. The main results of RLC on local village system were embodied in three aspects. Firstly, the living conditions have been significantly improved. The old houses were all dismantled, and more than 230 new houses were built up and put into use. The newly built houses with average cost of 120 thousand RMB¥ per household are nice and orderly. With the funds raised by local government, infrastructures were completed. The lanes are hardened and illuminated, and green plants were planted along the roadside. Moreover, a management mechanism for handling and treatment of garbage was developed. These improvements make clean and tidy village with high-quality residential environment come true. Secondly, the farmland area increased. The new settlement occupies only 8.67 ha, and newly increased farmland amounts to 23 ha. According to their planning, the newly increased farmland would be allocated for high-efficient agriculture such as greenhouse vegetable cultivation. Thirdly, because the RLC has been carried out based on villagers' spontaneous participation and effective organization of the VPB and VC, relationships between the cadres and villagers have been improved significantly, and thus increased the village's social capital that may benefit further development. Villagers in Chicun think highly of their settlement reconstruction. Most of them are satisfied with the improved living conditions (Table 11.7).

Table 11.7 Villagers' perception of the comprehensive effects of RLC in Chicun (Li et al. 2014)

Questions	Sampled households' answers
How about the transportation condition after the RLC?	Obviously improved—**24** households; improved a little—**0** household; no change—**0** household; getting worse—**0** household
How about the residential environment after the RLC?	Obviously improved—**23** households; improved a little—**1** household; no change—**0** household; getting worse—**0** household
How about the living quality of your family after the RLC?	Obviously improved—**23** households; improved a little—**1** household; no change—**0** household; getting worse—**0** household
Are you satisfied with the effects of RLC?	Totally satisfied—**20** households; rather satisfied—**4** households; unsatisfied—**0** household; totally unsatisfied—**0** household
How many people do you think would be satisfied with the effects of RLC?	Less than 50%—**0** household; 50–60%—**0** household; 60–70%—**0** household; 70–80%—**0** household; 80–90%—**1** household; above 90%—**23** households

Source Random sampling investigation of 24 households in Chicun

11.3.3.2 RLC in Wangcun: To Raise Land Use Efficiency and Develop Local Industries

(1) **Problematisation**

Wangcun is an administrative village under the jurisdiction of Dingcun Town, with a population of 1800 in 2009. Similar with the dilemmas of Chicun, the phenomenon of village hollowing got increasingly worse during the "going-out for work" period in 1990s. In 1998, residential land in Wangcun covered 56.67 ha, with 360 m^2 per capita. Wangcun carried out the second round of land contract adjustment in 1998. During the adjustment, a piece of farmland with the area of 5.07 ha had not been contracted for its low-lying terrain where would be waterlogged easily. How to make full use of this piece of land plagued the village cadres and villagers.

(2) **Strategic decision and planning**

Based on various investigations, Wang Jincheng, VPB secretary of Wangcun, held the opinion that villagers could gain little money from grain production, while off-farm work in cities was rather dirty, bitter and tired, and starting own business needs much money. In comparison, establishing village fair would be a rather good way, which need less up-front investments and through which villagers can not only do some business, but also change the old residential land into farmland combining with RLC, and thus solving the contradiction of increasing population versus decreasing farmland. Therefore, Wang put forward the suggestion of allocating unexpected land and establishing village fair after a comparison analysis of different developing strategies at a meeting of VPB, and his suggestion was approved by presented Party members. Then this strategy was confirmed at later meetings of VC. However, to establish a rural market needs scientific and detailed planning. The villagers of Wangcun started to prepare the RLC planning from 1998. They established a planning group composed of 18 individuals elected from members of VPB, VC and representatives of villagers, to be in charge of planning and correlation affairs. Then the planning group not only investigated typical rural market both in and out of the county and learned lessons from their experiences, but also visited each household to seek their ideas and suggestions, and after that, a planning scheme was formed. The planning scheme was revised and perfected at meetings of VPB and VC, and was finally adopted by way of democratic vote in the village meeting. The reconstruction began in 2000. Our investigation dada shows high levels of participation in Wangcun's RLC. The majority of selected households had been asked for advice during the course of planning (Table 11.8).

(3) **Settlement reconstruction, residential land consolidation and allocation**

Wangcun began to build houses with both commercial and residential functions according to RLC plan, which had made detailed regulations on location, area and structure of houses in the low-lying land with an area of 5.07 ha since 2000. A total of 100 new houses had been built until 2002, which met the basic scale of establishing a village fair. The village fair started trial operation at the end of 2002. Owe

Table 11.8 The basic characteristics of villagers' participation in Wangcun's RLC (Li et al. 2014)

Questions	Villagers' answers
How did you participate in RLC planning?	Participated in making important decisions—**8** households; put forward some advices—**7** households; indirectly participated in it—**7** households
Did the planners seek your advice?	Yes—**21** households; no—**1** household
How did you know about the RLC plan?	Completely know it—**8** households; know a lot—**9** households; know a little—**5** households

Source Random sampling investigation of 22 households in Wangcun

to village cadres' efforts, Wangcun was included in Inter-village Commuter Project (cuncuntong gongcheng), then a bus route between the county town and Wangcun was licensed, hence the transport infrastructure in Wangcun has been significantly improved and thus the development of village fair accelerated. Through visiting to various rural markets in typical regions for many times, the villagers of Wangcun found that comprehensive management system which makes the market well-ordered is essential for village fair development. Therefore, a series of institutional arrangement was designed. Village fair management committee has been established, which consisted of village cadres and prestigious and experienced Party members. They were in charge of daily management of the market and provided basic services such as order maintenance and dispute arbitration. The management mechanism of rubbish collection and handling was also set up and thus the rubbish could be collected and transported orderly. Moreover, the mechanism of appraising and electing "Creditable merchant" was established and carried out annually to build a better market environment. In terms of community management, Wangcun has established a complete set of measures about democratic decision-making, rewards and punishments. In general, the common affairs may be decided by meetings of VPB and VC; while the vital affairs should be decided by village meeting; and all the resolutions of various meetings should be announced by public notice. At the end of each year, a special village meeting would be held to honor "Good daughter-in-law", "Good mother-in-law", "Good teacher", "Good student", "Good Party member" and "Good cadre". Besides, the names of villagers who violate the local village's regulations and rules would also be criticized publicly at the meeting. These endogenous institutional arrangements benefit the transformation development of Wangcun significantly.

Wangcun's village fair construction and RLC have had positive effects on community development. Firstly, village fair develops rapidly. Wangcun's village fair has been a fixed-term rural market since 2003. Its total coverage area has enlarged from 5.07 ha in the early period to 14 ha at present, attracting the people of more than ten towns from east side of Dancheng. There are about 200 households in Wangcun engaging business directly or indirectly related to village fair, accounting for two thirds of the total households of the village. Besides, more than 80 commercial tenants from other villages joined them. Secondly, the appearance of village has been improved. A total of 306 households had moved into the new community until

August 2010, and roads, electric power, internet and other infrastructures have been completed already. Being different from serious village decline in TPAA, village fair construction and RLC made Wangcun more vigorous and thus made a positive contribution to slowing rural depopulation. Thirdly, agricultural production has been improved. After RLC, residential land decreased from 56.67 to 14 ha, and thus caused an increment of 42.67 ha of farmland, 29 ha of which have been consolidated and further allocated to each household. A total of 66.67 ha of farmland, including the newly increased 29 ha, has been rented out to the Dancheng's biggest agricultural industrialized enterprise. The enterprise is now taking advantage of the land to grow medicinal plant, and many villagers are employed to be in charge of field management. As such, villagers can benefit from both land rent and wages. Fourthly, RLC has brought about the emancipation of mind and the raising of development awareness. The majority of villagers consider that they are more active than before because of visiting, learning, experiences exchange and reconstruction activities. They used to drink and play cards during the slack season of agricultural production, but now they are busy with business or thinking about how they can run business better. Villagers in Wangcun think highly of their RLC practice, for most of them are satisfied with the improved living conditions (Table 11.9).

Table 11.9 Villagers' perception of the comprehensive effects of RLC in Wangcun (Li et al. 2014)

Questions	Sampled households' answers
How about the transportation condition after the RLC?	Obviously improved—**21** households; improved a little—**1** household; no change—**0** household; getting worse—**0** household
How about the residential environment after the RLC?	Obviously improved—**12** households; improved a little—**10** households; no change—**0** household; getting worse—**0** household
How about the living quality of your family after the RLC?	Obviously improved—**8** households; improved a little—**13** household; no change—**1** household; getting worse—**0** household
Are you satisfied with the effects of RLC?	Totally satisfied—**10** households; rather satisfied—**8** households; unsatisfied—**4** households; totally unsatisfied—**0** household
How many people do you think would be satisfied with the effects of RLC?	Less than 50%—**0** household; 50–60%—**0** household; 60–70%—**3** households; 70–80%—**4** households; 80–90%—**6** households; above 90%—**9** households

Source Random sampling investigation of 22 households in Wangcun

11.4 Conclusions

China's rural hollowing has been accelerated due to the rapid industrialization and urbanization under urban–rural dual structure since the mid-1990s. Taking Yangqiao community and Dancheng country as examples, this chapter analyzes the status quo of rural hollowing and the potential of RLC in eastern and central traditional farming areas of China, and considers implications for participatory RLC based on holistic analysis of the process, effects, and experiences of two typical RLC practices in Dancheng. The results show that Yangqiao and Dancheng have experienced a rapid process of rural hollowing characterized as the hollowing of rural industries, infrastructure, and residential population and settlements, thus present a challenge for catch-up development. However, from another point of view, hollowed villages in Yangqiao and Dancheng show tremendous potential for transformation development that would assist in achieving an optimal allocation of rural land resources, releasing and protecting farmland and helping to strike a balance between the demands of food security and urbanization. Rural restructuring driven by RLC is urgently needed for dealing with hollowed villages.

In recent years, dozens of RLC projects have been carried out in eastern and central traditional farming areas of China, and preliminary success has been made. Our case studies indicate that RLC, especially rural housing land consolidation, has positive effects on the improvement of living conditions, increment of farmland area and development of rural industries, especially agro-business. It is clear that RLC can help to revitalize hollowed villages. Overall, promoting RLC in TPAA scientifically according to local conditions could provide integrated platform for developing modern agriculture, stabilizing national grain production and promoting new countryside construction. The lesson we can learn from Yangqiao and Dancheng's consolidation practices is that RLC is a complex and systematic engineering, which needs comprehensive evaluation, overall planning, effective implementation, and most importantly, the close cooperation of various actors. In addition, Dancheng's experiences, including self-organized rural planning, democratic decision-making and endogenous institutional innovation can benefit the RLC in other plain agricultural areas, especially, may be helpful for revising the former local government's top-down way of RLC.

Furthermore, community-based RLC praxes in Yangqiao and Dancheng could also give us valuable implications for rural development in TPAA. Most importantly, villagers are main actors of rural development for their demands are vital driving forces, but they used to be ignored by local government or other powerful actors for various reasons. As such, mechanism on villagers' participation in rural development may be formed to keep villagers' willingness, rights and interests from being marginalized. Besides, village cadres are the nucleus of the internal power structure for rural development. Based on a systematic insight into local resources endowment development willingness, market demand, government's policy orientation, and external experiences and so on, they exerted themselves to stimulate the internal actors and integrate the external actors, and motivated these actors engage in collective action. As such, capacity building for village cadres in rural development is crucial. Moreover, participatory planning and design and endogenous institutional innovation initiated by village cadres or other rural elites were vital support to rural development. Future rural development praxes should pay more attention to these endogenous initiatives.

References

Li Y, Liu Y, Long H (2011) 黄淮海地区乡村发展格局与类型 (Study on the pattern and types of rural development in the Huang-Huai-Hai region). Geogr Res 30(9):1637–1647

Li Y, Liu Y, Long H et al (2014) Community-based rural residential land consolidation and allocation can help to revitalize hollowed villages in traditional agricultural areas of China: evidence from Dancheng county, Henan province. Land Use Pol 39:188–198

Liu Y, Long H, Chen Y et al (2011) 中国乡村发展研究报告: 农村空心化及其整治策略 (Research report on rural development in China: Hollowed village and its renovation strategy). Science Press, Beijing

Liu Y, Yang R, Li Y (2013) Potential of land consolidation of hollowed villages under different urbanization scenarios in China. J Geogr Sci 23(3):503–512

Long H, Li Y, Liu Y (2009) 中国空心化村庄演化特征及其动力机制 (Analysis of evolutive characteristics and their driving mechanism of hollowing villages in China). Acta Geogr Sinica 64(10):1203–1213

Long H, Li Y, Tu S et al (2018) 平原农区空心村整治关键技术研究与示范 (Key technologies research and demonstration of Hollowed village consolidation in plain farming areas of China). Science Press, Beijing

Long H, Tu S (2017) 论乡村重构 (On rural restructuring). Acta Geogr Sinica 72(4):563–576

Long H, Tu S (2018) 土地利用转型与乡村振兴 (Land use transition and rural vitalization). China Land Sci 32(7):1–6

Miranda D, Crecente R, Alvarez MF (2006) Land consolidation in inland rural Galicia, N.W. Spain, since 1950: an example of the formulation and use of questions, criteria and indicators for evaluation of rural development policies. Land Use Pol 23(4):511–520

Pasakarnis G, Morley D, Maliene V (2013) Rural development and challenges establishing sustainable land use in Eastern European countries. Land Use Pol 30(1):703–710

Vitikainen A (2004) An overview of land consolidation in Europe. Nord J Surv Real Estate Res 1(1):25–43

Part IV
Reflections and Futures

Chapter 12
Adjusting and Controlling Land Use Transitions

Abstract Land use transition is a two-edged sword for regional sustainability. Recently, there has been an increasing trend emphasizing land resource in formulating regional development policy and affecting rural sustainability. Optimal allocation and efficient management of land resource are favor of rural restructuring. As such, it is pivotal to establish a theoretical framework in order to develop land management policy that can promote future land use transitions capable of meeting multiple goals and satisfying demands from various stakeholders by incorporating abroad spectrum of disciplines. This chapter probes the mechanism of mutual feedback between land use transitions and land management based on a three-fold framework of natural system-economic system-managerial system, and to illustrate the mutual feedback based on the practices of land use management in China. Then, the influential mechanism of urban-rural integrated development promoted by adjusting and controlling land use transitions was analyzed. The direction of urban-rural integrated development affected by land use transitions lies on the urban-rural distribution of land incremental value. Under the background of rapid urban-rural transformation development in China, the recessive land use morphology and its changes are the key points of land use transitions and land management. Adjusting the changes of recessive land use morphology can be taken as an efficient way to innovate land management policies and institutions in China, also a way to promote land management mode changes from the one-dimensional quantity management to the multi-dimensional quantity-quality-ecology management as well as from the one-dimensional resources management to the multi-dimensional resources-assets-capital management.

12.1 Introduction

Land use transition is a two-edged sword for regional sustainability. Recently, there has been an increasing trend emphasizing land resource in formulating regional development policy and affecting rural sustainability (Long et al. 2016). Optimal allocation and efficient management of land resource are favor of rural restructuring and sustainable rural development. However, the process of land use transitions affected by the allocation and management of land resource is complicated, as the value of one

H. Long, *Land Use Transitions and Rural Restructuring in China*,
https://doi.org/10.1007/978-981-15-4924-3_12

land use type relative to that of its competing use changes over time. Sometimes, the actual values that are used to allocate land may be far from optimal, and undermine the rural sustainability by the distorted economic and political incentives due to the policy and institutional failures (Barbier et al. 2010). So, there is an urgent need to examine the interconnections between the environment, the social setting and resultant tensions in considering how land use transitions can achieve multiple benefits (Robinson and Carson 2013), i.e., how to control land use transitions by efficient land use management. Aiming at this point, a better understanding of the mutual feedback between land use transitions and land management is an important premise for promoting rural sustainability via land use transitions, by which land use transitions can be adjusted to appropriate trajectory by implementing land management measures in the direction of sustainable regional socio-economic development.

As such, it is pivotal to establish a theoretical framework in order to develop land management policy that can promote future land use transitions capable of meeting multiple goals and satisfying demands from various stakeholders by incorporating abroad spectrum of disciplines. This chapter tries to probe the mechanism of mutual feedback between land use transitions and land management, and to illustrate the mutual feedback between land use transitions and land management based on the practices of land use management in China, for the purpose of translating the research of land use transitions into the practical countermeasures of land use management to cope with the land use issues resulted from rapid urban-rural transformation development in China. Finally, the influential mechanism of urban-rural integrated development promoted by adjusting and controlling land use transitions was analyzed.

12.2 Mutual Feedback Mechanism Between Land Use Transitions and Land Management

Land management refers to implementing different kinds of managements on financial land use institution or rational land utilization by a country with legal and administrative measures. It is essentially a comprehensive activity of decision, plan, regulation, coordination and control, taken by a country under certain conditions via comprehensive use of administrative, economic, legal or technical methods, for improving socio-economic and ecological benefits, maintaining social dominated land ownership system, regulating land-based relationship, and supervising land use activities. As one of the important methods of land management, land use institution and policy plays an important role in sustaining the superstructure of one country over a certain period of time, which is a general term of the land-based relationships under certain socio-economic conditions as well as a legal form of the man-land interrelations of a country. Land use policy is the measurement system and the behavior norm for the implementation of land use institution. The Chinese land use institutions include farmland protection institution, land use regulation institution, land requisition institution, economical and intensive land use institution, etc., the implementation of which may have direct impacts on land use transitions.

Currently, China has stepped into an important period of urban-rural transformation development, facing severe challenges from transforming economic growth mode and adjusting industrial structure. Land is a spatial carrier for major socio-economic activities. Its traditional exploitative and extensive utilization mode has got a dim future due to aggravating conflicts between inevitable increasing demand and fixed land supply. With the advance of urbanization and industrialization, there will be a vigorous demand for and inertia low-efficient growth of living and industrial construction land, which will further drive high speed rural land (especially farmland) conversion as well as low land use efficiency in the urban and the rural. The transformation of urban-rural development in China will bring about a long-term existence of the de-agriculturalization of farmland (mainly farmland transition), the hollowing of rural settlements and the fragmentation of farmland management. So, there is an urgent need for the transformation of land use idea, mode and management measures. However, the sharply decrease of farmland in China during 1996–2011, decreased by 6.4% and 8.27 million ha, has formed a strong contrast to the world's most rigorous farmland protection institution and intensive land use institution implemented at the same period. The above phenomenon also reflects some problems existed in present land use management practices of China, reflecting the current land use management institution is no longer matched with the modern developing tendency, and an accelerating promotion for related innovation is in need. So, innovations on land use institution as well as improvements on land use functions and efficiency allow of no delay. The key point of land use policy and institution innovations rests on resolving the institutional conflicts that driving land requisition, pursuing for land finance and wasting land resource, and promoting a series of innovations of land use management institutions in the aspects of rights, planning, requisition and collective business construction land transfer, to further deepen and promote urban-rural development transformation in China.

National or regional land use morphology usually changes with its economic and social development phases (Long 2003, 2012). Lambin and Meyfroidt (2010) argued that land use transition emphasizes a non-linear progress of land use morphology changes related closely to social and bio-nature system changes, and it mainly derives from negative feedbacks caused by key resource depletion as well as socio-economic changes and innovations independent from ecological system. Land use transition occurs in a three-fold framework of natural system-economic system-managerial system in anytime (Barlowe 1986; Li 2002). Land use transition driven by socio-economic changes and innovations may have direct resources and environmental effects on natural system, which are always negative. And natural system may intensify the progress of land use transition with the occurrence of disaster and land degradation. Usually, changes of natural environment of the earth's surface represent as natural resources exhaustion (i.e. farmland decreasing and its quality deterioration) and environmental degradation (i.e. soil pollution or salinization). When it comes serious enough to attract the public concern, the managerial

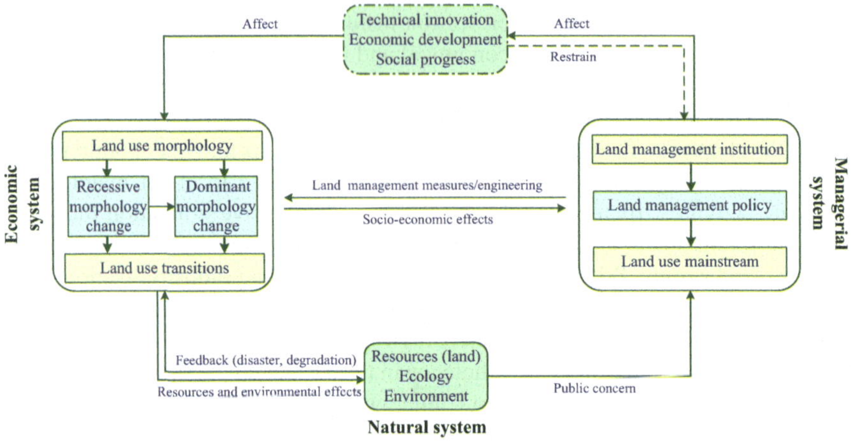

Fig. 12.1 The mechanism of mutual feedback between land use transitions and land management (Long 2015; Long and Qu 2018)

(institutional) system may start to regulate the behavior of land use mainstream by related legal or policy of land use management, also implement land use management policy and institution by land resource engineering (Liu 2015), to regulate land use economic system directly or indirectly, also to manage land use transition. Finally, management departments will adjust the relevant regulation measures based on its socio-economic effects, and make land use transitions meet the expected goal by influencing socio-economic changes and innovations as well as implementing adjusted management measures (Fig. 12.1).

12.3 Influences of Land Management on Land Use Transitions

Land use transition is a result driven by the mutual feedback and joint constraint between natural environmental conditions and socio-economic development factors. As external factors, socio-economic institutions and policies, especially land use management institutions and policies, may exert important impacts on land use transitions. Accordingly, it is necessary for land use management department to gather land use information timely and rapidly, and to formulate relative policies and macro planning decisions to regulate land use transitions.

12.3.1 Economic Measures of Land Management

The economic measures of land management refer to the methods or means to regulate and guide the land use activities according to objective economic laws, so as to achieve the management functions by land use managers. Among all the economic measures, land rent and land price are the most popular two, and they are the direct macroeconomic levers grasped by the government. The socialist public ownership of land is implemented in China, so there is no land ownership transfer among individuals. The price in existing urban land market is exactly a price for the state-owned land use right in a certain period, and the price in rural land market refers to the price for the collective farmland contractual operation right or the collective construction land utilization right in a certain period. These prices are actually the land rents in the same period conceptually and quantitatively. Land price is a useful tool to balance current land market. For example, to solve the irrational construction land expansion in China, the State Council issued the "Notice of the State Council on Issues Concerning Strengthening the Regulation of Land", specifying the minimum price standard for industrial land transfer. It curbed disordered construction land expansion to some extent, and forced the land users to improve land use efficiency and intensity, which caused the change of recessive land use morphology.

Meanwhile, governments take different types of land taxes (e.g. tax for farmland occupation) and different standards of land tax rates to guide the land exploitation and land utilization. For example, to stabilize the farmland quantity, the State Council implemented the new "Provisional Regulation for tax on farmland occupation" in January 1st, 2008. According to this regulation, the upper limit and lower limit of existing tax standard about farmland occupation were both increased by four times. Especially, the aforementioned tax standard should be increased by another 50% for basic farmland occupation. By the economic lever of farmland occupation tax, the government reduced the irrational farmland occupation by non-agricultural construction, and slowed down the speed of transferring cultivated land to construction land.

In addition, governments also take macro-regulation on land use by financial means. For example, in order to solve the current financial difficulties in the transfer of rural collective business construction land, the Chinese Banking Regulatory Commission and the Ministry of Land and Resources jointly issued the "Interim Measures for the Administration of Mortgage Loans to Rural Collective Business Construction Land Use Right" on May 13th, 2016, to promote the mortgage loans for rural collective business construction land in the pilot areas. The promulgation of this policy may revitalize the assets of rural collective business construction land, and promote effective land resource transfer, thus improve the intensive level of land use and achieves land use transition in the aspect of recessive morphology.

12.3.2 Land Engineering

Engineering and technical means of land management include systematic engineering techniques, land consolidation techniques, land use planning techniques, etc. Land engineering is a systematic project, referring to a cross integration of relevant disciplines consist of science, engineering, management, and agriculture (Liu 2015). Land engineering can help to utilize and consolidate production factors, integrating the fields, water, road and forest to better serve agricultural production, and to improve the utilization efficiency of the agricultural resources in the world (Liu 2015). Land consolidation, as one of the important regulation means of land use transitions, includes land exploitation, arrangement and reclamation, all of which essentially belong to the sphere of land engineering and techniques. China tends to build 26.67 million ha high-standard basic farmland to promote farmland use transition during the 13th five-year plan, which will vitally rely on a series of important engineering techniques on high-standard basic farmland construction, e.g., rural hollowed villages land consolidation, gully land comprehensive consolidation, sandy land comprehensive consolidation, etc.

Currently, many land use issues in China, such as suboptimal urban land use structure and non-economic agriculture land use scale, can be solved by land consolidation. So, China's future land management missions may rely more on land consolidation. The major objectives of land consolidation are to form rational, high-efficient and intensive land use structure, increase available farmland area, improve land use efficiency, and satisfy the land use requirements from socio-economic development. Land consolidation is helpful for realizing land use transitions by optimizing recessive land use morphology. For example, in rural area, land utilization ratio and input-output ratio can be enhanced by implementing land consolidation, regulating agricultural land use structure, and merging scattered plots; and farmland quality can be improved by changing recessive farmland morphology, i.e., low-yield field upgrading or saline-alkali land amelioration. While in urban area, internal land use potentials can be excavated via increasing investment and consolidating urban construction land, i.e., improving land use efficiency without increasing construction land, and promoting the recessive land use transition in urban area.

Land use planning is a long-term arrangement of regulating and allocating different land use structures and patterns. Lu et al. (2006) argued that land use planning should take into account the idea of land use transitions. A planning with the idea of land use transitions can make an overall allocation of land use types among different industries in different regions, and form rational land use structure. The Chinese government implement farmland protection institution and land use regulation institution. To realize these institutions, there are many methods, during which, land use planning is always an important one. By setting urban growth boundaries and ecological control lines, as well as defining permanent basic farmland, land use planning can take a powerful role in maintaining construction land scale, and controlling ecological land and farmland conversion scale, thus contribute to rigorous arable land and prime farmland protection. Consequently, the impacts of land use planning on land use transitions are mainly reflected on changing the dominant land use morphologies, e.g., the quantities and spatial patterns of land use.

12.3.3 Land Management Policy and Institution

Policy and institutional factors have strong restriction on regional land use mode and land use changes. Land use managers always formulate relative policies and rules from the point of whole society interest and macro-economic development, to guide land use transitions and improve general land use benefits. Land use policies and institutions may have significant impacts on both dominant land use transition and recessive land use transition.

12.3.3.1 The "Dual-Track" Structure of Urban-Rural Land Management Institution

According to "The Law of Land Administration", the utilization rights of collective-owned land is forbidden to remise, transfer and lease for non-agricultural construction, resulting in the phenomenon of "similar land with different rights" between urban and rural area. That means rural land cannot be transacted as urban land in the market for non-agricultural exploitation, and it constrains the transfer between different rural collectives as well as transfer from rural collective to the state. The restrictive law regulation on rural collective construction land is one of the major reasons for restricting the development of local economy severely in suburban area in China. In order to innovate rural land management institution, the central government issued an important document to separate the rural land rights into land ownership, contract rights and management rights, the so-called "Division of Three Rights" of rural land, on November 30th, 2016. It will have positive effects on improving rural land rights and contribute to "similar land with similar rights". The collective construction land circulation practices in Changshu county, southern Jiangsu province in China shows that, the document above may effectively revitalize the stock construction land in rural areas and improve total land use efficiencies. And will finally lead to remarkable land use transitions in regional recessive morphology.

12.3.3.2 Policies on the Transfer of Rural Land Contractual Operation Rights

The transfer of rural land contractual operation rights refers that farmers transfer their contractual operation rights to professional investors, family farms, farmers' cooperatives, or agricultural enterprises in the public market, maintaining their contract rights and transferring utilization rights, to develop various scale managements.

The transfer of rural land contractual operation rights represents the changes in land properties, and belongs to the recessive land use transition. In 2014, the State Council enacted the document on "Guiding the Healthy Development of the Market for Rural Property Rights Circulation and Trading", in which it is stipulated explicitly about the rights that can be transferred in the rural land market: the farmers' contracted land use rights, the collective forest land use rights, and the collective land contract rights with no relation to collective land ownership or rights that should be contracted only by farmers. With the normalizing and improving of land circulation market, land use transitions in rural area characterized by the transformation of land property rights will be carried out extensively.

12.3.3.3 The Economic and Intensive Land Use Institutions

Economic and intensive land use reflects the changes on land use intensity, which belongs to the recessive land use transition. Currently, China has issued several institutions and specific policies to promote economic and intensive land use. However, the actual effect is not ideal. Rural China, with the similar population size but four times construction land area compared with urban area, shows obvious lower land use efficiency than that of urban area. So, there is an urgent need for promoting economic and intensive land use, controlling construction land scale rigorously, improving land use efficiency in the rural-urban fringe zone and industrial areas, and guaranteeing the farmer's land-based livelihood.

12.3.4 Land Management Innovations

Land use policy and institutions are the core measures of land management. This section here takes policy-institution innovations as an example to illustrate how land management innovations affect land use transitions. Currently, the dominant land use transition in China is hard to achieve due to many rigid policy constraints under the existed macro land management policy framework. However, the recessive land use transition can be promoted through innovations on land management policy and institution, which may further cause the dominant land use transition.

In recent years, the implementation of hollowed villages consolidation as well as new community and rural industrial parks construction in Shandong province under the background of "increasing vs. decreasing balance" land-use policy concerning urban-rural construction land (Long et al. 2012), has promoted the rural construction land transition to a great extent. However, the implementation process did not work well in some towns for the short of funds. For example, Paizi village of Yucheng city in Shandong province has experienced the financial problem during the process of carrying out hollowed village consolidation and new-type community construction. The village land use morphology before consolidation is a hollowed village with 21 ha idle land and 15 ha rural housing land. According to the plan of hollowed villages consolidation and community construction, building a new rural community only needs 3 ha land, so the potential of increasing farmland via hollowed villages consolidation is 33, 12 ha from old rural housing land and 21 ha from idle land. Based on the 'increasing vs. decreasing balance' land-use policy, the key objective of which is to achieve equilibrium in the supply of land in China by balancing the increase in urban construction land (driven by urbanization) with the decrease in rural construction land (facilitated by out-migration). Former rural construction land could consequently be reclaimed as agricultural land, thus supporting food security goals and broader socio-economic development in rural areas. This policy ensures that the added 12 ha consolidated farmland from old rural housing land can get about two million RMB yuan subsidies from the local government. However, the added 21 ha consolidated farmland from idle land cannot get expected subsidies. Under this case, the added farmland has the same dominant morphology but the recessive morphology is different, and the activity of rural community construction is difficult to start due to lack of funds. If we can change the recessive morphology of the added farmland from idle land via policy and institution innovation, which means these added farmland can get the same or partial government subsidies. Then, we can get sufficient funds to initiate the rural community construction, and lead to expected land use transition, to form a new rural community with 33 ha farmland and 3 ha rural community (Fig. 12.2).

Currently, the new added farmland consolidated from old rural housing land can be converted into land use quota, which permits occupying farmland for urban construction in other areas, according to the 'increasing vs. decreasing balance' land-use policy. However, this kind of land use quota can only be traded at prefectural level instead of at provincial level. Yucheng has big potential of hollowed villages consolidation which can bring about many land use quotas for selling, but only part of these land use quotas can be consumed in what belonged Dezhou city due to its lagged economic development.

Actually, there are many developed cities along eastern coastal area of Shandong province with abundant funds while short of land use quotas for urban construction. So, if the additional construction land use quotas can be traded among different cities

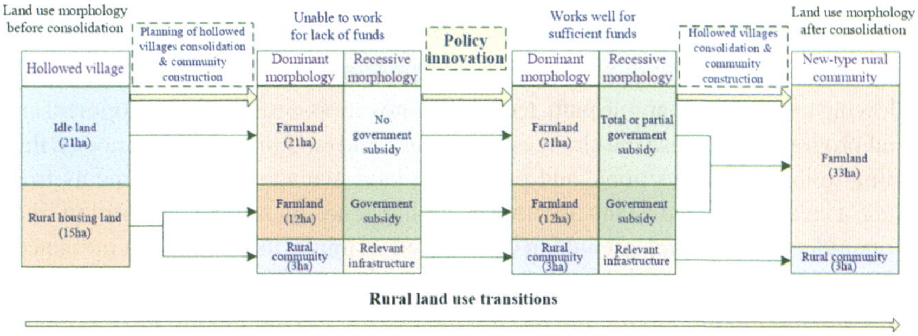

Fig. 12.2 Innovation of policy and institution promotes land use transitions (Long 2015; Long and Qu 2018)

in Shandong province by institution innovations, the Yucheng city will get sufficient fund for hollowed village consolidation and community construction. Besides, the additional farmland from rural idle land consolidation without subsidy shows a defect in current policy designing. If the government entitles the additional farmland from idle land consolidation with financial subsidies by policy innovations, the hollowed village consolidation and community construction of Paizi village will be pushed forward smoothly. Accordingly, there is an urgent need for land management policy and institution innovations to change the recessive land use morphology, and finally realize the dominant land use transition based on the changes of recessive land use morphology.

12.4 Influences of Land Use Transitions on Land Management

Aiming at the tendencies and requirements of socio-economic development, a nation or a region should adjust its land management measures timely according to its current land use morphology pattern and problems. For example, current land use morphology of the United States shows a similar proportion of forested land and grassland with the world's average level, but a significant difference from its former development phases (Long and Li 2002). In fact, land use management in the United States has been adjusted continuously according to the national land use morphology changes for more than 200 years since its founding, with a change from former irrational dealing with public land and predatory exploiting land resource, to paying attention to land resource protection, and formulating a new land ethics and government activism of food production resources protection (Long and Li 2000). Similarly, the land use management, especially the farmland and construction land management in China, is experiencing or will experience a series of changes and innovations under the background of urban-rural socio-economic development transformation.

12.4.1 Influences Following Socio-Economic Path

Following the socio-economic path, technical innovation, economic development and social progress may cause the changes of land use and land management modes, thus leading to land use transitions, and promoting land management adjustments from different levels. Taking the land use transitions happened on farmland operation since the open-door policy implemented in China as an example, the way how it influenced land management institution, policy and land use agent may be shown as follows (Fig. 12.3).

Before the open-door policy was implemented, there had been a people's commune system in rural China. In this system, returns for laborer had no connection

Fig. 12.3 Land management changes driven by land use transitions following the socio-economic path (Long and Qu 2018)

with their paid work. That made a low enthusiasm for farmers, also led to badly low agricultural production efficiency, and even resulted in severe food shortages. Under that background, farmers in Xiaogang village, Fengyang county of Anhui province, took the lead in performing 'fixing farm output quotas for each household', an operational farmland management mode, in 1978. Under this mode, farmers could get the full share of their efforts' marginal return. It greatly aroused farmers' enthusiasm of production and improved the grain production efficiency. And it opened a new stage of China's rural land system reform prelude. Adapting to this positive trend, the Chinese government issued several important documents continuously during 1983–2001, to put forward the household contract responsibility system and stabilize the farmers' land contractual operation right. The "Law of the People's Republic of China on Land Contract in Rural Areas" implemented on March 1st, 2003, authorized the legitimacy of rural land contract rights transfer. During this stage, there was an institutional level adjustment on land management.

With the development of productive force, the agricultural mechanization and the agricultural biotechnology improved constantly. Compared with manual operation, mechanical operation shows higher efficiency in agricultural planting and harvest, with lower cost. However, agricultural machineries with different horsepower may correspond to different configuration standards, so not every household can afford to high-powered machinery. Agricultural biotechnology shows an outstanding performance in saving manpower and improving land productivity. While it always has a high-tech threshold, which could not be mastered by the farmers directly. Under these backgrounds, rural professional service teams, especially the agricultural mechanization services, developed rapidly. With the transfer of labor-intensive agriculture production section from households to service teams, another type of land use transition based on farmland operation occurred. During this stage, land use transition only referred to the transfer of operation section in agricultural production process, and there was no change on rural land right. Meanwhile, land management adjustment only happened in the individual level, with no effect on policy level.

With the further development of the social economy and the productive forces, the secondary and tertiary industries show an increasing attraction to rural young labors, and the opportunity cost to perform agricultural production increases constantly. With more and more rural labors flowing into non-agricultural industry, farmers tend to give up farming and transfer their farmland to relatives, or even keep it abandoned. It caused a waste of land resource and serious threat to national food security. In order to stabilize the national grain production capacity, the Chinese government promoted the circulation of land contractual operation rights in two ways: encouraging farmers to lend more land for food production by providing agricultural subsidies, and smoothing large scale farmland transfers supported by detailed relative policies. During this stage, land use transition refers to land operation rights transferred among different households, and the land management adjustment has been up to the policy level.

At present, China is just at the third stage, the circulation of contractual operation rights shows an obvious higher speed and the moderate scale agricultural operation becomes an inevitable trend. As shown in the practice, rural land transfer and moderate scale operation is the only way for agricultural modernization. To adapt with this positive tendency and promote the transfer of contractual operation rights in an orderly way, the General Office of the CPC Central Committee and the State Council jointly issued a document entitled "The Opinions on Guiding Rural Land Operation Right Circulation in an Orderly Way to Develop Moderate Scale Operation". Although China has already got considerable progress in related institutional construction and modal innovation, there are still lots of tough problems seriously affecting the effectiveness and the sustainability of rural land use right transfer. And the problem abovementioned can be attributed to the contradiction between excessive agricultural population and limited arable land resource.

In the forthcoming stage, the driving force of land use transition and the subsequent management adjustment will be the improvement of social security. Currently, agricultural labor force gets such a low quality that they cannot be competent to the high-tech jobs in secondary and tertiary industries. While the number of rural surplus labors far exceeds the capacity of low-skilled jobs in the secondary and tertiary industries. Under the "dual-track" structure of urban-rural institution, the land in rural area needs to afford the function of social securities. Farmers won't leave their land before getting the same social security as the citizens. So the moderate scale management currently is just a "moderate scale" that corresponding to the current stage of socio-economic development, while not large enough to meet the "perfect scale" (Ni and Cai 2015). Small-scale management may hinder the development of agricultural modernization, as well as weaken the competitiveness in international market on agricultural product price, quality and structure. So, China is confronting with another land management adjustment, the process of which will be a long-term and the key point will be addressed on breaking the "dual-track" structure of urban-rural institution.

12.4.2 Influences Following Environmental Path

Following the environmental path, irrational land use behavior causes land use transitions and results in negative resources and environmental effects. These negative effects accumulate over time until they exceed the environmental carrying capacity and evolve into severe natural disasters or environmental degradation. The serious events may arouse public concern, and lead to land management adjustments and restoration measures. Compared with the socio-economic path, land management adjustment in this path always happens in a sudden way accompanied by a certain environmental event, so it is discontinuous. But these response measures may involve different aspects, and will perform collectively for the complexity and interactivity of natural system. Taking the forested land use transition and its environmental effects in China during the late 20th and early 21st century as an example, the way how it influenced land management is shown as follows (Fig. 12.4).

The large-scale sloping farmland reclamation in the late 20th century resulted in negative land use transitions in China, and badly destroyed the regional water and soil conservation functions. Meanwhile, it caused severe soil erosion and depletion. Furthermore, the decline of soil fertility led farmers to reclaim more new sloping land. This vicious circle posed a serious threat to the national ecological security. In response to this situation, the Chinese government put forward the Grain-for-Green program and carried out the pilot work in 1999 (Feng et al. 2005; Long et al. 2006). It was started all-round in 2002 with the issued document entitled "The State Council Opinions for Further Improving Sloping Land Conversion". The main work of this program is adjusting the farmers' land use behaviors by the means of subsidies, supervision and technical support. The Grain-for-Green program might be the largest eco-engineering projects in China and even in the world.

The natural forest resources in China mainly distribute in the northeast, Inner Mongolia and southwest area. It satisfied more than 50% of the timber demand in history. The forest mostly located in the core areas of major river sources, large reservoir surroundings and important mountainous regions, which plays an irreplaceable role in water saving, soil conservation, riverbed stabilization, river sediment reduction, and so on. But large numbers of natural forest cuttings seriously shrank the nature

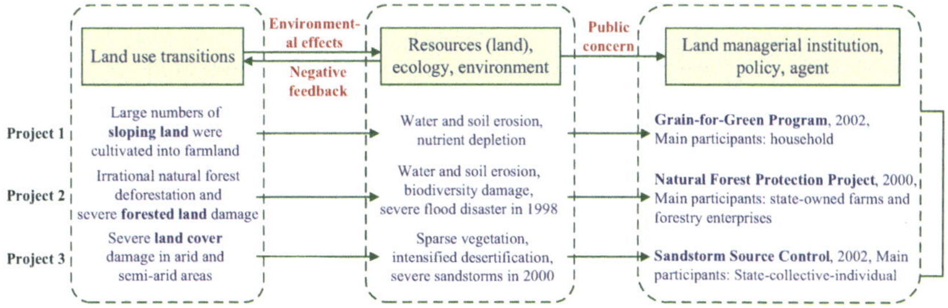

Fig. 12.4 Land management changes driven by land use transitions following the environmental path (Long and Qu 2018)

forested land areas, and badly damaged its soil and water conservation functions, which contributed to the catastrophic floods in 1998 to a great extent (Yin and Li 2001). To cope with this issue, the Chinese government proposed "The Natural Forest Protection Project" and carried out the pilot work immediately in the same year. This project accompanied with targeted guidance documents was officially launched in 2000 by the State Council. It was addressed on state-owned farms and forestry enterprises by the means of subsidies, supervision and technical support.

In the northwest and north China, there is a wide range of arid and semi-arid areas with fragile ecology and sparse vegetation, where there has suffered from frequent sandstorms for a long time. Date back to the year of 1978, there was "The Three North Shelterbelt Project" proposed for blocking sandstorms destructions in densely populated area. However, it could not cover the negative effects caused by irrational human behaviors. Long-term overgrazing and disordered herbal mining intensified the regional desertification, and further aggravated the sandstorm frequency and its influential scope. During the spring of 2000, 12 times sandy weathers of dust floating, sand blowing and sandstorm took place continuously in north China, with a high frequency, wide range, and high intensity. It caused serious economic losses in the capital of Beijing and aroused widespread public concern. To improve the ecological environment of the surrounding areas and alleviate this disaster, the CPC Central Committee and the State Council immediately launched a strategic ecological construction projects, so-called "The sandstorm source controlling project in Beijing-Tianjin-Hebei region". By the public participation of the government, collectives and individuals, this project has alleviated the environmental disaster effectively.

12.5 Influences of Land Use Transitions on Urban-Rural Integrated Development

In the process of rapid urbanization and industrialization, urban-rural integrated development has become an important socioeconomic phenomenon around the world (Liu and Li 2017). New changes have taken place in the division of industries and functions between urban and rural. Meanwhile, land use in the world has undergone a drastic transition, and land use types representing the interests of different sectors have conflicts in space (Bryan et al. 2018; Serra et al. 2008), which are highlighted by the rapid expansion of urban land and the acceleration of cultivated land conversion. Urban-rural integrated development is a key measure to share the fruits of building a moderately prosperous society in all respects by breaking urban-rural dual structure and shaping a new-type urban-rural relationship in China. At present, the urban-rural unbalanced development has become a concentrated manifestation of the contradiction between the people's increasing demands for a better life and unbalanced and inadequate development. Land is an important element and space carrier for urban-rural development (Long and Qu 2018; Long et al. 2016, 2019). As well, land use transitions are a mirror reflecting the stage of socio-economic development. Meanwhile, the various problems arising in the process of socio-economic development can also be solved through land use transitions (Long and Qu 2018). The urban-rural relationship in China has experienced a process from binary opposition and segmentation to overall planning and integration (Liu et al. 2013; Chen et al. 2018). Land use transitions and urban-rural development mutually evolve and develop. Under the current land institutional arrangement, the revolution of land use efficiency caused by land use transitions promotes the optimal allocation of land resources, while it is still worthy to discuss that whether the equivalent exchange of urban-rural land elements and urban-rural integrated development have been in the process of allocation. In the critical stage of socio-economic transformation development in China, it's the key scientific proposition that geography urgently needs to solve in the new era to clarify the mechanism and effect of land use transitions on urban-rural development and to give a full play to the role of land use transitions in promoting urban-rural integrated development.

12.5.1 Connotations of Urban-Rural Integrated Development

Urban is the spatial territory system with non-agricultural industries and non-agricultural population agglomeration as the main characteristics, while rural is all the spatial territory system except urban. There is no absolute spatial break point in urban-rural territory systems. Therefore, the difference of urban-rural territory systems doesn't lie in space, but mainly in function. The urban territory system mainly provides non-agricultural products, non-agricultural services and non-agricultural employment opportunities. Comparatively speaking, rural territory system mainly carries out agricultural production to furnish agricultural products and services, and gives play to its social and ecological functions (Zhang et al. 2018). Under urban-rural dual socio-economic structure, the functions of urban-rural territory systems are solidified and the transformation is blocked, which is mainly reflected in the poor circulation of development factors, unequal capability and power of development, and distorted development mechanism. Finally, urban-rural segmentation becomes inevitable.

Urban-rural integrated development is not to eliminate the boundary between urban-rural territory to achieve urbanization in rural, nor is it to assimilate urban-rural functions and landscape or homogenize urban-rural development, but to achieve urban-rural equivalent development. Urban-rural equivalent development means to promote the benign interaction and functional complementarity of urban-rural territory systems, so as to realize the impartial and equitable development rights of urban-rural territories (Liu et al. 2013). Urban-rural integrated development aims to eliminate the obstacles of urban-rural development through institution and system innovation. It helps to promote the free flow and equal exchange of urban-rural factors based on urban-rural common development to realize the return convergence of urban-rural development factors (Fig. 12.5). The final consequence of urban-rural

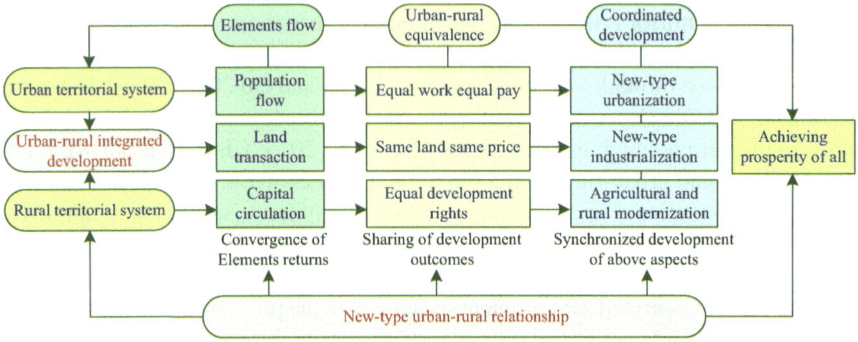

Fig. 12.5 The conceptual framework of urban-rural integrated development (Chen et al. 2020)

integrated development is to establish the new-type urban-rural relationship including the mutually promotion between industry and agriculture, the mutually complementary between urban and rural, the comprehensively integration and co-prosperity of urban-rural territories (Fig. 12.5). Urban-rural integrated development is embodied in the coordinated development of new industrialization, urbanization and agricultural and rural modernization. And the ultimate goal is for all the people to share the fruits of the building of a moderately prosperous society in all respects and the modernization development. Therefore, urban-rural integrated development does not pursue the equalization of space or administration, but strives to realize the gradual narrowing of the urban-rural labor productivity gap and urban-rural labor income gap until common prosperity is achieved.

The transformation development from urban-rural segmentation to integration is a process of urban-rural value remodeling and restructuring (Chen and Long 2019). Urban-rural integrated development will inevitably manifest itself in the transformations from the dual structure to the unitary structure, the narrowing of the urban-rural income gap, the advance of urban-rural productivity, and the improvement of urban-rural residents' living standards. The essence of urban-rural integrated development is to realize the overall functions optimization of urban-rural territory system through the interaction and complementarity between urban and rural territories.

12.5.2 *Influential Mechanism*

Land use transitions are inevitably accompanied by the land use function and value transformations. There are two most typical forms of land use transitions in China, i.e., cultivated land conversion and land nationalization. The cultivated land conversion is the transformation from agricultural land to construction land, and the land nationalization is the transformation from rural collective land to urban state-owned land, which is also the transformation from agricultural land or rural construction land to urban construction land. It is the high-efficient and high-value conversion of land use through land expropriation channel (Tan et al. 2009). Land use transitions play an important role in the operation of urban-rural territory system. Whether the allocation of land resources get optimized during land use transition is the foundation of promoting the level of urban-rural integrated development. More importantly, whether the rights and interests of urban-rural equal development, the free flow and equal exchange of urban-rural development factors and the urban-rural equivalent development has come true in the process of the reuse and redistribution of land resources during land use transitions is the concentrated embodiment of the positive impacts of land use transitions. Therefore, the direction and effect of land use transitions on urban-rural integrated development not only depends on whether land use transitions have achieved the improvement of the land use function and value, but also lies on the urban-rural distribution of land incremental value.

Land use transitions affect urban-rural integrated development through the ideal channel and distorted channel on the premise that the value of land use improves (Fig. 12.6). Land use transitions such as cultivated land converse and land nationalization can effectively improve the multiple values of economy, space and society (Tan et al. 2009). At present, the outstanding phenomenon hinders the rural transformation development that rural households in China have resources but no assets and capital. The land incremental value and land taxes and fees will increase by cultivated conversion and land nationalization (Xiao et al. 2017). In addition, land transfer after land nationalization will further improve the land economic value, and the land finance can alleviate the imbalance between central and local financial rights and administrative rights to some extent which can effectively promote regional development. Land nationalization further clarifies the property right attribute of land, and

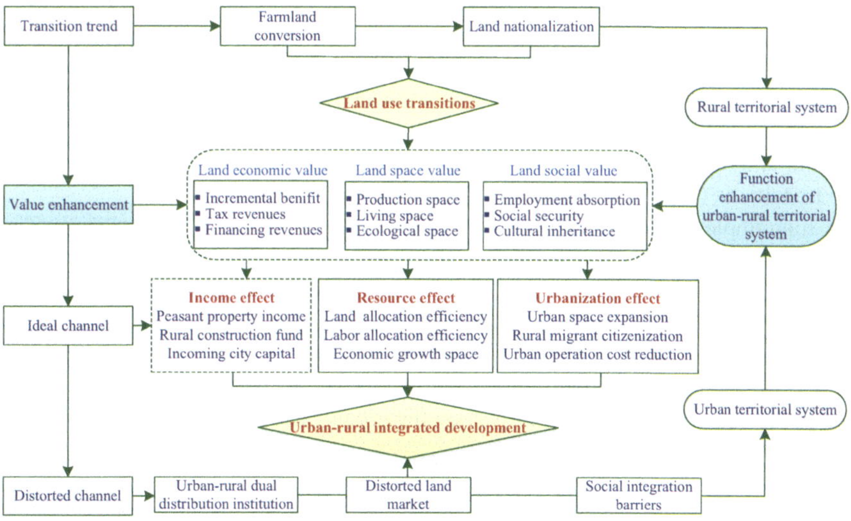

Fig. 12.6 The theoretical framework between land use transitions and urban-rural integrated development (Chen et al. 2020)

the resource-to-asset effect brought by land use transitions endows land the function of mortgage financing, which can increase the income of land financing. What's more important is that the land use transitions has accumulated funds for rural construction, effectively increased the property income of rural households. And to some extent, it is conducive to improve the living standard of rural households and promote the development of rural territory (Zhong et al. 2019), which will be the basis and premise of urban-rural integrated development in China. The multiplier effect of rural construction funds will also guide the injection of urban capital into the rural.

Land use transitions will accelerate urban-rural spatial development and utilization to improve the spatial value of land (Sun et al. 2020), which contributes to promote the urbanization development process and urban-rural integrated development. The improvement of the land elements value and land use efficiency from land use transitions drives regional economic development and urban-rural employment. With the help of surge of fiscal revenue brought by land use transitions, the level of government public service supply and public welfare will increase and enhance, which can optimize the social value of land. Of course, it is still worth warning that cultivated land converse and land nationalization cause large number of land-lost farmers to lose their social security functions (Skinner et al. 2001).

Land use transitions promotes land value. Under ideal channels, it is supposed to promote urban-rural integrated development in China through income effect, resource effect and urbanization effect. Land use transitions adjust the urban-rural man-land relationship, guide the non-agricultural employment and diversified development of farmers. It can also improve the efficiency of labor force allocation and expand the driving force and efficiency space of economic growth (Schleupner and Schneider 2010). Cultivated land converse and land nationalization themselves are accompanied by the expansion of urban construction space and the development of urbanization. And the development of urbanization would enhance the urban absorption level of rural migrants and further bring about the improvement of population urbanization. Moreover, the land use transitions through land expropriation and the migration move from rural into urban can effectively reduce the costs of urban development and operation, which will further strengthen the urbanization effect of land use transitions (Skog and Steinnes 2016).

China implemented urban preferential development strategy in a long-term. The land incremental value from land use transitions also has urban preference in distribution, resulting in the deprivation of the rural value. Under the three obstacles of urban-rural dual distribution institution, distorted land market and social integration barrier, the ideal channel land use transitions affect urban-rural integrated development is blocked. And the distorted channel becomes the main mechanism currently in effect, which puts a negative impact on the urban-rural integrated development in China (Fig. 12.6).

The scope of land expropriation expands beyond public interest purpose, and the formation mechanism of land expropriation price deviates from the marketization principle before the implementation of new *land administration law of PRC*. Land expropriation compensation is carried out according to the original use of land, while the price of land transfer is executed according to the market price (Hui and Bao 2013). The "dual-track" prices of land supply result in the first unreasonable distribution of land income and aggravate the deprivation of rural development value. The government's administrative intervention in land market is still mainly reflected in the land supply at different prices, the land supply with intervening types and the inefficient regional supply. Local governments prefer to sell industrial land at a low price and transfer commercial residential land at a high price to attract investment. The low-price preferential strategy to state-owned enterprises add government revenue to alleviate financial pressure, which objectively pushes up the ratio of approved transfer. Under the guidance of regional coordination strategy, the government takes the construction land index as a policy tool to support the development of the central and western regions. After 2003, the government was partial to supply land for the central and western regions and strictly controlled the construction land index in the eastern region (Liang et al. 2016). The distorted land supply strategies impair the efficiency and multiple values of land, and the positive role of land use transitions on urban-rural integrated development is damaged.

The land transfer revenue, tax revenue and financing income brought by land use transformation have relaxed the financial constraints of local governments. However, the urban preferential dual fiscal policy has formed the secondary unreasonable land income distribution for rural, affecting urban-rural integrated development. The abuse of land transfer revenue is an obvious representation of unreasonable distribution institution. China's land transfer revenue has long been outside of the government's financial budget management. It was not until December 31, 2006, when the ministry of finance issued "measures for the management of revenue and expenditure of state-owned land use right transfer", that the land transfer revenue was incorporated into the local government fund budget management system. In the actual land revenue distribution pattern, the ratio of rural households and village collectives is only 6.73%. In order to obtain construction funds and promote urban construction, there is great enthusiasm to increase land financing for local governments. Moreover, the income from land financing has an obvious construction tendency, which weakens the government's investment in people's livelihood. It is not conducive to urban-rural integrated development. The distorted land market pushes up the urban housing price (Wu et al. 2018), increasing the difficulty of urban integration for rural migrant. With the rapid development of rural land conversion, the household registration system hinders the transfer of rural population into urban, and the non-agricultural population is insufficient. Land use transitions liberalize the government financial constraints. But rural migrants are excluded from public services, public welfare and educational opportunities because of the household registration system. Social integration obstruction hinders the free flow of urban-rural population factors, and cuts off the essential condition of urban-rural integrated development.

12.6 Discussion

There are two key points concerning the concept of land use transitions, the socio-economic transition and the corresponding land use morphology changes. Technical innovation and economic development may lead to changes of land use. The resources and environmental effects caused by land use transitions may have reactions on land use. Furthermore, land management based on different measures can adjust land use. Thus, land use can be regarded as the center of these influential factors, and it is characterized by a certain land use morphology corresponding to a certain socio-economic condition. So, land use morphology is taken as a mirror to reflect the results influenced by all the above factors (Tuan 1971), and further reflect the inherent problems caused by the interaction of these factors.

Technical innovation and economic development belong to the sphere of productivity, while land management belongs to that of production relations. On the one hand, land management acts on the productivity by adjusting land use transitions. This effect of land management on improving productivity is direct and integrated. Land management can affect land use transitions directly by adjusting land use behaviors. And there are different kinds of managerial methods that need to be integrated. With the improving of productivity and the deepening of cognition on land use, land management measures will be improved continuously. Among which, land use policy and institution are the core measures, which are also the basis for implementing other land management measures. Therefore, innovations on policy and institution are the foundation for improving land management.

On the other hand, land use transitions may affect land management via both socio-economic path and environmental path. The key points of this influence process are the exogenous technical-economic factors, the origin of socio-economic path. However, the origin of environmental path is land use transitions, which are ultimately affected by the technical and economic development. For example, a wide range of sloping land reclamation aroused the Grain-for-Green Program by the environmental path. However, with the socio-economic development, the labor cost increases, and sloping land reclamation can no longer get the extra benefits (Shao et al. 2015). So, the farmers initiatively abandoned the sloping farmland, and changed the original driving factors of the environmental path. Then, the Grain-for-Green Program will be changed subsequently and gets a new meaning.

Different socio-economic development stages correspond to different regional land use morphology as well as different land use transition phases, thus inevitably result in certain land use transition process. For a long period of time in the future, the key issues of land management in China will still focus on the coordination of socio-economic development and agricultural resources protection. Therefore, the governments need to adjust their land management measures timely according to the changes of land use morphology, and promote the innovations on land management institutions. In this process, the formulation of land management policies and institutions needs to take the land use transition phase of targeted region into account, not only current land use transition phase but also its subsequent phase corresponding to regional socio-economic development transformation.

12.7 Conclusions

In this chapter, the mechanism of mutual feedback between land use transition and land management was probed based on a three-fold framework of natural system-economic system-managerial system. Generally, land use transitions are affected by land management via economic measures, land engineering, policy and institution. Land use transitions can also contribute to the adjustment of land management measures via socio-ecological feedback. Therefore, policy-makers need to adjust their land management policies by taking the changes of land use morphology and different phases of regional land use transitions into account. Land use transitions can contribute to improving land economic, space and social functions and the efficiency of land resource allocation. Under ideal channels, it is supposed to promote urban-rural integrated development in China through income effect, resource effect and urbanization effect. The direction of urban-rural integrated development affected by adjusting and controlling land use transitions lies on the urban-rural distribution of land incremental value. Urban-rural dual institution, distorted land market and obstacles to social integration will distort the positive role of land use transitions.

Land use transition is a forefront international research topic with high comprehensiveness, and is also a complex phenomenon. Under the background of rapid urban-rural transformation development in China, the recessive land use morphology and its changes are the key points of studying land use transitions and land management. Adjusting the changes of recessive land use morphology can be taken as an efficient way to innovate land management policies and institutions, also a way to promote land management mode changes from the one-dimensional quantity management to the multi-dimensional quantity-quality-ecology management as well as from the one-dimensional resources management to the multi-dimensional resources-assets-capital management. Only by this way can the optimal allocation and sustainable as well as efficient utilization of urban-rural land resource be achieved.

References

Barbier EB, Burgess JC, Grainger A (2010) The forest transition: towards a more comprehensive theoretical framework. Land Use Pol 27(2):98–107

Barlowe R (1986) Land resource economics: the economics of real estate, 4th edn. Prentice-Hall, Englewood Cliffs

Bryan BA, Gao L, Ye Y et al (2018) China's response to a national land-system sustainability emergency. Nature 559:193–204

Chen C, LeGates R, Min Z et al (2018) The changing rural-urban divide in China's megacities. Cities 81:81–90

Chen K, Long H (2019) 中国土地市场对城乡融合发展的影响 (Impacts of land market on urban-rural integrated development in China). J Nat Resour 34(2):221–235
Chen K, Long H, Liao L et al (2020) Land use transitions and urban-rural integrated development: theoretical framework and China's evidence. Land Use Pol 92:104465
Feng Z, Yang Y, Zhang Y et al (2005) Grain-for-green policy and its impacts on grain supply in West China. Land Use Pol 22:301–312
Hui EM, Bao H (2013) The logic behind conflicts in land acquisitions in contemporary China: a framework based upon game theory. Land Use Pol 30(1):373–380
Lambin EF, Meyfroidt P (2010) Land use transitions: socio-ecological feedback versus socio-economic change. Land Use Pol 27(2):108–118
Li X (2002) 土地利用变化的解释 (Explanation of land use changes). Prog Geogr 21(3):195–203
Liang W, Lu M, Zhang H (2016) Housing prices raise wages: estimating the unexpected effects of land supply regulation in China. J Hous Econ 33:70–81
Liu Y (2015) 土地综合研究与土地资源工程 (Integrated land research and land resources engineering). Resour Sci 37:1 8
Liu Y, Li Y (2017) Revitalize the world's countryside. Nature 548(7667):275–277
Liu Y, Lu S, Chen Y (2013) Spatio-temporal change of urban–rural equalized development patterns in China and its driving factors. J Rural Stud 32:320–330
Long H (2003) 土地利用转型: 土地利用/覆被变化综合研究的新途径 (Land use transition: a new integrated approach of land use/cover change study). Geogr Geo-Inf Sci 19(1):87–90
Long H (2012) 论土地利用转型与乡村转型发展 (Land use transition and rural transformation development). Prog Geogr 31(2):131–138
Long H (2015) 论土地利用转型与土地资源管理 (Land use transition and land management). Geogr Res 34(9):1607–1618
Long H, Heilig GK, Wang J et al (2006) Land use and soil erosion in the upper reaches of the Yangtze River: some socio-economic considerations on China's grain-for-green programme. Land Degrad Dev 17(6):589–603
Long H, Li X (2000) 美国土地资源政策演变及启示 (The evolution of land management policies in the United States and its implications). China Land Sci 14:43–47
Long H, Li X (2002) 区域土地利用转型分析: 以长江沿线样带为例 (Analysis on regional land use transition: a case study in transect of the Yangtze River). J Nat Resour 17(2):144–149
Long H, Li Y, Liu Y et al (2012) Accelerated restructuring in rural China fueled by 'increasing vs. decreasing balance' land-use policy for dealing with hollowed villages. Land Use Pol 29(1):11–22
Long H, Qu Y (2018) Land use transitions and land management: a mutual feedback perspective. Land Use Pol 74:111–120
Long H, Tu S, Ge D et al (2016) The allocation and management of critical resources in rural China under restructuring: problems and prospects. J Rural Stud 47:392–412
Long H, Zhang Y, Tu S (2019) Rural vitalization in China: a perspective of land consolidation. J Geogr Sci 29(4):517–530
Lu D, Shi P, Dong G (2006) 体现土地利用转型的重要理念 (Important ideas embodying land use transition). China Land Resour News. Accessed 17 Jul 2006
Ni G, Cai F (2015) 农户究竟需要多大的农地经营规模 (What is the proper land management scale really needed by farmers). Econ Res J 50(3):159–171
Robinson GM, Carson DA (2013) Applying landscape science to natural resource management. Ecol Soc 18(1):32. https://doi.org/10.5751/ES-05639-180132
Schleupner C, Schneider UA (2010) Effects of bioenergy policies and targets on European wetland restoration options. Environ Sci Policy 13(8):721–732
Serra P, Pons X, Saurì D (2008) Land-cover and land-use change in a Mediterranean landscape: a spatial analysis of driving forces integrating biophysical and human factors. Appl Geogr 28:189–209
Shao J, Zhang S, Li X (2015) Farmland marginalization in the mountainous areas: characteristics, influencing factors and policy implications. J Geogr Sci 25:701–722

Skinner MW, Kuhn RG, Joseph AE (2001) Agricultural land protection in China: a case study of local governance in Zhejiang Province. Land Use Pol 18:329–340
Skog KL, Steinnes M (2016) How do centrality, population growth and urban sprawl impact farmland conversion in Norway? Land Use Pol 59:185–196
Sun Y, Ma A, Su H et al (2020) Does the establishment of development zones really improve industrial land use efficiency? Implications for China's high-quality development policy. Land Use Pol 90:104265
Tan R, Beckmann V, DenBerg L et al (2009) Governing farmland conversion: comparing China with the Netherlands and Germany. Land Use Pol 26:961–974
Tuan Y (1971) Geography, phenomenology, and the study of human nature. Canadian Geogr 15(3):181–192
Wu Y, Mo Z, Peng Y et al (2018) Market-driven land nationalization in China: a new system for the capitalization of rural homesteads. Land Use Pol 70:559–569
Xiao Y, Wu X, Wang L et al (2017) Optimal farmland conversion in China under double restraints of economic growth and resource protection. J Clean Prod 142:524–537
Yin H, Li C (2001) Human impact on floods and flood disasters on the Yangtze River. Geomorphology 41:105–109
Zhang Y, Long H, Ma L et al (2018) Farmland function evolution in the Huang-Huai-Hai Plain: processes, patterns and mechanisms. J Geogr Sci 28(6):759–777
Zhong T, Zhang X, Huang X et al (2019) Blessing or curse? Impact of land finance on rural public infrastructure development. Land Use Pol 85:130–141

Chapter 13
Policies and Countermeasures Ensuring Rural Restructuring

Abstract With the influences of globalization and implementation of the national strategies on industrialization, urbanization, informatization and agricultural modernization, the changes of human resource, land resource and capital allocation have brought about many problems and challenges for the development of rural China. These problems and challenges are protruding in the following aspects: rural human resource allocation takes on the trend of low level and the mainstream of rural development is weakening; non-agriculturalization, non-grain preference and abandonment of farmland use together with the derelict and idle rural housing land result in low efficiency of rural land use; unfair urban-rural allocation of capital and its structural imbalance gradually weaken the self-development ability of rural China. Rural socio-economic development and optimized allocation of rural resources are vital to rural restructuring. As such, the author argues that it is necessary to restructure the contours of government intervention framework in rural economy and society to realize the optimal allocation and effective management of rural critical resource elements. This can be achieved through promoting the bidirectional flowing of urban-rural development elements, pushing forward the reform of rural land property rights system, deepening the reform of rural financial system, supporting land consolidation, and improving related policy and technical systems of rural restructuring planning. However, the smooth implementation of above policies and countermeasures lies in the scientific evaluation on the role of industrial and commercial capital invested into the countryside as well as the rational view on the socio-economic and ecological values of land resource allocation.

13.1 Introduction

Since the turn of the new millennium, with the influences of globalization and implementation of the national strategies on industrialization, urbanization, informatization and agricultural modernization, the changes of human resource, land resource and capital allocation have brought about many problems and challenges for the development of rural China, due to the constraints from the maintained rural-urban

H. Long, *Land Use Transitions and Rural Restructuring in China*,
https://doi.org/10.1007/978-981-15-4924-3_13

dualism of land ownership and household registration. These problems and challenges are protruding in the following aspects: rural human resource allocation takes on the trend of low level and the mainstream of rural development is weakening; non-agriculturalization, non-grain preference and abandonment of farmland use together with the derelict and idle rural housing land result in low efficiency of rural land use; unfair urban-rural allocation of capital and its structural imbalance gradually weaken the self-development ability of rural China.

Aiming at how to resolve the abovementioned problems and adapt to the challenges due to the changes of allocation of critical resources including human, land and capital, it is pivotal to restructure the rural development space, rural industry, and rural social organization and management mainstream. Rural restructuring, to some extent, is a process in which mainstreams optimize resources allocation by influencing rural development factors, whilst the government plays a leading role in promoting the flow of urban-rural factors and rural restructuring through policy-making, planning and resources allocation. This means rural restructuring process is directly or indirectly related to a series of resource allocation system and macro policy environment concerning land, finance, social security, etc. Meanwhile, thriving business features heavily in rural restructuring. Therefore, it is necessary to restructure the contours of state intervention in rural societies and economies (Woods 2012) and to allocate and manage the critical resources for rural economic development based on the recognition of the values and targets of rural restructuring.

13.2 Values and Targets of Rural Restructuring

To obtain a rational and scientific cognition of the values and targets of rural restructuring, it is necessary to have a systematic understanding of the value and function of the rural territorial system, and the interrelationship between the three core elements of rural development (including population, land and industry) and rural restructuring (involving spatial restructuring, economic restructuring, and social restructuring). As mentioned above, the rural territorial system is a complex composed of several elements such as nature, ecology, economy, society, etc., and has multiple value attributes of economy, society, resources, environment, and culture. Firstly, the rural territorial system supports a large population, which means that it has the social function of maintaining the population livelihood and ensuring social stability in a certain area, and the economic function of providing agricultural products, primary products as well as service products for urban and rural areas (Qiao 2008). Secondly, resources, including mineral, land, and biological resources, are the prerequisites and foundations for the operation of rural territorial system and are also the support for the cultivation and the optimization of the spatial distribution of industries in rural restructuring. In addition, because of its high vegetation coverage, rural areas have important ecological significance for energy storage and transformation, material synthesis and decomposition, degradation and purification of harmful substances, regulation of natural disasters and maintenance of biodiversity (Tu et al.

2015). Against the background of cultural convergence, rural areas, as the main accumulation and reserve of traditional culture and regional culture, have their unique cultural charm that is different from cities (Long et al. 2016). Rural restructuring is the coupling process of the key elements that affect rural development, in which the population, land and industry complement one another (Tu et al. 2015), and the three dimensions of spatial restructuring, economic restructuring, and social restructuring restrict one another (Long and Tu 2017).

The practice of rural restructuring should pay attention to several aspects. First, the multi-functional values of rural areas, such as economic, social, environmental, resource, and cultural values, should be given enough attention in the practice of rural restructuring (Fig. 13.1). Secondly, it aims at improving economic efficiency, gradually narrowing the gap between urban and rural areas, inheriting rural culture, protecting the ecological environment, and achieving sustainable use of resources. Meanwhile, it is also necessary to comprehensively consider the interaction among spatial, economic and social restructuring, the mutual effects among population, land, and industrial in rural development. In addition, to organically combine industrial development, spatial system optimization, human settlement environment construction, local culture inheritance, ecological value protection (Liu 2011), basic public service equalization, as well as the improvement the capacity of social development mainstream. In addition, the approaches to rural restructuring can be divided into

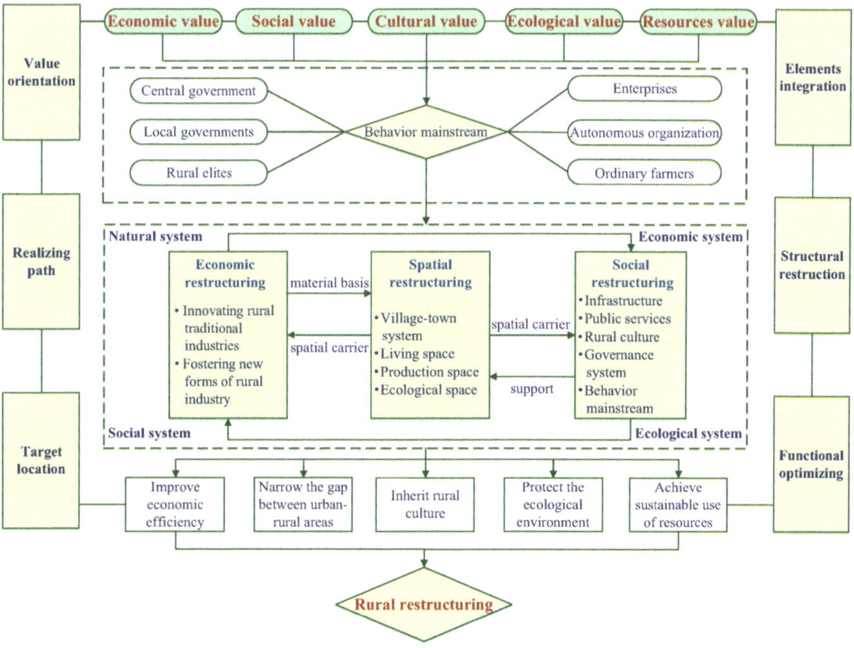

Fig. 13.1 Values and targets of rural restructuring (Adapted from Long and Tu 2018)

economic, spatial, and social restructuring. Economic restructuring includes readjusting and activating traditional rural industries and fostering new rural economy forms; spatial restructuring involves optimizing the urban-town-village spatial pattern and integrating rural living-production-ecological spaces; social restructuring means improving the rural social governance system, cultivating new rural development mainstream, strengthening infrastructure and public service networks, as well as protecting rural cultural landscape. Finally, it can promote the comprehensive development of the "natural-ecological-economic-social" system in rural areas and the enhancement of "production-living-ecological-cultural" functions (Long et al. 2016).

13.3 Restructuring the Contours of State Intervention

The inefficient, structural imbalance and unfair allocations of the critical resources are closely related to the defect of current institution system and policy restrictions. However, the restructuring of rural space, industry and society is not only directly or indirectly in relation to land use, finance, social security and other institutions of resources allocation, but also has connection with the macro policy environment. As a consequence, it is urgent to enhance the efficiency of resource utilization and promote the free flowing of the production elements, so as to accomplish the optimal allocation and efficient management of rural critical resources. For this purpose, it is necessary to restructure the contours of state intervention in rural societies and economies (Woods 2012), by coordinating urban and rural development, pushing forward the reform of rural land property rights system, deepening the reform of rural financial system, and improving related policy and technical systems of rural planning.

13.3.1 Promoting the Bidirectional Flowing of Urban-Rural Development Elements

The long-term urban-rural dual system in land, household registration and social welfare, etc. is regarded as the root of the structural imbalance of rural-urban development elements and low efficiency of resource allocation. On the one hand, urban-rural dual system has brought about the unidirectional flowing of production elements and the widening gap between the rural and the urban, which caused resources excessively concentrated in urban area and resulted in 'urban diseases'; on the other hand, it also gave rise to 'rural diseases' characterized by the rapidly aging and weakening of rural development mainstream, the increasingly vacant and derelict of rural construction land, the rapid non-agriculturalization of rural production elements, and the extreme shortage of rural development funds. Consequently, innovating the system and mechanism of coordinating urban and rural development will play an important role in breaking the dual track urban-rural structure, changing the situation of rural production elements out flowing and promoting the bidirectional flowing of urban-rural development elements. Under the background of the new-type urbanization, accelerating reform in the rural-urban dualism of land ownership and household registration and promoting the equalization of basic public services should be emphasized to balance urban and rural development. Meanwhile, the platform for equal exchanges of urban-rural development elements (Liu et al. 2013), which aims at providing institution guarantee for the optimal allocation and equal exchanges of land, laborer and public service resources, should be established timely to promote urban-rural mutual development.

In view of the problems of current rural development such as outflow of human resources, shortage of production funds and deficiency of endogenous development power, cultivating rural industry and strengthening public infrastructure construction should be highlighted in order to improve rural employment absorptive capacity and production elements gathering ability, and optimize the macro environment boosting inflow of laborers and capital. In addition, establishing the green channel through reinforcing the policy support is essential to attract high-quality human resources and funds from urban area. For instance, promoting rural land attending the market transactions, and allowing immigrants from the urban purchasing homestead and operating rural land by leasing; docking urban and rural public service system, and ensuring that the treatment like health care, social security, etc. is unchanged so as to eliminate the immigrants' worries; constructing reasonable flowing mechanism of urban-rural talents, e.g., teachers, medical and agricultural technical personnel, and giving them some incentives in wages, career development, etc.; and adopting some preferential policies in credit, taxation, land and other aspects to promote social capital flowing into the rural.

13.3.2 Pushing Forward the Reform of Rural Land Property Rights System

The current ambiguous rural land property rights and incomplete power bring about the unsmooth transfer of land as well as the serious influence on rural land use efficiency, spatial optimal allocation efficiency and the feasibility of transforming peasant's land assets to land capital. Especially, the obstacles existed in pushing forward scale management of farmland and the effective circulation of discarded rural housing land are becoming the bottleneck factors of stabilizing agricultural production, ensuring food security, protecting peasant's rights and interests and hindering production elements flowing into rural from urban. Therefore, there is an urgent need to push forward the reform of rural land property rights system. Rural land system reform should be carried out under the framework of collective ownership of rural land. Based on setting up respective rights system for land ownership, land use right and land profit right, an overall framework for property right system reform is expected to be established via confirming land rights, clarifying collective hierarchy and defining collective membership qualification to lay a solid foundation for optimally allocating land resource and endowing the peasant with actual land property rights. During the process of transforming land contractual management right and land use right into real right, the following three aspects are crucial to guarantee the right of the member of rural collective economic organization. Firstly, rural property value and profit assessment mechanism should be established. Besides, making a comprehensive evaluation of the productive value, ecological value and cultural value of the land is indispensable. Furthermore, correctly handling the allocation of added value income of capitalized land must be paid high attention. Aiming at the status quo of rural-to-urban migration and subsequent idle and derelict rural housing land, the establishment of rural homestead circulation and transfer mechanism in the context of urban-rural integration development is an effective way to accomplish the optimal allocation of land resource. Considering the rural laborers who transfer as migrant birds and the fact that rural collective land still undertakes the responsibility for ensuring peasant's basic life and satisfying residential need, land contract right and the withdrawing mechanism of rural housing land should be combined with related matching security system, so as to further improve current social security system, household registration system, educational system, medical insurance system, housing system and other related systems.

13.3.3 Deepening the Reform of Rural Financial System

In recent years, several top-priority annual number one policy documents of the Chinese government were devoted to rural financial reform. Nevertheless, rural financial system is still the weakest link in the whole national financial system. The difficulty in getting loans due to the lack of mortgage or guarantee contributes to the capital shortage issue constraining the further development of rural industry. In order to develop modern scale agriculture, the government should increase the agro-oriented special funds and lure more credit capital and social capital to be invested into rural area. At the same time, financial institutions are encouraged to issue bond specially used for farmers, agriculture and rural areas (so called "San Nong" in Chinese) and make the access control policy of rural finance relaxed to some extent. Rural commercial finance, policy-based finance, private finance and other financial components should be put on their proper place through making clear and complete related policies and regulations (Zhang 2014). Especially, more financial support should be inclined to rural small and tiny enterprises, and the support to the credit loan for agricultural scale management and intensive management should be strengthened, so as to satisfy the financial need of family farm, specialized household, farmer cooperation organization, leading enterprises of agricultural industrialization and other new-type agricultural managing mainstream. Aiming at the issue that it is difficult for farmers to get loan mortgage and guarantee, it is necessary to innovate mortgage guarantee mode by accepting rural land contractual management right, homestead land use right and forest ownership as loan mortgages. In addition, developing rural credit cooperation organization and establishing village-level cooperative financial guarantee fund are also feasible ways to resolve the problems.

13.3.4 Supporting Rural Land Consolidation

Land consolidation may establish a new platform for rural spatial restructuring, and is also an important approach for building new countryside and realizing urban-rural coordination development in China. However, ultimately accomplishing rural spatial restructuring relies on the driving forces of restructuring and corresponding strategy supported by land consolidation.

The mainstream of rural development is villagers, most of which gradually wake up to know that land consolidation is an indispensable way of rural production, living and ecological spatial restructuring, although it is just their irrational land-use activities resulted in their inconvenient living space, limited production space and deteriorated ecological space. Villagers' immediate need and common recognition of developing rural economy and improving production, living and ecological environment is the core driving force of carrying out land consolidation and restructuring rural space. Usually, a "bottom-up" restructuring strategy means that rural spatial restructuring is gradually accomplished relying on land consolidation, which are led by rural elites through strengthening the cohesive force of village, striving for outer aids mainly from governments, and motivating and integrating inner and outer needs. However, macro-policy transformation and corresponding changes of governmental functions involving pushing forward urbanization, building new countryside and advancing large-scale management of agricultural production is the major pushing forces of rural spatial restructuring relying on land consolidation. With the help of related policy platform and various projects favoring agricultural production financed by central or local governments, local governments advance rural spatial restructuring using a "top-down" strategy by integrating capital, technology and labor resources.

About the merits and demerits of these two strategies of 'bottom-up' and 'top-down', Bristow (2000) argued: a top-down emphasis on inward investment and job creation is important for generating employment and income growth but may fail to make the links with other rural activities that would establish resilience to the increasingly competitive pressures evident; however, a bottom-up emphasis based on local empowerment is likely to lead to more diverse and embedded activities, but may not be sufficient on its own to buttress the rural economy against the pressures of globalization. Many successful international rural restructuring projects own to experiencing a process of embeddedness and naturalization (Edwards 1998). The restructuring action has taken many forms, which ranged from the voluntary engagement of residents helping neighbours or co-ordinating local services; through deliberate strategies of co-operative self-help to organized partnership activities between community groups or other agencies (Yarwood and Edwards 1995). Each of these strategies involves the participation and engagement of local people in the process of restructuring action. Recently, many China's government-led rural development policies have not been smoothly implemented, the reasons for which mainly resulted from paying little attention to the process of embeddedness and naturalization. As such, a "bottom-up" restructuring strategy accompanied by a few "top-down" elements is helpful for smoothly pushing forward rural spatial restructuring in China.

13.3.5 Improving Related Policy and Technical Systems of Rural Restructuring Planning

In terms of regulation and policy, under the framework of National Urban and Rural Planning Law implemented in 2008, local regulations combining with the practice of rural construction can be formulated to provide operational policy basis for rural spatial restructuring. As for planning techniques, a scientific rural planning focusing on regional scale needs basing on the integrated study of the elements, function, mechanism and mode concerning rural development. Based on various territorial types, phases and capacities of rural development, the techniques and standard compiling rural planning system corresponding to different territorial types and developing phases can be developed involving some kernel fields, e.g., industry cultivation, village combination and consolidation, rural environment renovation, and resources allocation taking urban-rural integration into account. Additionally, it is necessary to perfecting current rural planning technical system focusing on town-village system and village layout.

13.4 Key Points Ensuring Rural Restructuring

13.4.1 Scientifically Appraising the Industrial and Commercial Capital into Countryside

In 2013, the top-priority annual number one policy document of the Chinese government argued that the industrial and commercial capital should be invested into rural China to develop the planting and breeding industries in an entrepreneurial manner. The CPC Central Committee's Decision on Major Issues Concerning Comprehensively Deepening Reforms adopted at the Third Plenary Session of the 18th Central Committee of CPC further emphasized the industrial and commercial capital should be encouraged and guided to be invested into the countryside to develop modern planting and breeding industries, introducing modern productive factors and management mode to agriculture. In fact, the industrial and commercial capital invested into agriculture is a double-edged sword and has brought about both positive and negative effects on agricultural development in China. On the one hand, it accelerated the transformation of traditional agriculture and promoted the development of modern agriculture. On the other hand, partial industrial and commercial capital invested into rural areas has engaged in non-agricultural industries through occupying large scale farmland and exacerbated the phenomenon of non-agriculturalization and non-grain preference of farmland use. In some areas, partial industrial and commercial capital took its advantages of capital, technology and scale management and produced a crowding-out effect so that the employment opportunities and market shares of local farmers were occupied, instead of adequately leading the farmer to develop industry (Zhang 2014); however, local resources were depleted. Aiming at the negative effects caused by industrial and commercial capital, strict access control and regulatory regime should be established at national level to manage the farmland lease behavior of commercial and industrial enterprises. As to the capital attraction and industry choice, local government should make an overall appraisal for the social, economic, and environmental effects of industrial and commercial capital on rural development, emphasizing on the leading role of enterprises in improving farmers' income and promoting local economic development. In addition, the mechanism on interests coupling between farmers and enterprisers should be built to prevent the phenomenon of revenue and profit leakage caused by the introduction of industrial and commercial capital, and ensure that it actually help to improve farmers' income and promote the rural economic development.

13.4.2 Rationally Treating the Socio-Economic and Ecological Values of Land Resource

Since the beginning of the 21st century, the phenomena of non-agriculturalization and non-grain preference of farmland use and village hollowing have gradually become the highlights in rural China. It is the result of land resource allocation among different industrial sectors and various planting structures, which is dominated by the rational behavior of market mainstream, also reflects the efficiency of resources allocation. However, accompanied with the increasing of economic value, the values of farmland in the aspects of social security and ecosystem services were lost. Meanwhile, it has brought about negative effects on national food security, ecological security and social stabilization. To address the issue of low-efficient utilization and allocation of farmland and rural construction land and to adapt to the changes of marginal productivity and market needs due to the variations in production elements, the central government implemented relevant policies, such as the "Transfer of Management Right of Contracted Land in Rural Areas" and the "Linking up Increased Urban Construction Land with Decreased Rural Construction Land". The government interventions aimed at promoting the orderly flow and optimal use of scarce resources. The policies concerning Transfer of Management Right of Contracted Land in Rural Areas and scale management are in favor of improving the agricultural production efficiency and establishing a platform for the development of modern agriculture. However, the large-scale production of agriculture will infringe the interests of small-scale households in some places. The policy of Linking up Increased Urban Construction Land with Decreased Rural Construction Land was designed originally to reconcile the contradiction between the insufficient supply of urban construction land and inefficient and extensive use of rural construction land. However, it led to the loss of development right of farmers and made the land transfer expense become an important financial source for local government (Zhang 2014). Considering the living security function of farmland and the habitat function of rural housing land, how to balance the economic value and social security function of rural land? Besides the allocation efficiency and economic value of rural land, the social security value and ecosystem services value of which also need to be paid more attention.

13.5 Conclusions

Rural territory is an open system consists of diversified elements including natural endowments, geographical conditions, economic base, human resource, cultural customs, etc. Meanwhile, rural territorial system is composed of kernel system and external system as well as object system and subject system, and takes on multi-functions in the aspects of living, production, ecology and culture. With the increasing complexity of man-land interrelations in rural areas in the process of rapid urbanization, rural restructuring has become an important frontier topic in geographical research. The practice of rural restructuring, based on the importance given to the multifunctional values of rural areas, should be aimed at improving economic efficiency, gradually narrowing the gap between urban and rural areas, inheriting rural culture, protecting the ecological environment, and achieving sustainable use of resources. Then, this kind of practice should comprehensively consider the inter-relationship between spatial restructuring, economic restructuring, and social restructuring, as well as the interplay between population, land, industry. Its ultimate goal is to promote the comprehensive development of the "natural-ecological-economic-social" system in rural areas and the comprehensive promotion of "production-living-ecological-cultural" functions.

The process of rural restructuring is directly or indirectly related to a series of resource allocation system and macro policy environment concerning land, finance, social security, etc. Rural socio-economic development and optimized allocation of rural resources are vital to rural restructuring, in which the government behavior plays the leading role. Thus, it is necessary to restructure the contours of government intervention framework in rural economy and society to realize the optimal allocation and effective management of rural critical resource elements. This can be achieved through promoting the bidirectional flowing of urban-rural development elements, pushing forward the reform of rural land property rights system, deepening the reform of rural financial system, supporting land consolidation, and improving related policy and technical systems of rural restructuring planning. In addition, the optimal allocation and effective management of critical rural resource elements also lies in the scientific evaluation on the role of industrial and commercial capital invested into the country side as well as the rational view on the socio-economic and ecological values of land resource allocation.

References

Bristow G (2000) Structure, strategy and space: issues of progressing integrated rural development in Wales. Eur Urban Reg Stud 7:19–33

Edwards B (1998) Charting the discourse of community action: perspectives from practice in rural Wales. J Rural Stud 14(1):63–77

Liu Y (2011) 中国新农村建设地理论 (Geography of new countryside construction in China). Science Press, Beijing

Liu Y, Lu S, Chen Y (2013) Spatio-temporal change of urban-rural equalized development patterns in China and its driving factors. J Rural Stud 32:320–330

Long H, Tu S (2017) 论乡村重构 (On rural restructuring). Acta Geogr Sinica 72(4):563–576

Long H, Tu S (2018) 乡村重构的理论认知 (Theoretical thinking of rural restructuring). Prog Geogr 37(5):581–590

Long H, Tu S, Ge D et al (2016) The allocation and management of critical resources in rural China under restructuring: problems and prospects. J Rural Stud 47:392–412

Qiao J (2008) 中国乡村地域经济论 (Rural territorial economy of China). Science Press, Beijing

Tu S, Long H, Li T et al (2015) 中国村镇建设和农村发展的机理与模式研究 (Study on the mechanism and models of villages and towns construction and rural development in China). Econ Geogr 35(12):149–156

Woods M (2012) New directions in rural studies? J Rural Stud 28(1):1–4

Yarwood R, Edwards B (1995) Voluntary action in rural areas: the case of neighbourhood watch. J Rural Stud 11(4):447–461

Zhang X (2014) 构建新型城乡关系 (Reconstruction of new-type urban-rural relationship). Social Science Academic Press, Beijing

Chapter 14
Research Prospects and the Future of Rural China

Abstract China is undergoing rapid socio-economic transformation development, at the same time the land use is experiencing a dramatic transition in both depth and breadth. Land use transitions and rural restructuring could lead to a breakthrough in the development of future rural China. Integrating multi-dimensional paths of land use transitions and rural restructuring to optimize the rural territory system is an important scientific proposition that needs to be focused on to deal with China's urban-rural integrated development. This chapter puts forward the research prospects of land use transitions and rural restructuring, and makes a preliminary conception concerning the future development blueprint of rural China under globalization. The future research fields concerning land use transitions are suggested to involve: the dynamic patterns and territorial types of land use transitions; the interactions and dynamic mechanism of land use transitions and rural transformation development; the resources and environmental effects of land use transitions; and the modes and paths of adjusting and controlling land use transitions. The research of rural restructuring in the future needs to pay more attentions to: process and pattern of long-term and multi-scale rural restructuring; mechanism of rural restructuring; regional modes of rural restructuring; rural planning technology system and standard; and policy and institutional innovations of rural restructuring. Rural China is constantly being integrated into the global socio-economic network. As such, it is necessary to fully understand the impacts of globalization on rural restructuring in China and to strengthen the relevant researches.

14.1 Introduction

China has made remarkable achievements in socio-economic development. China's rural is also undergoing historic changes. However, during the process of rapid urbanization, rural development issues are gradually emerging. Feasible supporting platform and hand grip are urgently needed to effectively solve the problems resulting from rapid urban-rural transformation development and to promote rural restructuring and vitalization. Land use transitions are a long-term and trend transformations of land use morphology and a reflection of China's socio-economic development.

H. Long, *Land Use Transitions and Rural Restructuring in China*,
https://doi.org/10.1007/978-981-15-4924-3_14

Meanwhile, various issues in the process of socio-economic development can also be resolved by adjusting and controlling land use transitions. Rural restructuring is a process of reshaping socio-economic morphology and spatial pattern in rural territory in response to the changes of elements both in kernel system and external system of rural development, by optimally allocating and efficiently managing the material and non-material elements in the two systems. Integrating multi-dimensional paths of land use transitions and rural restructuring to optimize the rural territory system is an important scientific proposition that needs to be focused on to deal with China's urban-rural integrated development.

Based on the abovementioned systematic analysis, this chapter puts forward the research prospects of land use transitions and rural restructuring, and makes a preliminary conception concerning the future development blueprint of rural China under globalization.

14.2 Research Prospects of Land Use Transitions

Currently, as an important measure to solve the issues relating to "agriculture, rural areas and farmers" in China, the double-wheel driving strategy of rural vitalization and new-type urbanization has put for word realistic demands of science and technology to realize the land use transitions successfully in rapid urbanization phase, which promotes an urgent requirement of theoretical innovation and academic achievements in agricultural geography and rural development as an strong support. Accordingly, the future research fields concerning land use transitions are suggested as follows (Long 2012):

(1) **The dynamic patterns and territorial types of land use transitions**

The dynamic patterns of regional land use transitions during the process of rapid urbanization could be depicted by establishing evaluation index systems and evaluation models. Focusing on the spatial allocation of factors and resources concerning rural development and the regional differentiations in the aspects of bio-physical geographical environment and socio-economic development, combining the evaluation results of regional dynamic patterns of land use transitions with relevant models is expected to identify the territorial types of land use transitions and to draw out the rules concerning the heterogeneous statuses of land use transitions related to regional socio-economic development differences. Meanwhile, based on the field investigations in typical areas, the natural, economic, social, political and cultural as well as the institutional factors affecting land use transitions in different territorial types, the stage characteristics, problems and their major influential factor of land use transitions are necessary to be analyzed.

(2) **The interactions and dynamic mechanism of land use transitions and rural transformation development**

Focused on the typical regional types of land use transitions and rural transformation development, the changing patterns and trends of the major driving factors (including land, capital, labor force and industrial development) could be analyzed and revealed in combination with field investigations and modelling simulations, from the perspectives of changes in regional natural-ecological structure, technological-economic structure and social-political structure, to qualitatively and quantitatively investigate the dominant driving force and interacting mechanism as well as the coupling relationship and its dynamic mechanism leading to the transformation of the stages of land use transitions.

(3) **The resources and environmental effects of land use transitions**

Under the background of rapid urbanization, there is an urgent need to study the value manifestation and proliferation effects concerning the flows of rural development elements in the process of land use transitions, and the systematical assessment of the influential process, intensity and effects of some new factors including rapid industrialization and urbanization development, and the formulation and deployment of national strategies and policies supporting and benefiting agriculture and rural areas on sustainable rural development at the scales of village and farmer. Furthermore, a series of indices system and integrated assessing models need to be established based on the rural physical and social environment elements to measure and evaluate the resources and environmental effects of land use transitions on local rural territorial system.

(4) **The modes and paths of adjusting and controlling land use transitions**

By integrating the time scale and historical background of social and environmental changes, the optimizing and adjusting modes of land use transitions during the process of rapid urbanization can be summarized and extracted. From the theoretical perspective of rural system growth and urban-rural interactive development, how to follow the rural transformation development in the new era and improve the capability of rural territories with different development types and degrees to adapt to the domestic and abroad competitive markets and their own development via adjusting and controlling land use transitions needs to be further discussed. Also, the paths to enhance rural productivity by spatial recombination of essential factors and resources favoring rural development and to facilitate urban-rural coordinated socio-economic development are also pivotal to the research of land use transitions.

In addition, the recessive morphology of land use and its changes should become the focus of future research on land use transitions and land resource management. Managing and controlling the changes of recessive morphology of land use contributes to innovating the policies, regulations and institutions of land resource management and accelerating the transformation of land resource management in China from the simple management of quantity to the comprehensive management with equal emphasis on quantity, quality and ecology as well as the change from the simple resources management to the comprehensive resource-asset-capital management, so as to optimizingly allocate and sustainably use land resource in both the urban and the rural.

14.3 Research Prospects of Rural Restructuring

Aiming at resolving the realistic dilemma brought about by the rapid urban-rural development transformation, the study of rural restructuring in the future needs to pay more attentions to the following aspects (Tu and Long 2017):

(1) **Process and pattern of long-term and multi-scale rural restructuring**

Geography emphasizes the research on the spatial distribution pattern and temporal evolution process of geographical elements (Cai 2011; Fu 2014), which involve time scale and spatial scale. Exploring the spatial distribution law and the temporal evolution law of geographical elements at different time and space scales, is conducive to provide scientific basis for simulating and predicting the development trend of geographical matter and taking appropriate human intervention to ensure its development along the expected path. The study of rural restructuring needs to focus on the process and pattern of restructuring from the perspectives of elements mobility, structural optimization and functional evolution, and explore the characteristics of spatial distribution and temporal evolution of the typical region, county, town and village, respectively, and reveal the laws of spatial distribution and temporal evolution at different time and spatial scales.

(2) **Mechanism of rural restructuring**

Interactively influenced by various factors such as natural environment, socio-economic development, historical and cultural features, macro-regional policies, etc., the mechanism of rural restructuring driven by the interaction of the internal and external factors in rural territory, is significantly diversified in different regions and at different periods. It is necessary to identify the main controlling factors and dynamic mechanism from the perspective of "elements-structure-function", based on analyzing different types of factors affecting rural restructuring, i.e., the endogenous factors and exogenous factors as well as subjective factors and objective factors, to explore the interactive rules of internal and external factors, to comparatively analyze the relationship between the process and the mechanism of rural restructuring at different spatio-temporal scales, and to summarize the general law and mechanism of rural restructuring.

(3) **Regional modes of rural restructuring**

The modes of rural restructuring refer to the theoretical summary of the development path with distinct characteristics and relative stability, which reflect the interaction law of elements and the mode of economic operation. The formation of rural restructuring is influenced by natural resources, economic development, productivity level, historical tradition, government behavior and other factors. Considering the terrain conditions, the economic development level and the modes of industrial development, e.g., the modes in plain areas, mountainous and hilly areas, metropolitan suburbs, developed areas or underdeveloped areas, as well as the modes led by modern agriculture or driven by industrial development, it is important to research different types of modes of rural restructuring by comparatively analyzing the internal and external factors, the operating mechanism, basic characteristics, obstacles and regulatory path so as to provide useful reference for the practice of rural restructuring in other similar regions.

(4) **Rural planning technology system and standard**

At present, the technologies at national level guiding rural restructuring in China involve "Standards for Town Planning" (GB50188-2007) and "Technical Specification for Village Consolidation" (GB50445-2008). Due to lacking of effective guidance for rural restructuring aiming at various territorial types, phases and capacities of rural development, the cohesion and complementarity should be emphasized in arranging the restructuring project, such as the rural residential system adjustment, industrial development and the public facilities allocation. As a consequence, aiming at the key technologies and fields such as rural spatial planning, public infrastructure configuration and industrial cultivation, it is an urgent need to carry out integration and demonstration research through multidisciplinary cooperation concerning geography, sociology, architecture and other disciplines, and ultimately form a series of the technology standards and operating specification of rural planning for different territory types and developing phases (Long et al. 2016), so as to provide technical support for the management of rural restructuring practice.

(5) **Policy and institutional innovations of rural restructuring**

Currently, a series of prominent issues concerning China's rural development, e.g., weakening behavior mainstream, low-efficient land use and insufficient infrastructure, are closely related to the defective institution and policy. Meanwhile, rural restructuring in the aspects of space, economy and society, has direct or indirect connection with land use, finance, social security and other institutions of resources allocation, which constitute the macro policy environment of rural restructuring (Long et al. 2016). Therefore, it is necessary to strengthen the research on policy support system for the purpose of enhancing the efficiency of resource utilization and promoting the orderly flow of the production elements in the practice of rural restructuring, and provide scientific basis for restructuring the contours of state intervention in rural societies and economies (Woods 2012; Long et al. 2016).

14.4 The Future of Rural China Under Globalization

14.4.1 The Future of Rural Places: A Global Countryside

At some point around the turn of the new millennium, the growth of world population passed an historic threshold with less than half the global population living in rural for the first time. This new age of global minority rurality has inevitably posed questions about the future of rural places, since the figure of the 'global countryside' follows the model of the 'urbanized countryside' in emphasizing the inter-connection and inter-dependency of rural and urban places, but it differs sharply in not privileging the city as the driver of change (Woods 2019). Affected by globalization, rural areas around the world are increasingly connected to each other and to cities through social, economic and cultural flows and networks, resulting in the restructuring of rural economies under changed market conditions and re-composition of rural development elements including populations, land and industry, as well as challenges to traditional rural cultural practices (Woods 2019).

Facing to the globalization processes, rural development mainstream such as rural institutions, residents and other actors has to actively respond and engage in negotiating, resisting, manipulating, and thus have the capacity to shape their own futures, albeit within constraints including geographical location, natural and cultural resources endowments, and local socio-economic development level (Long and Woods 2011; Woods 2013; Long et al. 2016). However, regional differentiations in the aspects of bio-physical condition, socio-economic development level, human capital and institutional capacity will determine differentiated interactions among these factors, subsequently causing various outcomes of globalization. As such, the global countryside is not a homogenous space, but a differentiated patchwork of hybrid places, in which some localities prosper and others struggle (Woods 2019).

Globalization hence presents challenges for rural planning, not only in the form of pressures on rural land use, but also the ambitious national/regional rural development strategy. Without exception, every country is constantly being integrated into the global socio-economic network, and the internationalization of production elements configuration like goods, capital and labor force, has exerted tremendous influences on the economy, society and culture in the rural areas. For instance, the influx of agricultural products with super quality have brought about gigantic lash to traditional agriculture, the globalization of capital market have further exacerbated the deagriculturalization of agricultural population and rural land, and the expansion of global culture and modern technology has led to a double effects on the countryside in most developing countries, which make their rural areas not only face the transformation from traditional agricultural society to modern industrial and urbanized society, but also face the risk of losing local culture and national culture (Long et al. 2016; Tu and Long 2017).

14.4.2 The Future of Rural China: Restructuring Under Globalization

Globalization, on the basis of the socialization of production and internationalization of division of labor and with worldwide flow and combination of production elements as its core, is a process of international interdependence and interconnection arising from increasingly frequent political, economic and cultural exchanges among different countries, regions and nationalities (Woods 2005, 2007, 2013). As one of the important developing countries, China has made full use of its cheap labor resources and abundant natural resources to participate in the global economic activities since the turn of the new millennium. The reconfiguration of production elements like laborer, capital and technology has exerted an increasingly profound impact on China's rural economy, society, culture, etc.

The flow of goods and capital at a global scale is a prominent feature of globalization. China's traditional agriculture used to have the features such as single planting structure, scattered distribution and small scale operation, and a low level of organization and standardization. Affected by the rising of the price of the means of agricultural production, land rent and laborer cost, the cost of agricultural products in China is increasing, especially for primary agricultural products like wheat and corn which are lack of competition compared with the international prices of the similar products. Driven by global commodity market system, the internationalization of agricultural product configuration and the influx of foreign bulk grain and oil products with super quality and competitive price have brought about unprecedented impacts on China's traditional agriculture, and at the same time have provided favorable opportunities for the structural adjustments and the standardization, management systematization and modernization of agricultural production. The globalization of capital market has accelerated the pace of industrialization and urbanization in coastal area of China (Long and Woods 2011), and the labor-intensive industrial clusters and foreign trade grounded attracted the trans-regional transfer of large number of rural surplus laborers, which exacerbated China's one-way flow of rural labor force and nonagriculturalization process of agricultural population transfer, and has further exerted profound influences on the transformation of agricultural development, utilization of land resources, construction of social organization, and inheritance of local culture in rural China. In addition, under the background of globalization, the synchronous propulsion of China's industrialization, urbanization, informatization and agricultural modernization has accelerated the penetration of modern culture and technology, and the radiation of urban production and life style, which effectively promoted the transformation of Chinese countryside from traditional agricultural society to modern industrial and urbanized society. Meanwhile, the local Chinese rural cultural order has suffered the double impacts from both the global culture and urban culture. In some rural areas, the national and regional culture is facing the risk of loss, and consanguineous relationship, ethical value, and the family cohesion and stability of traditional rural society have encountered unprecedented threat.

Globalization is an objective process of the development of human society. Facing to the challenges and opportunities brought about by globalization to China's rural socio-economic configuration, the Chinese government should take effective measures to actively participate in the global economic system and strengthen the rural spatial, industrial and social restructuring by taking the optimal allocation of resources as orientation, especially it should attach great importance to the construction of modern agricultural industrial system, and optimize the agricultural industrial structure, product structure and quality structure and speed up the construction of market information system, quality standard system, inspection and quarantine system of agricultural products. At the same time, the government should improve the systematization level of farmers and expedite the specialization, industrialization and large-scale production of agricultural management by taking leading enterprises and professional associations as the carrier. In recent years, in order to meet the needs of agricultural production and changes of international market, the Chinese government have carried out a series of measures aiming at promoting agricultural development, e.g., offering planting grain direct subsidy and other subsidies for purchasing agricultural machinery, the means of agricultural production and superior seeds. In the future, it should play an active part in international coordination and rule-making, gradually establish the agricultural support protective system in conformity with both the national conditions and international rules in order to create a good external environment for its agriculture to participate in international cooperation and competition.

14.5 Conclusions

China is undergoing rapid socio-economic transformation development, at the same time the land use is experiencing a dramatic transition in both depth and breadth. Land use transitions and rural restructuring could lead to a breakthrough in the development of future rural China. Integrating multi-dimensional paths of land use transitions and rural restructuring to optimize the rural territory system is an important scientific proposition that needs to be focused on to deal with China's urban-rural integrated development. This chapter puts forward the research prospects of land use transitions and rural restructuring, and makes a preliminary conception concerning the future development blueprint of rural China under globalization. The future research fields concerning land use transitions are suggested to involve: the dynamic patterns and territorial types of land use transitions; the interactions and dynamic mechanism of land use transitions and rural transformation development; the resources and environmental effects of land use transitions; and the modes and paths of adjusting and controlling land use transitions. Aiming at resolving the realistic dilemma brought about by the rapid urban-rural development transformation, the research of rural restructuring in the future needs to pay more attentions to: process and pattern of long-term and multi-scale rural restructuring; mechanism of rural restructuring;

regional modes of rural restructuring; rural planning technology system and standard; and policy and institutional innovations of rural restructuring. Rural China is constantly being integrated into the global socio-economic network. As such, it is necessary to fully understand the impacts of globalization on rural restructuring in China and to strengthen the relevant researches.

References

Cai Y (2011) 认识环境变化谋划持续发展: 地理学的发展方向 (Understanding environmental change planning sustainable development: development directions for geography in China). B Chinese Acad Sci 26(4):390–398

Fu B (2014) 地理学综合研究的途径与方法: 格局与过程耦合 (The integrated studies of geography: coupling of patterns and processes). Acta Geogr Sinica 69(8):1052–1059

Long H (2012) 论土地利用转型与乡村转型发展 (Land use transition and rural transformation development). Prog Geogr 31(2):131–138

Long H, Tu S, Ge D et al (2016) The allocation and management of critical resources in rural China under restructuring: problems and prospects. J Rural Stud 47:392–412

Long H, Woods M (2011) Rural restructuring under globalization in eastern coastal China: what can be learned from Wales? J Rural Community Dev 6(1):70–94

Tu S, Long H (2017) Rural restructuring in China: theory, approaches and research prospect. J Geogr Sci 27(10):1169–1184

Woods M (2005) Rural geography: processes, responses and experiences in rural restructuring. Sage, London

Woods M (2007) Engaging the global countryside: globalization, hybridity and the reconstitution of rural place. Prog Hum Geog 31:485–507

Woods M (2012) Rural geography III: rural futures and the future of rural geography. Prog Hum Geogr 36:125–134

Woods M (2013) Economic entanglements and the reshaping of place in the global countryside. In: Tamásy C, Diez J (eds) Regional resilience, economy and society: globalising rural places. Taylor and Francis, London, pp 11–32

Woods M (2019) The future of rural places. In: Scott M, Gallent N, Gkartzios M (eds) The Routledge companion to rural planning. Routledge, London, pp 622–632

The manufacturer's authorised representative in the EU is Springer Nature Customer Service Centre GmbH, Europaplatz 3, 69115 Heidelberg, Germany. If you have any concerns regarding our products, please contact ProductSafety@springernature.com

Printed and bound by CPI Group (UK) Ltd, Croydon, CR0 4YY
05/10/2026
02228124-0020